New Methods of Geostatistical Analysis and Graphical Presentation

Distributions of Populations over Territories

New Methods of Geostatistical Analysis and Graphical Presentation

Distributions of Populations over Territories

Roberto Bachi

Late of The Hebrew University of Jerusalem
Jerusalem, Israel

Published in cooperation with
The Israel Academy of Sciences and Humanities

Kluwer Academic / Plenum Publishers
New York, Boston, Dordrecht, London, Moscow

Library of Congress Cataloging in Publication Data

Bachi, Roberto, 1909–1995.

New methods of geostatistical analysis and graphical presentation: distributions of populations over territories / Roberto Bachi.

p. cm.

"Published in cooperation with The Israel Academy of Sciences and Humanities."

Includes bibliographical references and index.

ISBN 0-306-45544-7

1. Spatial analysis (Statistics) 2. Population geography—Statistical methods. 3. Geology—Statistical methods. 4. Statistics—Graphic methods. I. Akademyah ha-le 'umit ha-Yisre'elit le-mada im. II. Title.

HA30.6.B33 1999

304.6'01'5192—dc21

99-17060

CIP

ISBN 0-306-45544-7

233 Spring Street, New York, N.Y. 10013

10 9 8 7 6 5 4 3 2 1

A C.I.P. record for this book is available from the Library of Congress.

Printed in the United States of America

Roberto Bachi
1909–1995

Born in Rome in January 1909, Professor Roberto Bachi graduated from the University of Rome. He taught statistics and demography at several universities in Italy, and achieved the rank of full professor in 1937. He immigrated to Palestine soon after fascist Italy's "racial laws" went into effect in 1938 and worked first as a statistician for the Hadassah Medical Organization and then in the Mandatory Government's Department of Statistics. In 1940 he was invited to join the faculty of the Hebrew University of Jerusalem and became a full professor there in 1947.

When the State of Israel was founded in 1948, Professor Bachi was appointed to the post of Government Statistician. He established Central Bureau of Statistics, which he directed until 1971, and is credited with developing a modern and efficient service, remarkable for its reliability, independence, and professional integrity. During the early 1950s he headed the Hebrew University's Department of Statistics and Demography, and he played an essential role in creating the University's Faculty of Social Sciences, serving as its first dean from 1953 to 1956. Soon after he was appointed Pro-Rector of the Hebrew University. With the establishment of the University's Institute of Contemporary Jewry, he created the Division of Jewish Demography and Statistics, the leading research center in its field. In the course of a long and distinguished academic career, he was mentor and guide to generations of students, many of whom went on to prominence in academic and public life. Thus he became known as the "Father of Israeli Statistics."

After retiring in 1977, Professor Bachi continued to develop new research methodologies in the two fields that were his passion throughout his life—geostatistics and graphical representation of statistical data. Using his new methods to represent distribution of populations over territories, he worked with the U.S. Bureau of the Census, edited the monumental *Statistical Atlas of Italy*, and directed a major study of residential trends of Jews in major cities in the Western World. Until his last days he worked on the present volume, which describes these techniques and demonstrates their remarkable efficacy.

A member of the Israel Academy of Sciences and Humanities and an honorary member of the American Statistical Society, Professor Bachi was honored with many distinctions and awards over the years, including the most prestigious 1982 Israel Prize for lifetime achievement.

Foreword

The chief aim of this book is to present the reader with an integrated system of methods for dealing with geographical statistics (geostatistics) and their applications. It sums up the developments based on the vast experience accumulated by Professor Roberto Bachi over several decades of research.

Interest in the quantitative locational aspects of geography and the common ground of geography and statistics has grown rapidly, involving an ever-increasing spectrum of scientific disciplines. This expansion has gained momentum particularly since the advent of user-friendly statistical software facilitating research in these areas. The geostatistical methods described in this book make it possible to overcome many of the difficulties encountered in graphically presenting, processing, analyzing, and comparing spatial data over different territorial units. It is our hope that the present volume will fill a genuine need—as a textbook, as a reference work, and as a tool for geographers, applied statisticians, demographers, ecologists, regional planners, economists, professional staff of statistical agencies, and others.

The book's central theme is the presentation, treatment, and analysis of large collections of data subdivided into territorial units of irregular shape. Specifically, the methods presented enable one to: (a) graphically display the information supplied by the raw data in a way that permits a quick grasp of the geographical characteristics of the territorial distributions being studied; (b) produce accurate computerized maps comparing their features; (c) present geostatistical parameters summarizing the main characteristics of position, dispersion, and shape of a distribution under study and compare them even for different territories and at different points in time; (d) measure the geographical discrepancy between any two distributions.

The title of the book relates to distribution of populations over territories. However, the term population should be understood in its widest possible sense, to include events such as births, deaths, and accidents; establishments and institutions of different kinds; and various services and economic activities—in fact, any kind of phenomenon that can be described in spatial–geographical terms. More formally, the units under investigation here are geographical population sets or geosets, i.e., sets of triplets. Each triplet consists of the weight (or frequency) of a particular item of population in a territorial entity and the Cartesian plane coordinates of its location. These, then, are the basic building blocks and raw material with which this book deals, comparing them with each other and with the geographical features of the territory itself.

The impact of the methods described herein on applied geographical statistics constitutes a milestone in geostatistics. Professor Bachi was able to try out these methods in a series of extensive applications. Foremost among these were the studies carried out at the U.S. Bureau of the Census during the late 1970s, and the creation of the *Statistical Atlas of Italy*, a two-volume work of vast proportions produced by a team of specialists from the Italian National Institute of Statistics and coordinated by Professor Bachi himself during his frequent visits to Italy. More recently, he and his colleagues, supported by a grant from the US-Israel National Science Foundation applied the geostatistical methods described here to a study of the distribution of Jews in many cities of the Diaspora all over the world. One can therefore affirm with confidence that the methods outlined in the present volume have undergone extensive and thorough testing under a wide range of circumstances and conditions.

A few words should also be said about the beginning of Roberto Bachi's interest in, and involvement with, geostatistical research. His earliest papers in geostatistics were published over 60 years ago. The first, on a measure of distance between geographical distributions, was published in Italy in 1934. On the basis of this preliminary step he developed, chiefly during the 1950s and 1960s, a complete methodology for quantitatively analyzing and comparing geostatistical distributions by measuring the spatial position of the points constituting them, together with their thematic content, and then computing the shape, extension, and direction of the entire distribution, expressing them with the aid of a small number of both dimensioned and dimensionless parameters. These make it possible to draw quantitative as well as qualitative comparisons between different populations. Somewhat later, and then in parallel with the former, he devised a technique to facilitate the simple graphical presentation of geostatistical data, the graphical rational patterns (GRP) method, which allows geographical data over large territorial units to be displayed and grasped in a less area-generalized manner than they are in, for example, choropleth maps. The integration of these two fields of interest—the GRP and the earlier geostatistical method—led to the series of investigations that eventually resulted in the present volume.

As his granddaughter Clila notes in the Preface, Professor Bachi's book was almost finished at the time of his death in November 1995. A team of his colleagues undertook to assist her in completing the editing of the text. We all considered it an honor and a privilege to do this, feeling that we were fulfilling a debt of gratitude to the author, whose intellectual integrity and stature, unfailing commitment to this profession, respect for his colleagues, kindness to his students, devotion to his family, and gentleness toward all we learned to admire and esteem.

E. Peritz
G. Nathan
Jerusalem N. Kadmon

Preface

The death of Professor Roberto Bachi on November 26, 1995, brought to a close an unusually prolific career in the fields of demography and statistics. It also cut short his work on the present volume, the culmination of his research in geostatistics, to which he devoted his postretirement years. As his granddaughter, I had the pleasure of assisting him in his work on this book during the last two years of his life. This unique opportunity was a special privilege because he was truly a remarkable man, with outstanding scientific achievements to his credit and an innate graciousness and nobility.

It was my grandfather's dearest wish to see this book published. On his deathbed he told the family that it was almost finished and asked me to carry out the task of seeing it through to completion. Accordingly, I assumed responsibility for its publication, guided by my understanding of his intentions. For the most part, the text as he wrote it was left unchanged. In some cases, such as the last two sections of Chapter 10, the text was reconstructed from a draft; while in other instances, incomplete material had to be omitted. The maps and illustrations comprised the most problematic area by far: at the time of his death, many remained undone, and I had to rely on the figures as they were referred to in the text as my guide in their preparation.

I hope and trust that in fulfilling my grandfather's wish, I have made an important scientific document available, which will also constitute a step toward the adoption of his methods for theoretical and field studies.

This volume could not have been completed without the assistance of several people. I would like to thank Dr. Shai Ever-Hadani for his invaluable help in the preparation of the illustrations. Dr. Ever-Hadani drew upon his long years of collaboration with my grandfather. He dedicated both time and effort unstintingly to solving the many problems with which I was confronted.

I am also greatly indebted to three of my grandfather's colleagues for their assistance in preparing the text for publication. I am especially grateful to Prof. Gad Nathan for shouldering the largest part of the burden. He reviewed Chapters 8 through 16 and was particularly helpful in the reconstruction of Chapter 10 and Appendixes E and H. I would also like to express my appreciation to the late Prof. Eric Peritz for his work on Chapters 1 through 4 and for his pertinent comments. Special thanks must also be extended to Prof. Naftali Kadmon for his profound and insightful observations on Chapters 5 through 7.

I would like to thank Evelyn Grossberg for her excellent and very professional editorial work on the text as well as for her valuable comments and suggestions.

Finally, I wish to express my deep gratitude to my grandmother Vera, who, after standing by my grandfather's side for over 60 years as an unfailing source of encouragement in his academic pursuits, did her utmost to see this work published. Words cannot express my deeply felt appreciation for her assistance and support.

Clila Gerassi Tishby

Jerusalem

Contents

List of Plates xxv

Part I: Basic Concepts

Chapter 1
Introducing Geostatistical Methods

1.1. Primary Aims of this Volume 3
1.2. Introducing Rational Maps and Comparisons among Them 3
  1.2.1. The Distribution of Population over the Territory of Italy . . . 3
  1.2.2. Comparing Distributions of Populations to a Uniform Distribution 5
1.3. Formalizing Population Distribution as Geosets 10
  1.3.1. Primary Geosets (^gP, ^gU) and Geosets Derived from Comparing Them 10
  1.3.2. An Interpretation of the Methods above for Comparing ^gP and ^gU 11
1.4. Two or More Geosets over the Same Territory 12
  1.4.1. Further Examples of Geosets for Italy 12
  1.4.2. Geosets Derived from the Comparison of Two Primary Population Geosets 17
1.5. Examples of Geostatistical Parameters 22
  1.5.1. Order of Presentation of the Parameters 22
  1.5.2. Interpretation of the Crosses Representing Geosets Superimposed on the Maps 24
  1.5.3. The Mean Center 25
  1.5.4. Measuring Territorial Dispersion and Oblongity 27
  1.5.5. Measuring Territorial Coverage and Net Density 29
  1.5.6. Comparing Dispersion to Area Covered and Measuring Looseness 30
  1.5.7. Concise Measure of the Discrepancy between Two Geosets . . . 32

Chapter 2
Selected Applications of Geostatistical Methods

2.1. Generalizing the Approaches in Chapter 1 to Other Fields 35
2.2. Applying Geostatistical Methods in Different Fields 36
2.3. Scope of this Volume 38
2.4. A Brief Survey of the Development of Geostatistical Methods 40
2.4.1. Limitations of this Review 40
2.4.2. Centrography, Social Physics, and Geostatistics 40

Chapter 3
Geosets of Populations Distributed over a Predetermined Territory

3.1. Introduction 43
3.1.1. Why Geosets? 43
3.1.2. Sets and Set Operations 44
3.1.3. Introducing Population Geosets 45
3.2. Comparing the Actual Population Distribution to the Uniform Distribution over the Territory 48
3.2.1. The Uniform Distribution over the Territory 48
3.2.2. Comparing the Actual and Uniform Distributions Using the Cell Network 52
3.2.3. Distinguishing between Inhabited and Noninhabited Cells . . . 55
3.2.4. Transforming ${}^{g}P$ into ${}^{g}U$ by Means of Operations over Geosets . 56
3.2.5. Splitting the Habitat into Territories of Matched Elements and Surpluses 59
3.3. Comparing Two or More Geosets Distributed over the Same Territory 62
3.3.1. The Importance of this Topic 62
3.3.2. Introducing the Comparison between ${}^{g}P$ and ${}^{g}Q$ 63
3.3.3. Population Geosets Derived from Comparing ${}^{g}P^{\cdot}$ and ${}^{g}Q^{\cdot}$ . . . 64
3.3.4. Territorial Geosets Derived from Comparing ${}^{g}P^{\cdot}$ and ${}^{g}Q^{\cdot}$ 65
3.4. Comparisons between Geosets Representing Large Populations . . . 69
3.4.1. The Problems Dealt with in this Section 69
3.4.2. Reduction of the Number of Cells in the Geostatistical Analysis of Large Populations 69
3.4.3. Dealing with Input Formed by Aggregated Data 70
3.4.4. Determining the Reduced Number of Elements to Be Used in the Analysis of Large Populations 71
3.5. Operations on Geosets versus Operations on Sets 72
3.5.1. Similarities and Differences 72
3.5.2. Match of Geosets versus Intersection of Sets 73
3.6. A Compound Geoset Formed by Two Component Subsets 73
3.7. General Scheme of the Geosets Considered in this Chapter and the Symbols Used in this Volume 77

3.8. Extending the Approach Suggested in Section 3.6 to Other Settings . . 80
3.8.1. Selected Extensions 80
3.8.2. Geosets of a Compound Population Formed by Many Subpopulations 82
3.8.3. Geosets of a Compound Population Formed by Positive and Negative Subpopulations 84
3.8.4. Averaging or Compounding Two Geosets 84
3.8.5. Weighted Average of Two Geosets 84
3.8.6. Average of Many Geosets 85
3.8.7. Union of Geosets versus Average of Geosets 85
3.8.8. Intermediate Geosets 86
3.8.9. Decomposition of a Geoset 87
3.9. Location-Free Sets of Relative Cell Weights 87
3.10. Expressing Geosets of Surpluses Through Rates or Ratios 88

Chapter 4
Further Considerations on Geosets of Territories

4.1. Introduction 89
4.2. Defining the Territorial Gross Range for Unbounded Population Distributions 90
4.2.1. The Problem 90
4.2.2. Criteria for Solutions 92
4.2.3. The Convex Hull as Gross Territorial Range 93
4.2.4. Excluding Outliers from the Convex Hull 94
4.3. Tracts of Territory Covered by Individual Population Elements . . . 96
4.4. Ellipses of Equivalent Scatter 98
4.4.1. An Ellipse as a Substitute for Gross Territorial Range 98
4.4.2. The Ellipse of Equivalent Scatter as a Substitute for Habitat 99
4.5. The Geoset of a Simply Connected Territory Defined without Reference to a Population 105
4.5.1. A Geoset Formed by an Infinite Number of Infinitesimal Territorial Elements 105
4.5.2. Finite Territorial Geoset 106
4.6. Geosets of Compound Territories 107
4.6.1. A Territory Formed by Separate Fragments 107
4.6.2. Geosets of Subterritories 107
4.6.3. Territories with Internal Gaps 108
4.6.4. Territories with External Concavities 108
4.7. Operations on Territorial Geosets 108
4.8. Types of Scales Used to Determine Geosets of Populations and Territories 111

Part II: Graphical Presentation of Geographical – Statistical Data

Chapter 5
Statistical Mapping

5.1. Introduction . . . 115
5.2. Revolution in Statistical Graphics and Mapping . . . 117
5.2.1. Historical Highlights . . . 117
5.2.2. Automation . . . 118
5.2.3. Recent Scientific Developments . . . 120
5.3. The Present State of the Art . . . 121
5.4. Defining Principles and Setting Standards for Statistical Graphs . . . 123
5.5. Establishing Principles for Preparing Geostatistical Maps . . . 125
5.5.1. Basic Considerations . . . 125
5.5.2. General Properties of the Cartesian Graphical System . . . 126
5.6. Using the Cartesian System for Statistical Mapping . . . 127
5.6.1. The Inherent Difficulty in the Cartesian Method . . . 127
5.6.2. Using Dot Maps to Represent Locations of Individual Elements of a Population Distribution . . . 128
5.6.3. Extending the Use of Dot Maps to Represent Aggregates of Locations . . . 129
5.6.4. Stereogrammatic Statistical Maps . . . 130
5.6.5. Contour Lines . . . 132
5.6.6. Conclusions . . . 132
5.7. Using Non-Cartesian Graphical Methods to Construct Statistical Maps . . . 133
5.7.1. The Principal Alternative Graphical Systems . . . 133
5.7.2. General Criteria for Statistical Mapping . . . 133

Chapter 6
Graphical Rational Patterns

6.1. Introduction . . . 135
6.2. Graphical Rational Pattern Scale Representing Integers between 1 and 100 . . . 135
6.3. Other Types of Graphical Rational Patterns . . . 137
6.3.1. Alternative GRP Scale Showing Integers between 10 and 49 . . . 137
6.3.2. Alternative GRP Scale of "Black over White" and "White over Black" . . . 138
6.3.3. GRP Scales Showing Integers over 100 . . . 139
6.3.4. Circular GRP . . . 140
6.3.5. GRP Representing Positive or Negative Values . . . 140
6.3.6. Repeated Linear GRP . . . 140
6.3.7. Repeated Areal GRP . . . 142

6.3.8. Multiple GRP . . . 143
6.3.9. Adding Information over GRP . . . 144
6.4. Comparing Graphical Rational Patterns and the Cartesian System . . 144
6.5. Using Graphical Rational Patterns for Simple Statistical Maps 148
6.5.1. General Considerations . . . 148
6.5.2. GRP in Maps Representing Values that Correspond to Individual Elements of a Population . . . 148
6.5.3. Representing Data Referring to Individual Localities by GRP . . 150
6.6. Using Graphical Rational Patterns to Represent Values Corresponding to Territorial Units . . . 151
6.6.1. Principal Types of Territorial Units . . . 151
6.6.2. Transforming a Detailed Network of Irregular Territorial Units into a Network of Regular Cells . . . 152
6.6.3. Dealing with a Moderate Number of Meaningful Territorial Units . . . 157
6.7. Comparing the Graphical Rational Pattern Scale to Other "Regular" Scales of Symbols Used in Statistical Mapping . . . 158
6.8. Choropleth Maps . . . 162
6.9. Ensuring Correspondence between the Data and the Graphical Rational Patterns . . . 166
6.9.1. General Considerations . . . 166
6.9.2. Key to GRP in Simple Maps . . . 166

Chapter 7
Comparative Maps

7.1. Using Graphical Rational Patterns for Maps of Geosets Derived from Comparisons between Primary Geosets . . . 169
7.1.1. Selected Examples . . . 169
7.1.2. Emphasizing the Dissimilarities between Two Primary Geosets by Using Inflated GRP . . . 171
7.1.3. Presenting Geosubsets of Surpluses on the Same Two-Color Map . . . 172
7.2. Presenting Two or More Sets of Geographical–Statistical Data on the Same Map . . . 174
7.2.1. General Considerations . . . 174
7.2.2. Two-Color Maps Representing Primary Geosets over Cells of the Same Territory . . . 176
7.2.3. Maps in Three or Four Colors . . . 178
7.3. Representing Rates and Ratios by Means of Graphical Rational Patterns . . . 178
7.3.1. General Remarks . . . 178
7.3.2. Representing ${}^1R$ and ${}^2R$ by GRP . . . 179
7.3.3. Representing ${}^3R$ by GRP . . . 179
7.3.4. Keys for Showing Rates ${}^3R_j$ or ${}^3R_c$ on GRP Maps . . . 181

10.7.2. Identifying the Principal Axes 259
10.7.3. The Major Axis as an Indicator of the Principal Orientation of the Geoset 262
10.7.4. Identifying Equating Axes 262
10.7.5. Oblong and Nonoblong Geosets 263
10.7.6. Scatter 263
10.7.7. Measuring Oblongity 265
10.8. Selected Parameters of Dispersion and Oblongity 267
10.8.1. Types of Examples Considered 267
10.8.2. Parameters of Dispersion and Oblongity for Selected Population Geosets 268
10.9. Decomposition of Distance-Deviance of Compound Geosets 270
10.9.1. General Equations 270
10.9.2. Simple Territorial Models of Decomposition of Distance-Deviance 271
10.9.3. Decomposition of an Actual Fragmented Territory 272
10.9.4. Decomposition of Distance-Deviance of a Territory into that of the Habitat and the Nonhabitat 273
10.9.5. Decomposition of Distance-Deviance of Primary Geosets into Determinants of Derived Geosets 274

Chapter 11
Parameters of Form: Shapes of Territories and Patterns of Populations

11.1. Introduction 277
11.2. Shape and Parameters of Looseness of Territories 278
11.2.1. Shape of Territories 278
11.2.2. Parameters of Looseness of Territories 279
11.3. Relative Scatter of a Territory: General Considerations 282
11.4. Relative Scatter of Regular Geometrical Figures 283
11.4.1. Relative Scatter of Regular Polygons 283
11.4.2. Relative Scatter of Figures Affinely Equivalent to Regular Polygons 284
11.4.3. Relative Scatter of Isosceles Trapezoids 285
11.4.4. Relative Scatter of Other Simply Connected Figures 286
11.5. Parameters of Looseness for Territories with Concavities: Examples for Italy 286
11.5.1. Parameters of Looseness for Territories with Internal Concavities 286
11.5.2. Analysis of Factors of Looseness of Habitats and Nonhabitats for 385 Geosets of Italy 289
11.5.3. Extension of the Findings Concerning Looseness to Other Subterritories 291
11.5.4. Parameters of Looseness for Territories with External Concavities 292
11.5.5. Parameters of Looseness for Fragmented Territories 294

11.5.6. Parameters of Fragmentation Insensitive to the Location of the Fragments 296
11.6. Parameters of Looseness for Primary and Derived Population Geosets 297
11.6.1. Looseness of Primary Population Geosets 297
11.6.2. Analysis of Factors of Looseness for 385 Primary Population Geosets for Italy 298
11.6.3. Analysis of Factors of Looseness for Derived Population Geosets 299
11.7. Parameters of Looseness for Primary and Derived Geosets for the United States 299
11.8. Parameters of Looseness for Cities 300
11.8.1. An Almost Circular City: Rome 300
11.8.2. Other Cities 304
11.9. Determination and Analysis of Clusters of Surpluses of Population Geosets over the Uniform Distribution 305
11.9.1. Identifying Clusters in a Simple Setting 305
11.9.2. Extending the Study of Clusters to Large Populations 308
11.9.3. Examples of Clusters: Surpluses of the Population of the United States over the Uniform Distribution 309
11.9.4. Overall Parameters of Clustering and Its Impact 310
11.10. Elementary Methods for Partitioning a Geoset 311
11.11. Some Remarks on Parameters of Asymmetry 312

Chapter 12
Parameters of Position of Geosets: Coincidence, Congruence, and Similarity

12.1. Introduction 313
12.2. Coincident and Congruent Geosets 314
12.3. Preparatory Steps for Comparing the Locations of the Elements of Congruent Geosets 316
12.4. Principal Position of a Geoset 318
12.5. Parameters Describing the Position of a Geoset Uniquely 320
12.6. Similar Geosets and Reduced Principal Coordinates 321

Part IV: Summary Comparison of Geosets via Calculation of the Minimal Distance between Them

Chapter 13
Minimal Distance between Geosets: Objectives, Methods of Calculation, and Selected Examples

13.1. Comparing a Pair of Geosets by Means of a Summary Parameter 325

13.2. Minimal Distance between Two Aligned Geosets 327
13.2.1. Analogy to the Distance between Two Arrays of Cograduated Values 327
13.2.2. Arithmetic and Quadratic Average of Differences between Cograduated Values 329
13.3. Minimal Distance between Two-Dimensional Geosets 329
13.3.1. Introducing Two-Dimensional Minimal Distance in a Simple Setting 329
13.3.2. Adapting the Transport Problem Algorithm 331
13.3.3. Minimizing Totals of Distances and Squares of Distances . . . 333
13.3.4. Generalization of the Calculation of the Minimal Distance among All Types of Paired Population Geosets 334
13.3.5. Remarks Concerning the Notation for Minimal Distance . . . 334
13.3.6. Graphical Presentation of Lines of Minimizing Distances . . . 335
13.3.7. Expressing Minimal Distances as Pure Numbers 337
13.4. Examples of Minimal Distance between Actual Geosets and the Uniform Distribution 337
13.4.1. Minimal Distance from the Uniform Distribution of Selected Geosets for Italy 337
13.4.2. Minimal Distance from the Uniform Distribution of 202 Geosets for Italy 338
13.4.3. Minimal Distance between Populations and the Uniform Distribution for Various Countries, Cities, and Time Periods . 339
13.5. Examples of Minimal Distances between Pairs of Geosets: Comparisons to Populations or Other Paradigmatic Geosets 340
13.5.1. Comparisons between Pairs of Geosets 340
13.5.2. Minimal Distances between Geosets of Populations and Subpopulations Classified by Sex, Age, and Marital Status . . 341
13.5.3. Minimal Distance between the Labor Force and Individuals Engaged in Commerce in Italy 343
13.5.4. Minimal Distances between the Population and Ethnic Subpopulations in the United States 343
13.6. Minimal Distance between Territories 344
13.7. The Influence of Aggregation of Input Data over Minimal Distance 346

Chapter 14
Properties of Minimal Distance: Triangle of Distances, Joint Analysis of Several Minimal Distances, and Measuring Inconsistency

14.1. Minimal Distance as a Metric 347
14.2. Minimal Distances between Three Geosets 347
14.2.1. Triangularity or Collinearity and Measures of Inconsistency . . 347
14.2.2. Triangle of Minimal Distances 348
14.2.3. Consistencies and Inconsistencies in Minimal Distances of Ethnic Subpopulations in the United States 350

14.2.4. Minimal Distances of a Geoset from Two Paradigmatic Distributions . . . 351
14.2.5. Minimal Distances between Three Ordered Geosets . . . 353
14.2.6. Minimal Distances between Many Ordered Geosets . . . 356
14.3. Collinearity of Three Geosets, ${}^gK$, ${}^gG$, and ${}^gJ$, Related through ${}^gK = v_G^K\,{}^gG + v_J^K\,{}^gJ$ . . . 357
14.4. Triangle of Minimal Distances between Two Primary Geosets and Three Geosets Derived from Them . . . 361
14.5. Minimal Distance between Two Geosets as a Product of the Proportion of Surpluses and the Distance between Them . . . 364
14.5.1. The General Rule ${}^m\bar{D}(G, J) = \Delta\,{}^m\bar{D}(SGJ, SJG)$ . . . 364
14.5.2. Interpreting ${}^m\bar{D}(SGJ, SJG)$ and Δ in the Study of ${}^m\bar{D}(G, J)$ . . . 364
14.6. Minimal Distances between Two Components and the Compound Geoset . . . 365
14.6.1. Triangle of Minimal Distances between Two Components and Their Compound Geoset . . . 365
14.6.2. An Excursus: Equations for the Triangle of the Minimal Distances between Components and Compound Geosets . . . 367
14.6.3. Practical Rules Concerning Minimal Distances between Components and Compound Geosets . . . 368
14.6.4. A Special Configuration: Nonintersecting Components . . . 370
14.6.5. Extensions to Other Types of Geosets . . . 370
14.7. An Excursus: Indirect Calculation of Minimal Distance between Any Two Geosets Based on Their Distances from Paradigmatic Distributions . . . 370
14.7.1. Possible Approaches to the Problem . . . 370
14.7.2. Calculating the Limits to the Minimal Distance . . . 372
14.7.3. Calculating the Coordinates of Dots Representing the Geosets Studied on the Basis of Minimal Distances from Paradigmatic Distributions . . . 373
14.7.4. An Experiment for Evaluating the Minimal Distances Calculated Indirectly . . . 375
14.8. Minimal Distance between Net Geosets . . . 376
14.8.1. Geosets Netted from the Influence of a Common Determinant . . . 376
14.8.2. Netting Each Geoset from the Influence of a Different Determinant . . . 376

Chapter 15
Decomposition of the Average of the Squares of Minimizing Distances

15.1. Introduction . . . 379
15.2. First Approach to Decomposition of ASMD . . . 381
15.2.1. Developing ASMD . . . 381
15.2.2. Interpreting $d_G^2 + d_J^2$ . . . 382

15.3. Decomposition of ASMD into Nonnegative Components 383
15.3.1. Absolute and Percentage Components of ASMD 383
15.3.2. Ratios of ASMD and Its Components to the Joint Dispersion of ${}^gG$ and ${}^gJ$ 384
15.3.3. Decomposition of the ASMD between Surpluses SGJ and SJG 386
15.3.4. An Excursus: Algebraic Relationships between Components of ASMD for ${}^gG$ and ${}^gJ$ and ${}^gSGJ$ and ${}^gSJG$ 388
15.3.5. Graphic Presentation of ${}^m\ddot{D}^2(G, J)$ and Its Components 389
15.3.6. The Quadratic Average of Minimizing Distances 391
15.3.7. Decomposition of ASMD for Component and Compound Geosets 392
15.4. Examples of Decomposition of ASMD 394
15.4.1. Types of Examples Presented 394
15.4.2. Components of ASMD between Subpopulations in the United States 394
15.4.3. Components of ASMD According to Sex and Marital Status 397
15.4.4. ASMD for a City Growing Concentrically (Rome) 397

Chapter 16

Minimal Distance between One Geoset and Another Obtained by Transforming the Coordinates of the First: Eliminating Influences of Position and Dimensions from Minimal Distance

16.1. Introduction 399
16.2. Minimal Distance between One Geoset and Another Obtained by Parallel Translation of the First 399
16.2.1. Parallel Translation 399
16.2.2. Translation of a Geoset to Its Center 401
16.2.3. Comparing Two Geosets Translated to Their Centers 401
16.3. Minimal Distance between One Geoset and Others Obtained from It Through Rotation and Reflection 402
16.4. Comparing Geosets Found over Different Territories 403
16.4.1. Minimal Distance between Geosets in the Same Position . . . 403
16.4.2. Checking Coincidence and Congruence of Two Geosets 403
16.5. Minimal Distance between Similar, Reduced, and Standardized Geosets 404
16.5.1. Comparing a Geoset to a Similar Set Obtained from It by Dilatation 404
16.5.2. Minimal Distance between Geosets Transformed in the Same Way 404
16.5.3. Checking Whether Two Reduced Geosets Are Similar 405
16.5.4. Comparing Standardized Geosets and Checking Affine Equivalence 405

16.6. Converting the Minimal Distance to a Pure Number 406
16.6.1. Studies in Which Dimensionless Minimal Distance May Be Required 406
16.6.2. An Ideal Solution 406
16.6.3. Problems Arising from the Distance between Centers and a Solution 407
16.6.4. Rendering ${}^{m}\bar{D}(G, J)$ and ${}^{m}\ddot{D}^{2}(G, J)$ Dimensionless Through Comparison with the Average Distance between Pairs of Elements of the Basic Set 408

Part V: Epilogue

Chapter 17
An Overview and Some Concluding Remarks

17.1. Introduction 413
17.2. Principal Guidelines for the Geostatistical Methods Described in this Volume 414
17.3. Types of Territorial Containers 415
17.3.1. Gross Territorial Containers 415
17.3.2. The Habitat as a Net Territorial Container 415
17.3.3. Other Types of Territorial Containers 416
17.3.4. Enumerative or Meaningful Units into which the Gross Territorial Container Is Divided 416
17.4. Geostatistical Maps 417
17.4.1. General Considerations 417
17.4.2. Presenting Geosets by Means of Graphical Rational Patterns 417
17.4.3. Other Graphical Presentations Used in Geostatistical Research 418
17.5. Geostatistical Parameters 418
17.5.1. Parameters of Territorial Coverage and Density 419
17.5.2. Parameters of Position 420
17.5.3. Parameters of Dispersion and Oblongity 421
17.5.4. Parameters of Shape 422
17.6. Measuring the Terrritorial Discrepancy between Geosets by Means of the Minimal Distance between Them 424
17.6.1. Basic Concepts of Minimal Distance 424
17.6.2. Properties of Minimal Distance 425
17.6.3. Some Applications of Minimal Distances 426
17.7. Differences in Detail, Generalization, Abstraction, and Comparability of Maps and Geostatistical Parameters 427
17.8. Mainframe Computer Programs Used in the *Statistical Atlas of Italy* 427

17.9. Software for Processing Geosets, Constructing Maps, and Calculating Parameters . . . 428
17.9.1. The Software Package . . . 428
17.9.2. Maps and Parameters Comparing Geosets with a Uniform Distribution . . . 428
17.9.3. Maps and Parameters Comparing Two Geosets Distributed over the Same Gross Territorial Container . . . 428
17.9.4. Minimal Distance between Two Geosets: Parameters and Maps . . . 429
17.9.5. Additional Features . . . 429

Appendixes

Appendix A: Distance-Variance per Unit Area of a Circle . . . 431
Appendix B: Selected Demographic Findings: Washington, D.C., Atlanta, and Austin . . . 431
Appendix C: Selected Findings of Comparative Studies Regarding the Characteristics of the Territorial Distribution of the Jewish Minority in Various Cities and Time Periods . . . 437
Appendix D: Definitions and Symbols Used in this Book . . . 447
Appendix E: Relationships between the Mean Centers of Primary and Derived Geosets . . . 449
Appendix F: Iterative Process for Calculating the Two-Dimensional Median Point . . . 451
Appendix G: Minimal Standard-Distance Set with Given Oblongity . . . 452
Appendix H: Measuring Central Asymmetry through the ASMD . . . 454
Appendix I: Numerical Example of Decomposition of the ASMD for Nonintersecting Geosets . . . 454
Appendix J: Geostatistical Methods and Related Computer Software . . . 456

Bibliography . . . 459

Index . . . 471

List of Color Plates

Plate 1. Physical map of Italy. Color scheme: mountains, brown; hills, yellow; and plains, green. (Source: *Statistical Atlas of Italy*, 1988).

Plate 2. Jerusalem. Territorial geosubsets for the distribution of hotels, ^{g}P, and uniform distribution, ^{g}U, 1989. Colors of the hexagons show three geosubsets: white, nonhabitat (NH); red, surplus of P over U (SPU); and black, match $M(P, U)$. (Source: Israel Ministry of Tourism).

Plate 3. Jerusalem. Territorial geosubsets for the distribution of hotels, ^{g}P, and tourist agencies ^{g}Q, 1989. Colors of the hexagons show four geosubsets: white, nonhabitat (NH); green, surplus of P over Q (SPQ); blue, surplus of Q over P (SQP); and red, match $M(P, Q)$. (Source: Israel Ministry of Tourism).

Plate 4. Italy. Area of citrus groves, 1981. (Source: ISTAT).

Plate 5. Selected values from a scale of positive and negative GRP.

Plate 6. Rome. Surplus of children aged 0–14 over persons aged 65+ (blue) and surplus of persons aged 65+ over children aged 0–14 (red), 1981. (Source: Commune di Roma, Segretariato Generale il Direzione, Ufficio di Statistica e Censimento).

Plate 7. Hexagon cell with four positions representing two primary geosets.

Plate 8. Hexagon cell with four positions representing four primary geosets.

Plate 9. Rome. Distribution of children aged 0–14 (red) and the distribution of persons aged 65+ (blue), 1981. (Source: Commune di Roma, Segretariato Generale il Direzione, Ufficio di Statistica e Censimento).

Plate 10. United States. Individuals of Asian origin (blue) and of Hispanic origin (red), 1980. (Source: Strategic Mapping—Counties Data Files, USA).

Plate 11. Rome. Four periods of dwellings. Color of the GRP indicates the period: red, dwellings built up to 1919; blue, dwellings built between 1920 and 1945; green, dwellings built between 1946 and 1960; and orange, dwellings built between 1961 and 1970 (Source: Commune di Roma, Segretariato Generale il Direzione, Ufficio di Statistica e Censimento).

Plate 12. United States. Rates per 100,000 students (4+ years in college) by states, 1987. Color of GRP: red, above average; blue, below average. (Source: Strategic Mapping—Counties Data Files, USA).

Plate 13. Rome. Comparison between the distribution of persons aged 45–64 (1981) and the distribution of dwellings built during 1946–1960. Color of GRP: red (71.10%), locations where both distributions are above average; black (27.10%), locations where the first distribuion is below average and the second is above average; blue (1.75%), locations where both distributions are below average; and green (0.05%), locations where the first distribution is above average and the second is below average (Source of data: Commune di Roma, Segretariato Generale il Direzione, Ufficio di Statistica e Censimento).

Plate 14. United States, 1980. A residual spatial analysis by counties of income per capita as a function of four population groups: Whites, Afro-Americans, Hispanics, and Asians. Color of GRP indicates residual: red, positive; blue, negative. (Source: Strategic Mapping—Counties Data Files, USA).

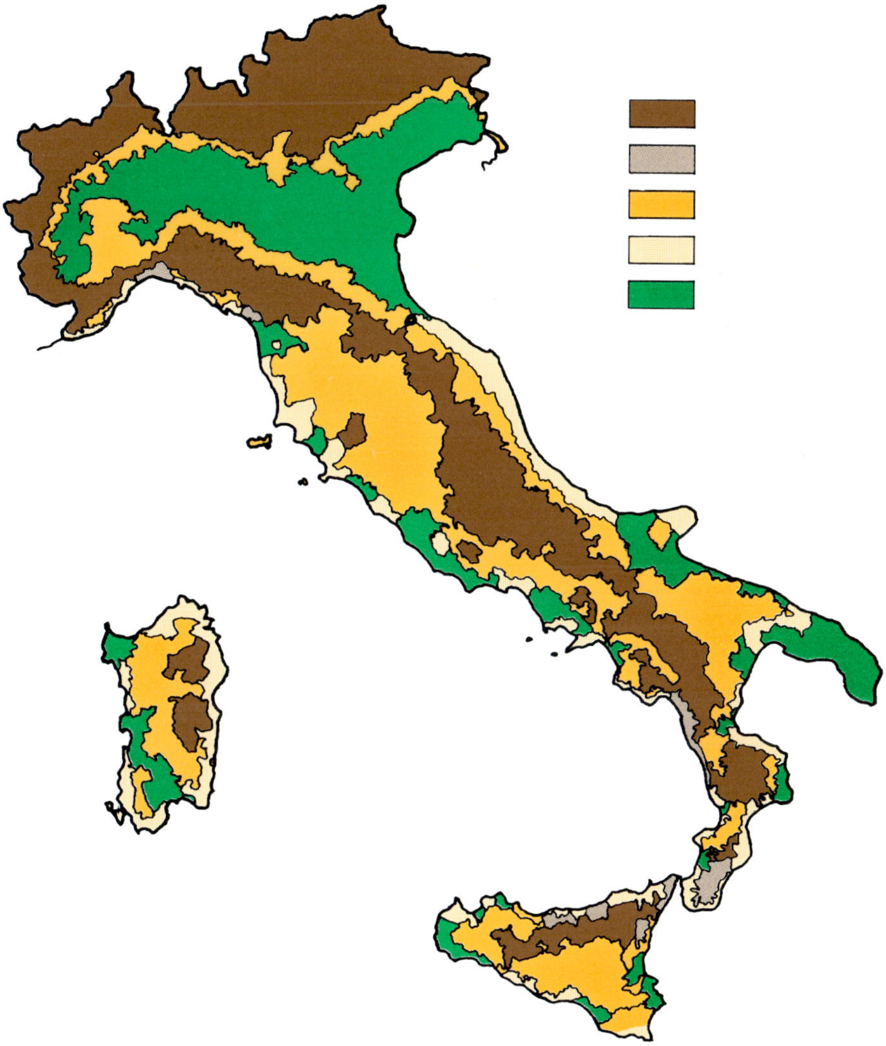

Plate 1

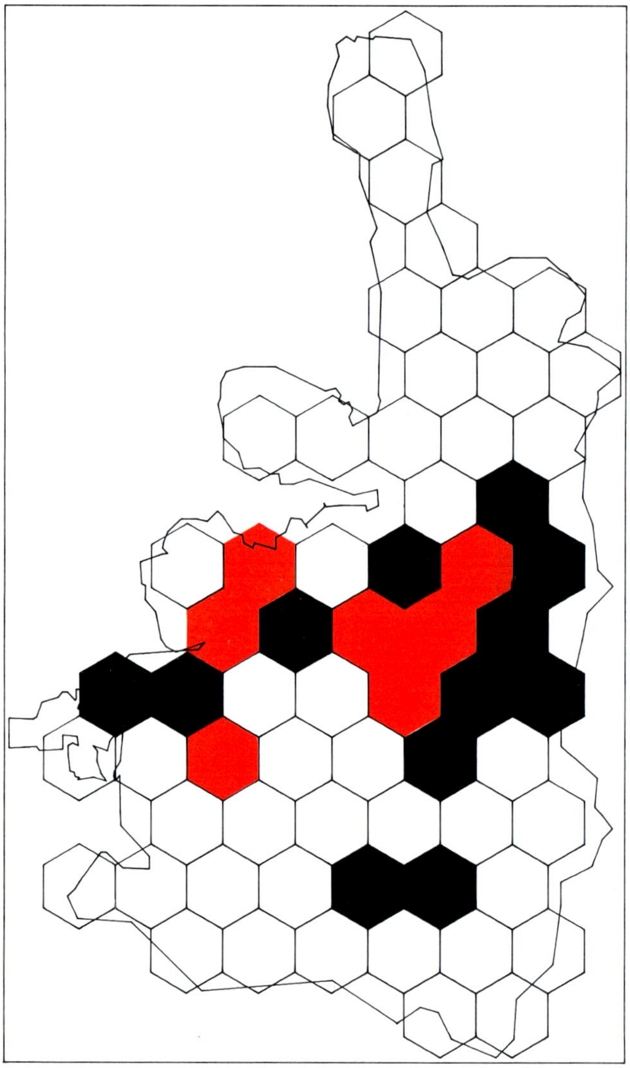

Plate 2

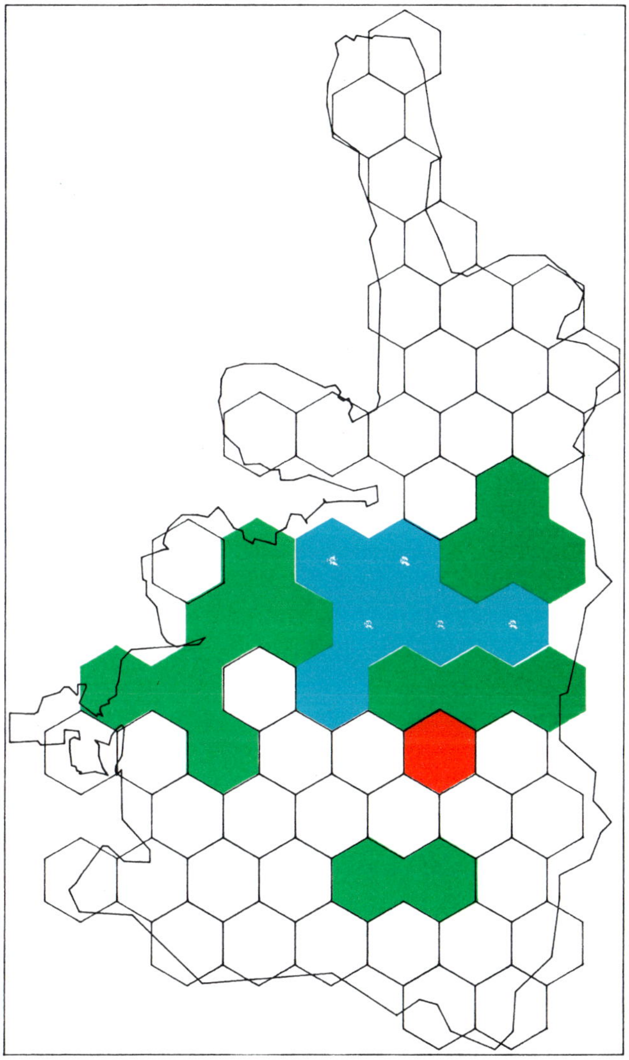

Plate 3

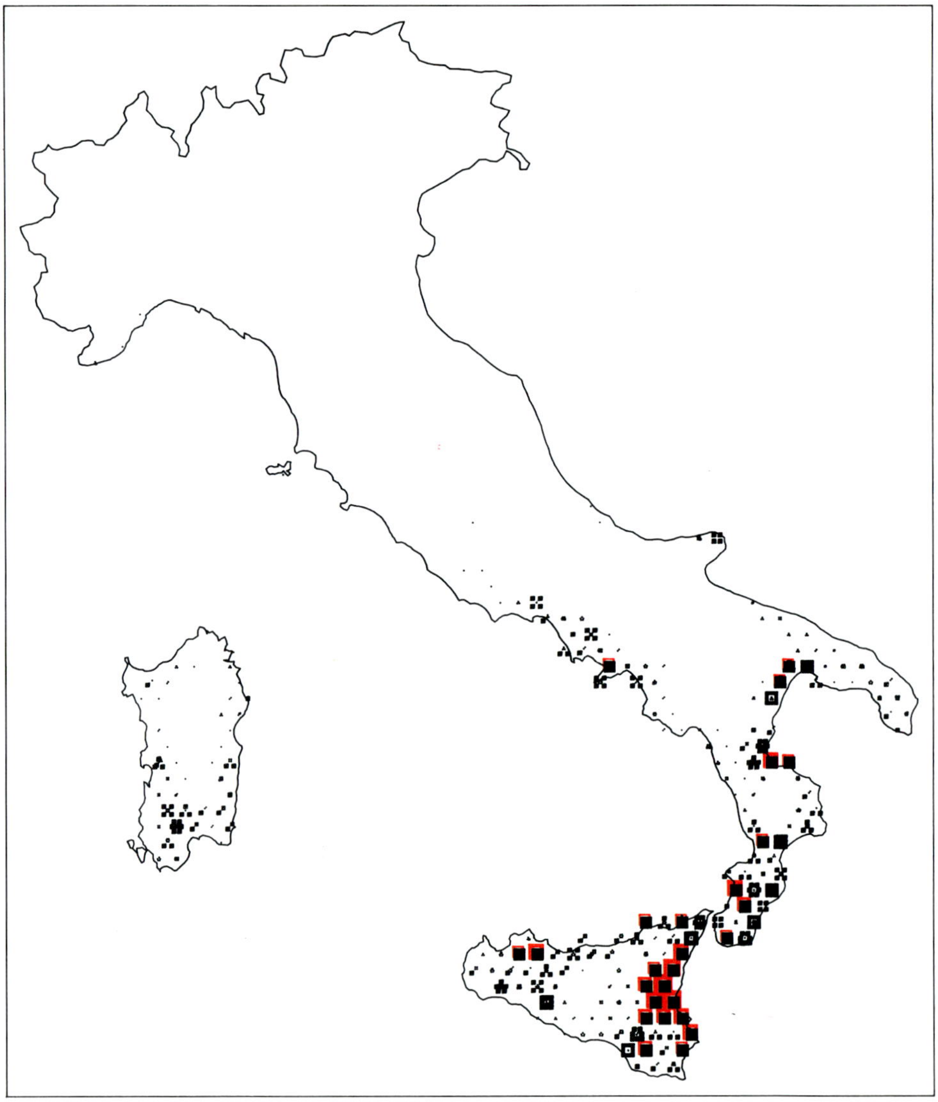

Plate 4

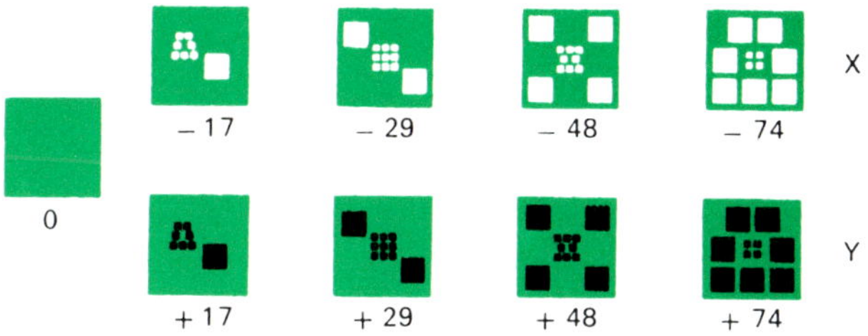

Single GRP

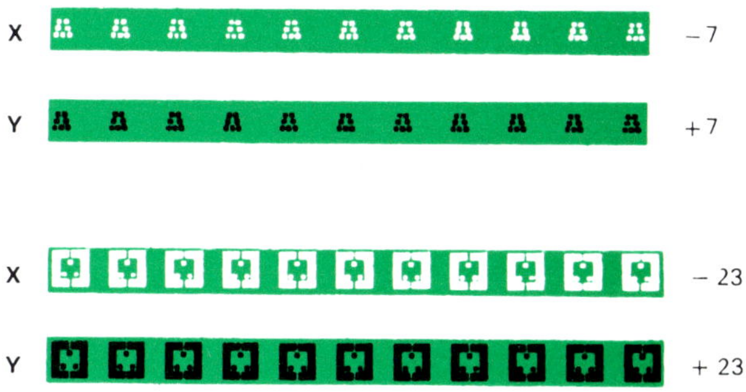

Linear GRP

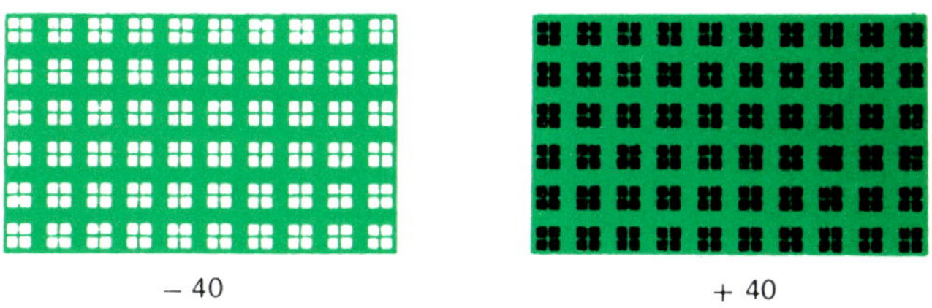

Areal GRP

Plate 5

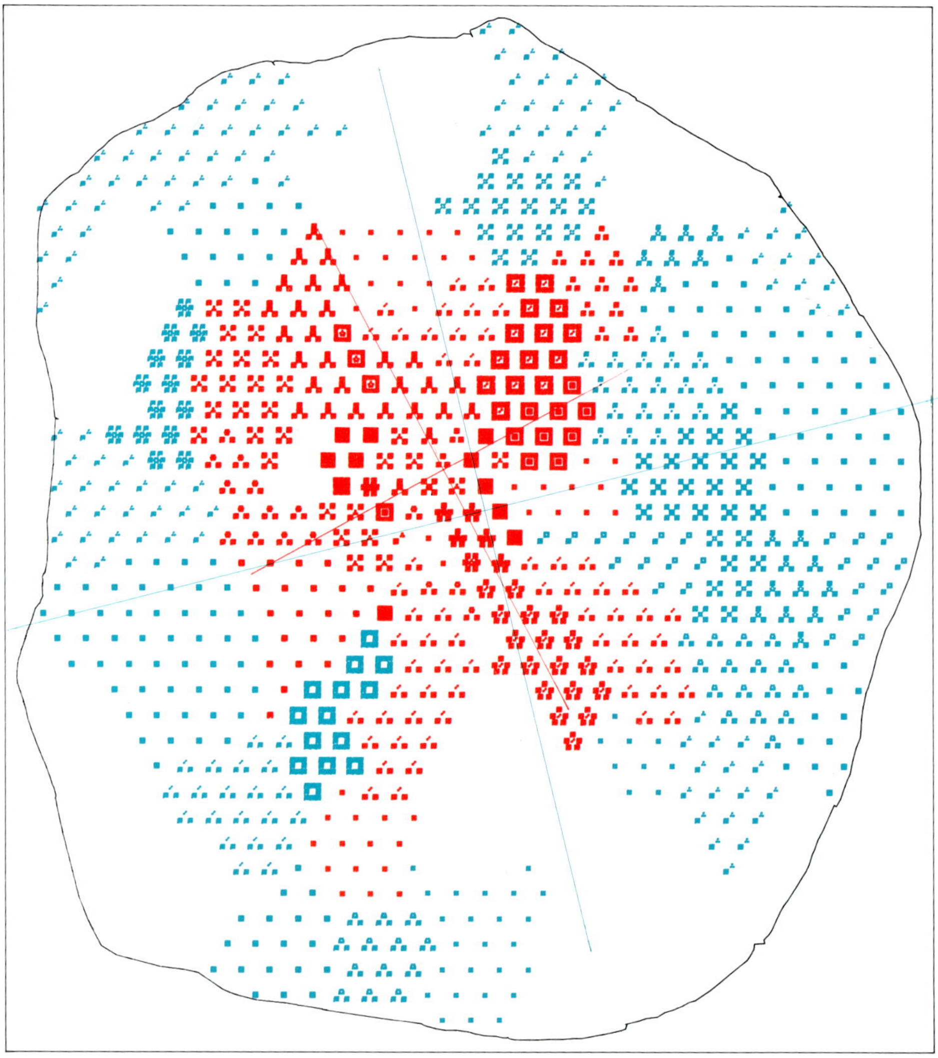

Plate 6

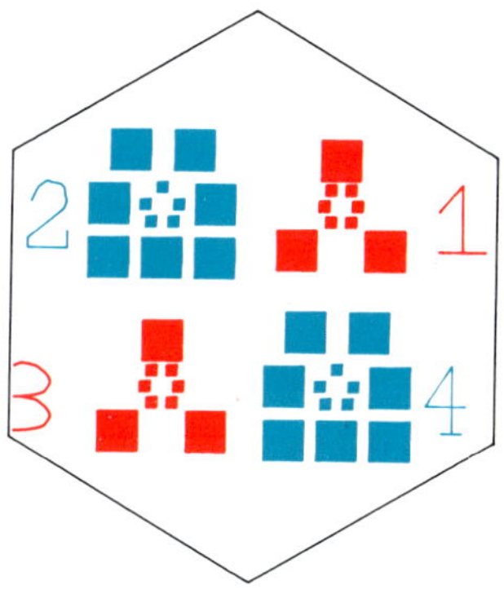

Plate 7

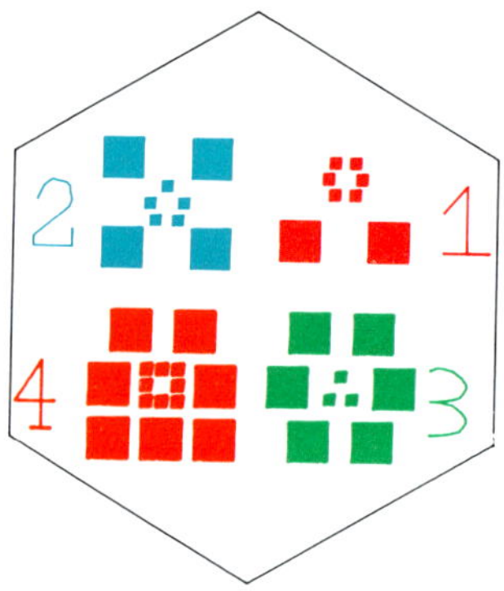

Plate 8

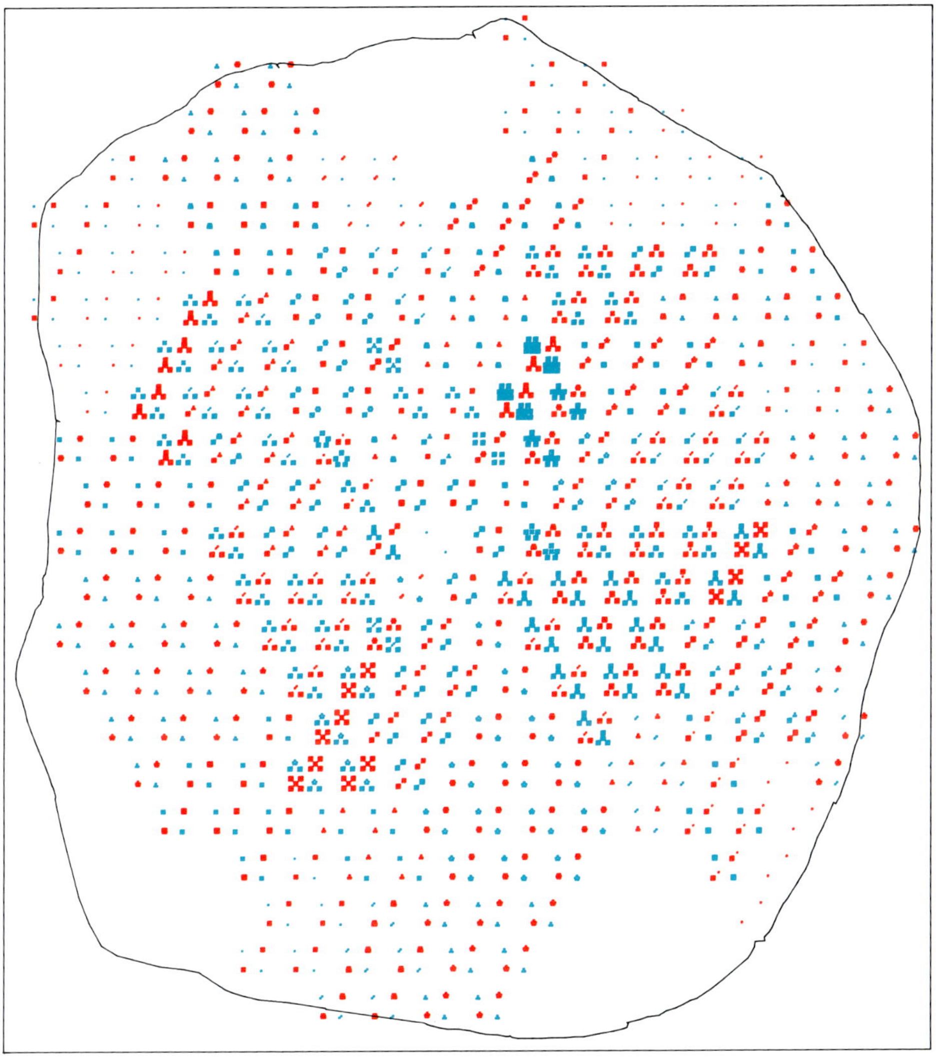

Plate 9

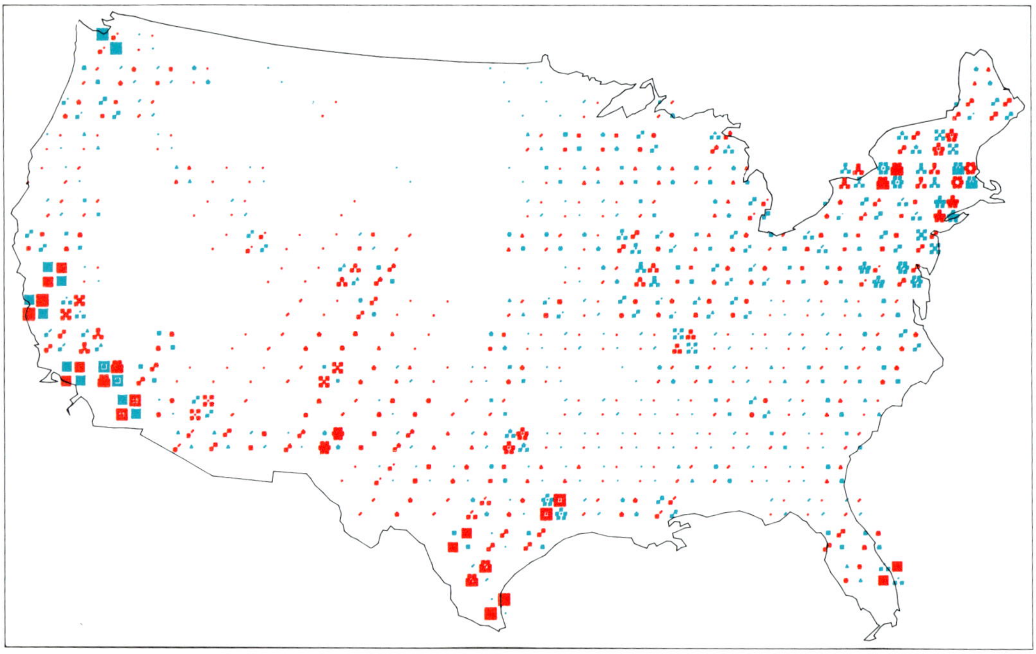

Plate 10

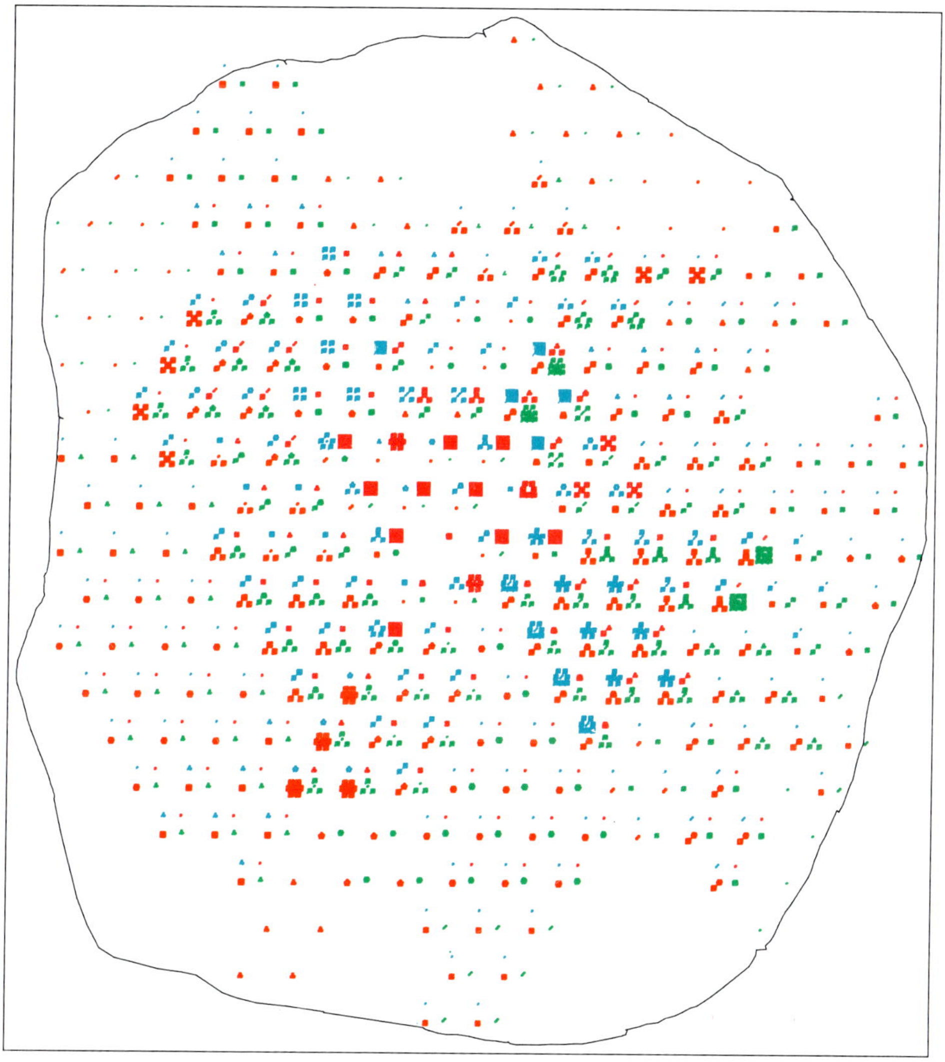

Plate 11

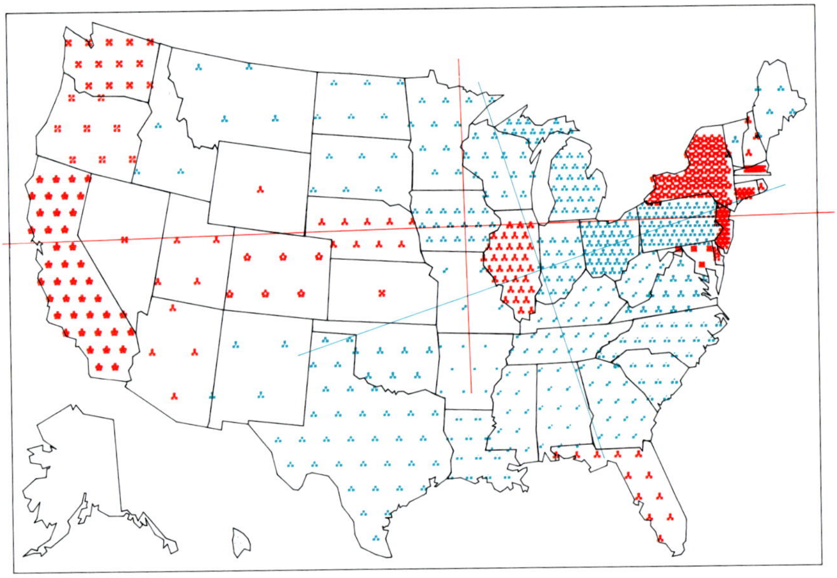

Plate 12

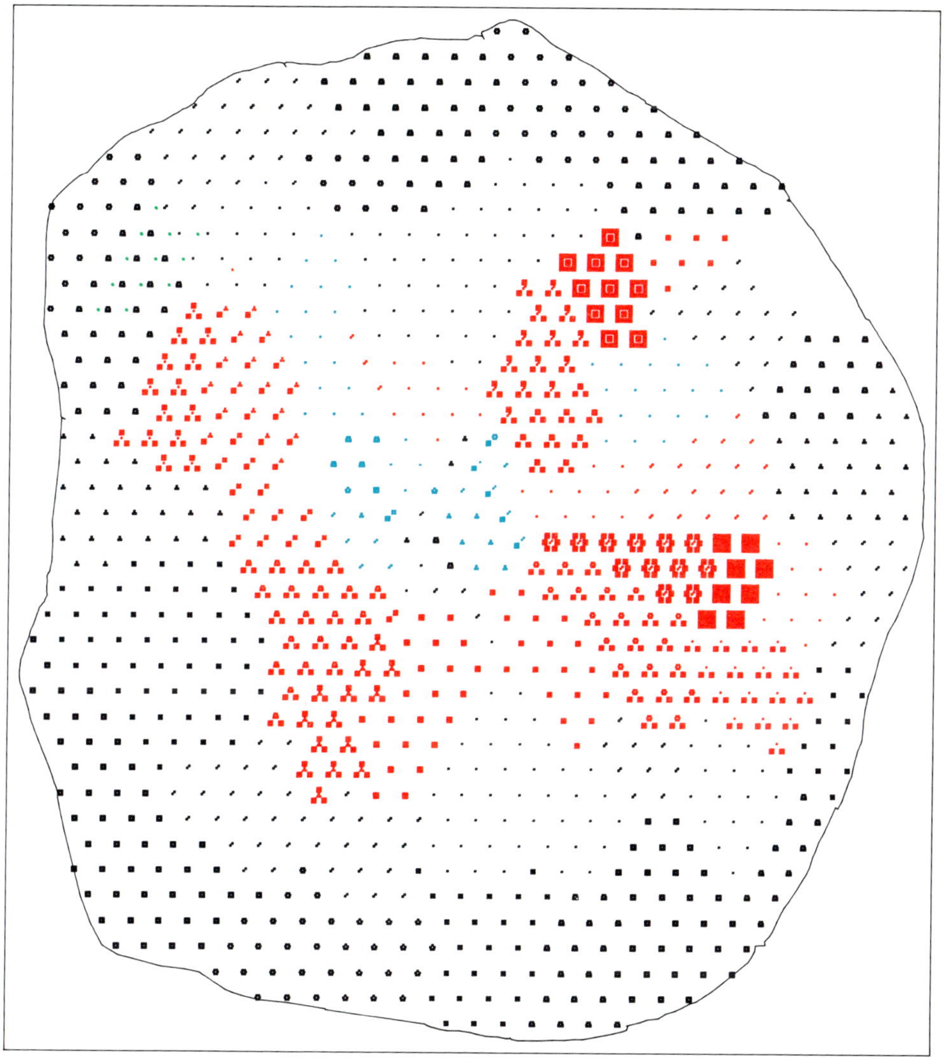

Plate 13

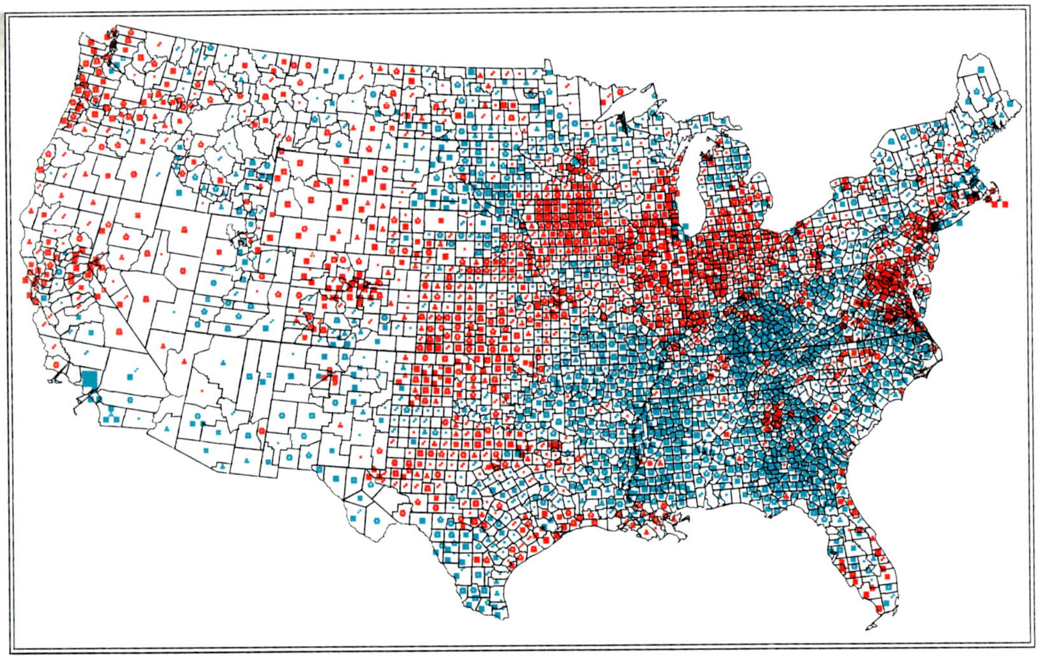

Plate 14

Part I
Basic Concepts

Chapter 1

Introducing Geostatistical Methods

1.1. Primary Aims of this Volume

The primary, but not the sole, aim of this book is to make an integrated system of methods, described as "geostatistical," available to statistical agencies and to researchers in geographical statistics. Extensive field experience has shown that these methods enable us to overcome many of the difficulties usually encountered in the graphical presentation, treatment, and analysis of bulky data that originate from large numbers of small and irregular territorial units. They permit us to: (a) represent graphically the information supplied by these data in ways that make it possible to (i) grasp the macrogeographic characteristics of the territorial distributions studied quickly, (ii) identify (when required) microgeographical features, and (iii) make quick comparisons between maps; (b) use geostatistical parameters to quantify the main aspects of position, dimensions, and shape of the distributions studied; (c) compare these parameters for different distributions, different territories, and different time frames; (d) measure, geographical discrepancy ("minimal distance") between pairs of distributions synthetically; and (e) obtain averaged values from many of these distances taken together.

1.2. Introducing Rational Maps and Comparisons among Them

1.2.1. The Distribution of Population over the Territory of Italy

As an introduction to geostatistical methods, we shall present some simple examples taken from a recent large-scale application of geostatistical methods, published in the two volumes of the *Statistical Atlas of Italy*, called hereafter the *Atlas* (ISTAT, 1988, 1992).

The National Statistical Institute of Italy (ISTAT) collects large quantities of data drawn from the 8086 municipal circumscriptions (*comuni*) into which Italy is divided and publishes them in its censuses and other reports. While these data meet administrative, political, and socioeconomic needs, it was decided to use them as well as input to obtain and present information on the characteristics of the distributions of various demographic, economic, social, and other phenomena over the territory of Italy. However, enormous difficulties are encountered in trying to utilize the detailed data for these purposes, stemming *inter alia* from the following:

1. The enormous length of these series of data: no statistician or geographer, even if very well acquainted with the names and locations of the *comuni*, can make sense out of data that is attached to a list of 8086 names of circumscriptions.
2. The fact that the various *comuni* are of very different sizes: e.g., the *comune* of Rome extends over an area larger than 1500 sq km, while other *comuni* have areas of less than 1 sq km.
3. The territories of the *comuni* have very different shapes.

Consequently, the *comuni* do not constitute manageable statistical units. The larger circumscriptions often used in the Italian official statistics (95 provinces or 20 regions) do have administrative or political meaning, but have proved unsuitable for furnishing sufficiently detailed information for the preparation of maps and for systematic geostatistical research. Thus, new ways had to be sought. We describe below the main principles adopted in the first stage of the work, namely the preparation of maps:

1. The territory of Italy was divided—as shown in Figure 1.1—into a network of cells of a fixed area (about 300 sq km) and shape (regular hexagons). The number of cells (1000) was selected after experimentation, so as to give sufficient territorial detail and still be convenient for easy manipulation.
2. The *Atlas* was conceived as an organic collection of easily comparable maps. Moreover, its maps were constructed within the framework of a general program for standardization of the graphs currently produced by the ISTAT (Napolitano, 1989). Accordingly, whenever possible, the distributions presented were reduced to a convenient standard size (10,000 elements).
3. The computer transformed absolute frequencies detailed by *comuni* into relative frequencies per 10,000, distributed over the 1000 cells.
4. A graphical symbol proportional to the relative frequency was superimposed over each cell. The symbols used are *graphical rational patterns* (GRP) (see Figure 1.2), which make possible the presentation of any integer ($n = 10t + u$) by a pattern formed by u small unitary squares of area a (which denote units) and t squares of area $10a$ (which denote tens).

Figure 1.3 shows an example out of the 385 maps of territorial distributions in the *Atlas*, which shows the distribution of the population of Italy in 1987. It is immediately evident from the map that the actual population distribution differs tremendously from a uniform distribution over the territory. To people conversant with the geography of Italy it shows—among other characteristics—important concentrations of population in and around the cells containing the cities, which are found throughout Italy.

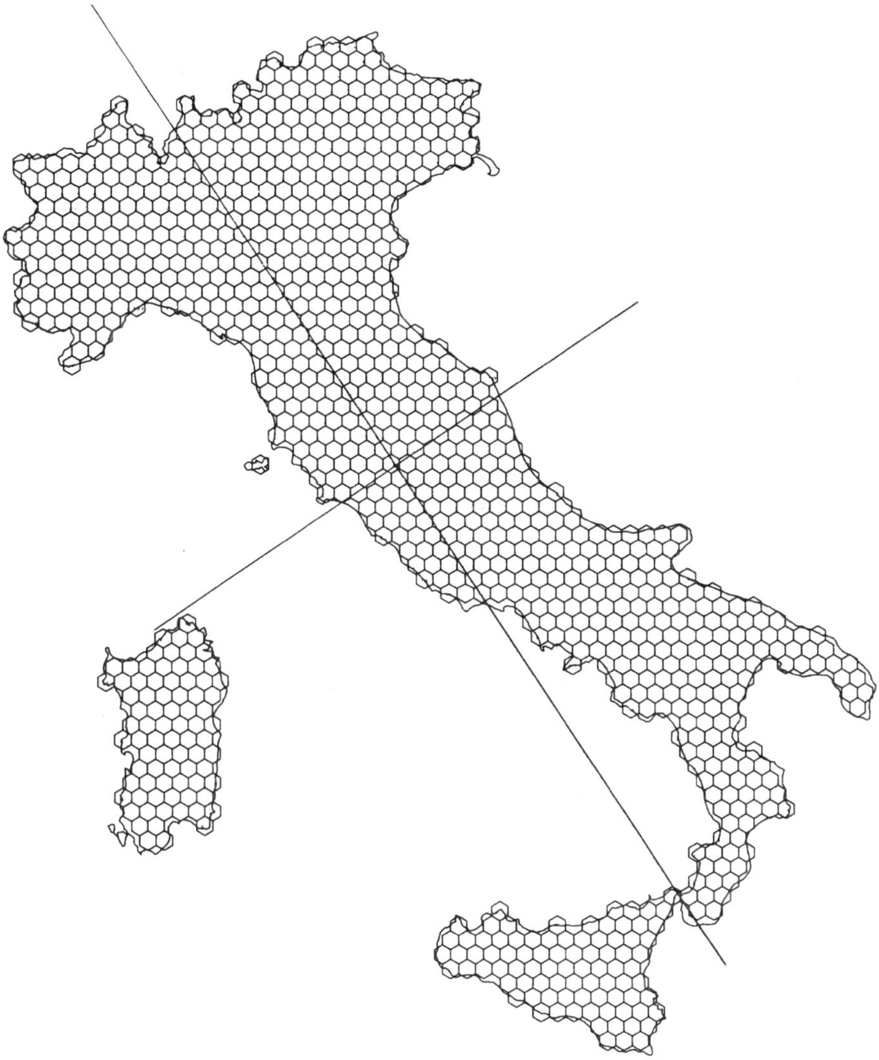

Figure 1.1. Cell network for the territory of Italy (Source: *Statistical Atlas of Italy*, 1988).

1.2.2. Comparing Distributions of Populations to a Uniform Distribution

In practice there are considerable discrepancies between the actual and the uniform distributions over the territory of all demographic, economic, social, and other phenomena taken into consideration in official Italian statistics. Thus there is no need in this case to test whether an actual distribution differs significantly from a uniform or random distribution. Rather we want to envisage and measure the extent of the discrepancies from a uniform distribution, their geographical characteristics, and their determinants.

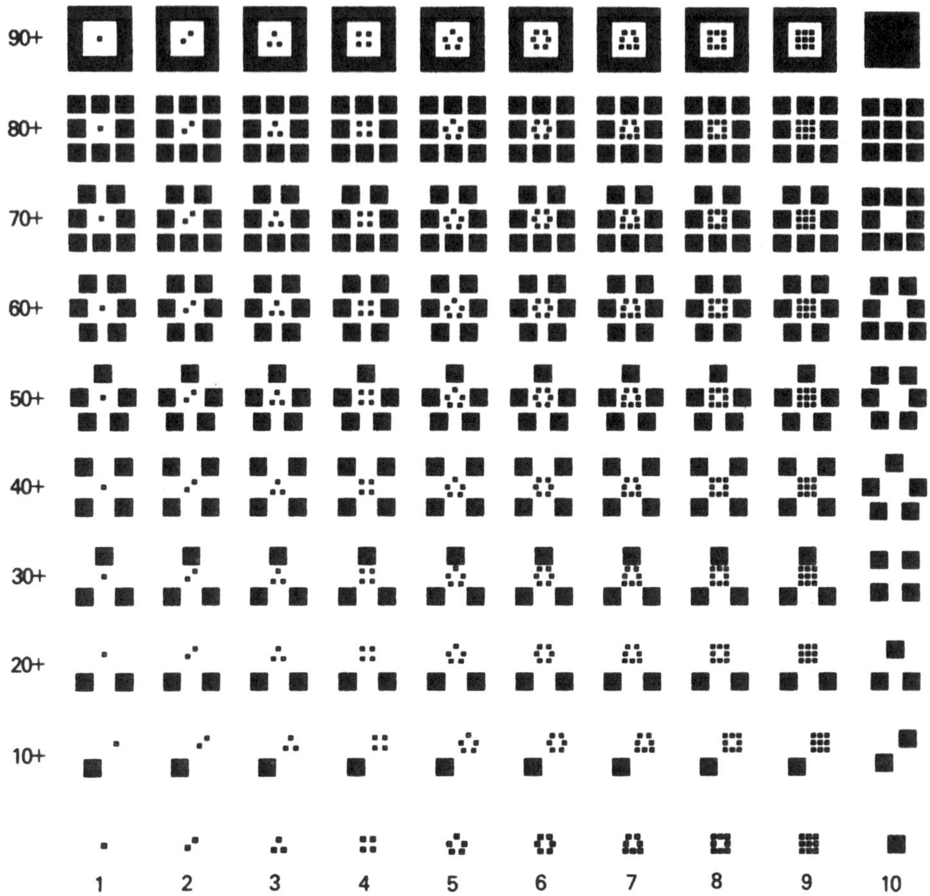

Figure 1.2. Graphical rational patterns (Source: *Statistical Atlas of Italy*, 1988).

An important step can be taken in this direction by calculating the comparison between the actual and the uniform distribution. Let us indicate by w_c^P the relative weight per 10,000 of the population distribution (shown in Figure 1.3) in each cell c. The corresponding relative weight per 10,000 in the uniform distribution is 10. We can easily visualize the uniform distribution by assuming that we put a GRP of 10 in the center of each of the cells of Figure 1.1. If in a cell c we have $w_c^P \geqslant 10$, then this cell has a surplus of population over the uniform distribution amounting to $(w_c^P - 10)$. In cells in which $w_c^P < 10$, there is a surplus of $(10 - w_c^P)$ elements of the uniform distribution over the actual population distribution. In both cases we can consider that there is a match between $\min(w_c^P, 10)$ elements of the actual and the uniform distribution.

Figure 1.4 shows a GRP map representing the surpluses of the actual population of Italy over the uniform distribution. To interpret this it may be useful to have a look at the sketch of the physical map of Italy given in Plate 1: We can see there

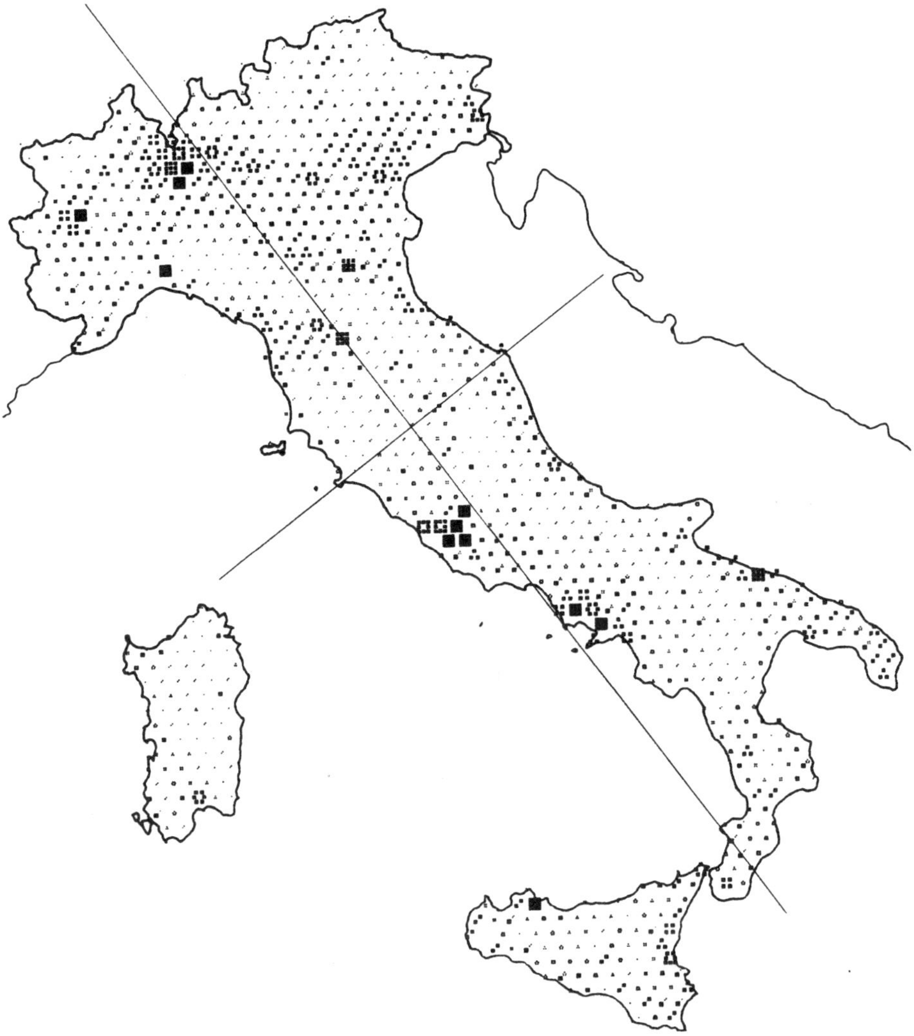

Figure 1.3. Total population of Italy, 1987 (Source: Statistical Atlas of Italy, 1988). Number of individuals: . 5700 ▪ 57,300 ▪ 114,600 ⁙ 286,500 ■ 572,900 + .*

the division of the territory of Italy into three major segments (Continental Italy, Sicily, and Sardinia), the enormous length of Italy's maritime coasts, the principal mountain ranges (Alps, Apennines), the extensive hilly areas, and the broad plain in the Po Valley.

It appears from Figure 1.4 and Plate 1 that actual population surpluses are found in the urban and semiurban areas, in many of the coastal regions, and in the

*There are several ways to represent the key to the GRP maps (see Section 6.9.2). The computer software provides a key that includes all the symbols that appear on a specific map. However, in order not to overburden the reader, we have decided in this volume to use a representative key with only five selected symbols for each GRP map.

Figure 1.4. Italy. Surplus of population over uniform distribution, 1987; Number of individuals: .1 ▪10 ▪▪20 ⁘ 50 ■ 100 + (Source: *Statistical Atlas of Italy*, 1988).

very fertile Po Valley. The GRP map of surpluses of uniform distribution over the actual population is given in the *Atlas* (Vol. II, p. 33) and is not reproduced here. However, the sizes $\min(w_i^P, 10)$ and the location of the matches are shown graphically in a different and most significant way in Figure 1.5. These matches can only assume the values $1, 2, \ldots, 10$ per 10,000, and so are represented by hexagons having areas 1/10, 2/10, ..., 10/10 of the standard hexagonal cell of Figure 1.1. These hexagons are blackened in Figure 1.5. Hexagons fully covered by solid black indicate the part of territory that has a population density equal to, or exceeding, the average

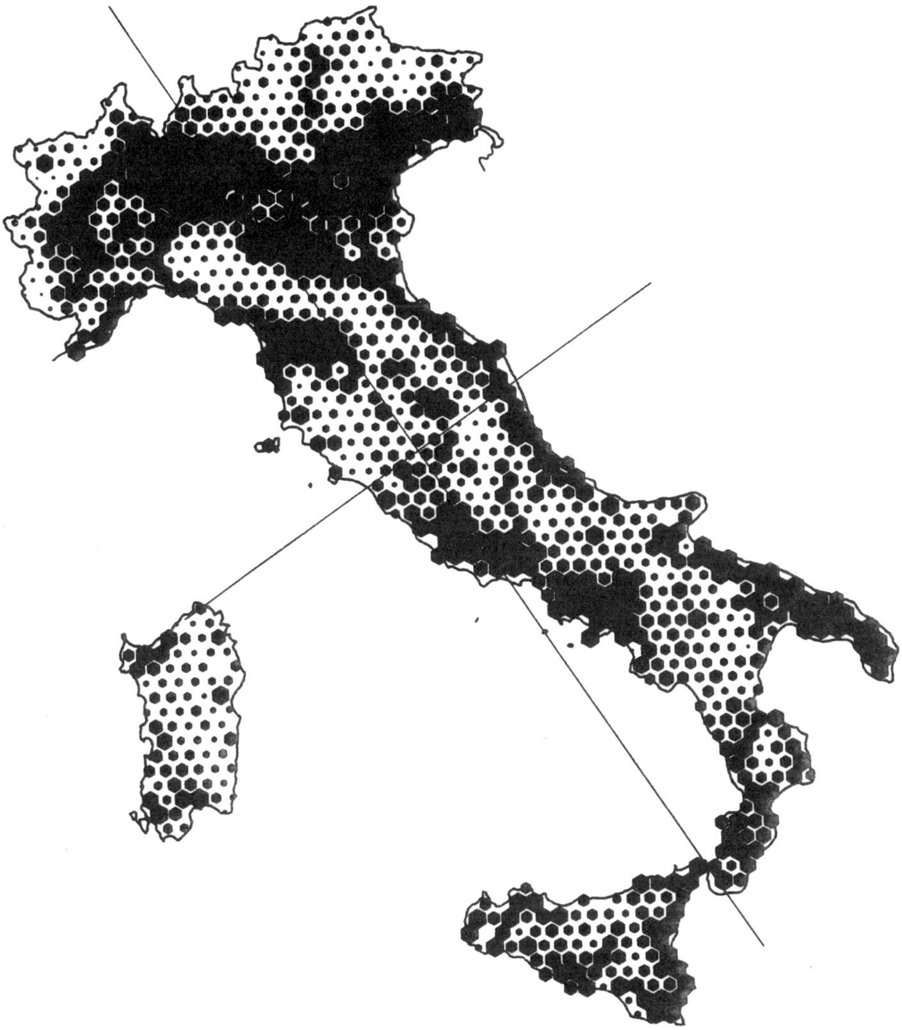

Figure 1.5. Italy. Habitat and nonhabitat of population, 1987 (Source: *Statistical Atlas of Italy*, 1988).

density. Other hexagons correspond to areas where population elements that are less dense than the average are found. The entire territory of Italy can be considered as the gross container of the population distribution. The black hexagons show the distribution of the net territory actually inhabited. We call this net territory as the "habitat." Similarly, the white areas represent the "nonhabitat," namely the parts of the territory over which no population is found.

Figure 1.5 shows clearly (as does Figure 1.4) the areas with density equal to, or higher than, the average density. It also makes clear the large zones of nonhabitat, which correspond mainly to the Alps and the Apennine range and the surrounding hills and areas in southern Italy and Sardinia known to be less developed in comparison with many zones in central and northwest Italy.

1.3. Formalizing Population Distribution as Geosets

1.3.1. Primary Geosets (${}^{g}P$, ${}^{g}U$) and Geosets Derived from Comparing Them

In the simple example given above, we dealt with the distribution of the relative weights of the elements of the population in the locations studied. Both to generalize this example, and to make the presentation of operations over a distribution of this type more accurate, it has been found convenient to formalize our work by introducing the concept of *geostatistical set* (abbreviated *geoset*). A geoset is a set of triplets that indicates the two Cartesian coordinates of each location and the relative weight to be attached to each location, in order to construct the map or calculate geostatistical parameters.

In our work we shall generally consider large populations (e.g., that of Italy, discussed above, which in 1987 was 57,290,519 people). However, to introduce the concept of geoset in a simple and accurate way it will be convenient to start with a small population, as we shall do in Chapter 3, where we deal with the distribution of hotels over the municipal area of Jerusalem (see Fig. 3.3 in Section 3.2).

Let set H be the list of all hotels h, $F^{H} = 63$ the size of this "population" of hotels, and (X_h, Y_h) the coordinates of the points indicating the location of each hotel on the map. We consider the following as geoset ${}^{g}H$ corresponding to set H:

Geoset ${}^{g}H$ = {Triplets (X_h, Y_h, w_h) are the coordinates of each hotel h; w_h is the relative weight to be attributed to these coordinates. If we assign the same weight to each hotel, $w_h = 1/F^{H}$. If we take into account the different sizes of the hotels, $w_h = B_h/B$, where B_h is the number of beds in hotel h and B is the total number of beds in all the hotels of Jerusalem}.

In the case of a large population P, such as that of Italy, individual locations are not assigned to each element, and even if they were that information could not be used. Information is given only in aggregate form (by 8086 *comuni*). We have found it convenient to work on data indicating the number of individuals in each of the hexagonal cells c and to reduce the total size of the population to 10,000 elements. We assign the coordinates (X_c, Y_c) of the center of cell c to all the individuals found in that cell. Then w_c^P is the relative weight per 10,000 to be attached to these coordinates. We consider the geosets ${}^{g}P$ of the population of Italy in 1987 to be formed by triplets indicating the coordinates (X_c, Y_c) and the relative weight w_c^P (referred to a total of 10,000).

The operations described in Section 1.2.2 can be generalized as follows: We start by considering two primary geosets (${}^gP, {}^gU$), where ${}^gU$ is the geoset of the uniform distribution over the territory T (see Figures 1.3 and 1.1). In practice, geostatistical parameters obtained directly from the contours of the territory T or from ${}^gU$ coincide (see Section 1.5).

Comparing ${}^gP$ with ${}^gU$ we obtain: (a) Three "derived" population geosets, which can be denoted, respectively, by ${}^gSPU$, ${}^gSUP$, and ${}^gM(P, U)$: ${}^gSPU$ (see Figure 1.4) is the geoset of the surpluses of the actual population over the uniform distribution; ^{g}SUP is the geoset of the surpluses of the uniform distribution over the population; ${}^gM(P, U)$ is the set of the matches between the elements of the uniform distribution and the distribution of the actual population. (b) Two territorial geosets ${}^gH$ and ${}^gNH$, which are, respectively, those of habitat and nonhabitat (see Figure 1.5).

1.3.2. An Interpretation of the Methods above for Comparing ${}^gP$ and ${}^gU$

The methods for comparison between ${}^gP$ and ${}^gU$ suggested above can be presented in a somewhat different form. Adopting an approach often used in geographical statistics, we assume that in order to compare ${}^gP$ with ${}^gU$ we transform ${}^gP$ into ${}^gU$: We shift elements of population from cells in which they are in surplus with respect to the uniform distribution to cells containing surpluses of elements of the uniform distribution. Returning to our specific example of the distribution of the population of Italy ${}^gP$ *vs.* ${}^gU$ (both reduced to a total of 10,000), we consider the following numerical values (slightly adjusted) from the *Atlas*:

1. Out of the 10,000 elements that constitute the reduced geoset of population, 5326 remain in their cells as they are matched by 5326 elements of the uniform distribution.
2. The remaining 4674 elements of population in surplus of those of the uniform distribution have to be shifted from their original location in the habitat and moved to a new location in the nonhabitat. There they will match the surplus elements of the uniform distribution. The locations of these elements constitute the destinations of the shifted elements.
3. The total number of these destinations is also 4674 out of 10,000.

The ratio 0.4674 indicates the proportion of elements to be shifted in order to transform ${}^gP$ into ${}^gU$. It can also be interpreted: (a) as a ratio of the total size of derived geosets ${}^gSPU$ and ${}^gSUP$ to the size of ${}^gP$ (and ${}^gU$); and (b) as the ratio of the area of nonhabitat to the total area of the territory, which corresponds to the well-known Duncan concentration index, usually denoted by Δ (see Chapter 8). Accordingly, we denote it by the symbol Δ_P^U. The ratio $(1 - \Delta_P^U)$ can be interpreted as the proportion of the elements of ${}^gP$ and ${}^gU$ that match and as the proportion of the area of habitat to the total area of the territory. We define as geostatistical parameters values such as Δ_P^U and $(1 - \Delta_P^U)$ measure characteristics of a geoset. Later, we shall introduce other types of geostatistical parameters.

1.4. Two or More Geosets over the Same Territory

1.4.1. Further Examples of Geosets for Italy

Up until this point we have been discussing the presentation and interpretation of a geoset ${}^{g}P$ or of ${}^{g}P$ as compared to ${}^{g}U$. We now go on to consider the more general case in which we deal simultaneously with two or more primary geosets (${}^{g}P, {}^{g}Q, {}^{g}R, \ldots$) over the same territory T.

We shall again take some examples of geosets distributed over the territory of Italy, which can shed light on several important aspects of that country's economic and social geography:

1. Figure 1.6 shows the distribution of the economic sector, measured by the number of individuals employed in industry, trade, and services, according to the economic census of 1981. Comparing this distribution with that of the population (see Figure 1.3), we see that: (a) as expected, there is a general similarity between the two distributions; (b) however, economic activities are even more concentrated in urban and semiurban areas than the population; (c) the share of more developed zones, especially in the northwest and in part of the center, is considerably larger for economic activities; and (d) by contrast, less developed southern and insular areas have a smaller share in economic activities than in the population.
2. An important determinant of some of the characteristics noted above is the concentration in the metropolitan areas in the northwest of many branches of industries (particularly construction, metal, and basic food industries); an extreme example of this is given in Figure 1.7 (showing the distribution of individuals employed in construction and the production of heavy equipment and construction materials).
3. Tertiary activities connected with modern economic development also tend to be concentrated in more progressive, metropolitan areas. Figure 1.8 illustrates by way of an example the distribution of individuals employed in advertising and public relations. By contrast, other tertiary activities (such as retail trade) gravitate around the population, and still others have an intermediate position.
4. A very large number of geosets of industry, trade, and services can thus be grouped into a relatively small number of classes, the territorial characteristics of which can be determined and explained. Some of these activities have peculiar distributions. As an example, Figure 1.9 shows the distribution of the petrol industry, which is found over small areas near to petrol layers or to harbors widely dispersed along the coasts.
5. With regard to agriculture, Figure 1.10 shows that the total farm area is spread over a very large part of the territory. Other maps (not reproduced here) show that the density of individuals engaged in agriculture is especially high, in the intensively developed Po Valley and in some coastal zones and, on the other hand, in poorer areas of the south and the islands.

Upon examining crop production, we find that most crops have peculiar distributions, which can be accounted for by such factors as soil characteristics,

Figure 1.6. Italy. The economic sector, measured by the number of individuals employed in industry, trade, and services, 1981: .1700 ▪ 16,900 ▪▪ 33,800 ⁘ 84,000 ■ 168,800+ (Source: *Statistical Atlas of Italy*, 1988).

water supply, climatic conditions, and connections with other economic activities. For some crops this also implies that their habitat is concentrated over small areas. Two extreme examples of this are given in Figures 1.11 and 1.12, which show, respectively, the distribution of area devoted to citrus groves and rice paddies: The latter is found in parts of the Po Valley particularly rich in water and the former in coastal zones of southern Italy and the islands, which have favorable soil and climatic conditions.

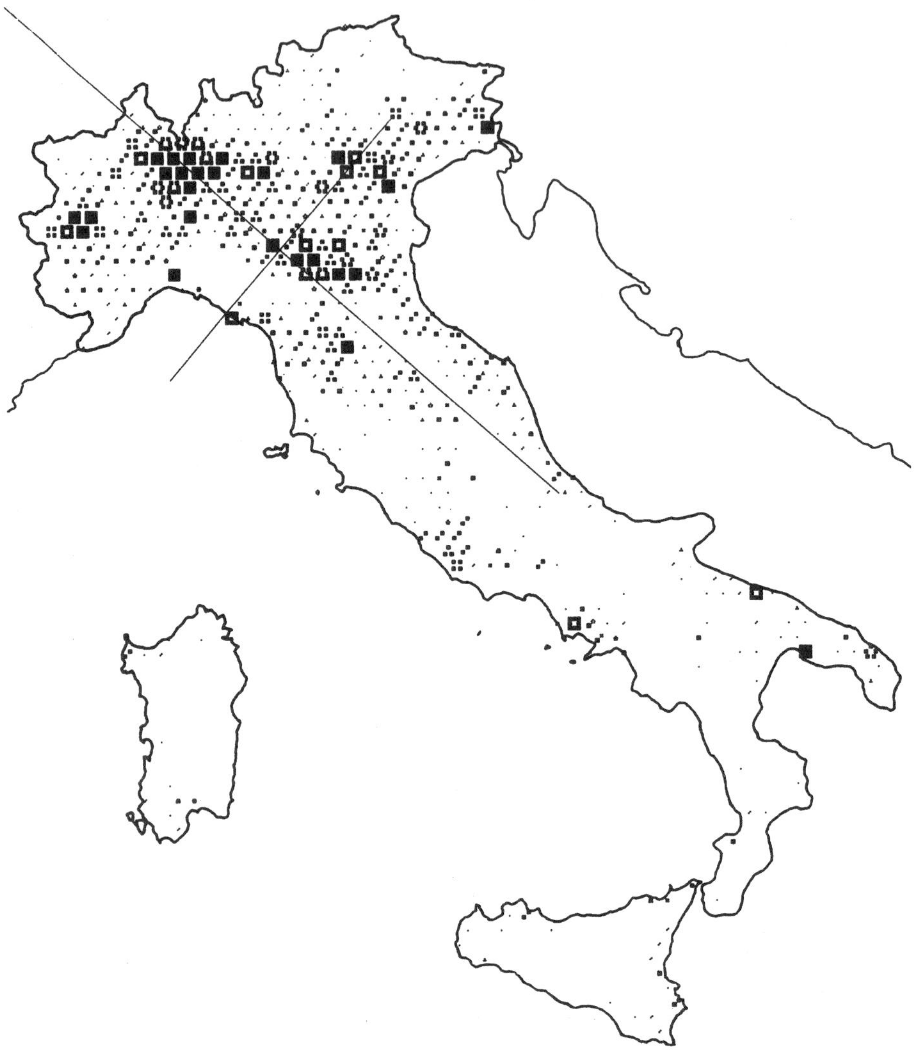

Figure 1.7. Italy. Individuals employed in heavy industry and construction 1981: . 51 ▪ 513 ▪▪ 1025 ⁘ 2563 ■ 5125+ (Source: *Statistical Atlas of Italy*, 1988).

Figure 1.8. Italy. Individuals employed in advertising and public relations, 1981: . 2 ▪ 22 ▪▪ 44 ⁘ 111 ■ 222+ (Source: *Statistical Atlas of Italy*, 1988).

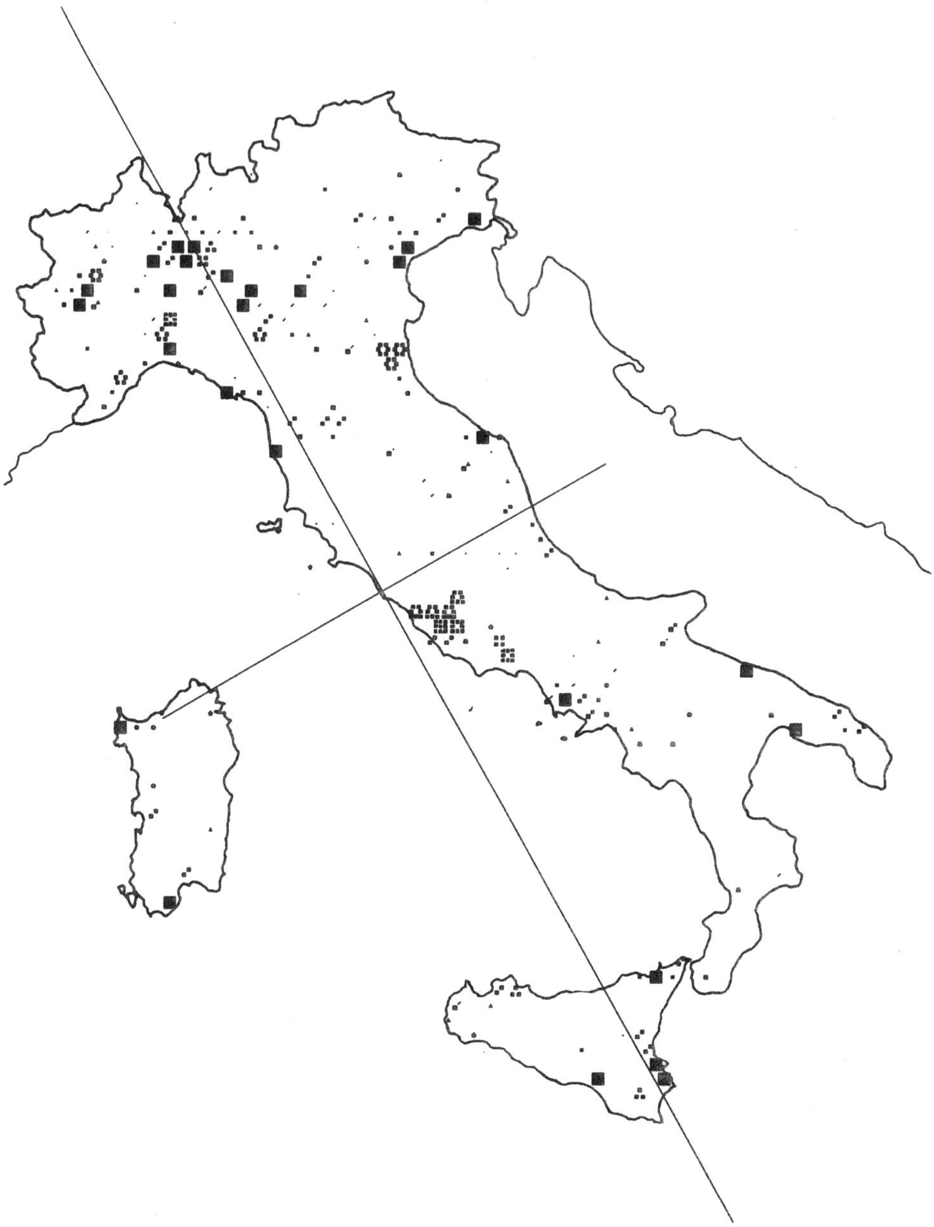

Figure 1.9. Italy. Individuals employed in the petrol industry, 1981: . 3 ▪ 27 ▪ 53 ⁘ 137 ■ 274+ (Source: *Statistical Atlas of Italy*, 1988).

Figure 1.10. Italy. Farmland, 1982 (in 10,000 square meters): . 2400 ▪ 23,600 ▪▪ 47,300 ⁛ 118,200 ■ 236,300+ (Source: *Statistical Atlas of Italy*, 1988).

1.4.2. Geosets Derived from the Comparison of Two Primary Population Geosets

In Section 1.4.1 we discussed some general characteristics of geosets, familiarizing ourselves with overall, macrogeographical, visual impressions obtained by looking at the maps. However, this approach is often insufficient. To enhance and deepen our comparative study of pairs of population geosets (such as ${}^{g}P$ and ${}^{g}Q$), we can adapt an approach analogous to that followed in Section 1.3.1 for comparing a geoset ${}^{g}P$ to the geoset ${}^{g}U$ of the uniform distribution.

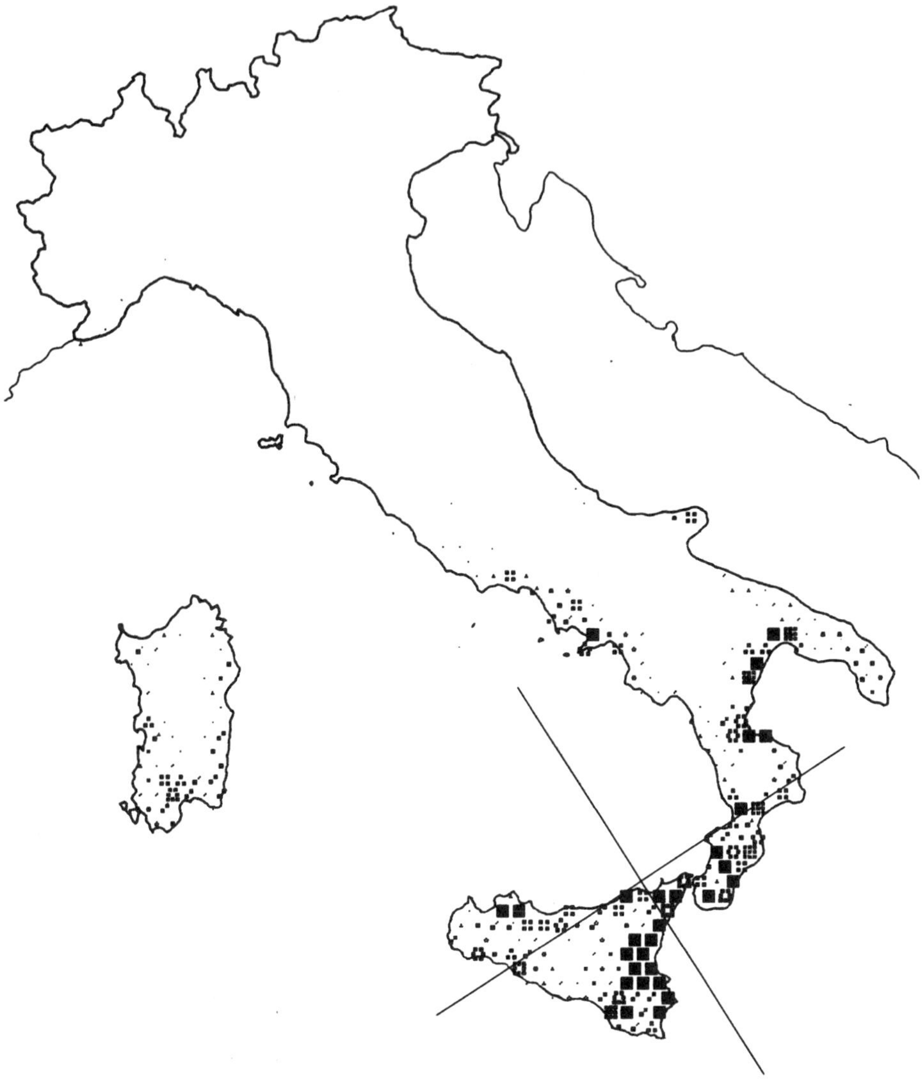

Figure 1.11. Italy. Area devoted to citrus groves, 1982 (in 10,000 square meters): · 20 ▪ 170 ▪▪ 340 ⁘ 850 ■ 1700+ (Source: *Statistical Atlas of Italy*, 1988).

Let us denote, respectively, by ^gP and ^gQ, the geosets of two populations (both reduced to the same size, say, $M^P = M^Q = 10{,}000$, and distributed over the same network of, say, $N^c = 1{,}000$ cells). We calculate the relative weight (per 10,000) of each population in each cell. We then obtain three derived population geosets: (a) the geoset $[^gM(P, Q)]$ of the matching elements of ^gP and ^gQ; (b) the geoset ^gSPQ of the surpluses of ^gP over ^gQ; and (c) the geoset ^gSQP of the surpluses of ^gQ over ^gP.

To compare ^gP and ^gQ, we propose transforming ^gP into ^gQ so that we can refer to the territorial distribution of the elements of ^gP that remain in place and the

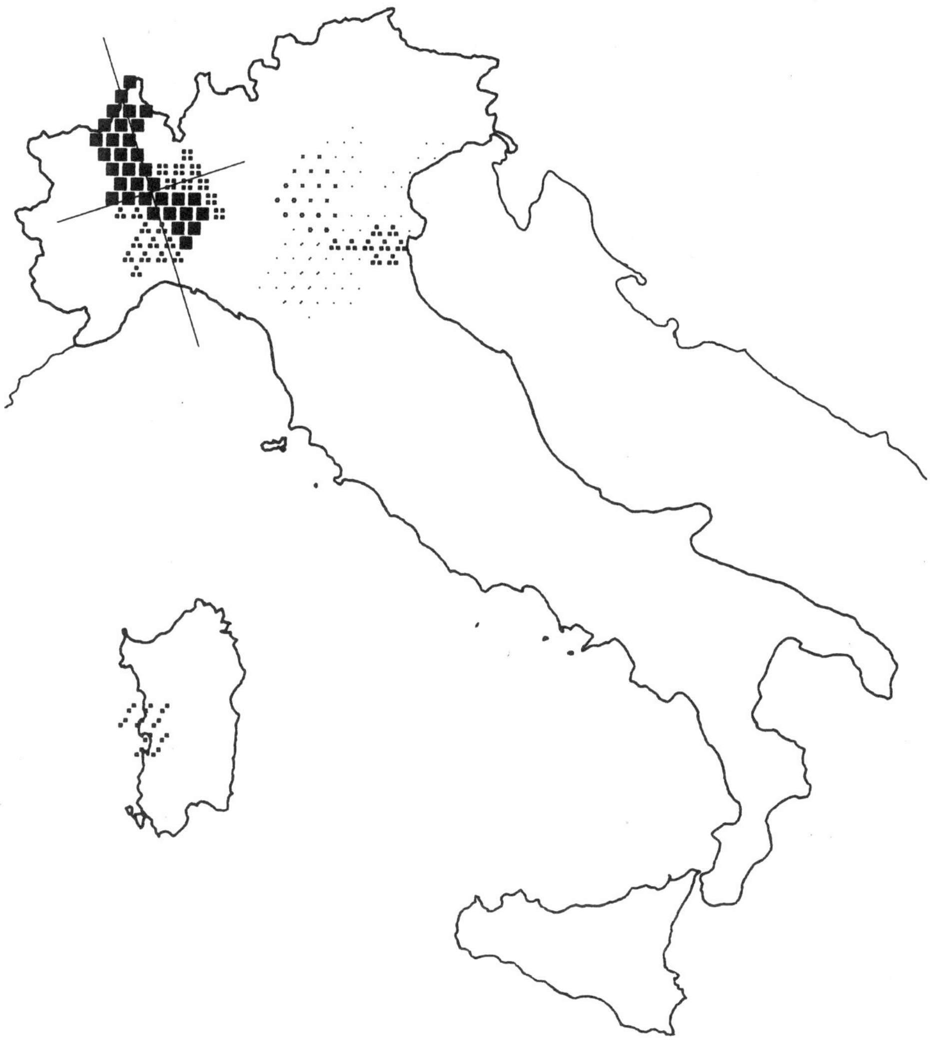

Figure 1.12. Italy. Average rice production, 1984–1986 (in 10,000 square meters): . 100 ▪ 1100 ▪ 2200 ⁘ 5500 ■ 11,000+ (Source: *Statistical Atlas of Italy*, 1988).

origins and destinations of the elements that are shifted. Comparing ${}^{g}P$ with ${}^{g}Q$, we can also divide the territory into four parts corresponding, respectively, to matches, surpluses of ${}^{g}P$, surpluses of ${}^{g}Q$, and "void" areas (where there are no elements of either ${}^{g}P$ or ${}^{g}Q$).

Derived geosets were found to be very useful in the detailed geostatistical analyses carried out in the *Atlas*. Some examples follow:

1. Comparisons have been carried out between pairs of distributions of the population of Italy. Successive analyses included the years from 1871 to 1987. Let us denote the geoset of the population at a given census by ${}^{g}P$ and the distribution at the following census by ${}^{g}Q$. In many comparisons it was found that ${}^{g}P$ and ${}^{g}Q$ seemed to be about equal, thus suggesting almost no change during the period between the two censuses. However, a deeper study of ${}^{g}SPQ$ and ${}^{g}SQP$ revealed important changes (such as growing concentration in cities versus rural exodus and positive versus negative changes in central metropolitan areas versus suburbs). ${}^{g}SPQ$ and ${}^{g}SQP$ enable us to analyze changes, while the geoset of matches permits us to study the part of the population distribution that remained constant over time.

2. Let ${}^{g}P$ be the geoset of the population (aged 6 or more) and ${}^{g}Q$ be the geoset of individuals aged 6 or more who are illiterate (both geosets being reduced to a total of 10,000 elements). Figure 1.13 shows the geoset ${}^{g}SQP$ of the above surpluses

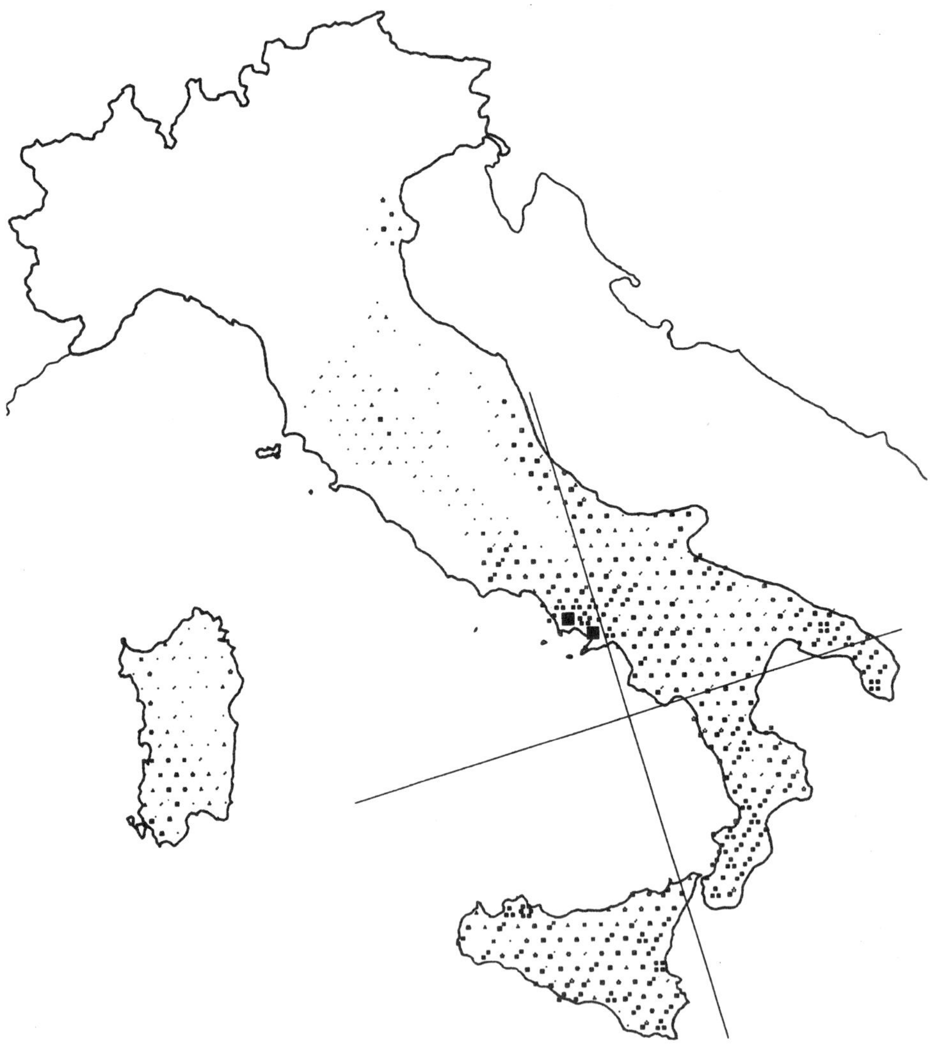

Figure 1.13. Italy. Surplus of illiterates over the population aged 6+, 1981: . 1 ▪ 10 ▪ 20 ⁘ 50 ■ 100+ (Source: *Statistical Atlas of Italy*, 1988).

of illiterates as compared to the number expected under the hypothesis of a distribution equal to that of the population. It is seen from the figure that the residual illiteracy in Italy is mainly concentrated in the southern zones of the country, which for historical, economical, and social reasons are known to be comparatively less developed.

3. Let ${}^{g}P$ be the geoset of individuals in the labor force, and ${}^{g}Q$ the geoset of individuals unemployed or seeking a first job. Figure 1.14 shows the geoset ${}^{g}SQP$. The similarity between this geoset and that of illiteracy is striking: many of the maps in the *Atlas* show analogous aspects, which are connected to the socioeconomic development of many southern areas.

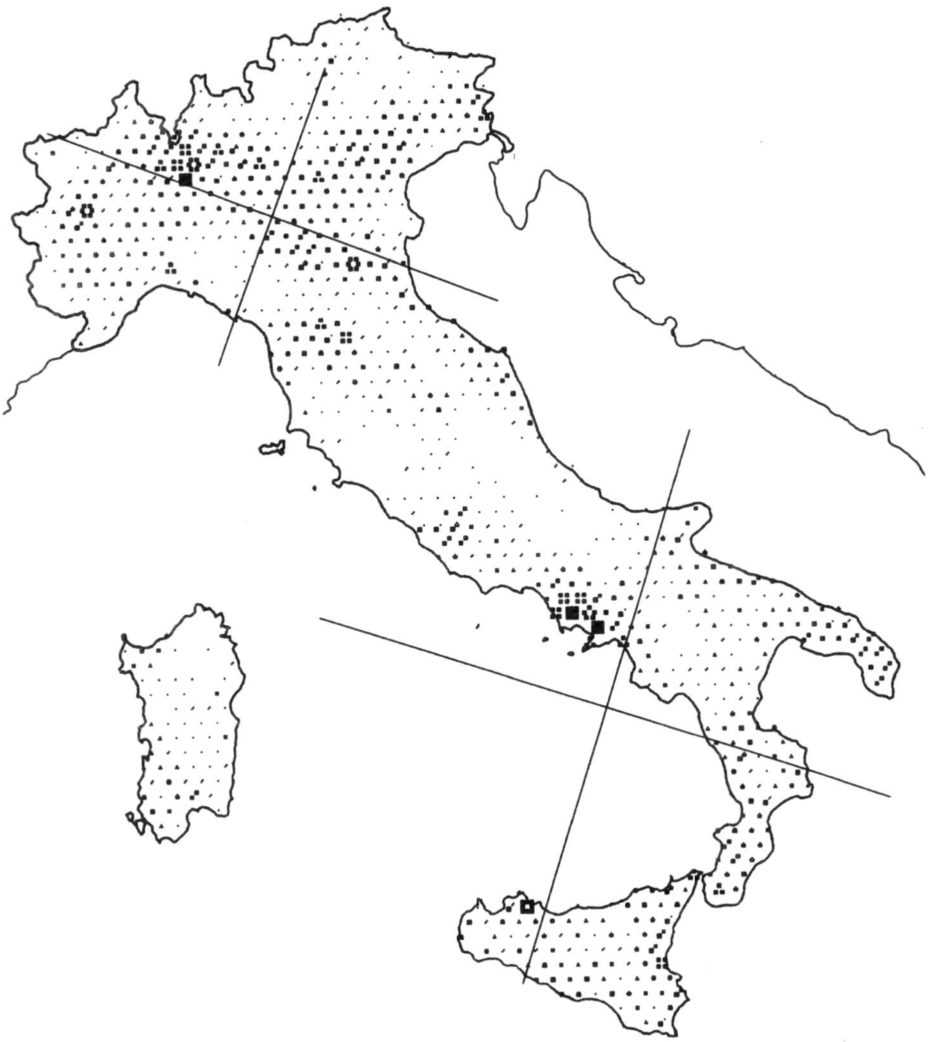

Figure 1.14. Italy. Surplus of unemployed over the labor force, 1981: . 1 ▪ 10 ▪ 20 ⁙ 50 ■ 100+ (Source: *Statistical Atlas of Italy*, 1988).

Table 1.1. Examples of Parameters for Selected Geosets (Italy)[a]

		Standardized parameter			Relative parameters		
Geosets considered	Figure No.	Scatter (s_Z^2/s_U^2) (1)	Oblongity $({}^rb_Z^2/{}^rb_U^2)$ (2)	Distance variance (d_Z^2/d_U^2) (3)	Scatter ${}^rs_Z^2$ (4)	Oblongity ${}^rb_Z^2$ (5)	Distance variance ${}^rd_Z^2$ (6)
1. Uniform distribution over territory	1.1	1	1	1	2.31	1.26	2.91
2. Population 1987	1.3	0.84	1.14	0.96	3.64	1.45	5.26
3. Surplus of population 1987 over uniform	1.4	0.72	1.26	0.91	15.13	1.58	24.14
4. Habitat of population 1987	1.5	0.94	1.08	1.01	4.05	1.36	5.50
5. Nonhabitat of population 1987	1.5	1.05	0.94	0.99	5.19	1.18	6.14
6. All economic activities	1.6	0.71	1.13	0.80	3.53	1.43	5.05
7. Mechanical industry	1.7	0.36	1.03	0.37	2.91	1.31	3.80
8. Advertising and public relations	1.8	0.46	1.17	0.54	5.12	1.48	7.58
9. Petrol industry	1.9	1.04	1.25	1.30	18.44	1.57	29.02
10. Farmlands	1.10	1.01	0.98	0.99	2.75	1.24	3.41
11. Citrus groves	1.11	0.32	0.79	0.25	4.97	1.0	4.98
12. Rice paddies	1.12	0.09	0.90	0.08	2.40	1.13	2.70
13. Surplus of illiterates over individuals aged 6+	1.13	0.56	0.80	0.45	6.65	1.01	6.73
Parameters for 385 geosets							
Average		0.75	1.10	0.82	4.20	1.39	5.82
Standard deviation		0.20	0.13	0.22	2.38	0.16	3.24

[a]*Notes:* SYMBOLS: Z is any primary or derived geoset considered; U uniform distribution over territory; A_z the area of the territory corresponding to geoset Z; and A_T total territory. EXPLANATIONS OF PARAMETERS: Columns 1–3 show the ratios of parameters for geosets considered to corresponding parameters for the uniform distribution: $s^2 = 2\sigma_m\sigma_M$ is the scatter; ${}^rb^2 = d^2/s^2$ is the relative squared oblongity; and d^2 is the average of squares of distances of the elements for the center, which equal to distance-variance. Columns 4–6: ${}^rs_Z^2 = 2\pi s_Z^2/A_Z$ is the relative scatter; and ${}^rd_Z^2 = 2\pi d_Z^2/A_Z$ is the relative distance-variance. Column 7: Territorial coverage: A_Z/A_T is the proportion of total area covered by territory corresponding to geoset Z. For geosets 1, 2, 4–12 it corresponds to $1 - \Delta_Z^U$. Column 8: Proportion of elements of Z in surplus versus elements of uniform distribution Δ_Z^U.

1.5. Examples of Geostatistical Parameters

1.5.1. Order of Presentation of the Parameters

Geostatistical parameters—measuring important characteristics of the primary geosets ${}^gP$, ${}^gQ$, ${}^gU$ and of geosets derived from them—constitute a central theme of this volume. Part III is devoted to the systematic analysis of their definitions and equations, properties, and primary applications. Here we introduce some of them by means of examples and we start by considering the parameters implicit in the crosses

Table 1.1. (Cont.)

Territorial coverage A_z/A_T (7)	Proportion of surpluses of population over uniform distribution Δ_z^u (8)	Standard mean center $(\bar{X}_Z^T/d^T)\cdot 100$ (9)	$(\bar{Y}_Z^T/d^T)\cdot 100$ (10)	Standardized eccentricity $e_Z^T=\{\sqrt{[(\bar{X}_Z^T)^2+(\bar{Y}_Z^T)^2]}/d^T\cdot 100\}$ (11)
1	0	0	0	0
0.53	0.47	3.08	2.94	4.25
0.11	—	2.05	8.26	8.51
0.53	0.47	3.98	−1.76	4.35
0.47	0.53	−4.55	2.03	4.98
0.46	0.54	2.62	28.00	28.12
0.28	0.72	0.74	71.18	71.18
0.21	0.79	−4.39	59.28	59.44
0.13	0.87	−15.92	−12.13	20.01
0.85	0.15	0.60	−3.94	3.99
0.15	0.85	−19.65	−162.13	163.31
0.09	0.91	−21.80	105.96	108.18
0.19	0.39	9.56	−103.87	104.30
0.45	0.55	2.25	7.93	8.24
0.15	0.15	7.25	40.38	41.03

superimposed over the maps of the geosets presented above (namely center, main direction, dispersion, and oblongity). We shall then go on to consider measures of concentration, "segregation," territorial coverage, and density. The comparison of dispersion to territorial coverage will enable us to measure a basic aspect of the form of geosets: "looseness," i.e., lack of compactness.

In Table 1.1 we introduce the parameters with numerical examples taken from the geosets for Italy shown in the maps presented above.

Interested readers may find the following exercise useful: Consider the detailed maps of each geoset given above together with the line of Table 1.1 indicating the numerical values of the parameters of the same geoset. A good correspondence is found between the general characteristics suggested by the numerical parameters and the geostatistical distribution seen in the map. For an interpretation of the findings from a substantive viewpoint (demographic, economic, or social) the reader is referred to Sections 1.2.1, 1.2.2, 1.4.1, and 1.4.2.

Reading the entries of Table 1.1 along its columns will provide some information on the behavior of each of the parameters considered. The last two lines of the table show the average and standard deviation of the values of each of the parameters for 385 geosets of demographic, economic, and social phenomena considered in the *Atlas*. These averages summarize the information supplied by $385 \times 8086 = 3{,}113{,}110$ primary data detailed by *comuni* for each parameter.

1.5.2. Interpretation of the Crosses Representing Geosets Superimposed on the Maps

Let us consider a collection of maps prepared with the same GRP method, illustrating the distribution of many geosets over the same territory, as in examples in the *Atlas* given above. A cross has been superimposed on each map. These crosses are formed by two perpendicular axes, known as the principal axes. We indicate the axes through the mean center along which the dispersion is maximal and minimal by major (M) and minor (m), respectively. The longest and shortest arms of the cross have a length of $4\sigma_M$ and $4\sigma_m$, respectively, where σ_M and σ_m are the standard deviations of the elements of the geoset from their center, measured along axes M and m. The major axis can be considered as the line nearest to the elements of the geoset in the sense of least squares, and can be interpreted as an indicator of the principal direction of the distribution. To identify this line, we calculate the angle α between the major axis and the closest conventional axis.

Let us start by considering the cross superimposed on the map of the uniform distribution. This map can be envisaged by putting a GRP of 10 at the center of each of the 1000 hexagonal cells of Figure 1.1. The major axis forms an angle $\alpha = 33.5°$ with the conventional Y-axis. The length of the major axis is 1337 km, while that of the minor axis is only 660 km. These values express basic characteristics of the territory of Italy: strong oblongity and prevailing direction roughly from northwest to southeast.

Most of the crosses superimposed over the 385 maps in the *Atlas* have an angle between the conventional Y-axis and the major axis very near to that of the territory, with the median of these angles being $+36.5°$. For 80% of the distributions, the angle is between $+32.5°$ and $+42.2°$, which suggests that the determinants of the principal direction of the territory of Italy have a strong influence on the determination of the principal direction of many of the geosets. However, there are geosets that have a very different principal direction: this is immediately shown by the cross, which may prompt the reader to look at the influence of its specific determinants.

For many geosets it is found that their crosses differ from that of the uniform distribution by having shorter or longer principal axes and/or a different location of the mean center. Whenever these differences are immediately evident at first inspection, it should be a warning to the reader to look for the specific determinants of the discrepancy. In the examples given above, this is the case for Figures 1.7, 1.8, and 1.11–1.14.

It is always true that marked differences in the position of the center or in the length of the axes between the map of any geoset and that of the uniform distribution suggest strong discrepancies between these distributions. However, the reverse is not always true. Similarity or apparent similarity of crosses does not necessarily imply congruent distributions. For example, the cross for the population of Italy (Figure 1.3) appears to be almost the same as the cross of territory (Figure 1.1). However, we saw in Sections 1.2.1 and 1.2.2 that the population distribution differs considerably, for many reasons, from the uniform distribution, and these factors balance each other in the determination of centers and dispersions of these distributions.

Among the parameters corresponding to the crosses, we have singled out for further discussion the mean center and the measures of territorial dispersion, which are analogous in some respects to the elementary parameters usually calculated as a first step in the analysis of unidimensional variables (means and measures of variability).

In the following, ${}^{g}Z$ will represant any primary geoset (as ${}^{g}P, {}^{g}Q, \ldots, {}^{g}U$) and any geoset derived from the comparison of ${}^{g}P$ to ${}^{g}U$ and from ${}^{g}P$ to ${}^{g}Q$.

1.5.3. The Mean Center

From the locations of each of the elements of any geoset ${}^{g}Z$, we obtain the mean center as an average of these locations. For a primary or derived geoset giving the information in aggregate form, by means of triplets (X_c^Z, Y_c^Z, w_c^Z), the center of gravity can be conceived as a dot on the map having coordinates $\bar{X}_Z = \Sigma w_c^Z X_c$ and $\bar{Y}_Z = \Sigma w_c^Z Y_c$. Here, w_c^Z is the proportion of elements of Z found in any hexagonal cell c, while (X_c, Y_c) are the coordinates of the center of hexagonal cell c. The coordinates can be expressed in any system of orthogonal axes. Figure 1.15 shows the mean centers for all the geosets listed in Table 1.1, which correspond to the centers shown by the crosses in each of the figures shown previously. Following a procedure adopted in the *Atlas*, the coordinates have the mean center of the territory as origin $(0, 0)$. The very significant principal axes of the territory now serve as new conventional axes (see Section 1.5.2). Accordingly, the mean centers of ${}^{g}Z$ are indicated as $(\bar{X}_Z^T, \bar{Y}_Z^T)$, where the symbols X^T and Y^T are used to stress the fact that we are dealing with deviations from the axes along principal axes of the territory T. The numerical values of the mean centers shown in Figure 1.15 are given in columns (9) and (10) of Table 1.1. To make these values comparable to those of geosets for other countries, they have been standardized by dividing the deviations from the center of the territory by the general measure of territorial dispersion, i.e., the standard distance d_T of the elements of territory (see Section 1.5.4).

Apart from $\bar{X}_Z^T/d^T$ and $\bar{Y}_Z^T/d^T$ Table 1.1 also gives, in column (11), the standardized eccentricity of ${}^{g}Z$ with respect to the mean center of the territory, $e_Z^T = \sqrt{([\bar{X}_Z^T)^2 + (\bar{Y}_Z^T)^2]}/d_T$. We see that the population and the geosets derived from the comparison between population and uniform distribution have very small eccentricities. This amazing situation has remained unchanged over almost 120 years. Both farmland (which is spread over almost the entire territory) and the petrol industry (which is concentrated in small scattered areas) have low values of eccentricity. The center of individuals employed in the financial sector is, as might be expected, more to the northwest than to the center of the population. The other geosets have very strong eccentricities, as expected: illiteracy and citrus are very southern, heavy industry is very northern. Eccentricity e_Z^T of the geosets ${}^{g}Z$ is mainly determined by the position of the mean centers of ${}^{g}Z$ along the major axis of the territory Y^T. This finding (which is due to the peculiar geographic characteristics of the territory of Italy) has been largely utilized in the second volume of the *Atlas*: $\bar{Y}_Z^T$

Figure 1.15. The mean centers for all the geosets in Table 1.1.

has been found to be an extremely sensitive and meaningful descriptive parameter. The dispersion of the 385 centers presented in the *Atlas* along the Y^T axis is much larger than that found along the X^T axis.

We shall see that the study of mean centers is particularly useful in the following settings:

1. For any geoset ${}^gP$ compared to ${}^gU$, the mean centers of ${}^gSPU$, ${}^gP$, ${}^gM(P, U) = {}^gH$ are collinear and so are the centers of territories ${}^gH$, ${}^gT$, and ${}^gNH$.
2. From the comparison between any two geosets ${}^gP$ and ${}^gQ$ over the same territory, we find collinearity of the centers of ${}^gSPQ$, ${}^gP$, and ${}^gM(P, Q)$ and of

${}^{g}M(P, Q)$, ${}^{g}Q$, and ${}^{g}SQP$. In both (1) and (2) the analysis of the angle between the two lines of centers and of the "triangle of centers" (which shows the centers of ${}^{g}NH$, ${}^{g}T$, ${}^{g}H$, ${}^{g}SPU$, ${}^{g}P$; see Section 9.4.2) has been found to be very useful for analytical purposes.

3. Comparison of three (or more) geosets bound by special relationships: e.g., (a) they refer to the same population at different times or to subpopulations of the same population distinguished by different levels of a variable (such as income or age); (b) they refer to a population and to two or more of its subpopulations; or (c) they are a compound geoset and its components. In all of these cases, their centers may be bound by algebraic relationships that facilitate analysis.

1.5.4. Measuring Territorial Dispersion and Oblongity

The study of the dispersion of geosets ${}^{g}Z$ over the territory T is an essential aspect of geostatistical analysis. We have already introduced a way to express the degree of a dispersion by means of the length of the axes of its cross, which are proportional to $4\sigma_M$ and $4\sigma_m$, respectively. These standard deviations of the elements along the principal axes have been found to be very meaningful for a first appraisal of the dispersion of the geoset studied.

We denote by oblongity the property of a geoset having two principal axes through the center with different standard deviations ($\sigma_M > \sigma_m$). Oblongity has almost always been found not only in the hundreds of geosets studied empirically in the *Atlas* but in many other studies of countries and cities.

In more extensive research on dispersion we may prefer to use squared measures, which have important algebraic properties. One such measure is the distance-variance d^2, which is the average of squared distances of the elements from their mean center. For data of the aggregated type

$$d_Z^2 = \Sigma w_c^Z[(X_c - \bar{X}_Z)^2 + (Y_c - \bar{Y}_Z)^2]$$

where w_c^Z is the proportion of elements of ${}^{g}Z$ in cell c and $(\bar{X}_Z, \bar{Y}_Z)$ are the coordinates of the mean center of ${}^{g}Z$.

Obviously, d_Z^2 is connected to (σ_Z^M, σ_Z^m) by the equation $d_Z^2 = (\sigma_Z^M)^2 + (\sigma_Z^m)^2$. In the analysis of dispersion we can also use equations that measure the extent of oblongity and its impact on dispersion separately. These have been found to be important ingredients in geostatistical analysis. In this book, we introduce various measures of oblongity, to be selected according to the type of data studied, as well as to the aims of the research. We note two of them here:

1. Absolute Squared Oblongity:

$$\Omega_Z^2 = (\sigma_Z^M - \sigma_Z^m)^2 = d_Z^2 - 2\sigma_Z^M \sigma_Z^m$$

where $s_Z^2 = 2\sigma_Z^M \sigma_Z^m$ is a measure of dispersion, called *scatter* for short. We shall see that s_Z^2 has very important properties and applications. The equation $d_Z^2 = s_Z^2 + \Omega_Z^2$ expresses the split of the distance-variance into two additive components: s^2 measures the dispersion free of the influence of oblongity and Ω^2 measures the oblongity.

2. Dividing d_Z^2 by s_Z^2, we obtain a parameter ${}^r b_Z^2 = d_Z^2/s_Z^2 = 1 + \Omega_Z^2/s_Z^2$, which indicates the relative increase of dispersion determined by the oblongity with respect to the scatter. Parameter ${}^r b_Z^2$ can also be expressed as

$$ {}^r b_Z^2 = 0.5[(\sigma_Z^M/\sigma_Z^m) + (\sigma_Z^m/\sigma_Z^M)] $$

which is the average of the ratios between the standard deviations along the principal axes. If there is no oblongity, ${}^r b^2 = 1$, while $(\sigma_Z^M/\sigma_Z^m) = k$, then ${}^r b^2 = 0.5\ (k + 1/k)$. The values of ${}^r b_Z^2$ given in column (5) of Table 1.1 suggest that for most of the geosets considered, the relative oblongity is strong, as might be expected from the corresponding maps. Exceptions are found for citrus (Figure 1.11), rice (Figure 1.12), and surpluses of illiterates over population (Figure 1.13).

In many instances, the strong oblongity is due to the shape of the territory of Italy. To evaluate parameters of oblongity, scatter, and distance-variance of the geosets studied, we have standardized them by dividing these parameters by the corresponding parameters for the uniform distribution over the territory [see columns (1), (2), and (3) of Table 1.1].

To study the dispersion of the geosets ${}^gZ$ listed in Table 1.1, column (3) gives values for the ratio d_Z^2/d_U^2, which compares the distance-variance of each ${}^gZ$ to the distance-variance of the uniform distribution over the territory. Since $d^2 = s^2\,{}^r b^2$, both d_Z^2 and d_U^2 have been split into their two factors, the scatter s^2 (i.e., the dispersion free of the effects of oblongity) and relative oblongity ${}^r b^2$. We thus obtain

$$ (s_Z^2/s_U^2)\,({}^r b_Z^2/{}^r b_U^2) = d_Z^2/d_U^2 $$

Wide-ranging research on s_Z^2/s_U^2 suggests that this is a very meaningful and sensitive measure, for which minimal and maximal limits can be determined (under certain conditions). Values of s_Z^2/s_U^2 considerably less than 1 would suggest that substantial proportions of the elements of ${}^gZ$ are distributed over territorial cells rather close to one another. These elements may be isolated or form one or many clusters (provided that these clusters are not very far from each other).

In this way, we can interpret the very low values of this ratio found [according to column (1) of Table 1.1] for rice (Figure 1.12), citrus (Figure 1.11), heavy industry (Figure 1.7), advertising and public relations (Figure 1.8), and surplus of illiterates over individuals aged 6+ (Figure 1.13). All economic activities (Figure 1.6) and surpluses of population (1987) over uniform distribution (Figure 1.4) have moderate values of s_Z^2/s_U^2.

Scatters lower than that of the uniform distribution are found for more than 90% of the 385 geosets studied in the *Atlas*. The rather infrequent occurrence of values of $(s_Z^2/s_U^2) > 1$ suggests the existence of considerable proportions of elements of ${}^gZ$ in peripheral zones. Among the geosets listed in Table 1.1, values of 1.05 have been found for the nonhabitat of the population (Figure 1.5) and for the petrol industry (Figure 1.9).

The large majority of the geosets in the *Atlas* have been found to be more oblong than the uniform distribution. Values of $({}^r b_Z^2/{}^r b_U^2) > 1$ have also been found in several of the geosets listed in Table 1.1 [see column (2)], while four of them have values varying between 0.9 and 1.0. Very low values have been found only for citrus (Figure 1.11) and for the surplus of illiterates over individuals aged 6+ (Figure 1.13).

The relatively high values of ${}^{r}b_Z^2/{}^{r}b_U^2$ render d_Z^2/d_U^2 generally higher than s_Z^2/s_U^2. Despite this fact, d_Z^2/d_U^2 is less than 1 for all the geosets in Table 1.1, with the single significant exception of the petrol industry (Figure 1.9), which is found in very scattered and largely peripheral locations.

Owing to its algebraic properties, the distance-variance d_Z^2 is particularly useful in the analysis of the same settings [(1), (2), and (3)] that were described in Section 1.5.3 with regard to the mean center. Nonetheless, it is often preferable to use its square root d_Z, the quadratic average of the distance of the elements of ${}^{g}Z$ from their center. This parameter, called the standard distance, is a linear measure of dispersion, while d^2, Ω^2, ${}^{r}b^2$, and s^2 can be considered as area measures.

1.5.5. Measuring Territorial Coverage and Net Density

We have seen above that for any geoset ${}^{g}Z$ we can determine the area of the territory corresponding to it. The ratio between the area A_Z of this territory and the area A_T of the total territory often constitutes a very interesting tool for geostatistical research, as it answers the basic question: "In what proportion of the territory is the geoset effectively found." Table 1.1 gives, in column (7), examples of $(1 - \Delta_Z^U) = (A_Z/A_T)$ for the habitat of the geosets studied. Obviously this ratio varies between 0 and 1. In the table it is nearer to the upper limit only for farmland and the other geosets have habitats lower than 54%. This is in line with the findings of the *Atlas*, according to which 50% of the 385 distributions studied have a habitat under 46% of the territory of Italy and 90% have a habitat lower than 63%.

Some of the geosets that have very small habitats, such as citrus (Figure 1.11) and rice (Figure 1.12), also have small dispersions; by contrast, the petrol industry has an extremely small habitat and a large dispersion, as its elements are scattered over small areas or clusters that are very far from one another.

We are accustomed to calculating gross density of population by dividing the population by the area of the gross territory over which it is distributed. We shall see that knowing of the area of the habitat makes it possible to calculate the "net density" by referring the size of the population to the area of the habitat. We shall also be able to calculate the density of surpluses of population by relating them to the area of their territory.

In comparing two geosets ${}^{g}P$ and ${}^{g}Q$ we can apply similar reasoning. As shown by the example of surpluses of illiterates over individuals aged 6+, given in line (13) of Table 1.1, we can determine the proportion of surpluses of P over Q, obtained from ${}^{g}SPQ$, and compare it to their corresponding territories. The same applies to surpluses ${}^{g}SQP$ and ${}^{g}M(P, Q)$.

We have already remarked on the interest that can be found in the study of Δ_P^U and Δ_P^Q, indicating, respectively, the proportions of elements ${}^{g}P$ in surplus of those of ${}^{g}U$ or ${}^{g}Q$ (see Section 1.3.2). Examples are given in column (8) of Table 1.1. In introducing Δ_P^U, we noted its meaning as a measure of concentration; similarly, Δ_P^Q corresponds to a measure frequently employed for the study of "segregation" or dissimilarity in distribution of P and Q. We shall see that presenting these measures together with the detailed maps of origins and destinations of designated shifts of surplus elements overcomes serious criticism often made in regard to Δ and—together with other geostatistical parameters—makes it possible to broaden concepts and measures of concentration and segregation.

1.5.6. Comparing Dispersion to Area Covered and Measuring Looseness

In the geostatistical analysis of the shape of territories, we shall measure some of their main characteristics, such as: (1) compactness, (2) oblongity, (3) external or internal concavity, (4) fragmentation, and (5) asymmetry. These measures can be extended, with proper adaptation, to the study of analogous characteristics of patterns of distribution of populations over territories (where in place of fragmentation we shall consider clustering). We shall introduce compactness here only cursorily; oblongity already been discussed; and concavity, fragmentation, and asymmetry will be described later (see Chapter 11).

A parameter proposed by various authors for measuring lack of compactness (or looseness) of a territory T is $2\pi d_T^2/A_T$, where A_T is the area of T and d_T^2 is the distance-variance of the elements of T. This parameter can be calculated as follows: $2\pi s_T^2\,{}^r b_T^2/A_T = 2\pi d_T^2/A_T$, where ${}^r b_T^2$ is the relative squared oblongity of T and s_T^2 is its scatter. It can be shown that $\min(s_T^2/A_T) = \frac{1}{2}\pi$, which is found when T has the shape of an ellipse (see Appendix A). Thus, $2\pi s_T^2/A_T$ is the ratio of the scatter per unit of area of any territory T to its minimum value. In more concise terms, we can denote this parameter as the "relative scatter" ${}^r s_T^2$ and $2\pi d_T^2/A_T$ as the relative distance-variance ${}^r d_T^2$.

Affinal equivalent figures or territories always have the same relative scatter and the the same shape. For example, all ellipses have ${}^r s_T^2 = 1$, all parallelograms have ${}^r s_T^2 = 1.047$, and all triangles have ${}^r s_T^2 = 1.209$. For the most compact figure, the circle, ${}^r s_T^2 = 1$, ${}^r b_T^2 = 1$, and ${}^r d_T^2 = 1$.

Actual territories do not, generally, have regular shapes. However, for some the values of ${}^r s_T^2$ and ${}^r b_T^2$ may suggest an analogy to regular figures or some important characteristics, e.g., convex territories have an ${}^r s_T^2$ that varies between 1 and 1.209; a territory with ${}^r s_T^2 \leqslant 1.209$ is not necessarily convex, but its departure from convexity (owing to fragmentation, external concavities, or internal gaps) cannot be large, so that we can consider it as compact or rather compact; territories with large external concavities, large internal gaps, or strong fragmentation are expected to have high values of ${}^r s_T^2$. Consider for example the territory of Italy and its three major fragments (as shown in Plate I). Their parameters of looseness and its determinants are as follows:

Looseness	Continental Italy	Sicily	Sardinia	All Italy
Scatter ${}^r s^2$	1.61	1.28	1.08	2.32
Oblongity ${}^r b^2$	1.67	1.20	1.27	1.25
Both ${}^r d^2$	2.69	1.53	1.38	2.91

The relative scatters suggest that:

1. The structures of Sicily and Sardinia are rather compact and there may be some vague similarity between the structure of Sicily and that of a triangle (as implied in the name of Trinaria given to the island by the Greeks in ancient times).
2. The structure of Continental Italy has a considerable departure from convexity (which is due to its wide external concavities).
3. The departure is even larger for the fragmented territory of all Italy.

We can obtain the relative scatter for any population geoset ${}^gZ$ distributed in territory T_Z of area A_{TZ} as follows:

$$ {}^r s_Z^2 = 2\pi s_T^2 / A_{TZ} = (s_Z^2/s_{TZ}^2)\,{}^r s_{TZ}^2 $$

where A_{TZ} for the equivalent distribution is the product of the fixed area of one cell times the number of inhabited cells. Thus the relative distance-variance of ${}^gZ$ is,

$$ {}^r d_Z^2 = [(s_Z^2/s_{TZ}^2)\,{}^r s_{TZ}^2]\,{}^r b_Z^2 $$

Table 1.1 shows in columns (4), (5), and (6), respectively, the values of ${}^r s_Z^2$, ${}^r b_Z^2$, and ${}^r d_Z^2$ for the geosets Z considered. It may be noted that if we compare ${}^r s_Z^2$ to ${}^r s_T^2 \simeq {}^r s_U^2$ for the entire territory of Italy (${}^r s_T^2 = 2.31$) we obtain ${}^r s_Z^2/{}^r s_T^2 = (s_Z^2/s_U^2){:}(A_{TZ}/A_T)$.

Parameters s_Z^2/s_U^2 and A_Z/A_T, shown in columns (1) and (7), indicate ratios of the structural dispersion of ${}^gZ$ (compared to that of the entire territory) and territorial coverage, respectively.

Consider, for example, three geosets with somewhat similar high structural dispersion: (a) the petrol industry with $s_Z^2/s_T^2 = 1.05$; (b) farmlands with 1.03; and (c) population with 0.84. These geosets have very different territorial coverage: (a) 0.13; (b) 0.85; and (c) 0.53. Consequently, ${}^r s_Z^2$ is very different for each: for (a) it is 18.44, suggesting an enormous fragmentation of the distribution; for (b) it is only 2.75, suggesting that farmland has a degree of compactness similar to that of the entire territory of Italy (2.31); for (c) it is 3.64, suggesting that population is less compact than that of the uniform distribution over the territory of Italy, but not drastically different.

Let us now compare two geosets with low structural dispersion: advertising and public relations with $s_Z^2/s_T^2 = 0.46$ and heavy industry with 0.36. The territorial coverage is smaller for the first (0.21) than for the second (0.28); however, ${}^r d_Z^2$ is larger (7.58) for the first than for the second (3.80), suggesting that the geoset of publicity is less compact than that of heavy industry. A look at the relevant maps confirms that these findings correspond to the visual impression.

In the analysis of the geosets examined in the *Atlas*, the parameters noted above were found to be very useful for both descriptive and analytical purposes. The relative scatter was found to be a very significant and sensitive parameter. For the 385 primary geosets examined in the *Atlas* its range varies between a minimum of 1.34 and a maximum of 38.45. For 354 geosets of surpluses it varies between 11.35 and 257.31.

1.5.7. Concise Measure of the Discrepancy between Two Geosets

Apart from the methods introduced above for comparing the geosets, in this book we generally use a summary parameter of territorial discrepancy between any pair of primary or derived geosets, which is based on the approach suggested in Sections 1.3.2 and 1.4.2. To compare, say, ${}^gP$ to ${}^gU$, we transform ${}^gP$ into ${}^gU$ by deriving geosets of surpluses ${}^gSPU$ and ${}^gSUP$ considered, respectively, as origins and destinations of the elements to be shifted. Then we link each origin to a single destination, so that the sum of distances covered by the designated shifts is minimized.

In the same way we compare ${}^gP$ to any other primary geoset ${}^gQ$. A simple example of this approach is shown later in Figure 13.3. Let us indicate: the total number of elements of the reduced geosets considered (e.g., $M = 10{,}000$) by $M^T = M^Q$; the number of surplus elements by M_s; and the total distances covered by the shifts by Σ. Dealing, for example, with distances between elements of ${}^gP$ and ${}^gQ$, we obtain: ${}^m\bar{D}(P, Q) = \Sigma/M_s$ = average minimal distance; and ${}^m\bar{D}(SPQ, SQP) = \Sigma/M_s$ = average minimal distance between surpluses. We then see that ${}^m\bar{D}(P, Q) = \Delta^m\bar{D}(SPQ, SQP)$. This equation is of great help in simplifying calculations and in analysis.

The minimal distance has many important properties, among which is that of being a metric. Consequently, we can build "triangles of distances" (as shown later by the examples of Figures 14.2, 14.4, 14.5, and 14.6 in Section 14.2), and we find functional relationships between minimal distances between pairs of geosets. If we take the average of the squares of the distances between linked elements in ${}^gP$ and ${}^gQ$, we can separate it into components, such as the square distance between the centers of ${}^gP$ and ${}^gQ$, the square difference in dispersion of ${}^gP$ and ${}^gQ$ measured by $(d_P - d_Q)^2$, and a residual. This separation enables us to link the analysis of the minimal distance to that of the center, measures of dispersion, and other parameters. These and other properties have been found to be of great help in geostatistical analysis.

We give below a few examples of calculations of minimal distances: (1), (2), and (3) are based on the comparison between selected pairs of the geosets for Italy that have been presented here:

1. Minimal distance from uniform distribution over the territory: farmland: 20.5; population (1981): 54.2; all economic activities: 119.8; petrol industry: 125.2; publicity and public relations: 248.0; heavy industry : 281.7; and citrus: 551.9.

2. M!inimal distances from the population (1981) (Figure 1.16): all economic activities: 87.8; petrol industry: 115.5; publicity and public relations: 217.4; heavy industry: 252.

3. Minimal distances from all economic activities: publicity and public relations: 133.7; heavy industry: 165.1; petrol industry: 175.5. The minimal distances listed above are absolute values, expressed in kilometers. We will see later that these parameters can be rendered dimensionless, thus enabling comparisons for population geosets distributed over different territories. As expected, a very low minimal distance is thus found for farmland over the uniform distribution. The population differs more from the uniform distribution, but is not very far from it. Also as

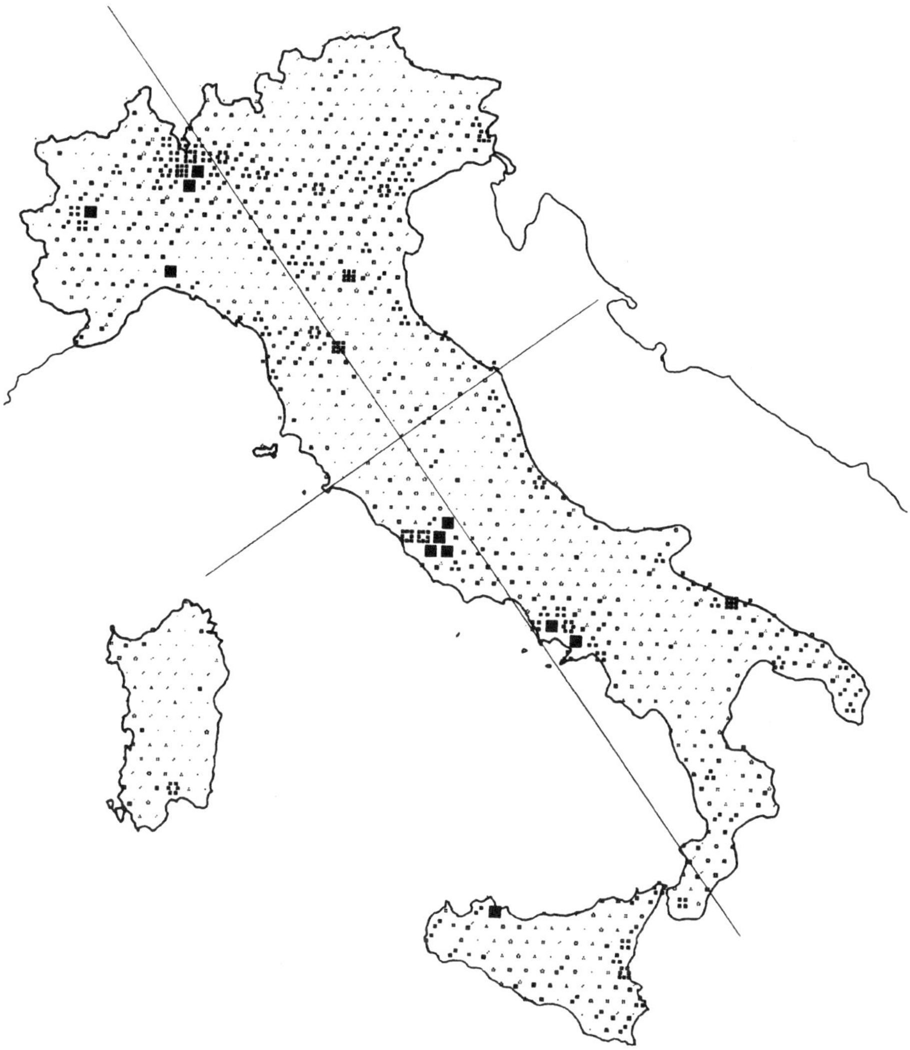

Figure 1.16. Total population of Italy, 1981: . 5700 ▪ 57,300 ▪▪ 11,4600 ⁘ 286,500 ■ 572,900+ (Source: *Statistical Atlas of Italy*, 1988).

expected, all other geosets differ rather strongly from the uniform distribution, the largest distance being found for citrus. Minimal distances from population and economic activities correspond well to expectations.

4. Later we will show examples in which, given three geosets generally indicated as ${}^gP$, ${}^gK$, and ${}^gQ$, we find that ${}^m\bar{D}(P, K) + {}^m\bar{D}(K, Q) = {}^m\bar{D}(P, Q)$.

5. Findings of this type are always true for: (a) geosets related to sets in which P and Q are components of a compound set K; (b) geosets (${}^gP, {}^gQ$) that are compounded in ${}^gK$; (c) other primary or derived geosets that are related functionally.

Consistency may also be found empirically for geosets that have a special nonfunctional relationship.

6. We shall see that it is most interesting to discover possible consistency or to measure inconsistency in the study of geosets relating to "ordered" sets, such as a population at various times, or subpopulations grouped according to a particular measurable or ordered variable.

7. Following this or other approaches, it is found that minimal distances can be of great help, e.g., in the analysis of long-time series of geosets; in forming "families" of analogous distributions and in analyzing their determinants; and in testing hypotheses.

Chapter 2

Selected Applications of Geostatistical Methods

2.1. Generalizing the Approaches in Chapter 1 to Other Fields

We trust that the examples given in Chapter 1 are sufficient to suggest directions in which the integrated use of graphical rational pattern (GRP) mapping and geostatistical parameters can be of use. In the following, we shall generalize the conclusions that can be drawn from these examples by suggesting other fields of application, channels for integration, ways of selecting goals, and stages of research.

All of the examples described thus far related to the use of geostatistical methods in a specific field, namely for agencies that deal with national statistics, where enormous quantities of data specified by small areas are accumulated by demographic, economic, and other censuses. We have found that GRP maps and related summary parameters can be of help in the presentation of the information in an easily understandable form as well as in making it available to specialists in formats that enable serious research. Obviously, the suggestions we have made can be extended to other fields that involve detailed statistics resulting from special surveys, current vital and migration data, and agricultural and industrial statistics.

All the examples given in Chapter 1 were based on statistics for Italy in order to utilize the extensive information in the *Statistical Atlas of Italy*. However, we have applied these geostatistical methods in various studies, some published, some unpublished, in Israel, France, Argentina, Turkey, Germany, and particularly the United States, among others (see Bachi, 1955–1989). Among the works regarding the United States, we shall note in this book mainly: (a) extensive calculations prepared in the Bureau of the Census (primarily from 1977 to 1979) on the basis of data detailed by counties, taken from various censuses (see Appendix B), and (b) calculations prepared more recently for this book, also on the county data. We have not had the opportunity to undertake systematic international comparisons on the basis of these studies, as such an enterprise would need the extensive backing of research institutions or international agencies. However, the material we have obtained is sufficient to suggest that reducing the geostatistical parameters so as to render them dimensionless may offer the possibility of useful international comparisons and overviews.

2.2. Applying Geostatistical Methods in Different Fields

Geostatistical methods have been widely and effectively used in urban research in dealing with official data for cities, originally correlated from census tracts, districts, and wards. Some of these studies have dealt with individual cities and others were designed for the purpose of interurban comparisons.

The most extensive study in comparative geostatistical research, carried out in cities in many parts of the United States and countries of Latin America, Europe, and South Africa was the one done on urban ecology of the Jews in the Diaspora (Bachi *et al.* 1988). Sponsored by the United States–Israel Binational Science Foundation, this project was devised to examine ecological characteristics of Jewish minorities, as compared to other minorities and the general population, and to provide an empirical evaluation of hypotheses and models regarding the distribution of Jewish populations over urban and metropolitan areas. Data obtained from a great variety of sources (mainly other than censuses for the Jews) were systematically mapped and summarized by means of geostatistical parameters. The research pointed to peculiarities in the territorial distribution of the Jewish population, for which the following was found: (a) less dispersion, special position, a persistence of clustering, even within a context of modern suburbanization; (b) some broad similarities between ecological characteristics of the Jews in different cities, together with divergences in various contexts and by dissimilarities in community development. Detailed monographs were prepared for Rome (1527–1985), Warsaw (1839–1938), and the Toronto metropolitan area (1981). For other cities the study was more limited. Another objective of the research was to evaluate the utility of geostatistical methods in comparative urban research and to examine their efficacy in comparison to other methods currently employed to measure segregation and other aspects of urban population distribution (see Appendix C).

In this book we shall largely utilize the findings of this research and especially its methodological conclusions. We shall also make reference to the use of geostatistical methods in urban studies such as the following: (a) Research undertaken in the U.S. Bureau of the Census concerning Washington D.C., Atlanta, and Austin within the framework of the more general research referred to above (see Bachi and DeAre, 1978). (b) Research on various Israeli cities. (c) Recent research on two "eternal cities": Rome and Jerusalem.

The studies referred to above are generally based on settings in which we consider a well-defined territory, over which a population, or many populations or subpopulations, are distributed. We use here, and throughout the book, the term "population" in a sense much broader that the demographic one. For example, geostatistical methods have been applied to: (a) events such as births, deaths, crimes, accidents; (b) sites or establishments such as factories, farms, shops, educational and other institutions, tourist resorts; (c) services or other economic activities; and (d) total values referred to statistical units, such as quantities or monetary value of production, sales, and income. Presumably, geostatistical methods could also be applied to animal populations and plantlife.

In most of the applications described in this book, the gross territory T over which the geosets ${}^gP$ are distributed is clearly predefined according to political, administrative, or other criteria. The clear definition of T as container of ${}^gP$ allows us to obtain the geoset ${}^{g}U$ of the uniform distribution over *T*. This, in turn, enables us to compare ${}^{g}P$ to ${}^{g}U$ systematically, as well as to obtain maps and parameters for derived geosets. However, geostatistical methods can also be used if we know the locations of a set of elements, even if the gross territory over which these elements are distributed is either not defined or irrelevant. We shall introduce methods that enable us to overcome this difficulty: knowing the locations of the elements, we shall determine the boundaries of the gross territory over which all the elements, or a large majority of them, are included (see Chapter 4). Having estimated the gross territory, we can then obtain the net territory (habitat) from it and apply the same geostatistical methods that are used for the analysis of ${}^{g}P$ over a predetermined territory.

It has not yet been possible to test systematically how well and how quickly laymen are able to understand GRP maps and the information given by geostatistical parameters. Informal experience, gained by the presentation of maps and parameters to many individuals and to audiences, seems to be encouraging. However, geostatistical methods are not suggested as a panacea, nor can they be expected to completely solve complex research problems in a simple way. They can be used extensively in research, but generally geostatistical findings are integrated with interdisciplinary efforts of geographers, general statisticians, and scientists conversant with the specific fields (e.g., demography, economics, sociology) to which the data refer.

Also, from this viewpoint, the experience gained in the preparation of the *Atlas* may help. In that work, the 385 geosets studies were divided into 36 subclasses according to their contents [e.g., (a) geosets of the distribution of the population in 12 censuses between 1871 and 1987; and (b) natural and migratory movements of the population between 1951 and 1986]. The subclasses were grouped into six classes according to: (a) the evolution of population distribution and its determinants; (b) the structure of the population by, e.g., sex, age, marital status, and education; (c) housing characteristics; (d) financial sector; (e) agriculture; and (f) tertiary occupations and services.

All of the significant geostatistical parameters for the geosets included in each subclass were shown graphically in ways that facilitate comparisons and were distributed to experts, together with numerical values and relevant maps from the first volume. Each of the experts studied the material relevant for him thoroughly and wrote a detailed report on substantive findings and on possible explanations and implications of results regarding the geographical characteristics, similarities, and differences of the distributions studied.

The main lesson learned from this monumental work was the utility of an integrated approach. Studying together the peculiarities of each distribution as it resulted from the detailed GRP maps and the comparative values of geostatistical parameters in the light of expert knowledge in each field has permitted us to obtain, and explain, an enormous number of findings, a considerable proportion of which were, until then, unknown or known only vaguely.

It is conceivable that integrated geostatistical methods might be of help in solving practical problems in industrial and commercial planning, in the distribution of services, and in the work of other government agencies. However, these possibilities have apparently not yet been explored.

In Chapter 17 we explain the graphical–geostatistical software that is available for the preparation of GRP maps and of geostatistical parameters, and the number of such tools that can be provided on the basis of each pair of geosets $({}^gP, {}^gU)$ or $({}^gP, {}^gQ)$ is very large. There is a danger that unsophisticated users might feel lost in the jungle of these maps and parameters or might misinterpret some of them. Thus it is important to learn to be selective.

For example, it is desirable to: (a) start with some preliminary exploration and use of previous expert research in the specific field of concern to try to define as clearly as possible the aims of the study; (b) select the geosets to be computed properly; (c) examine provisionally on the screen alternative GRP maps that can furnish some orientation; (d) select the proper parameters to be used by taking into consideration the aims of the research, the characteristics of the geosets under study, the properties, and the meaning of the available parameters.

A considerable part of this book is devoted to these problems. Prospective users of geostatistical programs are advised that before embarking on an application of geostatistical methods they first become familiar with the general principles of geostatistics as explained in Chapter 1 and eventually with those in subsequent more detailed specific sections that are relevant for their work.

2.3. Scope of this Volume

The two last chapters of Part I are devoted to the formalization of geostatistical methods by means of geosets of populations and territories. Part II deals with mapping methods. Part III describes geostatistical parameters showing or measuring characteristics of primary and derived geosets. Part IV elucidates minimal distance and its properties and applications. Part V explores special problems, presentations of geostatistical computer packages, and some general conclusions.

As we note earlier, the main aim of this book is to put geostatistical methods at the disposal of prospective users. As these users may belong to different disciplines and not necessarily be professional statisticians, the book has been written in a simple style, confining demonstrations and somewhat more advanced discussions to appendixes. Whenever possible, apart from suggestions on how to employ geostatistical computer programs, we have also provided indications for direct calculation without use of the program for many (but not all) of the tools suggested.

Despite the emphasis on the practical aim of application, the book has not been designed as a simple list of "recipes," and we have given systematic explanations of the criteria and properties upon which the proposed methods are based.

As this book introduces geostatistical tools mainly as methods for graphical presentation, description, and summarizing of information on distributions of actual populations, the following fields are completely outside its scope: (a) inferential research regarding the parameters proposed; (b) study of spatial samples; (c) study of stochastic models related to spatial processes and analysis of point, line, and area patterns based on theoretical foundations (see, e.g., Rogers, 1974; Getis and Boots, 1978; Diggle, 1983). We have also omitted the study of models of distributions of populations or economic activities over cities, regions, or countries, despite the fact that geostatistical tools might be of use in the empirical verification of such models.

Population distributions and territories studied here are given over maps drawn with projections in which actual distances between points on the face of the Earth are recorded. Distances used in the calculation of geostatistical parameters are Euclidean. Extension of the parameters over the entire globe or considerable parts of it and the use of non-Euclidean distances, albeit sometimes possible, are not included in this volume. Virtually all consideration of the methods for defining the "regions" that serve as a basis for many of the primary data used as input for rational maps and geostatistical parameters has been left out.

Lack of uniformity of territory is one of the aspects that has, legitimately, drawn the most attention in the geographical–statistical literature. In order to compensate one of the basic research techniques is to define "regions" in a way that keeps at least some of the determinants that render the territory multiform under control. There is an extensive literature dealing with the methods for defining regions, e.g.: (a) administrative, political, historical, or other criteria can be used; (b) regions can be defined in ways that ensure an acceptable degree of uniformity within the region with regard to natural, socioeconomic, ethnic, or other factors; (c) regions can be defined so that each of them feels primarily the influence of only one of the nodal points in the territory.

We saw no need to summarize this literature or to add to it, because here we are not relating to regions as foci of research. They are rather only a means by which to convey statistical information whenever it is not possible to present detailed information concerning the location of each element of the distribution. Thus we generally utilize the most detailed information available as input, namely, the smallest possible regions. Rational maps and geostatistical parameters later transform this information into macroscopic form, which facilitates the comparison of findings obtained by data based originally on different regional networks.

Classification of small regions by qualitative characteristics, by a quantitative variable, or on the basis of multivariate analysis helps to divide them into classes, enabling comparison of geographical characteristics of each class on the basis of geostatistical treatment. However, again, we have not included here any consideration of the criteria of classification, although we have given examples of geostatistical analyses of such classes.

There are several manuals available to assist geographers dealing with statistical research and methods in quantitative geography, which cover many different types of statistical methods (see, e.g., Duncan *et al.*, 1961; Gregory, 1963; King, 1969; Yeates, 1974; Matérn, 1986). Insofar as these deal with macrogeographic overviews based on statistical information, tools such as centers and standard distance are often referred to. It may be noted that the main concerns of quantitative geography (such as distributions of points, lines, and bounded areas) can, with certain qualifications, be integrated with the geosets discussed in this volume.

While this book has been designed mainly to provide assistance with practical applications of the methods it describes, it may also encourage authors of textbooks in applied statistics to introduce or expand chapters dealing with elementary methods for graphic representation and explication of territorial series. The attention (and space) devoted to such series in statistical textbooks is generally very limited in comparison to that devoted, say, to time series. Moreover, various manuals of applied statistics that give examples of data specified by regions or localities treat this information as if it were based on a nominal scale (in which each place or region is considered as a separate category). As indicated later in Section 4.8, this approach ignores the possibility of utilizing the knowledge of geographical characteristics of regions or localities (e.g., position on the map, size, shape) in the analysis. Spatial data can often be considered as being based on a double interval scale, which requires special methods or treatment and these features can broaden the scope of applied statistics manuals.

2.4. A Brief Survey of the Development of Geostatistical Methods

2.4.1. Limitations of this Review

Some of the methods discussed in this volume have been in use for a long time, while others have been developed during the past several decades. Some historical references to the literature are indicated in the Bibliography, but in the following we give only a very sketchy outline of some general developments.

Integration of rational mapping, analysis of interconnected geostatistical parameters, and application of both approaches to primary and derived population geosets and related territories appear here for the first time although we did publish some preliminary indications (Bachi, 1989*a*,*b*). As this is the main theme of the book and limitations of space had to be taken into account we found it expedient in planning this volume to limit discussions comparing the specific tools used here and possibly analogous methods employed in other contexts, which can be found in the literature. To give just one example, let us refer to the methods of quantification of shapes: the need to express the shape of a figure quantitatively has been widely felt in many disciplines, including physical and human geography, geology, mineralogy, biological sciences, psychology, and scene analysis. The aims of these studies might be e.g., the reproduction of the individual figures (as in factual methods), pattern classification and recognition, or a comparative study of general characteristics of sets of figures. A comparative analysis of the literature in these fields would have been an impossible task, not only because it might have been beyond our area of competence but also because of limited space. We have therefore reduced our presentation in Chapter 3 to a consideration of quantitative parameters of shape for the study of the distributions of population and related territories.

A similar limitation holds for various other topics and to the enormous literature discussed in recent treatises on the analysis of spatial data (such as, say, those of Ripley, 1981; Upton and Fingleton, 1989; Haining, 1990; Cressie, 1991). These works are devoted primarily to the study of problems fundamentally different from those discussed here. The limitations we have imposed have resulted in unavoidable sins of omission of which we are deeply and regretfully aware and for which we beg the forgiveness of scholars who have not been duly cited.

2.4.2. Centrography, Social Physics, and Geostatistics

The first attempts to obtain concise parameters of population distributions were directed to the calculations of centers or centrography. Determination and analysis of centers of population and territory have a long and honorable history, going back for more than a century. Excellent reviews of this work, with full bibliographies are available in the literature (see, e.g., Jones, 1960 and later; Neft, 1966; Sonnino, 1968; Coccossis, 1980) and it need not be repeated in detail. Here we only wish to supply some basic background information:

1. Calculation of population centers started in the United States in 1874, and was reconstructed back to 1790 by the U.S. Bureau of the Census, and for the Colonial Period, from 1754, by Neft.
2. Extensive applications of centers were carried out in Russia between 1906 and 1934, and then in many other countries, including France, Australia, India, Japan, Germany, and Poland.
3. Most early applications were mainly *ad hoc*. Gini and his collaborators in Italy in the late 1920s and early 1930s emphasized for the first time that the calculation of centers was a step in the statistical analysis of "territorial statistical series," which were considered one of the basic types of primary statistical series including time series and quantitative and qualitative series.
4. In this context, a little later we (Bachi, 1934) suggested that statistical treatment of these data should also deal with the measure of dispersion (see also Risser, 1937). Other suggestions were put forward around the same time for a theoretical approach to the study of dispersion (see Gini, 1934, and Galvani, 1933). Actually these first attempts in the study of dispersion constitute the first steps that lead from the narrow study of location only (by means of the analysis of centers) to a broader view of what we define here as geostatistics.
5. The name centrography to mean the study of centers appeared much later in the literature, and was apparently suggested by researchers who dealt with the history of the lively debates around the meaning and way of calculating various types of centers. The intense activity devoted to the calculation of centers was termed by Neft (1966), the "centrographic fever of a generation ago," which term was probably rather unfortunate, as it suggested the existence of a branch of science based solely on the consideration of centers.

Attempts to describe geographical distributions in terms of physical laws have been based primarily on the use of gravity models to describe spatial interaction. Models of potentials have been designed as general descriptions of the interdependencies existing simultaneously among all locations in large areas. As noted in Berry and Marble (1968), "Such models have been argued by their proponents to correlate highly with the spatial distributions of a wide variety of social and economic phenomena..." (pp. 3–4). Gravity models were first introduced in geography by Ravenstein (1885, 1889) in the study of migrations. "Thereafter those models found wide applications, for example, in marketing geography (Reilly's laws on retail gravitation in 1929), or in urban transformation studies describing interzonal travel... The gravity analogs were generalized by the astronomer J. Q. Stewart, who developed the idea of potential surface that simultaneously describes the interactions of every place and every other." The method of potentials has been used extensively in many fields.

Geostatistical methods and social physics share the approach of obtaining an overview of macrogeographical features of the population distributions from detailed microgeographical–statistical data. However, the social physics approach is based on the utilization of theoretical laws of gravity, which may or may not be adaptable to the various types of distributions. If viewed from a purely descriptive point of view, the methods of potentials are far more complex than geostatistical parameters. For this reason and because of limitation of space as well as the fact that they are amply covered in the literature, we have not included the potentials in this book. Only Appendix B is devoted to a bridge between them and some geostatistical parameters and to suggestions for the elimination of a serious shortcoming of potentials as a descriptive method.

In the late 1950s and early 1960s, we (Bachi, 1958, 1962) proposed methods measuring territorial dispersion. Contributions to this topic were also made by Warntz (1956). In 1966, Neft published a volume entitled *Statistical Analysis for Areal Distributions*, in which he made a systematic attempt to apply statistical methods for the analysis of two-dimensional distributions to geographical data. Since then, there have been many applications of geostatistical methods, which we refer to in the Bibliography. We note in particular only the important contributions made at Cornell University, starting with the pioneer work of Jones and including works by Peters, Lee, Loo, Manson, Caccossis, and others, which deal mainly with applications to urban research and have brought about the development of new geostatistical tools.

The term "geostatistics" was probably suggested for the first time, "with some trepidation," by Hart (1954), as a useful way to describe statistical techniques such as centrography, which emphasize location within real distribution rather than their value in linear distributions. We used the term somewhat later in a wider context (Bachi, 1962*b*) and our example was followed by other scholars. Subsequently and completely independently, the term geostatistics was employed in a rather loose way to denote applications of statistics to geological problems, primarily in the United States and the United Kingdom, and an example of this usage is found in Merriam (1970). The word also assumed a much more specific meaning, which was applied for the first time by Matheron (1962) to the analysis of a special type of two- and three-dimensional, nonstationary, spatial series. It was also used in English for the first time by Matheron (1963), and since then its use in this sense has been rather wide (see the bibliography in the *Review of the International Statistical Institute* and Bachi, 1973). The term "geomathematics" is sometimes used in place of geostatistics in its wider geological meaning but despite the regrettable risk of confusion, we have continued here to use geostatistics in its original specific sense.

There is an extensive literature in the general field of the preparation of statistical maps, which is scattered among statistical, geographical, and cartographic books and papers. However, we shall review only selected developments in dealing with "rational" mapping (see Chapters 5–7). We carried out some early research on the possibility of developing scales of symbols that make the accurate representation of values over choropleth maps possible (Bachi, 1932, 1937, 1955, 1962, 1965). A book published in 1968 introduced the theory and applications of GRP, but papers based on the computerization of GRP were published only later (see Bachi, 1975, 1976, 1978, 1988).

Chapter 3

Geosets of Populations Distributed over a Predetermined Territory

3.1. Introduction

3.1.1. Why Geosets?

The preparation of rational maps and geostatistical parameters is based on information concerning the locations of the elements of the populations and territories considered. In this chapter we formalize and conceptualize this information as geostatistical sets *geosets*, by which we mean the distribution of relative frequencies of the locations of the elements of a population (or a territory). More specifically, the geoset is a list of locations of elements of population or of territory to which we attach the relative weight assigned to each location in order to prepare maps and geostatistical parameters. Locations can be expressed by longitude and latitude, by Cartesian coordinates, or in any one of several other ways, but in this volume we use only Cartesian coordinates. For some hints on calculations of geostatistical parameters based on longitudes and latitudes, see Shimony (1963).

Reference to geosets has proved to be useful for many reasons: (a) it helps to clarify input procedures and the relationships among the various types of input, which are given either individually for each element or in aggregate form; (b) it introduces simple equations that enable us to express ourselves in concise form and to calculate the various parameters, e.g., centers, dispersion measures, and distance between geosets; (c) with suitable adaptation it extends ordinary set operations to geostatistical research and helps derive other geosets from the primary ones; (d) this, in turn, makes the relationships between primary and derived geosets and their geostatistical parameters more easily accessible; (e) it clarifies the relationships between maps representing primary and derived geosets; (f) it can expedite comparisons between actual and theoretical population distributions.

We start our presentation of geosets with a short discussion of ordinary sets (Section 3.1.2) and introduce a very simple example of a population geoset (Section 3.1.3). The example given deals with the distribution of the elements of a small population treated as if it were found over an undetermined unbounded territory. These restrictions are later removed. The populations considered throughout this book are generally large (Section 3.4), and the assumptions are that a territorial range is determined for unbounded population distributions (Section 4.2) and that individual population elements cover sizable tracts of territory (Section 4.3).

The concept of geoset is extended from elements of population to elements of territory (Section 3.2 and Chapter 4), which is generally useful and, in particular, very helpful for the comparison between the actual and the uniform distribution over the territory (Section 3.2). Comparisons between the geosets of two populations over the same territory are discussed in Section 3.3. Special emphasis is placed on the comparison of geosets of compound populations formed from various components (Section 3.6).

We then describe a general scheme based on these ideas for the relationships between primary and derived geosets of populations and territories (Section 3.7), which is followed throughout the book. Section 3.9 deals with sets of frequencies in which the locations of the cells are not taken into account.

3.1.2. Sets and Set Operations

We shall take as our starting point the presentation of sets and of operations on sets, as indicated by the usual Venn diagrams. One such diagram is given in Figure 3.1, and we can assume that it refers to any one of the "populations" mentioned in Section 1.1 as an object of geostatistical research. More specifically, we refer to a human population, as enumerated in a census taken over a designated territory (country, region, or city) at a given time. Following a widely accepted convention, we indicate population sets by P, A, B, etc. To avoid confusion, corresponding geosets are denoted as ${}^{g}P$, ${}^{g}A$, ${}^{g}B$, etc.

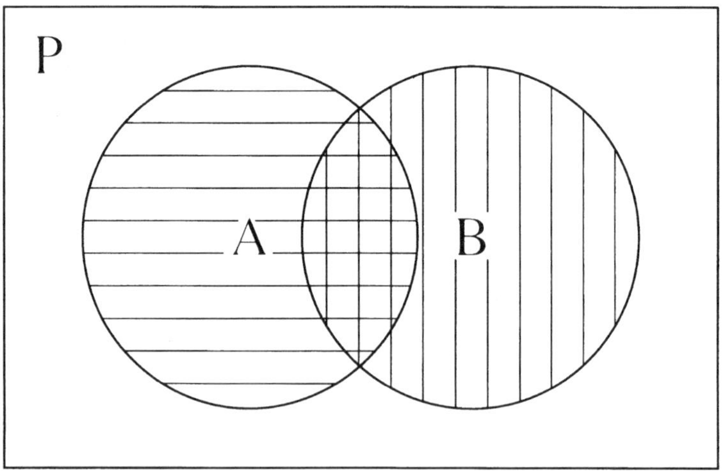

P – population
A – aged 25+
B – single

Figure 3.1. A Venn Diagram.

The rectangle of Figure 3.1 represents the universal set of all the individuals or "all the elements of the population" enumerated. The universal set of the figure is indicated by P and its elements by p_i: $P = \{p_i | p_i$ is an individual enumerated in the determined territory at a given time$\}$. The circles A and B, covered, respectively, by horizontal and vertical lines, represent sets $A \subset P$ and $B \subset P$ formed by specified elements of the population P. For example, A might be the set of individuals aged 25 or older and B the set of single people. We can carry out some of the usual set operations. As there are individuals who are single and aged 25 or more, the intersection $A \cap B$ is not empty. This intersection is shown by the cross-hatched part of the figure. The difference set $(A - B)$, including persons aged 25 and more who are not single (married, divorced, widowed) is shown by the part of circle A that is not cross-hatched. The other difference set $(B - A)$, single people who are younger than 25, is shown by the part of circle B that is not cross-hatched.

By complementation we can obtain, respectively, the set of all individuals who are not single $(P - B)$, which is shown by the part of the rectangle not covered by B; the set of all individuals who are younger than 25 $(P - A)$ is shown by the part of the rectangle not covered by A.As sets A and $(P - A)$ are complementary, $A = (P - A)'$ and $(P - A) = A'$. Similarly, $B = (P - B)'$ and $(P - B) = B'$.

The union of A and B is obtained by taking $(A \cap B)$, $(A - B)$, and $(B - A)$ together. In our example, the union of A and B would indicate individuals who are either single, or aged 25 or older, or both.

For disjoint or mutually exclusive sets $P_j (j = 1,2,3,\ldots)$ it is sometimes acceptable to designate the operation of union with the symbol $+$. For example, if P_j is the population included in any age group j, combining all age groups, we can write $P = \Sigma P_j$. In this book we sometimes consider populations $P = \Sigma \varepsilon_j F_j^P$ (where $\varepsilon_j = \pm 1$) formed by taking both positive $(+)$ and negative $(-)$ sets (provided $F_j^P > 0$, where F_j^P is the size of the set). For example: $j = 1 =$ the population at the beginning of a period; $2 =$ births, $3 =$ deaths, $4 =$ immigrants, and $5 =$ emigrants during the period, and $P =$ the population at the end of the period; 1, 2, and 4 are positive; 3 and 5 are negative.

Let us denote the absolute size (i.e., frequency = number of the elements) of the sets considered above by the symbols F^P, $F^j = F(P^j)$, F^A, F^B, $F^{A'}$, $F^{B'}$, $F(A - B)$, $F(B - A)$, $F(A \cap B)$, and $F(B \cap A)$. Some examples of relationships between these frequencies are:

$$F^P = \Sigma \varepsilon_j F^j \tag{3.1a}$$

$$F^P = F^A + F^B \tag{3.1b}$$

$$F^P = F^A + F^{A'} \tag{3.1c}$$

$$F^A = F(A \cap B) + F(A - B) \tag{3.1d}$$

3.1.3. Introducing Population Geosets

We now go on to consider population geosets. To introduce them, we will use a very simple example, dealing with an actual small population. Despite the fact that this book deals in general with large populations, we have selected this particular example because it enables us to check the small numbers of elements involved in the geoset operations on the relevant maps.

The population set we will deal with is a list of the hotels operating in 1989 within the municipal territory of Jerusalem, Israel. This list, which is denoted by the generic symbol P (population), was prepared for the use of tourists and includes apart from the name of each hotel information such as the number of beds, the rating, and the address. If we identify each hotel by a conventional symbol h, there are $F^P = 63$ hotels h in the set P.

A geoset corresponding to P (denoted by ${}^gP$) was constructed by transforming the address of each hotel h into coordinates (X_h, Y_h). These are expressed as distances (in km) along the X- and Y- axes crossing at a well-defined origin—a familiar spot known to natives and tourists alike as David's Tower.

We can now construct a map, such as that shown in Figure 3.2, in which each hotel is represented by a dot with the coordinates (X_h, Y_h). For orientation purposes, we have indicated the outline of the Jerusalem city walls. Built in the 16th century, these walls enclose the historic Old City, a major tourist attraction as it has many sites holy to Jews, Christians, and Moslems as well as many archaeological excavations, and we can see on the map that there is a concentration of hotels in and around this area.

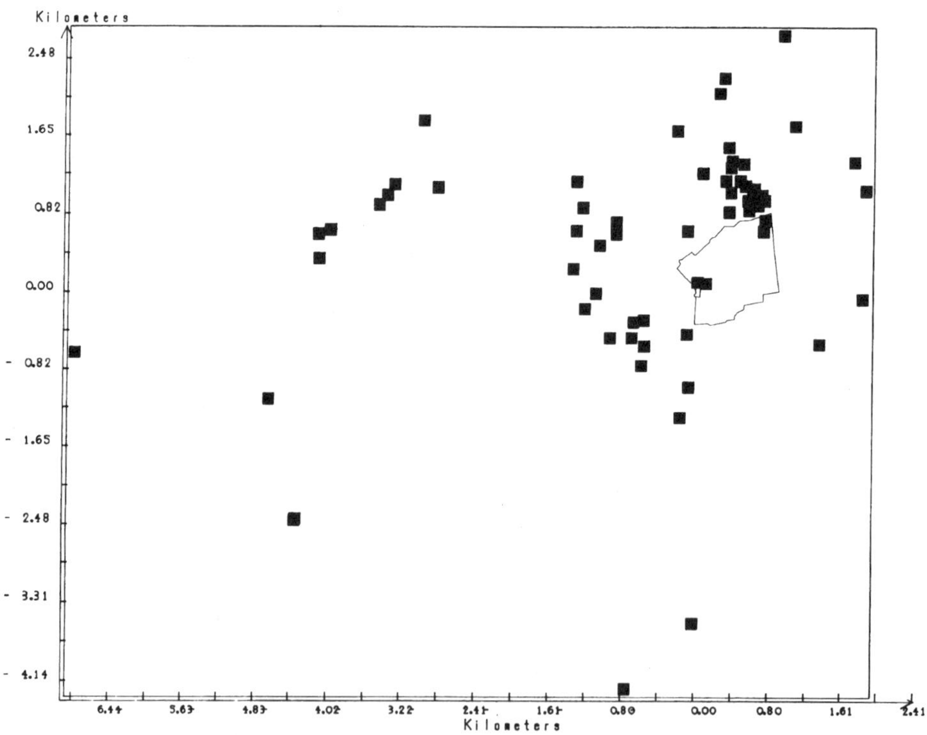

Figure 3.2. Jerusalem. Hotels within a rectangle, 1989 (Source of data: Israel Ministry of Tourism).

To proceed from vague generalities apparent from the map to exact parameters such as average location and dispersion, comparable to parameters for other relevant geosets, we need to take further steps. The first is to attach a relative weight to the coordinates of each population element h. For example:

1. We can attribute the same value to each hotel and take $1/F^P = 1/63$ as the relative weight for each h. Then the geoset ${}^gP$ would be: ${}^gP^\cdot = \{X_h, Y_h, w_h^P | (X_h, Y_h)$ are the coordinates of element h of set P; the relative weight to be attached to these coordinates in the calculation of geostatistical parameters is $w_h^P = 1/F^P\}$. The geoset of the hotels is indicated here by the symbol ${}^gP^\cdot$. The point that is included in this symbol has been added to indicate the specific way followed here to form the geoset by showing each element h of the population considered. As most of the geosets given in this book refer to large populations, the primary information available for them does not make reference to individual population elements. We shall see that this information is given in aggregate form by indicating the relative frequency of the relevant population in each of the cells of fixed area and shape into which the gross territorial container will be divided. Such "cell geosets" will be denoted without the point superscript (as for example ${}^gP$).
2. Alternatively, we may prefer to take into account—in calculating the geostatistical parameters—the different sizes of the hotels, as measured, e.g., by the number of beds. Let us indicate by F_h^B the number of beds in each hotel h, and by F^B the total number of beds (in our numerical example $F^B = 7702$). Then we can obtain a geoset of hotel beds as follows: ${}^gB^{H\cdot} = \{X_h, Y_h, w_h^B | (X_h, Y_h)$ are the coordinates of hotel h; $w_h^B = F_h^B/F^B$ is the relative weight to be attached to these coordinates in the calculation of geostatistical parameters$\}$

In regard to Figure 3.2 we note the following:

1. To introduce the geoset ${}^gP^\cdot$ visually we have used the most simple and common method of a dot on the map to represent each element of the geoset. We can accept this as representative of geoset ${}^gP^\cdot$ under the convention that the same relative weight $1/F^P$ is attributed to each element h.
2. The dot map has the advantage of giving a one-to-one correspondence between an element of ${}^gP^\cdot$ and a dot on the map.
3. However, hotels are not dimensionless entities: they are found in buildings that cover a tract of territory (see Section 4.3). We might discuss whether the coordinates to be attributed to this tract should correspond to the mean center of the ground floor, to the mean center of all floor space, to the reception desk, or perhaps to the entrance to the hotel from the road. However, in general, this is irrelevant because any solution would lead to the same results if the hotels are viewed over the scale of the map representing the territory T. The small variations found cannot seriously affect either the map or the geostatistical parameters based on (X_h, Y_h).
4. Theoretically, a dot is a dimensionless symbol that is placed on the map to

correspond to the location of an element of ${}^{g}P^{\cdot}$. However, for a "dot" to be visible in practice, it must be a small circle or square covering some finite space on the map. If the number of elements to be represented is large, symbols may overlap, and this makes the use of dot maps inadvisable. Other representations, such as the graphical rational patterns (GRP), are then preferred.

5. Use of GRP is also preferable for showing sets of type ${}^{g}B^{\cdot}$ and for presenting data aggregated into cells for large populations, as we shall see in Chapters 6 and 7.

3.2. Comparing the Actual Population Distribution to the Uniform Distribution over the Territory

3.2.1. The Uniform Distribution over the Territory

Figure 3.2 shows a distribution of dots (corresponding to the elements of ${}^{g}P^{\cdot}$) over an undetermined territory. Much progress in the analysis of the geoset can be achieved if we compare the actual distribution to a theoretical uniform distribution over the gross territorial container of the population. This container is often predetermined, as it is a function of the definition of the population considered. For example, the set P to which the geoset ${}^{g}P^{\cdot}$ refers has been defined as the list of the hotels operating in 1989 within the municipal territory of Jerusalem (denoted by the symbol T). Thus we are justified in presenting the geoset ${}^{g}P^{\cdot}$ over the background of T, as we have done in Figure 3.3, where T is shown by its contour line, elements h are again represented by squares (as in Figure 3.2), and the origin of coordinates (David's Tower) is repeated for reference purposes. It is immediately obvious from this map that most of the hotels are heavily clustered and that the distribution is far from uniform over T.

To highlight and measure the discrepancy between actual distribution ${}^{g}P^{\cdot}$ and uniform distribution over T, it is convenient to construct the geoset of the uniform distribution (to be denoted as ${}^{g}U^{T^{\cdot}}$) and to show it graphically, as we have done in Figures 3.4 and 3.5. The territory T (with some adaptations) has been divided into $N^c = F^P = 63$ cells of regular hexagonal shape, and area $A^T/F^P = 1.722$ sq km, with $A^T = 108.5$ sq km. To facilitate the comparison between ${}^{g}P^{\cdot}$ and ${}^{g}U^{T^{\cdot}}$ we have taken a number of cells N^c equal to the number of elements F^P of the population studied. Let us indicate the cells by $c = 1,2,\ldots, F^P$, and the coordinates of the center of each cell c by X_c, Y_c. The uniform distribution has a geoset ${}^{g}U^{T^{\cdot}} = \{X_c, Y_c; w_c^U = 1/F^P\}$, as shown by Figure 3.5 in which there is a frequency of 1 in each cell. To simplify the notation, whenever there is no danger of misunderstanding, we use U instead of U^T.

${}^{g}U^{T^{\cdot}}$ is implicitly based on the hypotheses that: (a) the population P, might be distributed over any spot in T; (b) the overall "density" (hotel per sq. km of T) is the same in ${}^{g}P^{\cdot}$ and in ${}^{g}U^{T^{\cdot}}$; (c) the density is the same in all the cells in ${}^{g}U^{T^{\cdot}}$, but different among cells in ${}^{g}P^{\cdot}$; (d) locations of individual elements are clearly determined in ${}^{g}U^{T^{\cdot}}$ (as they are in ${}^{g}P^{\cdot}$).

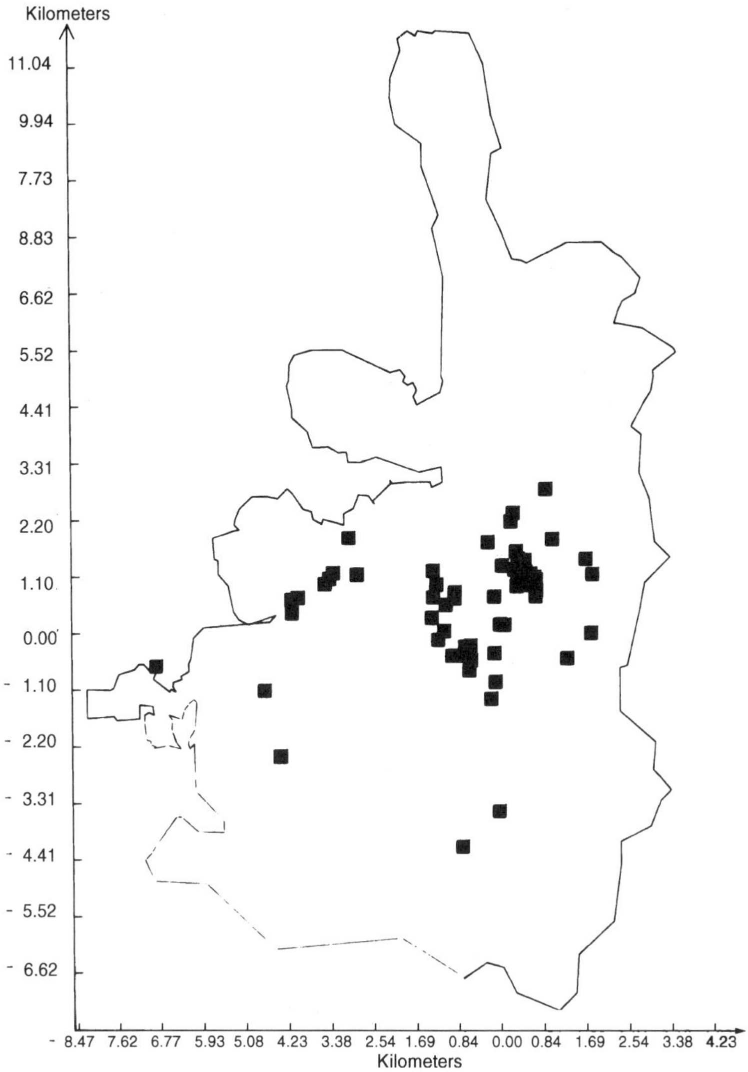

Figure 3.3. Jerusalem. Hotels within the municipal area, 1989 (Source of data: Israel Ministry of Tourism).

Hypothesis (a) may be in conflict with known characteristics of T. For example, it would be unthinkable to establish a hotel on tracts of land occupied by holy places (synagogues, churches, mosques), archaeological sites, public gardens, or public buildings. Such considerations may be of importance for microgeograhic research but are not relevant for the macrogeographic considerations, with which we are mostly concerned in this book.

Here the uniform distribution is just a method for showing the effects of the shape and size of T and discrepancies between ${}^{g}P$ and U^T. It is obvious that U^T is

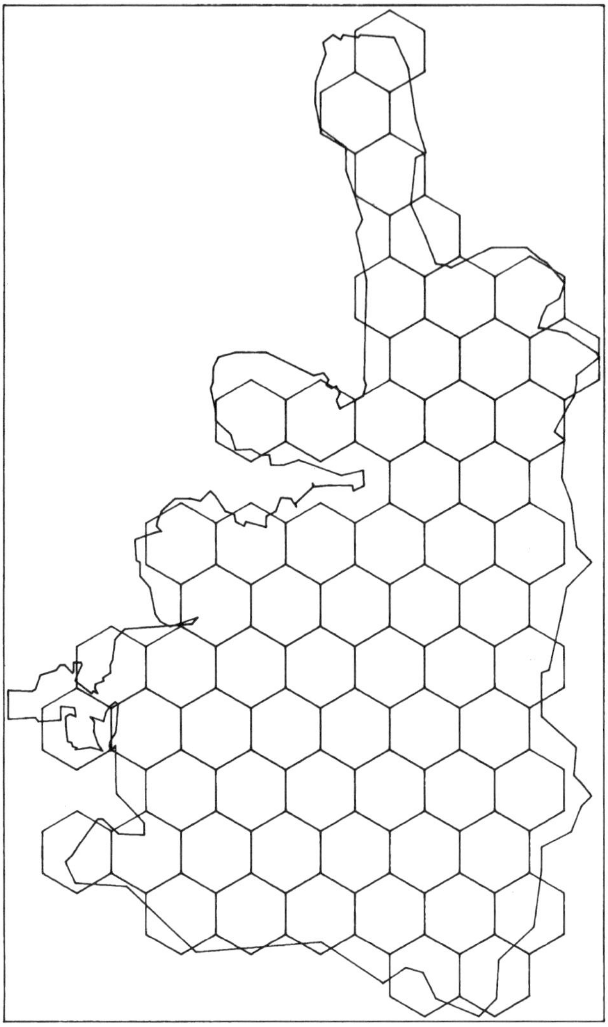

Figure 3.4. Jerusalem. Network of 63 hexagonal cells (Source of data: Israel Ministry of Tourism).

a good indicator of the effects of the shape and size of T if the number of cells is sufficiently large. Then the set of cells is generally found to correspond well to T. In our example, F^P is comparatively small and the correspondence of the set of cells to T is not very good, but we have already explained the reasons that prompted us—despite this—to select this example of a small population.

From a more general point of view, the presentation of a population with F^P elements over $N = F^P$ territorial cells can be envisaged as follows: We assume that a territory T (with area A^T) is "at the disposal" of a population of size F^P. If T is subdivided equally among the population elements, then each element would have

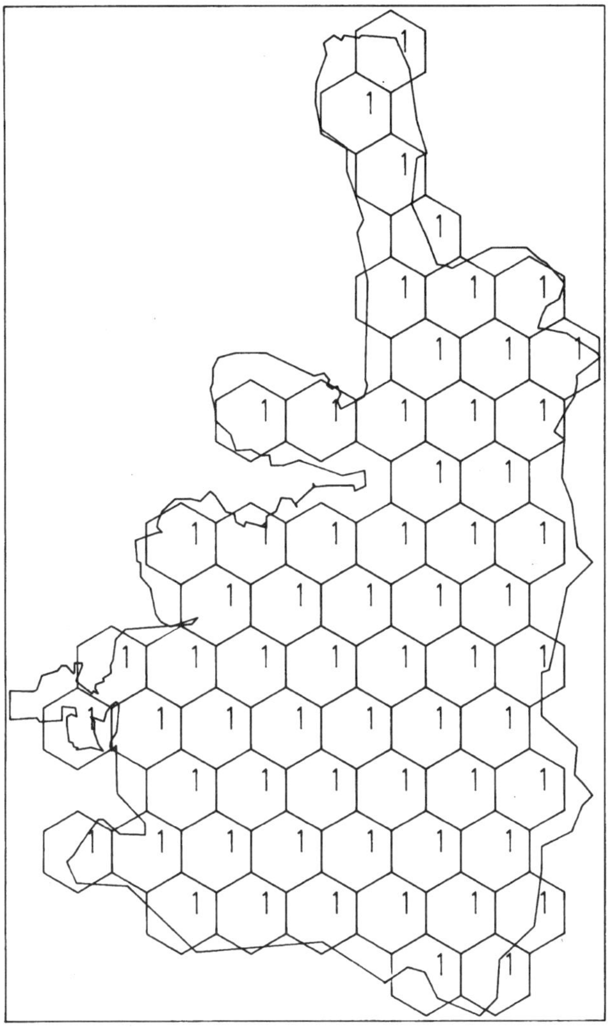

Figure 3.5. Jerusalem. Uniform distribution.

the same size tract, namely a cell of area $a = A^T/F^P$. The density in each cell, $1/a$, would be equal to the general density F^P/A^T. The actual populations usually have geosets ${}^gP$ different from ^{g}U, and our task is to compare ${}^gP$ and ${}^gU$, which is generally a rather complex affair. We shall accomplish it by deriving geosubsets from primary geosets, showing them graphically, and calculating and evaluating an array of geostatistical parameters for each of the primary geosets and for the geosubsets derived from them.

We start here by introducing a method that facilitates comparison between ${}^gP^{\bullet}$ and ${}^gU^{\bullet}$ and the carrying out of geoset operations over them.

3.2.2. Comparing the Actual and Uniform Distributions Using the Cell Network

We can facilitate the comparison between the actual and uniform distributions and pave the way for the presentation of geoset operations in the following way:

1. The network of hexagonal cells *i* of Figure 3.4 is superimposed over the dot distribution of Figure 3.2 as shown in Figure 3.6.
2. We count the number of elements of the actual population that falls in any

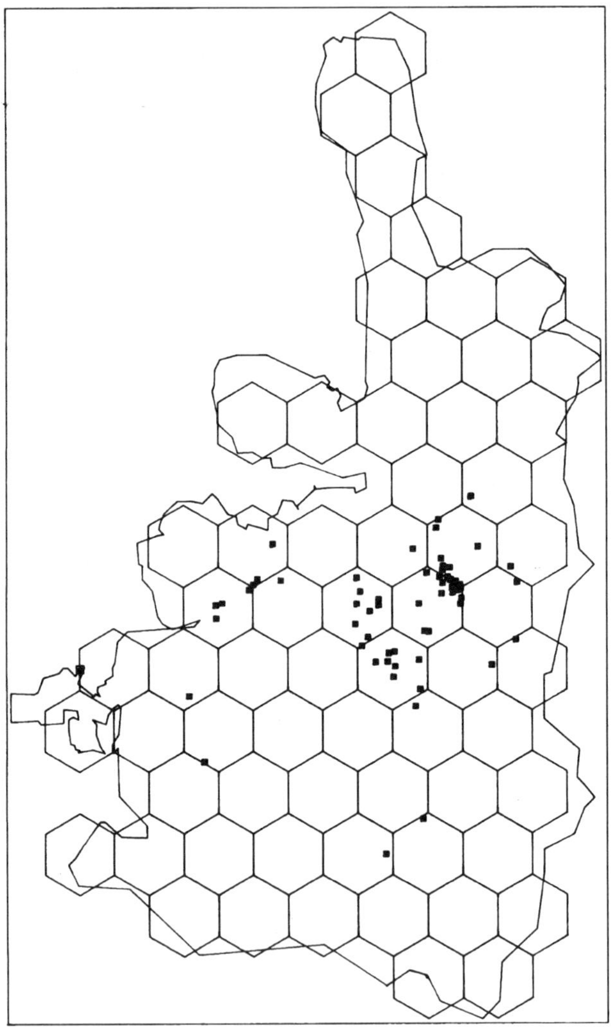

Figure 3.6. Jerusalem. Distribution of hotels over cell network, 1989 (Source of data: Israel Ministry of Tourism).

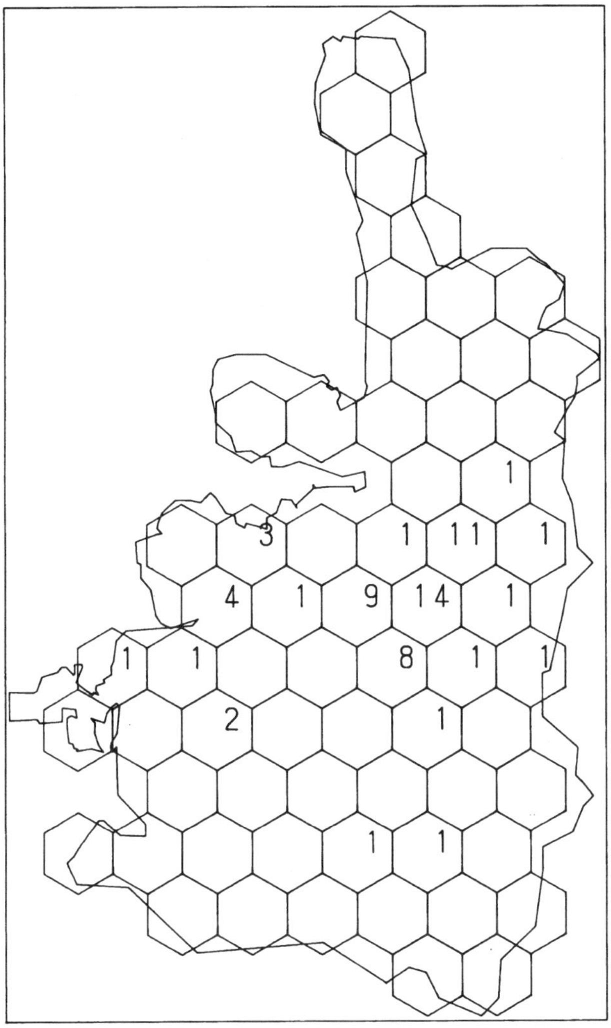

Figure 3.7. Jerusalem. Number of hotels falling in any cell i, 1989 (Source of data: Israel Ministry of Tourism).

cell i, as shown in Figure 3.7, and denote this number by F_i^P. Cells in which $F_i^P = 0$ are void in the figure.

3. We attribute the coordinates (X_c, Y_c) of the center of any cell c to all the elements of the actual population found in that cell.
4. The relative weight to be attributed to coordinates (X_c, Y_c) in the calculations of geostatistical parameters becomes $w_c^P = (F_c^P/F^P)$. We can thus build a cell geoset to be indicated by ${}^gP$ (instead of ${}^gP^{\cdot}$): ${}^gP = \{X_c^P, Y_c^P, w_c^P | (X_c^P, Y_c^P)$ are the coordinates of the center of cell c; w_c^P is the proportion of elements of P falling in cell c; $w_c^P = (F_c^P/F^P)\}$

This is the standard presentation of most of the population geosets in this book. In referring to them, the appellation "cell" geoset is omitted whenever it is not necessary. Concerning the method described above, we note the following:

1. The formation of ${}^{g}P$ by aggregating individual elements into cells corresponds to the familiar method used in statistics to present a large set of statistical data as a frequency distribution, according to classes of constant size.
2. For a long time now this method has been adopted frequently in geographical statistical research and serves as well as a basis for "quadrant analysis."
3. Two different kinds of regular grids are generally used in geographical statistical research: the square and the one based on equilateral triangles. Each is associated with a regular system of square or hexagonal cells, respectively, their centers coinciding with the points of the grid. Here we have used regular hexagons because, as was explained by Dr. P. Napolitano, regular grids can be created by translating a given parallelogram and characterized by the minimum distance between contiguous points. Among regular grids with the same minimum distance, the one based on equilateral triangles has the important property that the number of points of the grid distributed within a fixed area is a maximum.
4. Use of cell coordinates (X_c^P, Y_c^P) instead of exact coordinates (X_h, Y_h) of the elements h of the population introduces some inaccuracy. However, if F^P is not too small, neglecting differences of location of elements within the same cell generally has little importance. Moreover, if we know the actual locations of the elements (X_h, Y_h), we can still use them to calculate the exact parameters and to evaluate the actual magnitude of the error introduced by using (X_c^P, Y_c^P). If we do not know (X_h, Y_h), we can still evaluate the order of magnitude of errors for the various geostatistical parameters. We repeat here that we have selected the example of the small hotel population (with $F^P = N^c = 63$) for didactic purposes, in order to introduce geoset operations, and that, in practice, this method should be used only with larger populations.
5. From a conceptual point of view, keeping the number N^c of cells equal to the size F^P of the population is very important as it helps to distinguish habitat from nonhabitat easily and to understand the meaning of other geoset operations (see Sections 3.2.3 and 3.2.4). Moreover, it avoids arbitrariness in fixing the number of cells, which sometimes constitutes a problem in quadrant analysis.
6. Many of the populations discussed in this book are very large (including tens or hundreds of thousands, or even many millions of elements). For these populations it becomes impossible to abide by the principle of keeping $N^c = F^P$, because the primary information is provided in aggregate form (say, by inhabited places, counties, provinces, or census tracts. However, there are practical solutions for utilization of these data (see Section 3.4.1).

3.2.3. Distinguishing between Inhabited and Noninhabited Cells

Looking at Figure 3.7, we can distinguish two types of cells among the $N^c = F^P = 63$ cells into which the territory T has been divided: (a) cells in which one or more elements of the "population" studied are found, and (b) cells in which no such elements are found. Utilizing in a somewhat arbitrary way a terminology adapted for human populations, we designate the cells (a) as inhabited, and the cells (b) as noninhabited. The set of inhabited cells is indicated as habitat (denoted by H). In Figure 3.8 the cells (a) of the habitat are made even more prominent than in

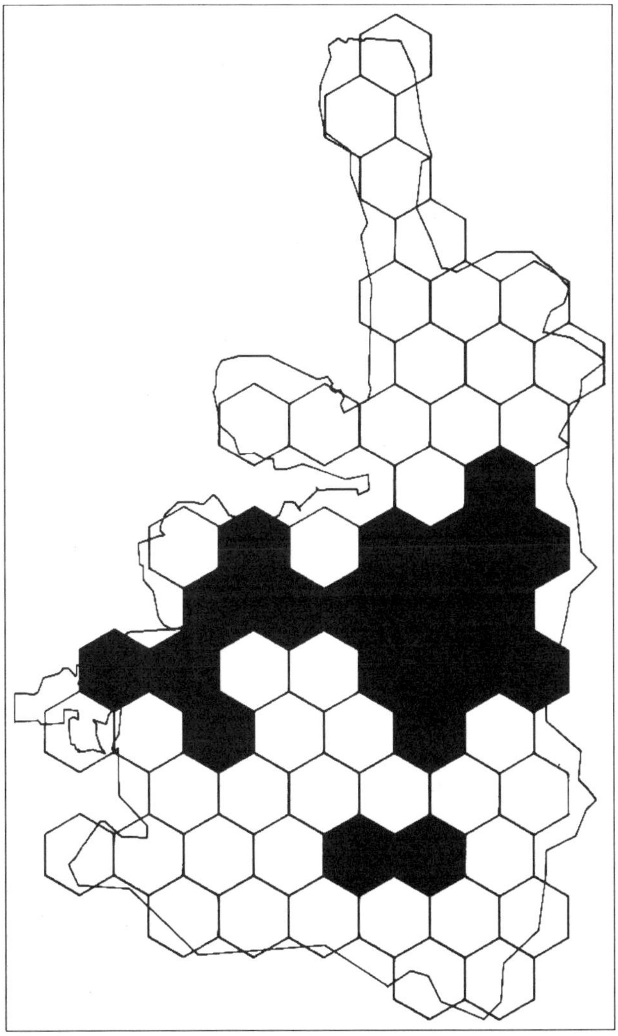

Figure 3.8. Jerusalem. Habitat and nonhabitat of hotels, 1989.

Figure 3.7 by being fully blackened. Cells (b) are shown in white to emphasize that they are empty. The set of these cells is designated as nonhabitat (denoted by NH). Extensive applications of H and NH and of their geosets in the *Atlas* and elsewhere have shown that the geostatistical analysis of H and NH can be very useful for research (see Chapter 8).

H can be regarded as the net territory over which the population studies is effectively found, which is often much smaller than the gross territorial container. Let us denote the coordinates of the center of any cell i by (X_i, Y_i), and the centers of the cells i that belong to the habitat by (X_i^H, Y_i^H), their number by N^H, their total area by A^H, and the corresponding values for the nonhabitat by (X_i^{NH}, Y_i^{NH}), N^{NH}, and A^{NH}. The geosets ${}^gH$ and ${}^gNH$ are:

$$ {}^gH = \{X_i^H, Y_i^H; w_i^H = (1/N^H)\} \tag{3.2a} $$

$$ {}^gNH = \{X_i^{NH}, Y_i^{NH}; w_i^{NH} = (1/N^{NH}\} \tag{3.2b} $$

The detailed study of the maps showing ${}^gH$ and ${}^gNH$ and of their geostatistical parameters often helps to suggest the reasons for the presence of the population in ${}^gH$ and its absence from ${}^gNH$. Obviously, we find $N^H + N^{NH} = N^T$ and $A^H + A^{NH} = A^T$.

3.2.4. Transforming ${}^gP$ into ${}^gU$ by Means of Operations over Geosets

Let us return to the comparison between ${}^gU$ and ${}^gP$, introduced in Sections 3.2.1 and 3.2.2 by means of Figures 3.5 and 3.7. Following an approach often used in statistical work, we can compare ${}^gP$ to ${}^gU$ by assuming that we transform ${}^gP$ into ${}^gU$. This transformation can be carried out stepwise by means of operations on geosets, described below.

In habitat cells (in which $F_i^p \geqslant 1$) we leave one element of ${}^gP$ unmoved and move $(F_i^P - 1)$ elements out. The elements that have been moved are distributed among empty cells of the nonhabitat, one per cell. We can thus examine separately the territorial distribution of (a) the unmoved elements and those of (b) origins, and (c) destinations of elements that are moved. To do so we introduce appropriate geoset operations, which enable us to show (a), (b), and (c) graphically, to calculate their geostatistical parameters, and to demonstrate the relationships between these parameters. We note the following concerning (a), (b), and (c):

1. In each cell of H we can consider that one element of population P "matches" the unitary element of the uniform distribution. Hence this element is not moved in the transformation of ${}^gP$ into ${}^gU$. We can construct the geoset of these matching elements, to be denoted as "match of ${}^gP$ and ${}^gU$," abridged as ${}^gM(P, U)$ as follows:

$$ {}^gM(P, U) = \{X_i^H Y_i^H; w_i^M = 1/N^M = 1/N^H\} \tag{3.3} $$

where (X_i^H, Y_i^H) are the locations of centers of inhabited cells. As the number of matched elements, N^M, is equal to the number of inhabited cells, the relative weight to be attributed to each element of P matched by one element of U is $1/N^M = 1/N^H$. Figure 3.9 shows the distribution of these matched elements, which are indicated by the numeral 1 in the relevant cells.

2. The elements of ${}^gP$ that do not match an element of ${}^gU$ and are to be moved out of the cell in which they are found are denoted as "surpluses of ${}^gP$ over ${}^gU$." They are found in cells indicated by the superscript SPU, in which $F_i^P > 1$; their number in each of these cells is $(F_i^P - 1)$; their total is denoted as F^{SPU}. The geoset of surpluses of P over U can be expressed as ${}^gSPU$:

$$ {}^gSPU = \{X_i^{SPU}, Y_i^{SPU}; w_i^{SPU} = (F_i^P - 1)/F^{SPU}\} \tag{3.4} $$

Figure 3.10 shows the distribution of these surplus elements, according to the cell constituting their "origin."

3. Let us consider now the opposite type of surpluses, to be denoted generically as SUP. These occur in cells in which there is one element of the uniform distribution not matched by an element of P. These cells are not inhabited by elements of P; their geoset has already been expressed as nonhabitat of P (denoted by ${}^gNH$); and their total number is indicated by $N^{NH} = N^c - N^H$. They can be regarded as the destinations to which surplus elements SPU are to be moved in order to transform ${}^gP$ into ${}^gU$.

The geoset of SUP, as shown by Figure 3.11, is

$$ {}^gSUP = \{X_i^{NH}, Y_i^{NH}; w_i^{SUP} = 1/M^{NH}\} \tag{3.5} $$

Geostatistical parameters of ${}^gM(P, U)$ are equal to those of ${}^gH$. However, it is important to make a clear distinction between the two concepts: ${}^gM(P, U)$ refers to a subpopulation, while ${}^gH$ refers to a subterritory. Geostatistical parameters of ${}^gSUP$ and ${}^gNH$ are the same, but here, too, it is important to distinguish between subpopulation SUP and subterritory NH. We have thus obtained from *primary geosets* ${}^gP$ and ${}^gU$: (a) three *derived population geosets*, ${}^gSPU$, ${}^gSUP$, ${}^gM(P, U)$ and two *derived territorial geosets*, ${}^gH$ and ${}^gNH$. Further territorial geosets will be introduced in Section 3.2.5.

Geostatistical parameters, such as mean centers and measures of dispersion, can be calculated in the same way for each of these primary and derived geosets. The parameters of primary and derived geosets are functionally related (see Table 3.1, Figure 3.12, and Appendix E). This facilitates their interpretation, as we will show by many examples. To give some indication of these relationships, we note below some values of the total sizes of the geosets we are dealing with here. From Figures 3.5–3.11 we can easily obtain the following:

1. Total number of elements considered in the actual population and in the uniform distribution: $F^P = F^U = 63$.

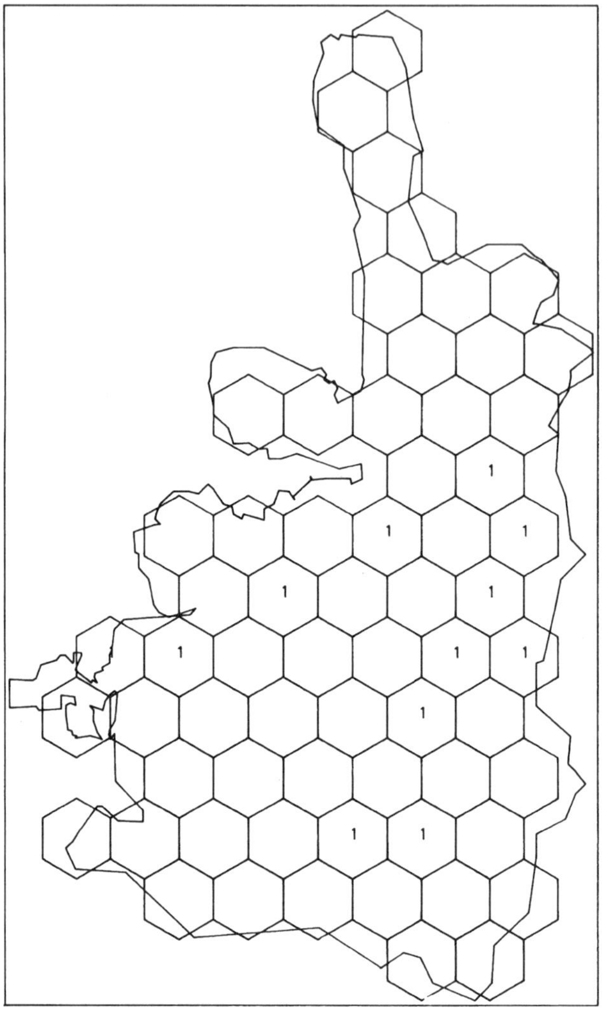

Figure 3.9. Jerusalem. Matched between hotels and uniform distribution, 1989 (Source of data: Israel Ministry of Tourism).

2. Number of elements matched in ${}^gP$ and ${}^gU$ = number of elements in the habitat = $F^{M(P,U)} = N^H = 19$.
3. Number of surpluses of hotels over uniform distribution = number of surpluses of uniform distribution over hotels = number of cells in the non-habitat: $F^{SUP} = N^{NH} = 44$.
4. Proportion of surpluses: $(44{:}63) = \Delta_P^U = 69.8\%$.
5. Proportion of matches: $(19{:}63) = (1 - \Delta_P^U) = 30.2\%$.

The symbol Δ is introduced here to stress conceptual identity with Duncan's well-known index of concentration (see Chapter 8). The term (1 − Δ) is also a

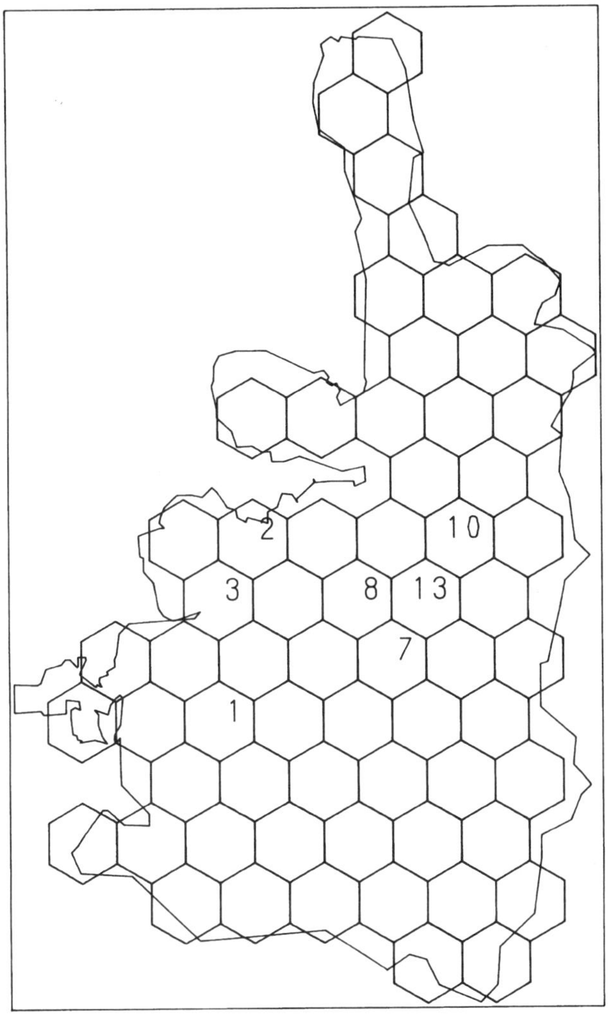

Figure 3.10. Jerusalem. Surplus of hotels over uniform distribution, 1989 (Source of data: Israel Ministry of Tourism).

measure of the proportion of the area of the habitat over the entire territory, while Δ is the proportion of the area of the nonhabitat over the entire territory. While $F^{SPU} = F^{SUP}$, the geographical distribution of the elements of ${}^{g}SPU$ (as shown in Figure 3.10) differs from that of the elements of the ${}^{g}SUP$ (as shown in Figure 3.11).

3.2.5. Splitting the Habitat into Territories of Matched Elements and Surpluses

In more advanced geostatistical research, it is sometimes desirable to introduce a more refined definition of territory than that given in Section 3.2.3. This can be obtained by subdividing the habitat into parts that correspond to surpluses of ${}^{g}P$

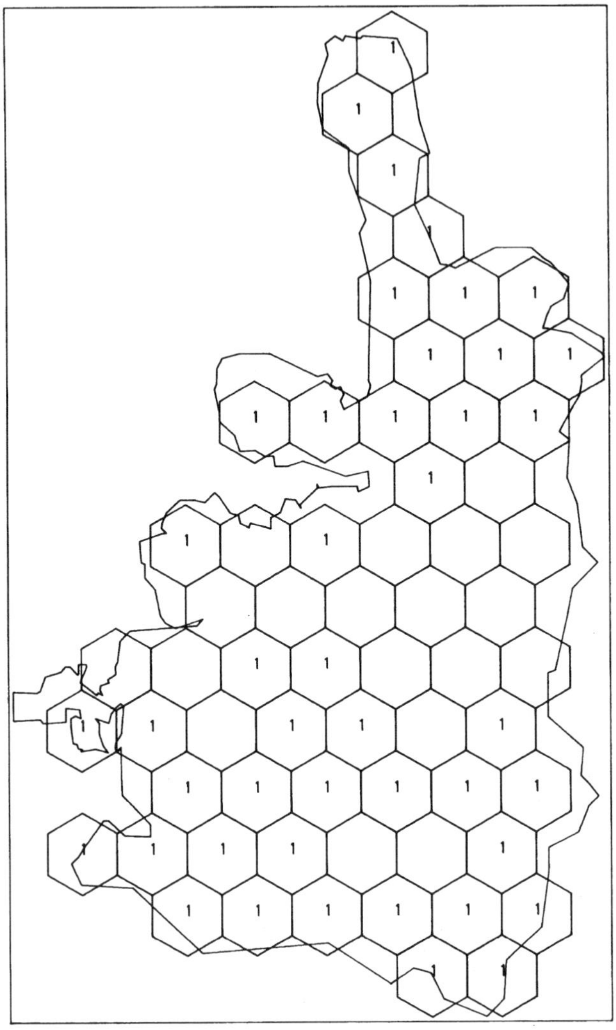

Figure 3.11. Jerusalem. Surplus of uniform distribution over hotels, 1989 (Source of data: Israel Ministry of Tourism).

over ${}^gU$ and residual parts (corresponding to matches between ${}^gP$ and ${}^gU$). Plate 2 shows this operation.

Let us compare Figures 3.7, 3.9, and 3.10. The first shows the number of F_i^P population elements found in each cell i of the habitat. Each of these numbers can be subdivided into: (a) the matches shown in the second figure and (b) surpluses $(F_i^P - 1)$ shown in the third. Conventionally, we can assume that there is a portion of the cell i whose area is $a(1/F_c^P)$, where a is the area of a cell, that corresponds to each of the F_i^P elements found in that cell (Figure 3.7). As there is one element of the match $M(P, U)$ in each inhabited cell, the territory $TM(P, U)$ corresponding to

Table 3.1. Hotels in Jerusalem (P) Compared to Uniform Distribution (U): A Numerical Example of Related Centers of Primary and Derived Geosets

	Symbol	Coordinates of mean center X	Y	Proportion of elements
Equivalent distribution of hotels	${}^{g}P$	8.12	8.51	1
Uniform distribution	${}^{g}U$	7.53	8.03	1
Surplus of hotels over uniform distribution	${}^{g}SPU$	8.12	8.65	Δ_P^U
Surplus of uniform distribution over hotels	${}^{g}SUP$	7.45	8.11	Δ_U^P
Matched between hotels and uniform distribution	${}^{g}M(P, U)$	8.17	7.52	$1 - \Delta_P^U$
Habitat	${}^{g}H$	8.17	7.52	$1 - \Delta_P^U$
Nonhabitat	${}^{g}NH$	7.45	8.11	Δ_U^P

match $M(P, U)$ is formed by the set of portions of inhabited cells whose area is $a(1/F_i^P)$. These portions are conventionally shown, as depicted in Plate 2, as black hexagons of area $a(1/F_i^P)$, concentric with the inhabited cell to which they belong. If $F_i^P = 1$, the entire inhabited cell is blackened. We can construct a geoset of the territory $TM(P, U)$ corresponding to the match as follows:

$$
\begin{aligned}
{}^{g}TM(P, U) &= \{X_i^H, Y_i^H; w_i^{TM}\} \\
&= (a/F_i^P)/\Sigma(a/F_i^P) \\
&= (1/F_i^P)/\Sigma(1/F_i^P)
\end{aligned}
\tag{3.6}
$$

The territory $TM(P, U)$ has a total area of

$$A[TM(P, U)] = a\Sigma(1/F_i^P) \tag{3.7}$$

In the cells in which $F_i^P > 1$ (as shown in Figure 3.10) an area

$$a - (a/F_i^P) = a(F_i^P - 1)/F_i^P \tag{3.8}$$

corresponds to surplus elements, which is shown conventionally by the marginal red portion of the cells in Plate 2. We can construct a geoset of the territory $TSPU$ corresponding to surpluses of P over U. The total area of this territory is

$$AT(SPU) = a\Sigma[(F_i^P - 1)/F_i^P] \tag{3.9}$$

Obviously,

$$AT(SPU) + A[TM(P, U)] = A(H) \tag{3.10}$$

In Plate 2, the nonhabitat is given in white, as in Figure 3.8. Thus we have divided the entire territory into three parts; corresponding, respectively, to matches, to surpluses SPU, and to nonhabitat.

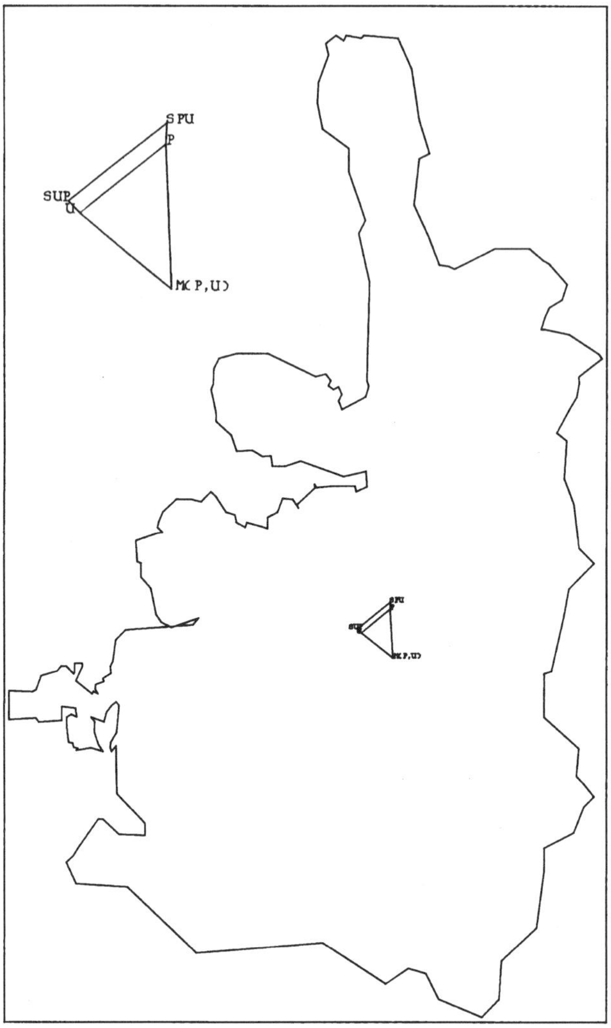

Figure 3.12. Jerusalem. The five centers of ${}^{g}P$, ${}^{g}U$, ${}^{g}SPU$, ${}^{g}SUP$, and ${}^{g}M(P, U)$ of hotels, 1989 (Source of data: Israel Ministry of Tourism).

3.3. Comparing Two or More Geosets Distributed over the Same Territory

3.3.1. The Importance of this Topic

Comparing the distribution of two or more geosets (${}^{g}P, {}^{g}Q, {}^{g}R, \ldots$) over the same territory T constitutes a central theme of this volume. These geosets might correspond, say, to different populations ($P, Q, R, \ldots$) for which the researcher wishes to study similarities or dissimilarities; to a particular population considered at various times; to subpopulations of one population; to one or more populations from which one may want to isolate the influence of a general determinant; or to component geosets from which one may want to obtain a compound, average, or intermediate geoset.

Examples of such comparisons are given in Chapter 7, where methods for their graphical presentation are explained in detail. Chapters 8–12 will include comparisons between parameters that measure particular characteristics of the geosets studied. In Chapters 13–16, examples are given of summary comparisons between a pair of geosets by means of the minimal distance between them.

In this and the following sections, we shall deal with some general principles of the comparison between geosets, such as ${}^{g}P$ and ${}^{g}Q$ distributed over T. This presentation will be analogous to the one in Section 3.2 for comparing an actual geoset ${}^{g}P$ to the uniform distribution ${}^{g}U$ over the territory.

We shall start, as we did for the comparison between ${}^{g}P$ and ${}^{g}U$, by considering a very simple setting, one in which both ${}^{g}P$ and ${}^{g}Q$ are small and we know the exact location of each of their elements. As each element can be shown on the map by a dot, we denote such geosets as ${}^{g}P^{\cdot}$ and ${}^{g}Q^{\cdot}$. Again, this particular presentation is for didactic purposes, as most of the comparisons between geosets distributed over in T in this volume are based on large populations. The conceptual modifications needed to take this into account are discussed in Section 3.4.

3.3.2. Introducing the Comparison between ${}^{g}P$ and ${}^{g}Q$

We introduced the study of a geoset ${}^{g}P$ and its comparison to the uniform distribution ${}^{g}U$ by means of the example of the small geoset ${}^{g}P^{\cdot}$ of $F^{P} = 63$ hotels in Jerusalem in 1989. Here we shall compare the geoset ${}^{g}P^{\cdot}$ of the hotels to an analogous geoset ${}^{g}Q^{\cdot}$ that refers to travel agencies in the same city in the same year.

We have at our disposal a list of 68 travel agencies operating in Jerusalem in 1989 and their addresses, and we wish to compare the territorial distribution of hotels and travel agencies. To facilitate the comparison by rendering the sizes of the two populations to be compared equal, we have excluded at random 5 out of the 68 agencies. Let us indicate by Q the set of the 63 remaining agencies and its size by $F^{Q} = F^{P} = 63$. We have constructed a geoset ${}^{g}Q^{\cdot}$ by counting the number F_i^{Q} of the travel agencies found in each of the 63 cells i into which the territory of Jerusalem was divided, and assigning the coordinates (X_i, Y_i) of the center of the cell i to all the travel agencies found in that cell. In order to stress that we work in practice on cells i in which there are elements of Q, we denote their coordinates as $(X_i^{Q}, Y_i^{Q}, w_i^{Q})$. The geoset ${}^{g}Q^{\cdot}$ is thus

$$ {}^{g}Q^{\cdot} = \{X_i^{Q}, Y_i^{Q}; w_i^{Q} = F_i^{Q}/F^{Q}\} \tag{3.11} $$

Figure 3.13 shows the numbers F_i^{Q} of travel agencies, which can be compared to the number F_i^{P} of hotels, shown by Figure 3.7. To facilitate the comparison, we can again assume that one distribution (say, ${}^{g}Q^{\cdot}$) is transformed into the other (say, ${}^{g}P^{\cdot}$), by shifting some of its elements. Here too, the assumed transformation is carried out through various geoset operations.

3.3.3. Population Geosets Derived from Comparing ${}^{g}\mathbf{P}^{\bullet}$ and ${}^{g}\mathbf{Q}^{\bullet}$

We compare, cell by cell, the absolute frequencies F_i^Q and F_i^P as follows:

1. We calculate $\min(F_i^Q, F_i^P)$ in each cell. This is the number of elements that are not moved in the transformation of ${}^{g}Q^{\bullet}$ into ${}^{g}P^{\bullet}$. These elements match one another. They are shown in Figure 3.14, with their total number being

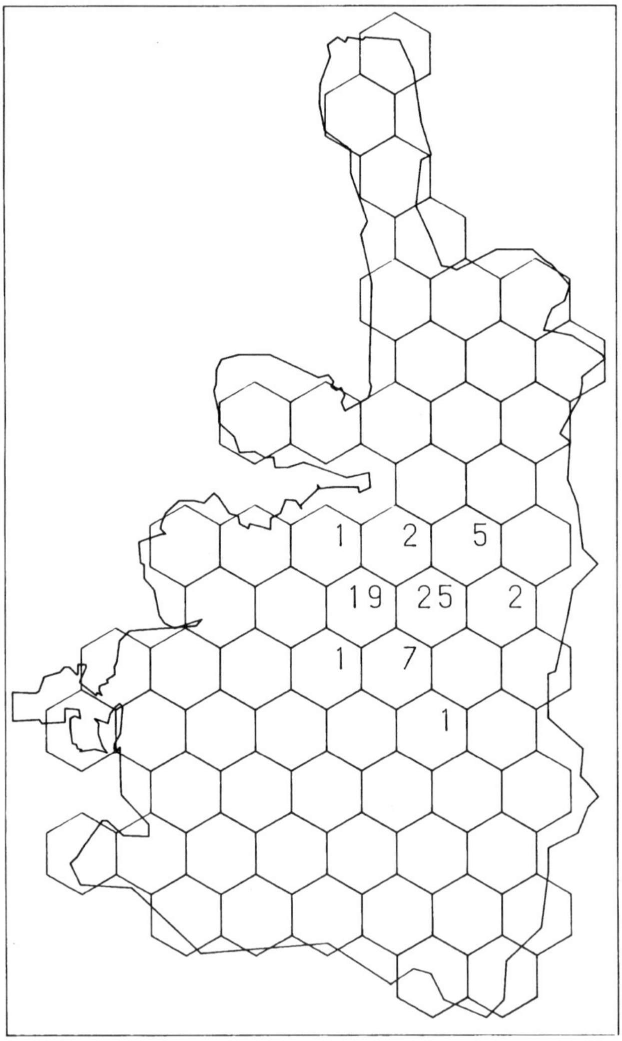

Figure 3.13. Jerusalem. Number of travel agencies falling in any cell i, 1989 (Source of data: Israel Ministry of Tourism).

denoted by $F^M(P, Q)$. We indicate the geoset of matches ${}^gM(Q, P)$ by:

$$ {}^gM(Q, P) = \{X_i^M, Y_i^M; w_i^M = \min(F_i^Q, f_i^P)/F^M(P, Q)\} \tag{3.12}$$

2. In cells in which $F_i^Q > F_i^P$, we determine this "surplus of Q over P," by calculating $SQP_i(F_i^Q - F_i^P)$. Figure 3.15 shows the territorial distribution of these surpluses. Their total, F^{SQP}, is the number of elements to be shifted in order to transform ${}^gQ^{\cdot}$ into ${}^gP^{\cdot}$. We can construct a geoset of the surpluses of Q over P, denoted by ${}^gSQP$, as

$$ {}^gSQP = \{X_i^{SQP}, Y_i^{SQP}; w_i^{SQP} = S(F_i^Q - F_i^P)/N^c\Delta_P^Q\} \tag{3.13}$$

where the superscript SQP means that these cells have a surplus of Q over P. This geoset gives the territorial distribution of the origins of the assumed movements needed to transform ${}^gQ^{\cdot}$ into ${}^gP^{\cdot}$.

3. In a similar way we can deal with surpluses of P over Q. Their territorial distribution is shown in Figure 3.16, and their geoset is given by

$$ {}^gSPQ = \{X_i^{SPQ}, Y_i^{SPQ}; w_i^{SPQ} = S(F_i^P - F_i^Q)/N^c\Delta_P^Q\} \tag{3.14}$$

which gives the territorial distribution of the destinations of the assumed movements needed to transform ${}^gQ^{\cdot}$ into ${}^gP^{\cdot}$.

We thus obtain a complete description of the geographical characteristics of the transformation of ${}^gQ^{\cdot}$ into ${}^gP^{\cdot}$ through the geoset operations ${}^gSQP$, ${}^gSPQ$, and ^{g}M(Q, P). Apart from the maps, we can also obtain geostatistical parameters from these geosets, many of which are functionally related.

3.3.4. Territorial Geosets Derived from Comparing ^{g}P$^{\cdot}$ and ^{g}Q$^{\cdot}$

From the comparison of ${}^gP^{\cdot}$ and ${}^gQ^{\cdot}$, we can also derive subterritories and corresponding territorial geosets, as we did above from the comparison of ${}^gP^{\cdot}$ and ${}^gU$. Let us refer to Plate 3, which represents territorial geosubsets for the distribution of hotels ${}^gP^{\cdot}$ and travel agencies ${}^gQ^{\cdot}$ over the municipal territory T of Jerusalem. The N^c cells into which the territory T has been divided can be distinguished as "void" and "nonvoid."

1. Void cells are those in which both $F_i^P = 0$ and $F_i^Q = 0$, and these are blank in Plate 3. Their number is N^v and their geoset ${}^gV$, to be designated as the geoset of void territory, is

$$ {}^gV = TV(P, Q) = \{X_i^v, Y^v; w_i^v = 1/N^v\} \tag{3.15}$$

with an area denoted by $A[V(P, Q)]$.

2. Nonvoid or inhabited cells (shown in color in Plate 3) are those in which there are elements of P or Q or both. Their number is $N^c - N^v$. The set

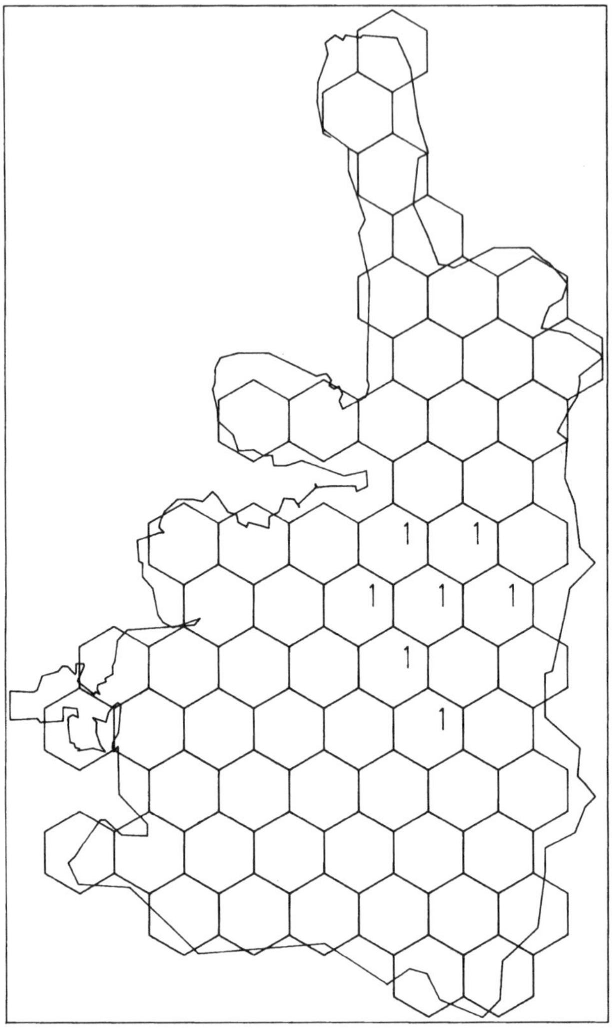

Figure 3.14. Jerusalem. Matched between hotels and travel agencies, 1989 (Source of data: Israel Ministry of Tourism).

$H(P, Q)$ of these cells can be considered the *union* of cells with elements of P and of Q. (a) Within each of these cells there is a concentric hexagon shown in red in Plate 3. The area of this hexagon is proportional to $a[\min(F_i^P, F_i^Q)/\max(F_i^P, F_i^Q)]$, where a is the area of a cell, conventionally considered to represent the territory of the match of P and Q. If $F_i^P = F_i^Q$, the entire cell is shown in red. (b) If $F_i^P > F_i^Q$, the remaining part of the cell has an area of $a(F_i^P - F_i^Q)/F^Q$, which is shown in green along the margin of the cell and represents the territory of the surplus $S(F_i^P - F_i^Q)$. (c). If $F_i^Q > F_i^P$, the area along the border of the cell is shown in blue, is proportional to $a(F_i^Q - F_i^P)/F_i^P$, and represents the territory of the surplus $S(F_i^Q - F_i^P)$.

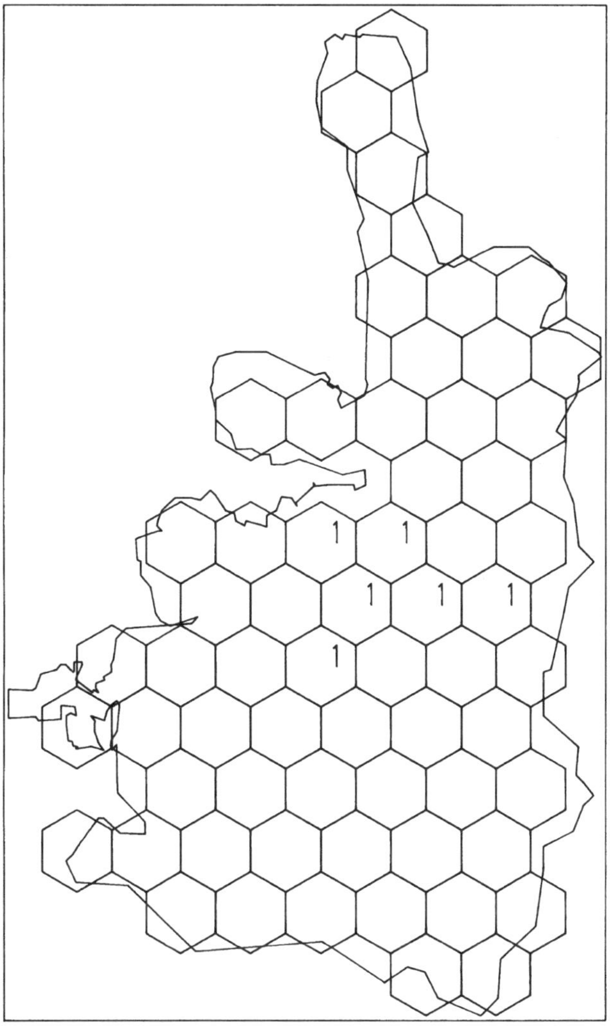

Figure 3.15. Jerusalem. Surplus of travel agencies over hotels, 1989 (Source of data: Israel Ministry of Tourism).

We can thus construct three territorial geosets: (a_1) *Geoset* $TM(P, Q)$ of the territory over which matched elements are supposed to be found:

$$
{}^{g}TM(P,Q)=\{X_i^M, Y_i^M; w_i^M = [(\min(F_i^P, F_i^Q)/\max(F_i^P, F_i^Q))]/\Sigma[\min(F_i^P, F_i^Q)/\max(F_i^P, F_i^Q)]\} \qquad (3.16)
$$

The total area of $TM(P, Q)$ is indicated as $A[TM(P, Q)]$. (b_1) Geoset ${}^{g}TSPQ$ of the territory over which elements of surpluses of P over Q are found:

$$
{}^{g}TSPQ = \{X_i^{SPQ}, Y_i^{SPQ}; w_i^{SPQ} = [(F_i^P - F_i^Q)/F_i^P]/\Sigma[(F_i^P - F_i^Q)/F_i^P]\} \qquad (3.17)
$$

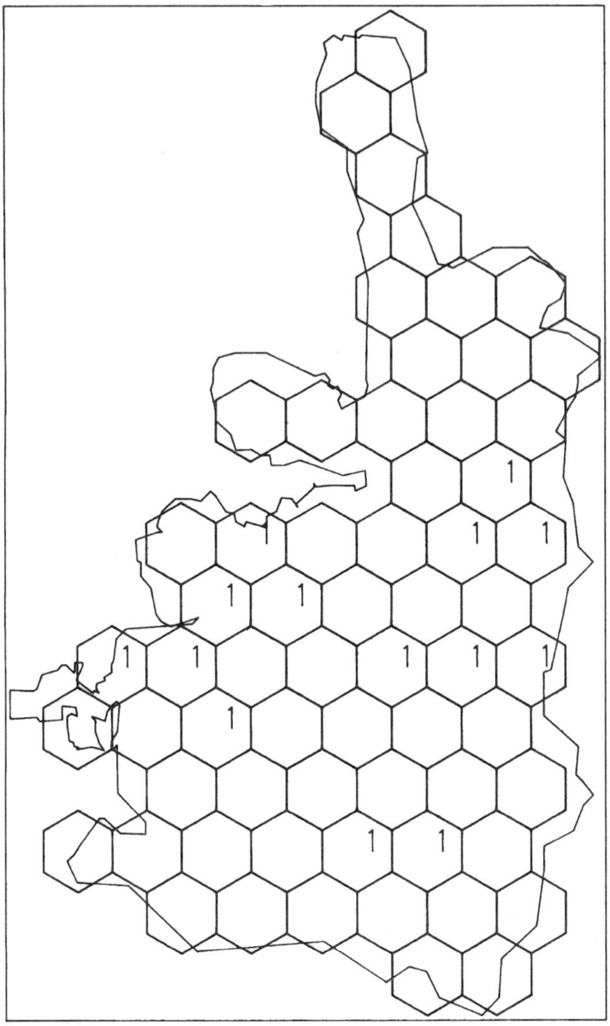

Figure 3.16. Jerusalem. Surplus of hotels over travel agencies, 1989 (Source of data: Israel Ministry of Tourism).

Here the superscript SPQ denotes cells with a surplus of P over Q. The total area of $TSPQ$ is $A[TSPQ]$. (c_1) In an analogous manner, we can construct the geoset ${}^gTSQP$ of the territory over which surpluses of Q over P are found, with $A[TSQP]$ denoting the total area:

$$ {}^gTSQP = \{X_i^{SQP}, Y_i^{SQP}; w_i^{SQP} = [(F_i^Q - F_i^P/F_i^Q)/\Sigma[(F_i^Q - F_i^P)/F_i^Q]\} \tag{3.18} $$

The area of $H(P, Q)$, namely of the territory over which elements of P, or Q, or both are found is

$$A[H(P, Q)] = A[TM(P, Q)] + A[TSPQ] + A[TSQP] \quad (3.19)$$

and is complementary to the area of void cells:

$$A(T) = A[H(P, Q)] + A[V(P, Q)] \quad (3.20)$$

Let us indicate the territory over which elements of P are found by $T(P)$ and by the territory over which elements of Q are found by $T(Q)$:

$$A[T(P)] = A[TM(P, Q)] + A(TSPQ) \quad (3.21a)$$

$$A[T(Q)] = A[TM(Q, P)] + A(TSQP) \quad (3.21b)$$

$$A[T(P)] + A[T(Q)] = 2A[TM(P, Q)] + A[TSPQ] + A[TSQP] \quad (3.21c)$$

3.4. Comparisons between Geosets Representing Large Populations

3.4.1. The Problems Dealt with in this Section

From a look at the methods of comparing the distribution of an actual population to the uniform distribution, it emerges that it is desirable, whenever possible, to abide by the rule of dividing the territory into a number of cells N^c equal to the number of elements of the population (see Section 3.2). Further, from the presentation of methods for comparing the distribution of two populations over the same territory, it was clear that whenever possible the two populations being compared should be assumed to be the same size ($F^Q = F^P = F$) and the territory should be divided into F cells ($N^c = F$).

However, these approaches are not always feasible. In this section we shall discuss: (a) how to select a proper number N^c of cells when F^P, or F^Q, or both are "large" or "very large"; (b) how to deal with the very frequent situation in which the primary data are not specified by the location of individual elements and are given in aggregate form; (c) how to find ways to equalize the number of elements in the comparison of ${}^gP$ with ${}^gU$ and of ${}^gP$ with ${}^gQ$.

3.4.2. Reduction of the Number of Cells in the Geostatistical Analysis of Large Populations

Most of the populations dealt with by geostatistical methods in this volume include thousands, millions, or even hundreds of millions of elements. Even if the individual location of each element were available, it would be technically impossible to construct maps with such a huge number of dots. It would also be impossible to distinguish so many cells of habitat and nonhabitat on a map. We are thus forced to accept a lesser degree of accuracy and to reduce our presentation to a distinguishable number of cells. Programs are available for the condensation of any number of individual locations to any desired network of N^c hexagonal cells. The size of N^c can be adjusted to simplify the calculations but one has to keep in mind that reducing N^c requires an effort, that accuracy increases with increasing N^c and the technical problems involved in the preparation of the maps. We shall see in Chapters 5 and 6 that dot maps have very limited use, and that much greater accuracy can be achieved by means of GRP. However, when using GRP, we often find that if we wish to obtain a "readable" map, of handy size, we are best off limiting the number of individual cells to, say, $N^c = 1000$. In practice, this size has been very widely and successfully used for the maps of the *Atlas* and for a great number of urban maps. In this book, it should be understood that the maps and geosets are based on $N^c = 1000$ cells unless otherwise noted.

3.4.3. Dealing with Input Formed by Aggregated Data

In recent years there has been much progress in obtaining, from censuses and surveys, exact locations for all of the elements of the populations under study. Hence, the method of condensing individual data into a limited, but still very generous, number of cells, such as $N^c = 1000$, can now be employed to an increasing extent. However, it is still true that most of the data on territorial distributions of large populations are made available by statistical offices in aggregated form, listed, e.g., by inhabited places, municipal areas, counties, provinces, or regions, and for cities by wards, districts, census tracts, or blocks.

There are computer programs to transform these data into values for networks of regular hexagonal cells. If the number of primary territorial units by which data are classified is large and these units have comparatively small areas, reduction to a network of 1000 cells does not present difficult conceptual or technical problems. For example, as we noted earlier in the extensive application of geostatistical methods in the *Atlas*, the basic data used were for 8086 *comuni*. The division of the territory of Italy into 1000 cells led to units of a rather modest area of about 300 sq km each, and fitting the *comuni* into the cells was a smooth operation. There were only a very few cases in which the territory of a *comune* had to be divided into two or more cells.

The fitting of territorial units having an area larger than $A(T)/N^c$ into the hexagonal cell network is based on the hypothesis that the density of the relevant population is uniform within the area of the unit. In general, if the number of the units is not too small, the inaccuracies introduced by the use of this device are not numerous. However, this may not be true if the number of territorial units by which the primary data are generated is small and the units themselves large. Transforming primary data specified, say, for some tens of regions of a country into a network of 1000 hexagonal cells may be a rather crude operation, which could involve presenting apparently accurate data that are basically inaccurate.

A different view may be taken whenever there is a legitimate research interest in the study of data classified by regions that have a clear political, social, economic, ethnic, or other meaning. In such cases we might prefer not to use the hexagonal cell network or, if we do use it, to make sure that the regional boundaries are properly marked on the maps.

In certain cases, it may be unavoidable to simply take a central point for each region and assume that the relative weight of the region is to be attached to it. Despite the criticism that can be raised against this approach, experience has shown that some of the geostatistical parameters calculated in this way are often not very different than those obtained by more refined methods.

With suitable modifications some geoset operations can be carried out on regional data. However, it may become too crude or even impossible to study, say, refined aspects of the geostatistical presentation such as habitat and nonhabitat in this way.

3.4.4. Determining the Reduced Number of Elements to Be Used in the Analysis of Large Populations

We have presented geosets generated by two kinds of information: (a) the list of coordinates of the cells, and (b) the relative weight to be attributed to each cell. After that, the way the territory is divided and the number and location of the cells have been fixed. Within the limits resulting from these divisions, the degree of accuracy of the geostatistical parameters depends on the degree of precision of the relative weights. By increasing the number of digits after the decimal point we can improve the accuracy, and this does not present any serious problem when the calculation is done by computer.

One important conceptual problem is that of making reference to the numbers of the supposedly unitary elements of the population(s) studied. In Sections 3.2 and 3.3 we generally referred to F^P and F^Q as the sizes of the population studied and noted the advantage of using an integer N^c of cells equal to F^P, or F^Q, or both. We have found, however, that this approach cannot be used for large populations or for aggregated primary data. Despite this fact, we should not give up on the idea that the populations studied are made up of individual elements that can be shown separately on maps, paired, and used in the explanation of geoset operations.

A very simple method for obtaining this result might be to assume that we are dealing with reduced populations having N^c elements each (N^c integer). To do so, it might be sufficient to multiply the relative weights of the geosets ${}^gP$ and ${}^gQ$ by N^c, rounding off each result to the nearest integer and making some additional adjustments, if necessary. However, experience has shown that distributing, say, 1000 population elements over 1000 cells often leads to very poor results. It is therefore preferable to use a larger multiplier for transforming the relative weights of ${}^gP$ and ${}^gQ$ into integer numbers of elements. In this book we have generally assumed that there are $\Phi^P = 10{,}000$ population elements over $N^c = 1000$ cells. The symbol Φ^P denotes the size of the "equivalent" population adopted (for "equivalent," see below). If the summary information is sufficiently detailed, we can generally expect that a geoset with $\Phi^P = 10{,}000$ elements distributed over $N^c = 1000$ cells has geostatistical parameters close to those that might be obtained from a geoset based on detailed information on individual elements of the population. Hence we sometimes refer to these reduced population distributions as "equivalent" to the detailed (and often unknown) distributions. The geosets used in this book have $N^c = 1000$ and $\Phi^p = 10{,}000$ unless otherwise noted. The transition from $N^c = F^P$ to $\Phi^p = 10N^c$ implies some modifications in the equations of geoset operations, such as surpluses, match, and determination of corresponding territories, but these are of a purely formal nature and do not imply any serious conceptual changes.

The approach suggested here leads to a very simple solution of the problem of rendering the sizes of populations P and Q to be compared equal, an equality that is essential for an understanding of geoset operations and for measuring minimal distance between geosets.

3.5. Operations on Geosets versus Operations on Sets

3.5.1. Similarities and Differences

The operations on geosets introduced above have similar common set operations; for example, determinations of ${}^gSPQ$ and ${}^gSQP$ are analogous to the determination of the differences $(P - Q)$ and $(Q - P)$ between population sets, and the determination of ${}^gM(P, Q)$ is analogous to the determination of intersection $(P \cap Q)$. However, in order to avoid misunderstandings, we should pay attention to the differences between the two kinds of operations and their consequences. They can be better understood through the differentiation introduced in the symbols for geoset operations versus the common symbols of set operations.

We shall deal here with sets of populations or subpopulations $P, Q, R, \ldots$ and with the corresponding geosets ${}^gP^\bullet, {}^gQ^\bullet, {}^gR^\bullet \ldots$:

1. Coming back to the example introduced in Section 3.1.3, we assume that P is the set of the hotels, Q is a subset of P including large hotels with over 100 beds, and R is another subset of P including highly rated hotels of four or five stars and deluxe. We assume further that we know the location of each of the elements h of P so that the corresponding geoset can be indicated as ${}^gP^\bullet$. Let us postulate that in a given location (X_h, Y_h) there can be only one hotel h. We assume that we have separate geosets ${}^gP^\bullet, {}^gQ^\bullet$, and ${}^gR^\bullet$. Further, we assume that we prepare corresponding dot maps, and mark over the dots of ${}^gP^\bullet$ a symbol such as $'$, mark over the dots of ${}^gQ^\bullet$ a symbol such as $+$, and over the dots of ${}^gR^\bullet$ another symbol such as $\times$. If we then superimpose the three maps, we shall be able, in the resulting map, to distinguish hotels that are large and highly rated (marked $*$), large with a low rating $(+)$, small and highly rated $(\times)$, and small with a low rating $(')$. Asterisks denote matching of geosets ${}^gQ^\bullet$ and ${}^gR^\bullet$; the symbols $+$ and $\times$ denote, respectively, surpluses of Q over R and of R over Q; the symbol $'$ denotes complementation of geosets. In this special setting, there is a perfect correspondence between the results of geoset operations and those of common set operations: ${}^gM(Q, R)'$ denotes the locations of elements included in the set $(Q \cap R)$; and ${}^gSQR'$ and ${}^gSRQ'$ denote, respectively, the locations of the elements included in $(Q - R)$ and $(R - Q)$.

2. If we now assume that P = hotels and Q = travel agencies, as in the example in Section 3.3.1 and examine the geosets ${}^{g}P^{\bullet}$ and ${}^{g}Q^{\bullet}$ we find that there are cases in which travel agencies are located inside hotels. While $(P \cap Q) = \phi$ (a hotel cannot be a travel agency), $M({}^{g}P^{\bullet}, {}^{g}Q^{\bullet}) \neq \phi$ (a travel agency can have the same location as a hotel).

3. Let us go on to consider geosets, such as ${}^{g}P$, ${}^{g}Q$, and ${}^{g}R$, corresponding to aggregates of elements over cells. Here, ${}^{g}M(P, Q)$ refers to elements of P and Q that are located in the same cell; however, we cannot deduce from this a clear relationship between P and Q. If we assume, for example, that the sets are P = individuals inhabiting a city at the end of 1991 and Q = individuals inhabiting the same city at the end of 1992, it is likely that many of the individuals included in ${}^{g}M(P, Q)$ are also in $P \cap Q$. However, if we compare P and Q referring to individuals inhabiting the city on dates that are increasingly far apart, the proportion of individuals included in ${}^{g}M(P, Q)$ who are also in $(P \cap Q)$ decreases. At the limit, this proportion becomes 0, as happens, say, when P refers to 1994 and Q to 1884.

3.5.2. Match of Geosets versus Intersection of Sets

To avoid misunderstandings, it seems worthwhile to compare the concepts of match of geosets and intersection of common sets. In set theory $(P \cap Q)$ is the set (or subset) including elements common to P and Q. Let us examine the analysis of geosets $({}^{g}P^{\bullet}, {}^{g}Q^{\bullet})$ defined by the locations of individual elements. We might consider the coinciding locations of an element of ${}^{g}P^{\bullet}$ and an element of ${}^{g}Q^{\bullet}$ as matching. One might think that if an element i of ${}^{g}P^{\bullet}$ and an element j of ${}^{g}Q^{\bullet}$ have matching locations, i.e., if $X_i^P = X_j^Q$ and $Y_i^P = Y_j^Q$, then their locations coincide. We could show this graphically by the coincidence (or overlapping) of the dots representing P_i and Q_i, respectively. However, such a coincidence of dots is not a concept as simple as it appears at first sight. We have already suggested that the consideration of a location as a dimensionless dot is rather elusive. Dealing with individual elements, we have found it more convenient, under the hypothesis of uniform distribution, to attribute to each element a nonzero area, namely a cell of area A^T/N^c. Coincidence of elements thus becomes location in the same cell. Neglecting "within cell" distances we assume that an element of P and an element of Q found in the same cell are located at "distance 0" or have coinciding locations. If we prefer to think of locations in terms of dimensionless dots we can say that the distance between two matching elements is less than $1.2408\sqrt{a_h}$, where $a = A^T/N^c$ is the area of the hexagonal cell. This is the maximum possible distance between two elements found in the same hexagonal cell. Generally, for N^c sufficiently large, these distances are negligible.

This approach is also useful for defining the territorial overlap of two geosets ${}^{g}P$ and ${}^{g}Q$. The degree of overlap is measured by $(1 - \Delta_P^Q)$ and its geographical characteristics are described by ${}^{g}M(P, Q)$.

3.6. A Compound Geoset Formed by Two Component Subsets

In practical geostatistical work, we often have to deal with "compound" geosets, which are formed by "components." The analysis of these geosets can be greatly enhanced by studying the relationships between maps and geostatistical parameters of the component versus those of the component geosets.

Let us denote the compound geoset by ${}^gK$ and the component geosets by ${}^gP$ and ${}^gQ$. We assume that the gross territory T over which the geosets ${}^gK$, ${}^gP$, and ${}^gQ$ are distributed has been divided into N^c cells i $(i = 1, 2, \ldots, N)$ of regular hexagonal shape and fixed area A^T/N^c. We denote the number of elements of K, P, and Q found in each cell i by F_i^K, F_i^P, and F_i^Q; their respective totals by F^K, F^P, and F^Q, the proportion of elements of the population found in each subpopulation by $v_P^K = F^P/F^K$ and $v_Q^K = F^Q/F^K$; and the proportion of elements of the population and the subpopulations, respectively, found in cell i by $w_i^K = F_i^K/F^K$, $w_i^P = F_i^P/F^P$ and $w_i^Q = F_i^Q/F^Q$.

We then see that

$$w_i^K = F_i^K/F^K = (F_i^P + F_i^Q)/F^K = [F^P(F_i^P/F^P) + F^Q(F_i^Q/F^Q)]/F^K$$
$$= v_P^K w_i^P + v_Q^K w_i^Q \tag{3.22}$$

The geoset ${}^gK$ can thus be expressed as

$${}^gK = \{X_i, Y_i; w_i^K = v_P^K w_i^P + v_Q^K w_i^Q)\} \tag{3.23}$$

The relative weight w_i^K of the compound geoset ${}^gK$ in each cell i can then be obtained as an average of the relative weights w_i^P and w_i^Q of component geosets ${}^gP$ and ${}^gQ$ in cell i. This is a weighted average in which we use the weights v_P^K and v_Q^K for ${}^gP$ and ${}^gQ$, respectively.

We can express the relationship between ${}^gK$, ${}^gP$, and ${}^gQ$ in short form as

$${}^gK = v_P^{Kg}P + v_Q^{Kg}Q \tag{3.24}$$

As an example of these relationships between the geosets of a compound population and those of two component subpopulations, we consider (with the help of Figures 3.17–3.21) the distribution of dwellings enumerated in Rome according to the date that they were built. Let us denote all dwellings by K; those built up to 1945 by P, and those built between 1946 and 1981 by Q.

Figures 3.17–3.19 illustrate the geosets ${}^gK$, ${}^gP$, and ${}^gQ$, respectively. Each has been constructed by distributing 10,000 dwelling elements over 1000 hexagonal cells. Comparing ${}^gP$ (Figure 3.18) to ${}^gQ$ (Figure 3.19), we see that dwellings Q built during the period between the end of World War II and 1981 are much more dispersed over the territory and much more in the periphery than the older dwellings P.

Figures 3.17–3.19 show the distributions of K, P, and Q independently of the actual sizes of the population and subpopulations to which they refer. If we are interested in studying the relative contribution made, respectively, by dwellings Q and P to the total K, we can take into consideration that the proportion of dwellings built before 1945 is $v_P^K = 23.81\%$, while the proportion of those built between 1946 and 1981 is $v_Q^K = 76.19\%$.

Figures 3.20 and 3.21 show the values of v_P^K and v_Q^K, respectively, in each relevant cell, and the sum of the values of Figures 3.20 and 3.21 is equal to those of Figure

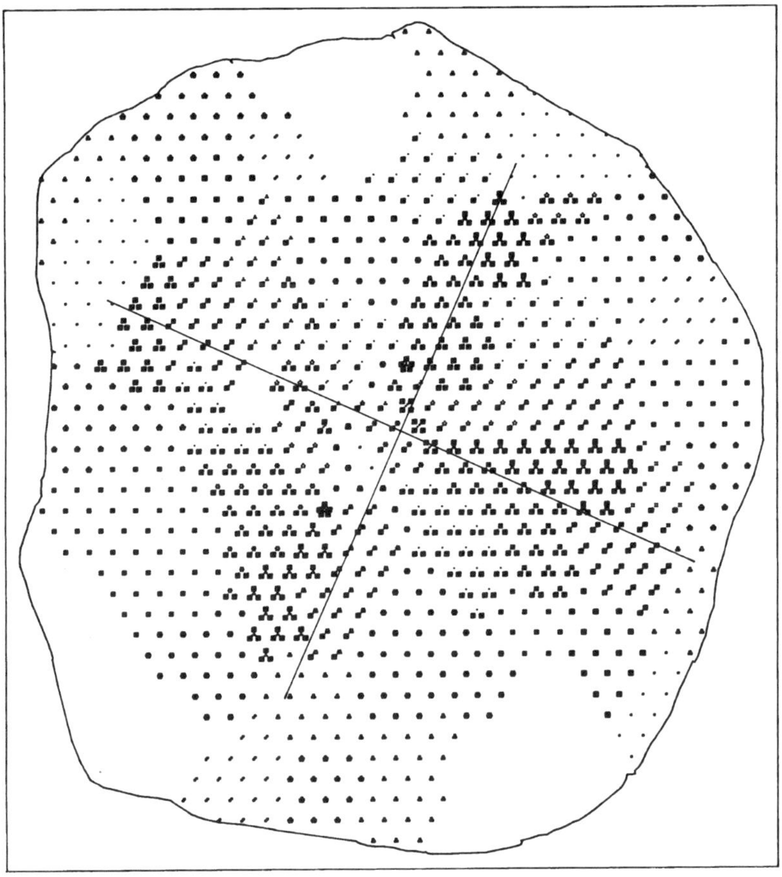

Figure 3.17. Rome. Total dwellings built up to 1981: . 81 ▪ 813 ▪▪ 1626 ▪▪ 3253 ▪▪ 4220 (Source of data: Comune di Roma, Segretariato Generale il Direzione, Ufficio di Statistica e Censimento).

3.17. Geostatistical parameters obtained from these geosets can reveal important relationships.

Let us denote the distance between the centers of ${}^gP$ and ${}^gQ$ by ${}^cD(P, Q) =$ 0.86 km and the minimal distance between ${}^gP$ and ${}^gQ$ by ${}^m\bar{D}(P, Q) = 2.532$ km (see Chapter 13). This is a measure of the territorial discrepancy between ${}^gP$ and ${}^gQ$, which indicates the average distance to be covered in order to transform each element of ${}^gP$ into an element of ${}^gQ$. We see from Figure 3.22 that

1. The centers of ${}^gP$, ${}^gK$, and ${}^gQ$ are collinear.
2. The distances between the centers of ${}^gK$ and those of ${}^gP$ and ${}^gQ$, respectively, are:

$$ {}^cD(K, P) = v_Q^{K_c}\,{}^cD(P, Q) = 0.66 \tag{3.25a} $$

$$ {}^cD(K, Q) = v_P^{K_c}\,{}^cD(P, Q) = 0.20 \tag{3.25b} $$

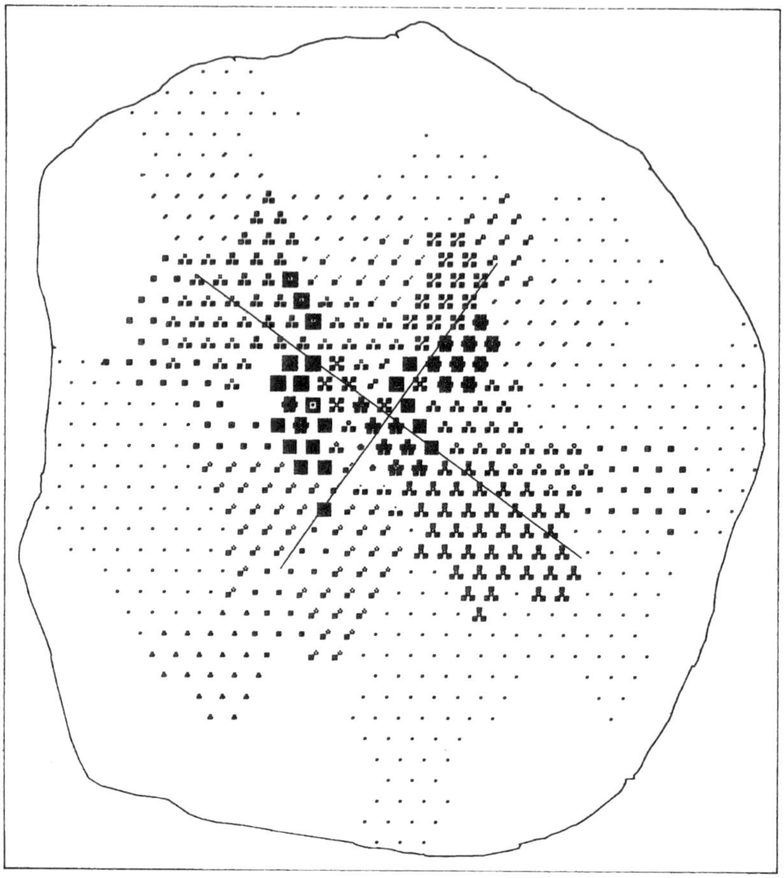

Figure 3.18. Rome. Dwellings built up to 1945: . 19 ▪ 193 ▪ 367 ⁘ 1047 ■ 3292 (Source of data: Comune di Roma, Segretariato Generale il Direzione, Ufficio di Statistica e Censimento).

3. The minimal distance between ${}^gK$ and ${}^gP$, and ${}^gQ$, respectively, is

$$ {}^m\bar{D}(K, P) = v_Q^{Km}\bar{D}(P, Q) = 1.929 \tag{3.26a} $$

$$ {}^m\bar{D}(K, Q) = v_P^{Km}\bar{D}(P, Q) = 0.603 \tag{3.26b} $$

4. To evaluate the meaning of (2) and (3), we give two general measures of dispersion, namely the standard distances of ${}^gK$ and of the territory of Rome ($d_K = 5.88$ km; $d_T = 7.66$ km), as well as the standard distances of ${}^gP$ and ${}^gQ$: ($d_P = 4.21$ km; $d_Q = 6.30$ km).
5. It is evident that the geosets ${}^gP$ and ${}^gQ$ differ considerably, despite the fact that they tend to be more or less concentric. The discrepancy between ${}^gP$ and ${}^gQ$ is due to a considerable extent to their different dispersions.

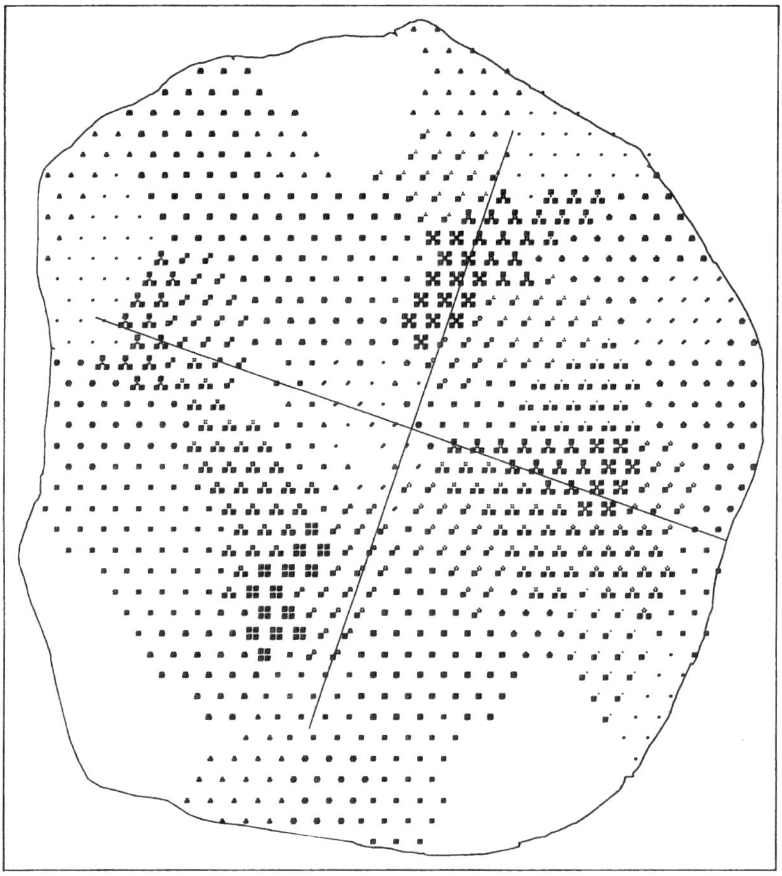

Figure 3.19. Rome. Dwellings built between 1946 and 1981: . 62 ▪ 619 ▪ 1177 ▪ 2045 ▪ 3036 (Source of data: Comune di Roma, Segretariato Generale il Direzione, Ufficio di Statistica e Censimento).

3.7. General Scheme of the Geosets Considered in this Chapter and the Symbols Used in this Volume

In order to help readers to find their way among the various types of geosets introduced in this chapter and among the symbols used in the volume, these have been listed schematically here and in Appendix D. Some additional types of geosets will be introduced in Chapter 4.

It should be remembered that a geoset is a list of individual or aggregate locations of elements of population or territory in which the relative weight to be attached to each location is also given. The geosets constitute the input needed for constructing the maps and calculating the geostatistical parameters. Knowledge of the relationships among the various types of primary and derived geosets greatly facilitates the interpretation of maps and parameters.

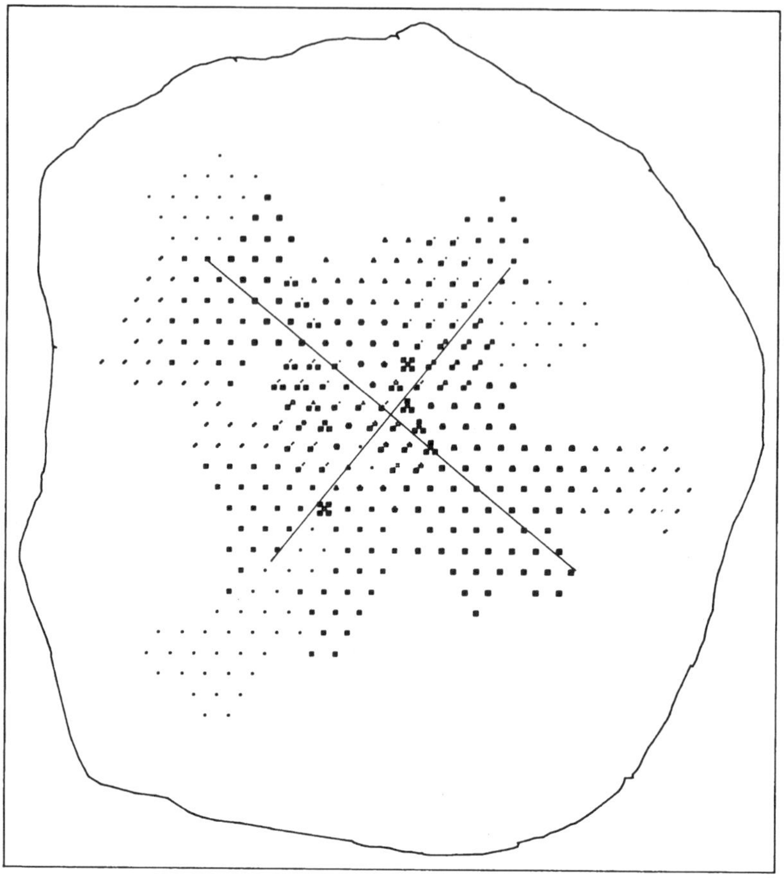

Figure 3.20. Rome. Dwellings built up to 1945 (total of 2298 elements): . 81 ▪ 813 ▪▪ 1626 ⌘ 3741 (Source of data: Comune di Roma, Segretariato Generale il Direzione, Ufficio di Statistica e Censimento).

Geoset is indicated by a superscript g. This symbol is followed by the indication of the population or territory to which the geoset refers. Populations are denoted by letters, such as $P, Q, \ldots$. The letter K generally denotes a compound geoset (see, e.g., Figures 3.17–3.19). The uniform distribution of elements of population over territory is denoted by the letter U (not to be confused with the symbol of union). Territories are denoted by the symbol T, followed by the letters of the population of which T is the container, or by a superscript number indicating subterritory.

Symbols such as ${}^{g}SPQ$, ${}^{g}SQP$, and ${}^{g}SKQ$ denote the geosets of surpluses of the first distribution over the second (see, e.g., Figures 3.23 and 3.24); symbols such as ${}^{g}M(P, Q)$ denote the geoset of the match obtained by comparing the geosets of the two indicated populations (see, e.g., Figure 3.25). Obviously, ${}^{g}M(P, Q) = {}^{g}M(Q, P)$, ${}^{g}H$ and ${}^{g}NH$ (or simply H, NH) denote geosets of the territory of habitat or nonhabitat of the population being considered. Here the T is dropped, as habitat and nonhabitat are, respectively, territories over which the population is found or

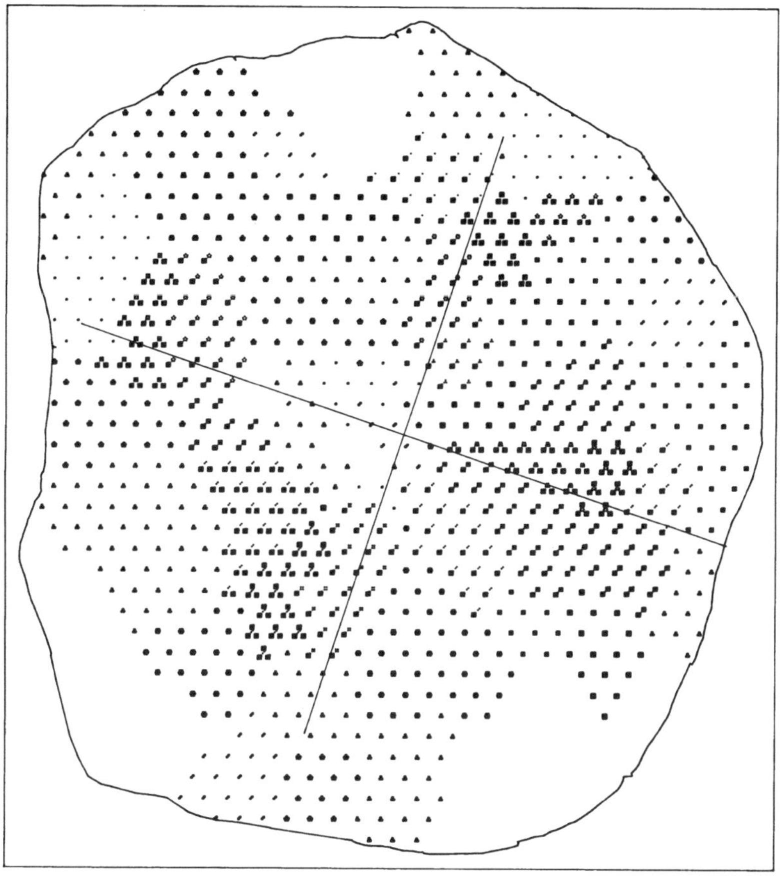

Figure 3.21. Rome. Dwellings built between 1946 and 1981 (total of 7702 elements): . 81 ▪ 813 ▪ 1545 ▪ 2358 ▪ 3090 (Source of data: Comune di Roma, Segretariato Generale il Direzione, Ufficio di Statistica e Censimento).

not. The symbol *TV* [as, e.g., ${}^{g}TV(P, Q)$] is used to denote a void territory in which there are no elements of either *P* or *Q*.

Parameters calculated on the basis of geosets will be denoted by the symbol of the parameter followed by the indication of the geoset(s) that served as the basis for the calculation. For simplicity the superscript g for geosets will be dropped in the symbols of parameters and, depending on what is more convenient, the symbol of the geoset used for the calculation of the parameter will be a subscript, a superscript, or set off in parentheses. For example:

1. Using the letter F to denote the frequency or total size of a population, we can write $F^P, F^Q, \ldots$.
2. Using the letter A to denote area, we can write A^T, A_T, or $A(T)$, where T is any territory.

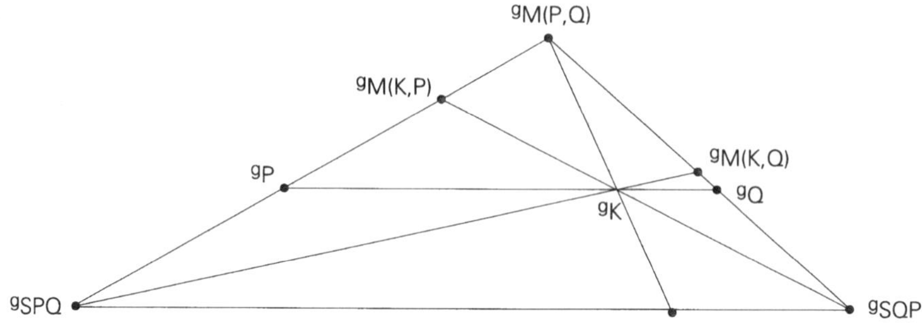

Figure 3.22. Rome. The minimal-distance triangle between two components (dwellings built up to 1945, P, and buildings built between 1946 and 1981, Q) and their compound geoset (all dwellings, K) (Source of data: Comune di Roma, Segretariato Generale il Direzione, Ufficio di Statistica e Censimento).

3. Using the letters $\bar{X}$, $\bar{Y}$ to denote the coordinates of the center, we can write, e.g., $\bar{X}_P$, $\bar{Y}_P$ or $\bar{X}[M(P, Q)]$, $\bar{Y}[M(P, Q)]$.
4. Using d^2 to denote distance-variance (a measure of areal dispersion), we can write, e.g., d_K^2 or $d^2[TSPQ]$.
5. Using $\sigma^{x'}$ to denote the standard deviation of the elements along the principal axis X', we can write $\sigma_{x'}P$ or $\sigma_{x'}[TM(P, Q)]$.
6. Using ${}^cD$ to denote the distance between the centers of a pair of geosets, we can write ${}^cD(P, Q)$.
7. Similarly, using ${}^m\bar{D}$ to denote minimal distance between two geosets, we can write ${}^m\bar{D}(P, Q)$.

We often distinguish between population geosets such as ${}^gP$, ${}^gK$, and ${}^gSPQ$ and territorial geosets such as ${}^gT$, ${}^gH$, and ${}^gVT(P, Q)$. In other cases, however, the distinction between population and territorial geosets is neither necessary nor desirable. In such cases we shall generally use letters such as G and J to designate two sets of any type. Accordingly, we can use the symbol ${}^gG$ to denote the geoset corresponding to set G. The use of the same letter—as a prefix superscript g to indicate geoset and as G to indicate a given set is unfortunate but inevitable, given the insufficient number of letters available for the specification of simple symbols.

The proportion of the size of a subpopulation (say, P) to the size of a compound population (say, K) is denoted by $v_P^K = F^P/F^K$. Similarly, the proportion of the area of a subterritory over the area of a territory is denoted by the letter v with the symbol of the territory as a superscript and the symbol of the subterritory as a subscript.

3.8. Extending the Approach Suggested in Section 3.6 to Other Settings

3.8.1. Selected Extensions

In Section 3.6 we gave the example of the geosets of a compound population (all the dwellings) divided into two subpopulations (dwellings built up to 1945 and

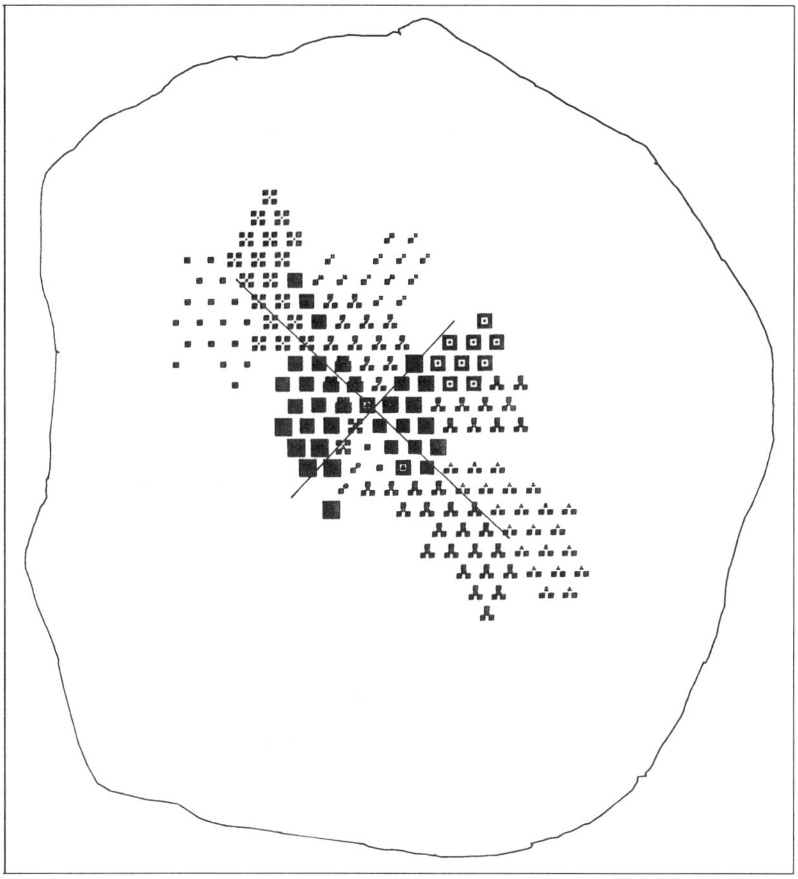

Figure 3.23. Rome. Surplus of dwellings built up to 1945 over dwellings built between 1946 and 1981: . 34 ▪ 313 ▪▪ 695 ⁘ 1252 ■ 3360 (Source of data: Comune di Roma, Segretariato Generale il Direzione, Ufficio di Statistica e Censimento).

those built from 1946 to 1981). This approach can be extended (with simple modifications) to other settings.

We consider first the generalization of the compound population by dealing with a population formed by many components (Section 3.8.2), some of which may be negative (Section 3.8.3). We then proceed to consider settings in which P and Q are not components of a compound population K. However, geosets ${}^{g}P, {}^{g}Q, \ldots$ are available and it may be justified and desirable to compound them into a geoset ${}^{g}K$, if, for example, we want to (a) calculate a simple or weighted average of geosets ${}^{g}P, {}^{g}Q, \ldots$ (see Sections 3.8.4, 3.8.5, and 3.8.6); (b) calculate the union of geosets (see Section 3.8.7); or (c) consider a geoset ${}^{g}K$ that is intermediate between ${}^{g}P$ and ${}^{g}Q$ (see Section 3.8.8). Examples of these settings will be given later.

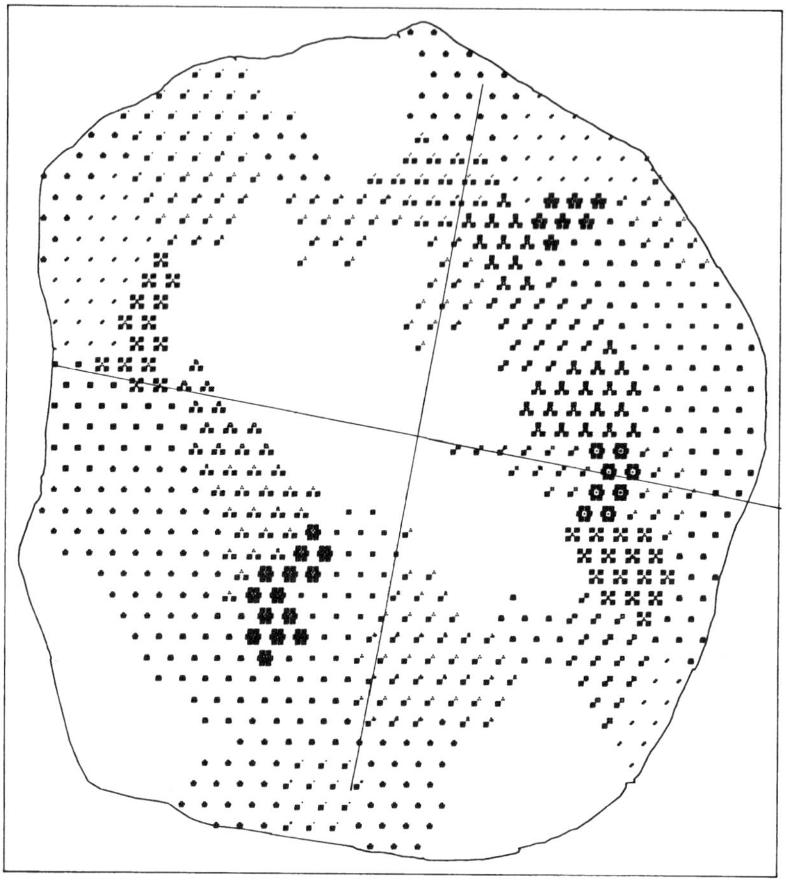

Figure 3.24. Rome. Surplus of dwellings built between 1946 and 1981 over dwellings built up to 1945: .10 ▪ 98 ▪▪ 196 ✿ 490 (Source of data: Comune di Roma, Segretariato Generale il Direzione, Ufficio di Statistica e Censimento).

3.8.2. Geosets of a Compound Population Formed by Many Subpopulations

The approach suggested in Section 3.8.1 can be extended to the settings in which a population K is formed in an exhaustive way by many subpopulations K^j (such as classes of age or ethnic groups). Let us designate the size of subpopulation K^j by F^j; the proportion of each subpopulation K^j over the total population ($v_j^K = F^j/F^K$) by v_j^K, and the proportion of elements of K^j in cell i by $w_i^j = F_i^j/F^j$. Then the geoset of ${}^gK$ can be given as ${}^gK = \Sigma\, v_j^K {}^gK^j$.

It frequently happens that many subpopulations K^j have various aspects of their territorial distributions in common, as indicated by each of the ${}^gK^j$. It is found,

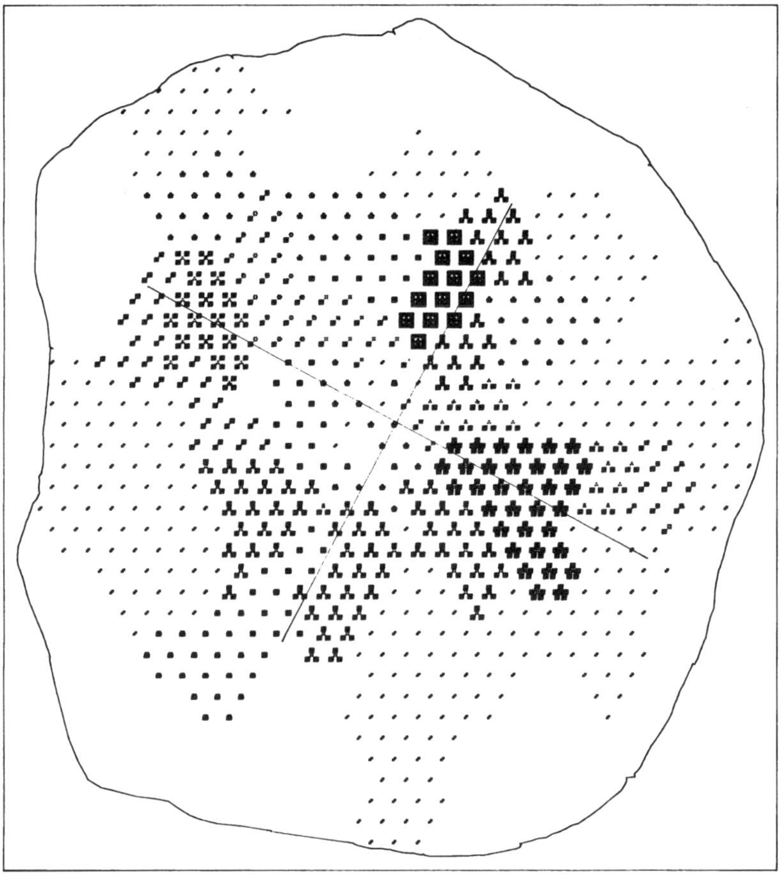

Figure 3.25. Rome. Matched between dwellings built up to 1945 and dwellings built between 1946 and 1981: .1 ▪ 10 ▪˙ 20 ✿ 50 (Source of data: Comune di Roma, Segretariato Generale il Direzione, Ufficio di Statistica e Censimento).

for example, that the distributions of the various age groups of the population of a country are rather similar and are also similar to ${}^{g}K$. However, we are interested in drawing attention to dissimilarities and peculiarities for each of the subpopulations. For this purpose, we often want to separate the general effects of K from each ${}^{g}K^{j}$. Thus, it is useful to derive three geosets from the comparison of each ${}^{g}K^{j}$ to ${}^{g}K$: (a) geoset ${}^{g}SK^{j}K$ of surpluses of subpopulations K^{j} over K; (b) geoset ${}^{g}SKK^{j}$ of surpluses of population K over subpopulation K^{j} (it is often found that the study of ${}^{g}SK^{j}K$ for the various subpopulations is of particular interest), and (c) geoset ${}^{g}M(K, K^{j})$ of the match between K and K^{j}.

3.8.3. Geosets of a Compound Population Formed by Positive and Negative Subpopulations

Sometimes it is convenient to consider a compound geoset that has both positive and negative components. Let us denote by $\varepsilon_j = +1$ or -1, the sign to be attached to each component, according to whether it is positive or negative. If all the w_i^j are nonnegative in any cell i, we can then write ${}^gK = \Sigma\varepsilon_j v_j^{K_g} K^j$. As an example of this setting, we consider the following: ${}^gP$ is the geoset of the distribution of a population at the end of a given period; $j = 0, 1, 2, 3, 4, 5, 6$, where $0 =$ population at the beginning of the period; $1 =$ births; $2 =$ immigrants; $3 =$ deaths; $4 =$ emigrants; $5 =$ origins of internal migrants; $6 =$ destinations of internal migrants. Here, components 0, 1, 2, 6 are given the sign $+$; components 3, 4, 5 are given the sign $-$.

3.8.4. Averaging or Compounding Two Geosets

Let us consider two geosets ${}^gP$ and ${}^gQ$ defining distributions over the same territory from which we wish to obtain an average geoset, to be denoted as $\mathrm{Av}({}^gP, {}^gQ)$, e.g.,

1. P might be a population at the beginning of a period, and Q the same population at the end of the period. We might want to examine the distribution of the population in the middle of the period, assuming a linear process (see Section 3.8.5).
2. ${}^gP$ and ${}^gQ$ might be unrelated populations, having, say, a very similar territorial distribution. We are interested in obtaining a geoset that can be considered as being influenced in the same measure by ${}^gP$ and ${}^gQ$. We can write:

$$\mathrm{Av}({}^gP, {}^gQ) = 0.5({}^gP + {}^gQ) = \{X_i^c, Y_i^c, w_i^c = 0.5(w_i^P + w_i^Q)\} \tag{3.27}$$

If we consider that

$$ {}^gP = \Delta SPQ + (1 - \Delta)M(P, Q) \tag{3.28}$$

and

$$ {}^gQ = \Delta SQP + (1 - \Delta)M(P, Q) \tag{3.29}$$

Then

$$\mathrm{Av}({}^gP, {}^gQ) = \Delta 0.5[SPQ + SQP] + (1 - \Delta)M(P, Q) \tag{3.30}$$

3.8.5. Weighted Average of Two Geosets

In Section 3.8.4, we considered the simple average of ${}^gP$ and ${}^gQ$, obtained by attributing the same weight ($\frac{1}{2}$) to both geosets. However, it is sometimes desirable to give them different weights. For example, ${}^gP$ and ${}^gQ$ might denote, respectively, the distribution of a population at the beginning and at the end of a particular period. Assuming a linear transformation of ${}^gP$ into ${}^gQ$, we can ask what the distribution would be after a proportion v of the period?

$$ {}^gR^v = (1 - v){}^gP + v{}^gQ = \Delta[(1 - v){}^gSPQ + v{}^gSQP] + (1 - \Delta)M(P, Q) \quad (3.31)$$

3.8.6. Average of Many Geosets

In previous sections we considered a geoset obtained from ${}^gP$ and ${}^gQ$. We can now generalize to a geoset obtained as a simple or weighted average of many geosets ${}^gP^j$. The need for such an operation might arise in situations such as the following:

1. Geosets ${}^gP^j$ might be analogous and might be considered as belonging to the same "family" of territorial distributions. In this case, it might be desirable to obtain an average geoset to be taken as a general indicator of the characteristics of the family. The average can be obtained either by giving the same weight to all geosets or by weighting them by attributing more importance to some than to others.
2. Similarly if ${}^gP, {}^gQ, {}^gR, \ldots$ are the geosets of a population on different dates of a given period, the average geoset can be taken to represent the distribution during the entire period.

The settings discussed in Sections 3.8.4, 3.8.5, and 3.8.6, can be generalized as follows: Let ${}^gP^j$ denote several geosets and v_j be the relative importance or weight attributed to each of them in the process of compounding or averaging ($\Sigma v_j = 1$). Then the average or compound geoset is

$$\text{Av}({}^gP^j) = \Sigma v_j{}^gP^j \quad (3.32)$$

3.8.7. Union of Geosets versus Average of Geosets

If we consider the average of two geosets ${}^gP$ and ${}^gQ$,

$$\begin{aligned}\text{Av}({}^gP, {}^gQ) &= v_P{}^gP + v_Q{}^gQ \\ &= v_P[\Delta{}^gSPQ + (1 - \Delta){}^gM(P, Q)] + v_Q[\Delta{}^gSQP \\ &\quad + (1 - \Delta){}^gM(Q, P)] \end{aligned} \quad (3.33)$$

There are cases in which, apart from or in addition to $\text{Av}({}^gP, {}^gQ)$, we might want to take the union of ${}^gP$ and ${}^gQ$ into account. In calculating the union we attribute a weight

$$[(1 - \Delta)/2]/[\Delta + (1 - \Delta)/2] = (1 - \Delta)/(1 + \Delta) \quad (3.34)$$

to the match, while the average of the geosets of surpluses is given a weight $\Delta/(1 + \Delta)$:

$$\begin{aligned}\text{Union}({}^gP, {}^gQ) = {} & [\Delta/(1 + \Delta)][v_PSPQ + v_QSQP] \\ & + [(1 - \Delta)/(1 + \Delta)]{}^gM(P, Q) \end{aligned} \quad (3.35)$$

In an analogous way, dealing with the geoset ${}^gK$ of a compound population K

formed by two component subpopulations, we obtain

$$ {}^gK = v_P^{K_g}{}^gP + v_Q^{K_g}{}^gQ \tag{3.36} $$

and

$$ \begin{aligned} \text{Union}({}^gP, {}^gQ) &= [\Delta/(1+\Delta)][v_P^K SPQ + v_Q^K SQP] \\ &\quad + [(1-\Delta)/(1+\Delta)]{}^gM(P, Q) \end{aligned} \tag{3.37} $$

For an example, see Section 3.6.

3.8.8. Intermediate Geosets

Let us consider two geosets ${}^gP$ and ${}^gQ$, denoting the distribution of two populations P and Q over territory T. The two populations (of equal size F) might or might not be complementary to one another:

$$ {}^gP = \{X_i^c, Y_i^c; w_i^P = F_i^P/F\} \tag{3.38a} $$

$$ {}^gQ = \{x_i^c, y_i^c; w_i^Q = F_i^Q/F\} \tag{3.38b} $$

The *intermediate geoset* $[{}^gI(P, Q)]$ is one of locations of elements over T, which in any cell i has relative weights w_i^I such that $\min(w_i^P, w_i^Q) \leqslant w_i^I \leqslant \max(w_i^P, w_i^Q)$.

Let us denote the absolute number of elements of I in cell i by F_i^I. We can write

$$ F_i^I = \min(F_i^P, F_i^Q) + F_i^{SI} \tag{3.39} $$

where $0 \leqslant F_i^{SI} \leqslant \max(F_i^P, F_i^Q) - \min(F_i^P, F_i^Q)$.

Here F_i^{SI} denotes the addition to $\min(F_i^P, F_i^Q)$ needed to reach F_i^I. We know that

$$ {}^gM(P, Q) = \{X_i^M, Y_i^M; w_i^M = \min(F_i^P, F_i^Q)/\Sigma \min(F_i^P, F_i^Q)\} \tag{3.40} $$

The geoset ${}^gSI$ is obtained as follows:

$$ {}^gSI = \{X_i^{Sj}, Y_i^{Sj}; w_i^{Sj} = F_i^{SI}/\Sigma F^{SI}\} \tag{3.41} $$

The total sizes of the populations considered are then

$$ F^I = \Sigma F_i^I, F^M = \Sigma \min(F_i^P, F_i^Q), F_{SI} = \Sigma F_i^{SI} \tag{3.42} $$

and the proportion of matched and added elements over intermediate elements becomes

$$ v_M^I = F^M/F^I \qquad v_{SI}^I = F^{SI}/F_I \qquad (v_M^I + v_{SI}^I) = 1 $$

The intermediate geoset ${}^gI$ can be expressed as:

$$ {}^gI = v_M^I{}^gM + v_{SI}^I{}^gSI \tag{3.43} $$

This equation is important because it shows us that the relevant geostatistical parameters of the intermediate geoset I are included in the corresponding parameters of ${}^gP$ and ${}^gQ$.

3.8.9. Decomposition of a Geoset

The computer package described in Chapter 17 enables us to carry out operations opposite to those considered in the preceding sections. Let us consider a geoset P for any population P distributed over a territory T, that is divided into N regions or cells i ($i = 1, 2, \ldots, N$). The center and the area of each i are known. We form groups or classes of cells, or regions. Each group or class is indicated by C. The classification of the i into classes may or may not be exhaustive; the territory of C may or may not be simply connected. The program makes it possible to map each C distinctly and to calculate geostatistical parameters for the territorial and population geosets of each C. This approach has many important applications, such as the following:

1. Classes C can be formed (a) according to the absolute size or density of the population in each i or (b) according to qualitative attributes or values that characterize each i.
2. Consider a city or conurbation that has enlarged its gross territory over the course of time. Suppose we know for each i the date on which it was incorporated into the gross territory. We can form C indicating the gross territory at each relevant date.
3. Suppose that for each i we have information on many of its attributes and/or levels of variables. We can at will form a group C of i, that possesses selected characteristics. For instance, in market research in a city we might want to single out cells or regions i in which we can find selected income levels and given proportions of several groups (formed according to age, ethnicity, or social standing), which are at selected distances from the business center and at an average distance from amusement areas. In an environmental epidemiological study, we might want to include or exclude regions or cells i in which the population has selected demographic, ethnic, and socioeconomic characteristics and the territory has selected environmental characteristics.

The flexibility of the program enables us to compare various alternatives easily.

3.9. Location-Free Sets of Relative Cell Weights

If for any of the geosets discussed in this chapter that are based on the division of territory T into N^c cells i of equal size and shape, we take into account only the relative cell weights w_i and disregard the location of each cell i, we call the list of the w_i a "location-free" set of relative cell weights. We have already seen an example of these sets in Section 3.2.4, where we considered a location-free set derived from the comparison of the actual distribution of a population with the uniform distribution. For cells in which $w_i^P > 1/N^c$, we calculated surpluses $(w_i^{SPU} - 1/N^c)$ and obtained $\Sigma(w_i^{SPU} - 1/N^c) = \Delta_P^U$ as the total of the surpluses. This was interpreted as the proportion of elements to be moved in order to transform ${}^gP$ into ${}^gU$. Obviously $\Sigma(1/N^c - w_i^{SUP})$ was also found to be $\Delta_U^P = \Delta_P^U = \Delta$, and the proportion of non-movers was $(1 - \Delta)$, a parameter seen to be sensitive to the geographical characteristics of the population distribution.

If we assume that we keep the values w_i^P fixed and assign each of them in all possible permutations to a different cell, we can obtain very different geographical configurations, but the value of Δ_P^V will not change. Hence Δ_P^V is not a proper parameter to be used in the description of geographical characteristics of ${}^gP$. The same can be said for all the location-free sets of relative cell weights (such as those denoting, e.g., w_i^{SPQ} and $w_i^{M(P,Q)}$), as well as for many of the commonly used measures of concentration, segregation, ecological correlation, among others.

However, this does not imply that Δ_P^U as well as the other parameters mentioned, cannot be of use. On the contrary, we shall see many important applications of these parameters (see Sections 8.3 and 8.4) but they can be used properly only for studies in which insensitivity to geography is not relevant.

3.10. Expressing Geosets of Surpluses Through Rates or Ratios

Geosets of surpluses ${}^gSPU$, ${}^gSPQ$, and ${}^gSQP$ are based on the differences between relative frequencies of P, U, and Q in each cell or region. Here, we create a bridge between this method and the approach frequently utilized to make comparisons by means of rates or ratios.

1. Let us consider surpluses of a population P over the uniform distribution. We denote the relative frequency of P in any region or cell i of the territory T over which P is distributed by $P_i = F_i^P/F^P$, the proportion of the area of i over the total area of T by $a_i = A_i/A$, the density of the population over the area of i by $h_i = P_i/A_i$, and the average density of the population over T by $\bar{h} = P/A$. The surplus of P over U in regions or cells in which $P_i > a_i$ is given by $(P_i - a_i) = a_i[(h_i/\bar{h}) - 1]$. If i denotes cells of equal area, $a_i = 1/N$, the map representing surpluses becomes practically identical to the map of $[(h_i/\bar{h}) - 1]$.

2. Let us compare ${}^gQ$ to ${}^gP$ for some phenomenon Q, such as income or births, occurring in a population P. Let us denote the relative frequency F_i^Q/F^Q of Q in i by q_i, the relative frequency of P_i^P/F^P of P in i by p_i, the birth rate or the ratio of income per capita in i by $g_i = Q_i/P_i$, and the average rate or ratio for the entire population by $g = (F^Q/F^P)$. The surpluses of Q over P in cells in which $(q_i + p_i)$ is given by $(p_i - q_i) = p_i[(g_i/\bar{g}) - 1]$. Thus, we see that this surplus depends on both the ratio of g_i to g and on the relative weight of the population in cell i.

Chapter 4

Further Considerations on Geosets of Territories

4.1. Introduction

In Chapter 3 we considered examples of geosets of territories that were constructed on the basis of the following approaches:

1. Dealing with a predetermined territory T of area A^T, we assume that a population P or two populations P and Q are distributed over T, which is divided into a convenient number N^c of cells i ($i = 1, 2, \ldots, N^c$), of regular hexagonal shape and area A^T/N^c; $N^c = F^P$ for populations of moderate size and $N^c < F^P$ for large populations; centers (X_i, Y_i) of cells i are determined.
2. If we wish to calculate the geostatistical parameters of T, as gross territory, we use the geoset ${}^gU$ as an approximate base:

$$ {}^gU = \{X_i, Y_i; w_i^c = 1/N^c\} \tag{4.1} $$

3. If greater precision is required, some geostatistical parameters, such as those measuring dispersion, can be adjusted in order to take intracell dispersion into account.
4. From the comparison between ${}^gP$ and ${}^gU$, we derive geosets ${}^gH$ and ${}^gNH$ of habitat and nonhabitat. Habitat can be further subdivided into (a) the subterritory corresponding to the match between ${}^gP$ and ${}^gU$; (b) the subterritory corresponding to surpluses ${}^gSPU$ of ${}^gP$ over ${}^gU$.
5. From the comparison of two population geosets ${}^gP$ and ${}^gQ$ over the same territory T, we derive territorial geosets ${}^gTSPQ$, ${}^gTSQP$, ${}^gTM(P, Q)$, and ${}^gTV(P, Q)$, which serve as a basis for calculating approximations to the geostatistical parameters of these territories.

In this chapter we generalize the concept of geoset ${}^gT$ of a territory T by introducing additional approaches:

1. We assume that we know the locations of the elements of a population P or

of two populations P and Q, but that we do not have at our disposal a relevant ${}^{g}T$ for a predetermined territory T to serve as a container of P or of P and Q. We shall see how we can determine a gross territorial range for these "unbounded populations," so that we can always compare ${}^{g}P$, or ${}^{g}P, {}^{g}Q$ to a ${}^{g}T$ (see Section 4.2).

2. In considering the relationships between ${}^{g}P$ or ${}^{g}P, {}^{g}Q$ and ${}^{g}T$, it is sometimes relevant to take into account settings in which each population element covers a meaningful and determinable tract of territory (Section 4.3).
3. We might wish to consider geosets of territories and/or subterritories independently of the geosets of population(s) distributed over them. For this purpose we start from a study of the division of a simply connected territory T into an infinite number of infinitesimal cells (Section 4.5.1).
4. For the actual calculation of geostatistical parameters, approximations to the model introduced under 3 are suggested (Section 4.5.2).
5. Apart from geosets of simply connected territories, we will also consider other types of territorial geosets, which correspond to territories formed by separated fragments and to territories with internal gaps, and external concavities (Section 4.6), as well as special aspects of geoset operations applied to territorial geosets (Section 4.7) and types of scales used in determining geosets of populations and territories (Section 4.8).

4.2. Defining the Territorial Gross Range for Unbounded Population Distributions

4.2.1. The Problem

In the study of geosets in Chapter 3, we assumed that the populations were distributed over a predetermined territory T with clearly defined boundaries. Here we consider settings in which we know the locations of elements of populations, but the territory over which these elements are found is either not defined or has virtually no relevance for our purposes. For example:

1. In studies of plant or animal ecology, the distribution of archaeological sites, or urban ecology, maps are sometimes designed to show locations of individual elements of a population P, but the gross territory T over which these elements are given remains undefined.
2. On the other hand, sometimes the boundaries of T are given, but in the shape of a rectangle or some other simple geometrical figure without any clear definition of its meaning. One example of such a setting was shown in Figure 3.2 (Section 3.1).
3. In urban ecology P might be defined by reference to a territory T (such as the municipal area of a city), but ${}^{g}P$ is actually found over a small part of T. In this case although T is "predefined," it may be almost meaningless for the study, and we might encounter difficulties even when we only use it as a tool for determining the habitat.

Let us consider an example taken from a study we published some years ago (Bachi, 1981). Figure 4.1 shows the distribution of 62 hotels in Tel-Aviv–Jaffa, Israel. As in the study described earlier of hotels in Jerusalem, each hotel is represented by a dot. Most of the dots are closely clustered along the shore of the Mediterranean and in the center of the city. In this particular study, the boundaries of the municipal area (shown in Figure 4.1) were found to be virtually irrelevant for the microgeographic description, so we chose to consider the distribution of the hotels of Tel Aviv–Jaffa as unbounded. This is illustrated in Figure 4.2, where the distribution ${}^{g}P$ of the dots is magnified and shown without reference to the municipal boundaries.

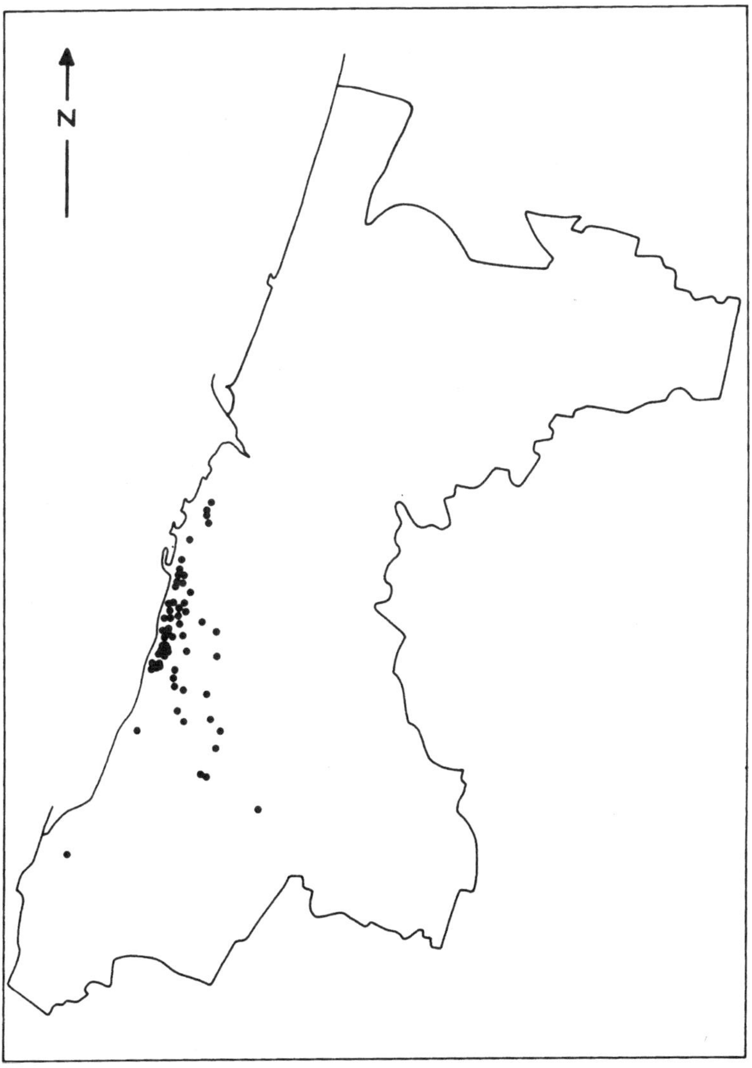

Figure 4.1. Tel-Aviv–Jaffa. Hotels within the municipal area (Source: Bachi, 1981).

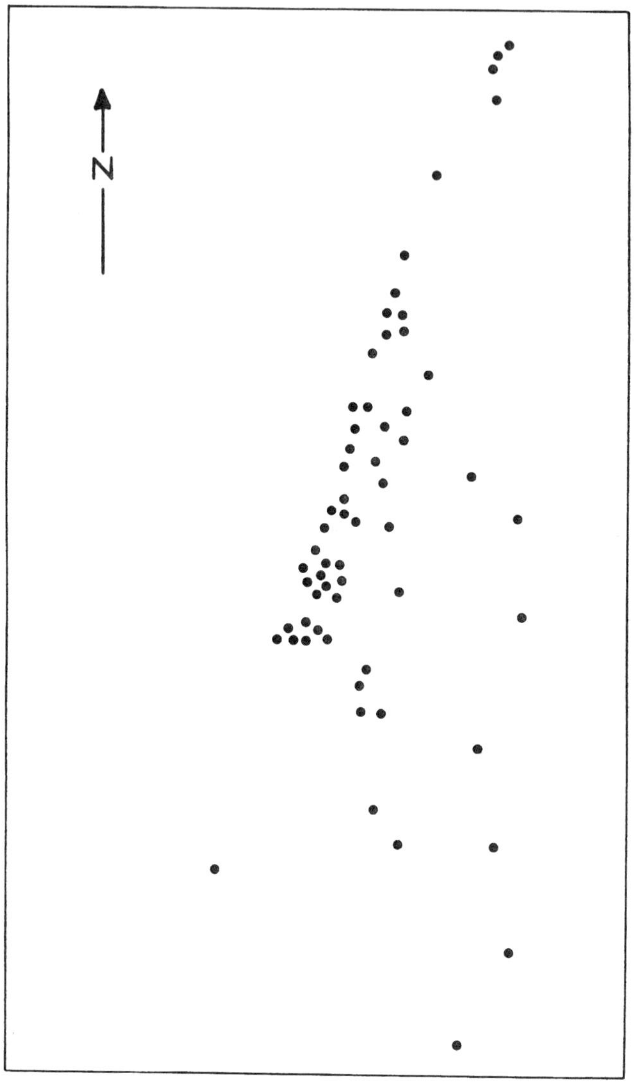

Figure 4.2. Tel-Aviv–Jaffa. Hotels distribution (magnified) (Source: Bachi, 1981).

In such cases, it is useful to define the gross container of the geoset ${}^{g}P$ in an "endogenous" way. We assume that the locations of the elements of P are known but that a predefined T is unknown or irrelevant, and we want to determine a gross container T_P of the elements of P solely on the basis of geoset ${}^{g}P$.

4.2.2. Criteria for Solutions

Simple "commonsense" criteria are as follows: (a) T_P should be convex; (b) T_P should include the entire population or at least a great majority of it (apart possibly from a small proportion of "outliers"); (c) the definition of T_P should correspond to an acceptable one-dimensional parameter; (d) under the above conditions, T_P should have the minimum possible area; (e) T_P should be easily describable. We justify these criteria as follows.

1. Territory T_P can be conceived as a gross container of the population. We can expect that the part of the plane taken as the container will be simply connected by a continuous contour. However, we can also request unhindered and easy accessibility from each point i to every other point j of the population; operationally, this may imply that every pair of points i and j should be connected by a straight line lying completely within T_P.
2. Theoretically, the definition of T_P should enable us to enumerate all the individuals in the population by spanning T_P, without going outside its boundaries. However, we might be prepared to neglect a few isolated cases of the population (outliers) that are apart from the main clusters that constitute the bulk of the population.
3. If we assume that the distribution is one-dimensional and that the dots i are identifiable through their abscissas X_i, our problem is then reduced to determining the part of the X-axis along which all, or most, of the dots are found. Two familiar ways for dealing with this are: (a) to use the range between the highest and the lowest value of X_i, and (b) to consider proper multiples of a parameter of dispersion around the center, which can be expected to include a proportion of the population sufficiently large for our descriptive aims. By applying these concepts to the two-dimensional distribution, we find two solutions to the problem of determining the territory corresponding to the population: "territorial range" (Section 4.2.3) and an ellipse with scatter equal to that of the population (Section 4.4).
4. The need to minimize the area of the container T_P is obvious: eliminating areas that do not include population points makes T_P a more efficient tool for describing the population, provided that it does not impinge on the conditions upon which the definition of T_P is based.
5. As T_P is a tool for describing an important aspect of the population distribution, the description of T_P itself must be easy.

4.2.3. The Convex Hull as Gross Territorial Range

Application of criteria (1)–(5) to an unbounded distribution of N points such as shown in Figure 4.2, leads us to define the convex hull drawn around the N points (or at least a great majority of the N points, after exclusion of outliers) as gross territorial range (or simply range). An example of this method is illustrated in Figure 4.3. The gross range can intuitively be conceived of as a territory of convex shape and minimal area that includes all or most of the elements of the population. Over this range each pair of population units can be connected linearly without going outside the range.

Having determined the convex hull as an indicator of gross territory T_P, we can carry out on it and on P all the operations described in Section 3.2 for comparing P to the uniform distribution over the predetermined T. As an example, Figures 4.4 and 4.5 show, respectively, the uniform and the equivalent distributions over a

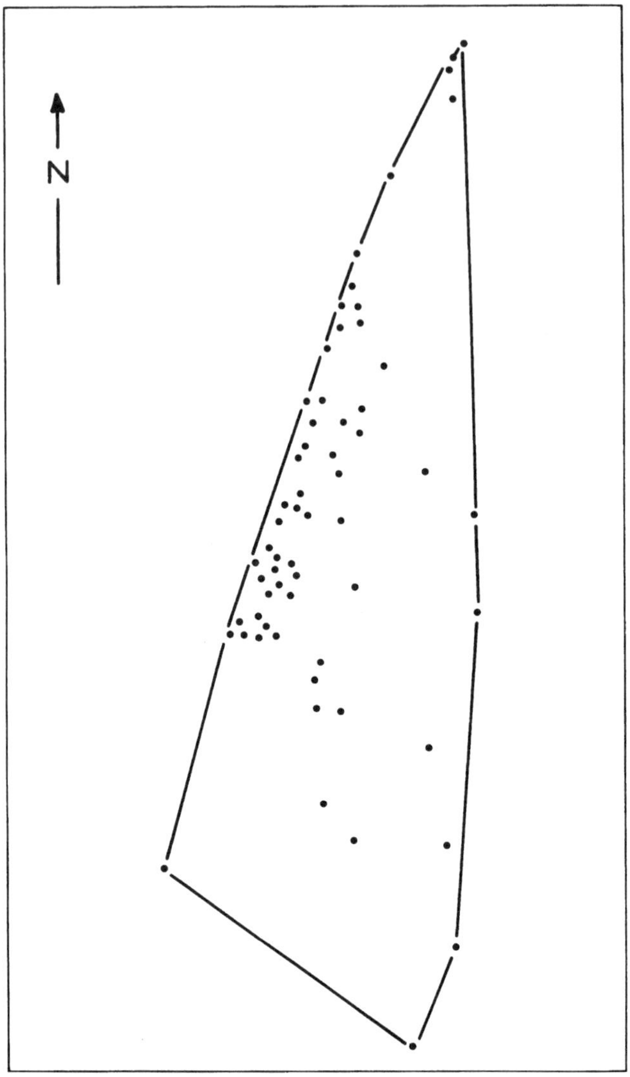

Figure 4.3. Tel-Aviv–Jaffa. Gross range of hotels (Source: Bachi, 1981).

convex hull. Figure 4.6 shows the division of T_P into habitat (black hexagonal cells) and nonhabitat (white cells).

4.2.4. Excluding Outliers from the Convex Hull

An obvious shortcoming of the use of a convex hull for determining T_P is that one or a few dots located far from the bulk of ${}^{g}P$ can cause it to be considerably distorted and enlarged. To avoid or minimize this problem, we can: (a) determine the convex hull around the population provisionally; (b) identify clusters of contigu-

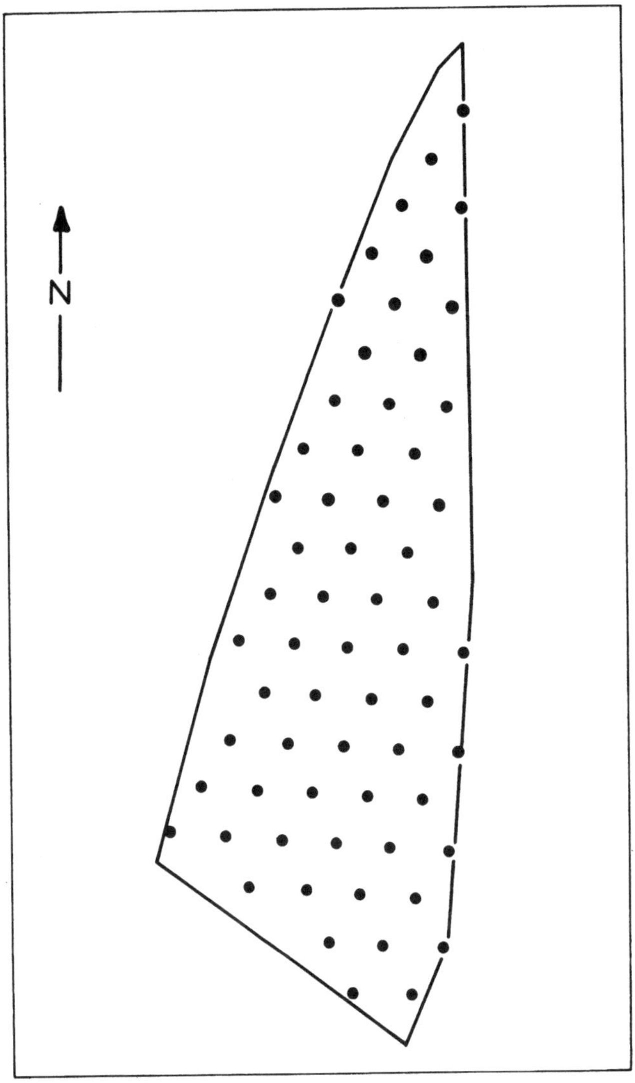

Figure 4.4. Tel-Aviv–Jaffa. Uniform distribution over gross range (Source: Bachi, 1981).

ous inhabited cells [contiguous cells being defined as cells having one boundary side in common (see Section 11.9.1)]; (c) define as "major" the clusters, with areas and/or populations not smaller than designated percentages of the total area and/or population, isolated inhabited cells can be included in this calculation and the requirements for contiguity of the cells of a cluster might be relaxed; (d) draw a corrected "reduced" convex hull in a way that ensures the inclusion of all major clusters.

A computer geostatistical program enables us to implement these methods. The criteria for the definition of major clusters depend on the decisions of the researcher,

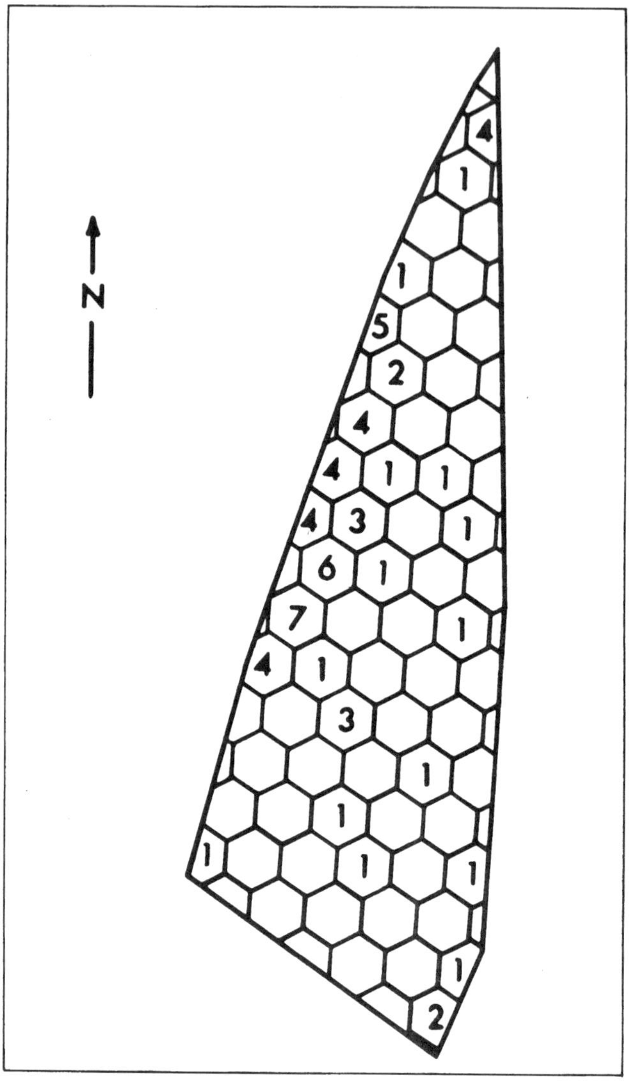

Figure 4.5. Tel-Aviv–Jaffa. Equivalent distribution of hotels (Source: Bachi, 1981).

taking into account the aims of the study, the nature of the relevant population, and/or results or alternative solutions obtainable from the computer.

4.3. Tracts of Territory Covered by Individual Population Elements

In this book we often employ the commonly accepted convention that the location of individual population elements can be represented by dimensionless dots. However, there are limitations to this approach, some of which will be discussed in

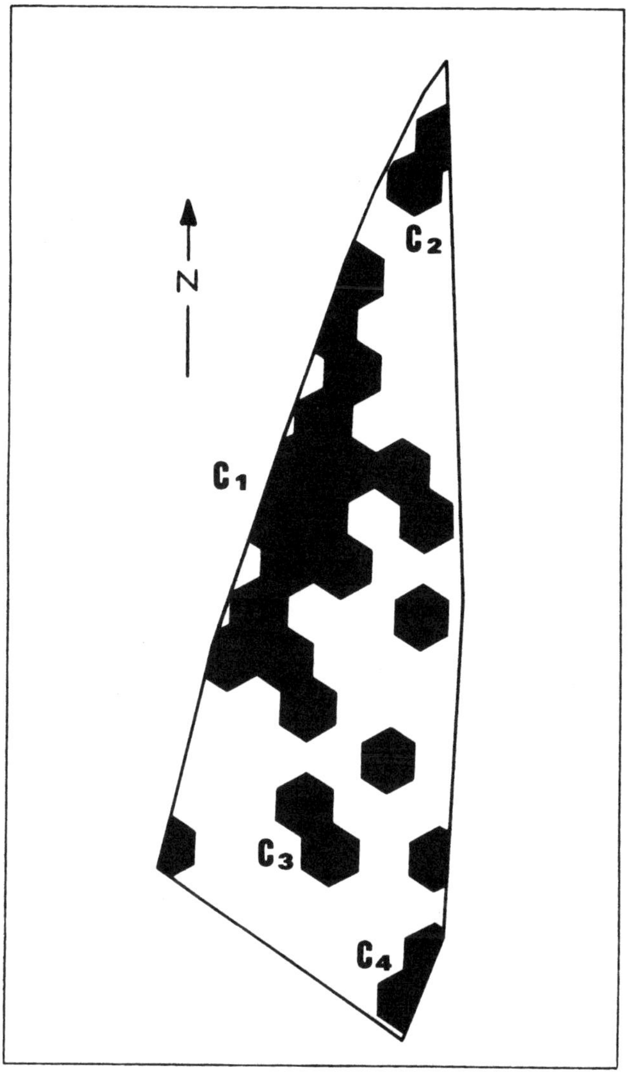

Figure 4.6. Tel-Aviv–Jaffa. Habitat and nonhabitat of hotels (Source: Bachi, 1981).

connection with the concept of concentration of populations and with the elusive concept of area covered by human beings or subject to their influence (see Section 8.5). Other limitations have already been alluded to in the example of the population of hotels in Jerusalem (Section 3.1.3). We note again that hotels are not dimensionless entities, but are found in buildings that cover a tract of territory. The same can be said of, e.g., commercial establishments, schools, and public office buildings. In urban research the area they cover is far from being null, is clearly determinable, and can sometimes be determined from statistical sources or maps. However, these areas, even if they can be delineated, are often of little interest for macrogeostatistical analysis. This is especially true when the areas covered by population units are small compared to the total area of the territory over which the population is distributed.

The situation is often different when geostatistical methods are used in other fields, such as for agricultural statistics, e.g.,

1. In many countries agricultural censuses furnish a wealth of geographical–statistical data on farms, considered as statistical units and classified by such criteria as tenure, characteristics of operator, type of organization, livestock, types of cultivation, and field crops harvested. The geographical specification of the data is generally given in aggregate form by territorial units such as regions, provinces, or countries, which provides us with population geosets similar to those discussed in Chapter 3. However agricultural censuses very often also supply information about the area itself for each type of population element being surveyed, e.g., the various crops being cultivated or whether the farms are run by individuals, family groups, or institutions. This additional information about the areas can be of great interest, and examples of studies of "compactness" of cultivation are given in Chapter 11.
2. The use of geostatistical parameters also appears to be possible and interesting in the study of the distribution of lands for farming, viewed in a deeper and more complex microgeographical context. If we assume that we are concerned with agricultural holdings in a region and that detailed maps of each holding, divided into separate parcels, are available, studies could then be carried out on, e.g., shape characteristics and fragmentation of the holdings, but we will not discuss this type of research any further in the present work.

4.4. Ellipses of Equivalent Scatter

4.4.1. An Ellipse as a Substitute for Gross Territorial Range

We might want to determine the gross territorial range by using an approach analogous to that described in Section 4.2.3 for the study of a one-dimensional distribution of a variable X, given its average $\bar{X}$ and its standard deviation σ_X, where we examine what proportion of the values of X are included within the limits $(X - h\sigma_X)$ and $(X + h\sigma_X)$, where h is a real number. This brings us to the idea of getting a rough estimate of the gross range by constructing an ellipse having the same center and principal semiaxes as the population with the lengths of the semi-axes being proportional to $h\sigma_{X'}$ and $h\sigma_{Y'}$, where $\sigma_{X'}$ and $\sigma_{Y'}$ are the standard deviations of the population along the principal axes.

A computer program can help us to calculate what proportion of the elements of the population is included within such an ellipse (see Section 17.9.5). In the selection of a suitable value for h, it is generally found that theoretical limits are not useful in the study of actual geosets because most of the spatial distributions do not obey simple laws. As we are interested in keeping the area of the territorial container as small as possible, a two-dimensional analogue to the Tchebycheff inequality would also not be of much use, as it would set excessively wide limits. We can consider gradually increasing the values of h and until we reach one such that a satisfactory proportion of the elements of the population is included within the ellipse and the area of the ellipse is substantially smaller than that of the gross territory. Having thus determined a convenient ellipse as the gross container, we can obtain the net habitat by the methods suggested above, adapted to the setting in which we deal with a predetermined territory.

If we take $h = 2$, we obtain an ellipse for which the lengths $(2\sigma_{X'}, 2\sigma_{Y'})$ of its principal semiaxes are equal to those of the arms of the cross superimposed on the map. This ellipse, which we call as an "ellipse of equivalent scatter," has the advantage of a clearly defined meaning: if the population is uniformly distributed over a territory of elliptic shape, the ellipse of equivalent scatter coincides with the territory.

4.4.2. The Ellipse of Equivalent Scatter as a Substitute for Habitat

The determination of the ellipse of equivalent scatter might be of use for a rapid evaluation of the habitat or for indicating the limits of its extent. This does not apply to settings in which the known actual gross territory has large concavities, internal gaps, or fragmentations. For example, in the study of geosets over the very concave territory of Italy, the use of ellipses of equivalent scatter was not found to be convenient.

However, exactly the opposite conclusion was reached in a very extensive study carried out at the U.S. Bureau of the Census in the late 1970s (see Appendix B), in which ellipses of equivalent scatter were obtained for 626 geosets. These were related to demographic, socioeconomic, and agricultural distributions (resulting from recent censuses) over the conterminous United States (detailed by over 3000 counties) and to the metropolitan areas of Washington D.C., Atlanta, and Austin (detailed by census tracts). All these territories are not far from being convex, and here it was found that the ellipses of equivalent scatter included the following percentages of the population of the geosets:

Up to	75%	75–79%	80–84%	85–89%	90–94%	95–100%
Per 100 ellipses:	—	2.2	24.9	46.6	25.2	1.0

If we denote the total size of the population considered by F^P and the size of the population included within the ellipse, by F^E, we find that for the majority of the ellipses F^E/F^P is close to 1. If A^T is the area of the gross territory and $A^E = 4\pi\sigma_{X'}\sigma_{Y'}$ is the area of the ellipse for $A^E/A^T < F^E/F^P$, there may be a descriptive advantage to considering the ellipse as the territorial container in place of the territory T. This implies that the density of the population per unit of area within the ellipse, F^E/A^E, is higher than the gross density F^P/F^T over the entire territory. For $A^E/A^T \ll F^E/F^P$, the descriptive advantage may be considerable. Under these conditions, we might use the ellipse as a rough substitute for the habitat as it has the clear advantage of quick determination and presentation by means of a few well-defined parameters of the population (center, angle between principal and conventional axes, standard deviations along principal axes).

Examples of applications are illustrated in Figures 4.7–4.15 (Bachi, 1981). Figures 4.7 and 4.8 compare: (1) an even distribution over the metropolitan area to actual distributions of (2) White population, (3) Spanish-speaking inhabitants, and (4) Afro-American population in Atlanta, Georgia, and Washington, D.C. To facilitate comparisons, the original scales have been changed so that the area included in ellipses (1) has been made equal in Atlanta and Washington. Moreover, the maps have been rotated in order to give the ellipses the same orientation. The analogy between the distributions shown by Figures 4.7 and 4.8 is evident: the areas of the ellipses decrease from (1) to (4).

Figure 4.9 shows that the aging population in Washington D.C., is less dispersed than the children, a feature common to many cities. Figures 4.10 and 4.11 give examples of agricultural distributions (cotton and tobacco harvested), the first being compared to the distribution of the Afro-American farm population and the second to the distribution of tobacco industry facilities.

Figures 4.12 and 4.13 illustrate use of ellipses to compare distributions of a single variable with changing parameters: farms of different sizes in the United States (Figure 4.12) and provinces of Italy classified by income per capita (Figure 4.13).

Figures 4.14 and 4.15 illustrate the ellipses for research on changes in the distribution of a population over the course of time. Figure 4.14 shows the enormous

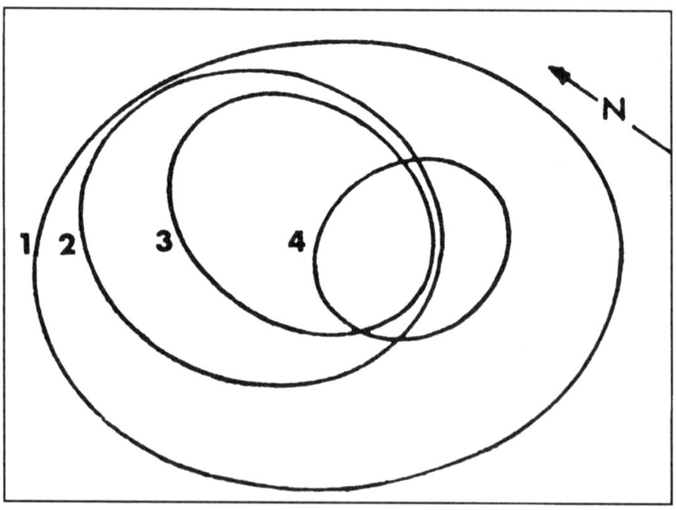

Figure 4.7. Atlanta, 1970: (1) Metropolitan area, (2) Whites, (3) Spanish, (4) African-Americans (Source: Bachi, 1981).

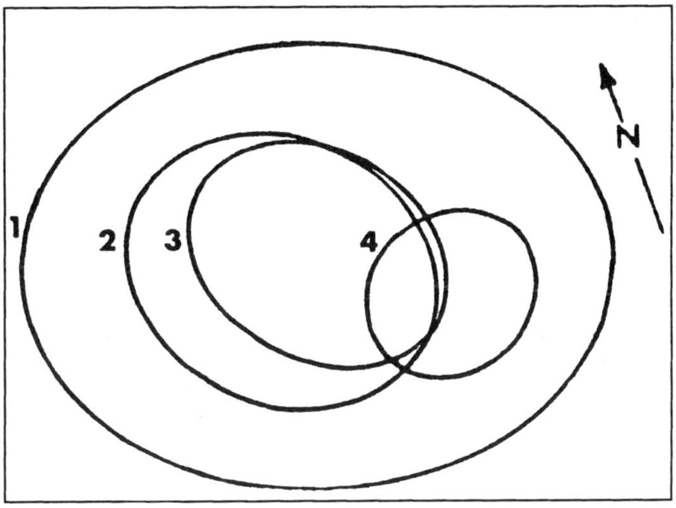

Figure 4.8. Washington, 1970: (1) Metropolitan area, (2) Whites, (3) Spanish, (4) Afro-Americans (Source: Bachi, 1981).

increase in the dispersion of the Chinese population in the United States and the eastward shift of its center between 1870 and 1970: an in-depth analysis would indicate that the 1970 distribution has two major and very distant clusters, one in the East and one in the West. Figure 4.15 shows the distribution of the population in Rome, Italy, in 1526, 1870, and 1961.

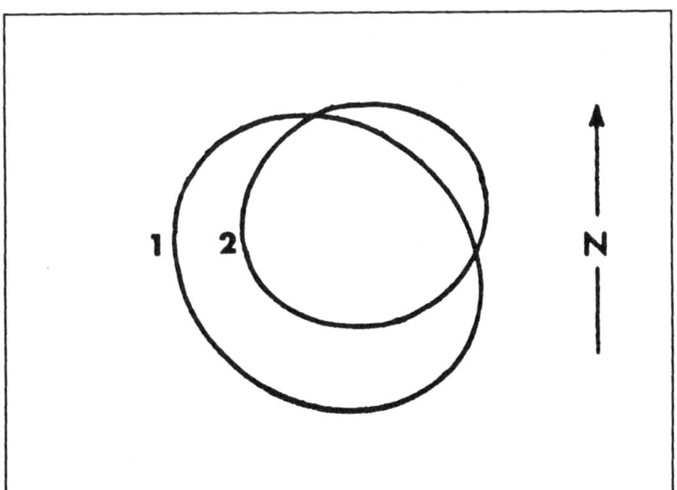

Figure 4.9. Washington, 1970: (1) under 18, (2) 65 and over (Source: Bachi, 1981)..

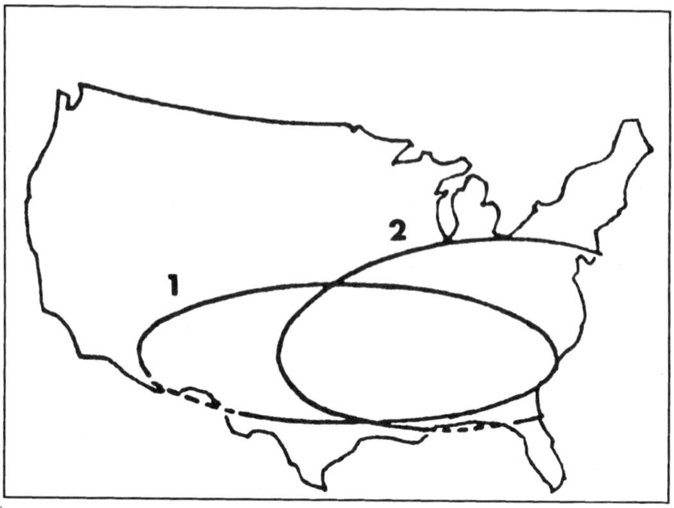

Figure 4.10. United States: (1) cotton harvested, 1969; (2) Afro-American farm population, 1960 (Source: Bachi, 1981).

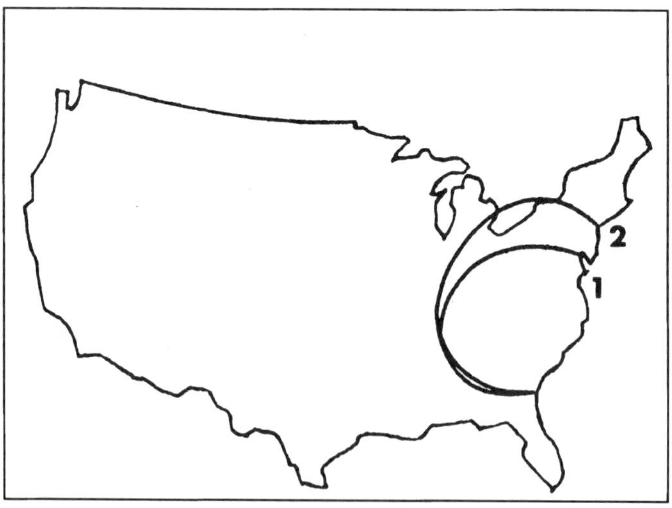

Figure 4.11. United States: (1) tobacco harvested, 1969; (2) tobacco industry, 1967 (Source: Bachi, 1981).

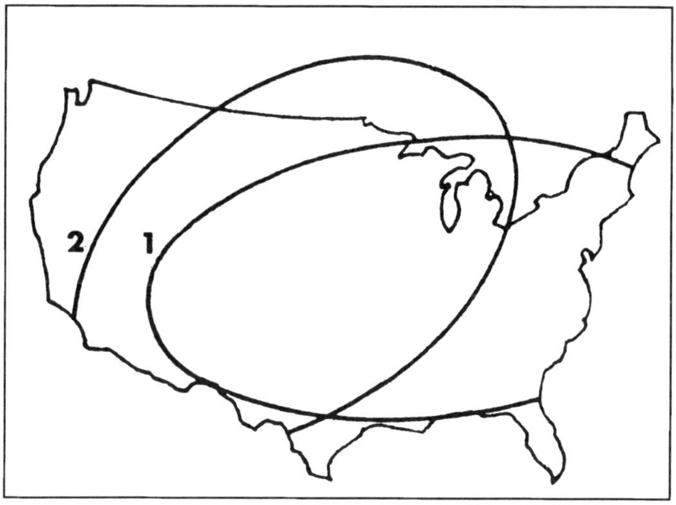

Figure 4.12. Farms in the United States, 1969: (1) 1–999 acres, (2) 1000 acres and over (Source: Bachi, 1981).

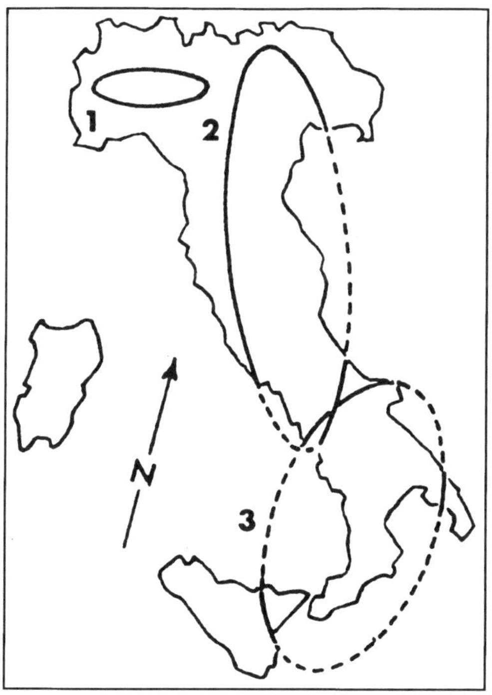

Figure 4.13. Italy. Income per capita, 1961: (1) highest decile, (2) middle decile, (3) lowest decile (Source: Bachi, 1981).

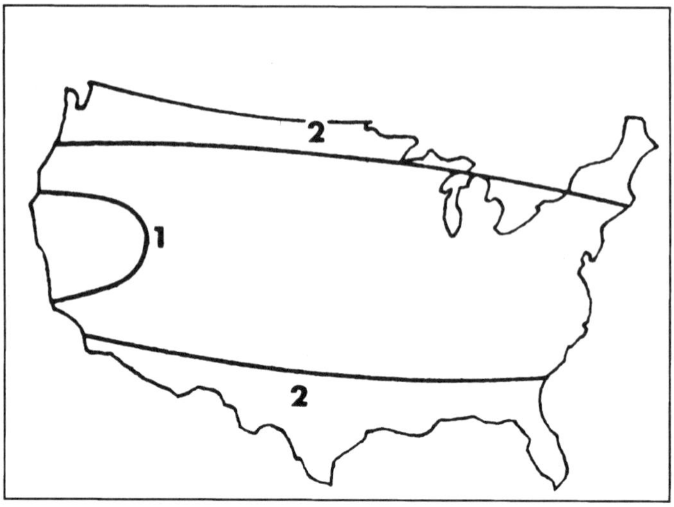

Figure 4.14. Chinese population in the United States: (1) 1870, (2) 1970 (Source: Bachi, 1981).

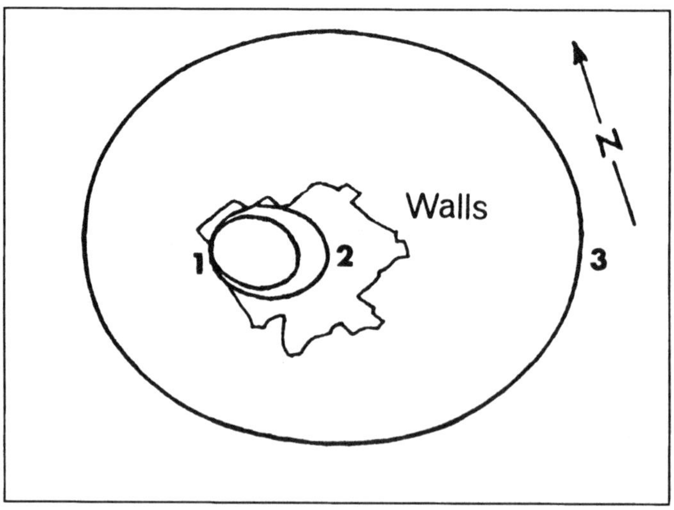

Figure 4.15. Population of Rome: (1) 1526, (2) 1870, (3) 1961 (Source: Bachi, 1981).

4.5. The Geoset of a Simply Connected Territory Defined without Reference to a Population

4.5.1. A Geoset Formed by an Infinite Number of Infinitesimal Territorial Elements

In Figure 4.16 the territory T of the Old City of Jerusalem, first shown in Figure 3.2 (Section 3.1) for purposes of orientation, is magnified and is itself the object of our study, without regard to the populations it contains (such as inhabitants, hotels, and travel agencies). This territory is determined by the boundary line B, which represents the walls of the Old City, and can be compared historically to the territories enclosed by the walls built by King David as well as other walls built around the city during its 3000 year history. Characteristics such as compactness, oblongity, and asymmetry in each epoch can be compared to those of walled cities in other countries dating from different eras. Any simply connected territory enclosed within a determined boundary line B can be conceived of as being formed by an infinite number of infinitesimal elements of area $dxdy$ and coordinates x, y. Such a territory is denoted by T^∞. In Figure 4.16 we show a finite number of squared elements.

For the area A^T of T^∞ we have $A^T \iint_T dxdy$. The principal geostatistical parameters of T are as follows:

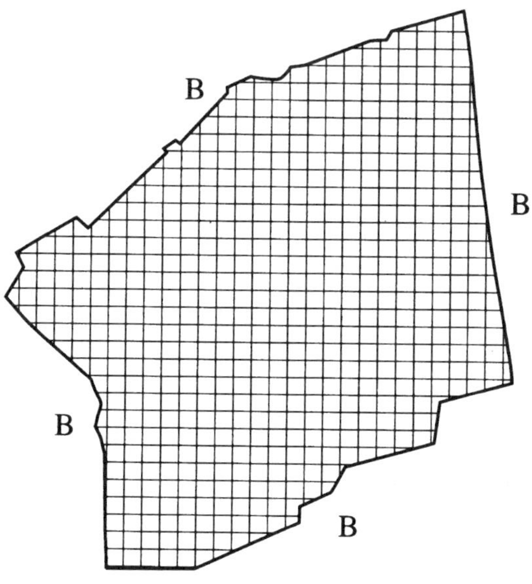

Figure 4.16. Jerusalem. The walls of the Old City (built by Suleiman the Magnificent 1537–1541).

COORDINATES OF THE CENTER:

$$\bar{X}_T = (1/A) \iint_T x dx dy$$
$$\bar{Y}_T = (1/A) \iint_T y dx dy \tag{4.2}$$

VARIANCES:

$$\sigma^2_{X,T} = (1/A) \iint_T x^2 dx dy - \bar{X}^2_T$$
$$\sigma^2_{Y,T} = (1/A) \iint_T y^2 dx dy - \bar{Y}^2_T \tag{4.3}$$

COVARIANCE:

$$\operatorname{cov}(X, Y) = (1/A) \iint_T xy dx dy - \bar{X}_T \bar{Y}_T \tag{4.4}$$

4.5.2. Finite Territorial Geoset

In order to utilize the equations (4.2), (4.3), and (4.4), we adopt the following approach (Bachi, 1973). We choose a series of points $\ldots i, i + 1, \ldots$, with coordinates $\ldots (X_i, Y_i)(X_{i+1}, Y_{i+1}), \ldots$ on the boundary B (equally or unequally spaced) going completely around the perimeter, using only enough points to trace the shape accurately (the larger the number of points the more cumbersome and expensive the calculations). We connect each i with $(i + 1)$ by a segment. Territory T^∞ is thus transformed into a polygon T, with the following parameters:

AREA:

$$A = \frac{\Sigma(X_{i+1} - X_i)(Y_{i+1} - Y_i)}{2} \tag{4.5}$$

COORDINATES OF ITS CENTER:

$$\bar{X}_T = \frac{\Sigma(X_i Y_{i+1} - X_{i+1} Y_i)(X_i + X_{i+1})}{3\Sigma(X_i Y_{i+1} - X_{i+1} Y_i)} \tag{4.6}$$

$$\bar{Y}_T = \frac{\Sigma(X_{i+1} Y_i - X_i Y_{i+1})(Y_i + Y_{i+1})}{3\Sigma(X_{i+1} Y_i - X_i Y_{i+1})} \tag{4.7}$$

VARIANCES:

$$\sigma^2_{X,T} = \frac{\Sigma(X_i Y_{i+1} - X_{i+1} Y_i)(X_i^2 + X_{i+1}^2 + X_i X_{i+1})}{6\Sigma(X_i Y_{i+1} - X_{i+1} Y_i)} - \bar{X}^2_T \tag{4.8}$$

$$\sigma_{Y,T}^2 = \frac{\Sigma(X_{i+1}Y_i - X_iY_{i+1})(Y_i^2 + Y_{i+1}^2 + Y_iY_{i+1})}{6\Sigma(X_{i+1}Y_i - X_iY_{i+1})} - \bar{Y}_T^2 \tag{4.9}$$

COVARIANCE:

$$\mathrm{COV}(X, Y) = \frac{\Sigma(X_iY_{i+1} - X_{i+1}Y_i)2[(X_iY_i + X_{i+1}Y_{i+1}) + X_iY_{i+1} + X_{i+1}Y_i]}{12\Sigma(X_iY_{i+1} - X_{i+1}Y_i)} - \bar{X}_T\bar{Y}_T \tag{4.10}$$

These equations are approximations of the true values of the parameters, and by choosing a larger number of points so as to trace the boundary line of B more exactly one can improve on these approximations.

4.6. Geosets of Compound Territories

4.6.1. A Territory Formed by Separate Fragments

Up to this point, we have dealt mostly with simply connected territories, both in our consideration of uniform distributions over territories (Section 3.2) and in our analysis of territorial geosets ${}^gT$. However, we often have to deal with other types of territories and these will be discussed below.

Let us consider a territory T, formed by separated fragments T^j ($j = 1, 2, \ldots,$), each of which is simply connected. As an example, let us look at Figure 4.17, which shows the territory of Italy as formed by major elements: 1 = Continental Italy; 2 = Sicily; 3 = Sardinia.

If we wish to determine the uniform distribution over T, we proceed as follows: (a) we calculate the proportion v_j^T of the area of T^j over the area of T; (b) having determined the number N^c of cells into which we wish to divide T, we attribute to each element T^j a number of cells N^j equal to the integer nearest to $N^c v_j^T$; (c) a special feature of the computer program enables us to distribute N^j cells over each T. The geoset of the uniform distribution over T is then obtained as a compound geoset: ${}^gU = \Sigma v_j^T {}^gU^j$, where ${}^gU^j$ is the geoset of the uniform distribution over the element T^j. The same approach can be used in the study of a territory consisting of separate parts by the method described in Section 4.5.2. If we denote the geoset of element T^j by ${}^gT^j$, then

$${}^gT = \sum v_j^T {}^gT \tag{4.11}$$

4.6.2. Geosets of Subterritories

Let us consider a territory T that is subdivided into simply connected patches, with each path being characterized by qualitative or quantitative features such as geomorphological, geological, altitudinal, meteorological, economic, social, and/or type of land. A set of patches with identical characteristics can be considered as a fragmented territory, which immediately allows us to extend the methods suggested in Section 4.6.1.

4.6.3. Territories with Internal Gaps

Equation (4.11) can be interpreted as showing the geoset ${}^{g}T$ of a compound territory formed by "positive" components. However, we can also use the approach that was introduced in Section 3.8.3 for compound populations for a compound territory, namely the possibility that some of the components are positive and others negative. We might deal, for instance, with a territory formed by separate elements (positive) that have negative components within them, e.g., a country that has lakes or other inland bodies of water or tracts of land such as glaciers and deserts that are irrelevant for certain kinds of studies. In cities there may be public parks or other areas that are not suitable either for housing or for commercial enterprise. From a formal point of view, we can transform equation (4.11) into

$$ {}^{g}T = \sum \varepsilon_j v_j^{T\,g} T^j \tag{4.12} $$

where ε_j can be either positive or negative.

4.6.4. Territories with External Concavities

Figure 4.17 gives examples of the types of territorial components that are sometimes of interest for geostatistical analysis. The boundaries marked CH^1, CH^2, and CH^3, are, respectively, the convex hulls drawn around continental Italy, Sicily, and Sardinia. We denote the territory enclosed by the hulls as TCH^1, TCH^2, and TCH^3. The territories included between TCH^1 and T^1, TCH^2 and T^2, and TCH^3 and T^3 can be thought of as the territories of the external concavities around each of the three subterritories considered with geosets ${}^{g}S(TCH^j, T^j)$. The analysis of their geostatistical parameters is sometimes of considerable interest (see Section 11.5.4 for the example given in Figure 4.17). We also might take into account "interelemental gap," which is obtained by subtracting the convex hulls around the component subterritories from the territory of the convex hull around the compound territory.

4.7. Operations on Territorial Geosets

1. *Compound territory formed by disjoint components:* Turning again to the example of the territory of Italy, we can say that ${}^{g}T^1$, ${}^{g}T^2$, and ${}^{g}T^3$ do not overlap but form the compound ${}^{g}T$ in a mutually exclusive and comprehensive way.

2. *Compound territory formed by overlapping components:* Considering Figure 4.17, we see that all the elements of T^j match elements of TCH^j, and we can write ${}^{g}M(T^j, TCH^j) = {}^{g}T^j$. The geoset of concavities around a territory T^j can be considered as the geoset of surpluses of TCH^j over T^j. We thus introduce the possibility of comparing two territories that have different areas and partially overlap. For one more example of this operation, let us look at Figure 4.18. Here boundary $B1$ delineates Suleiman's walls of Jerusalem's Old City. The corresponding territory TC is indicated by horizontal lines. Boundary $B2$ outlines Herod's wall and the corresponding territory TH is shown by vertical lines. The areas marked by vertical or horizontal lines or by hatching represent the territory of the union of TC

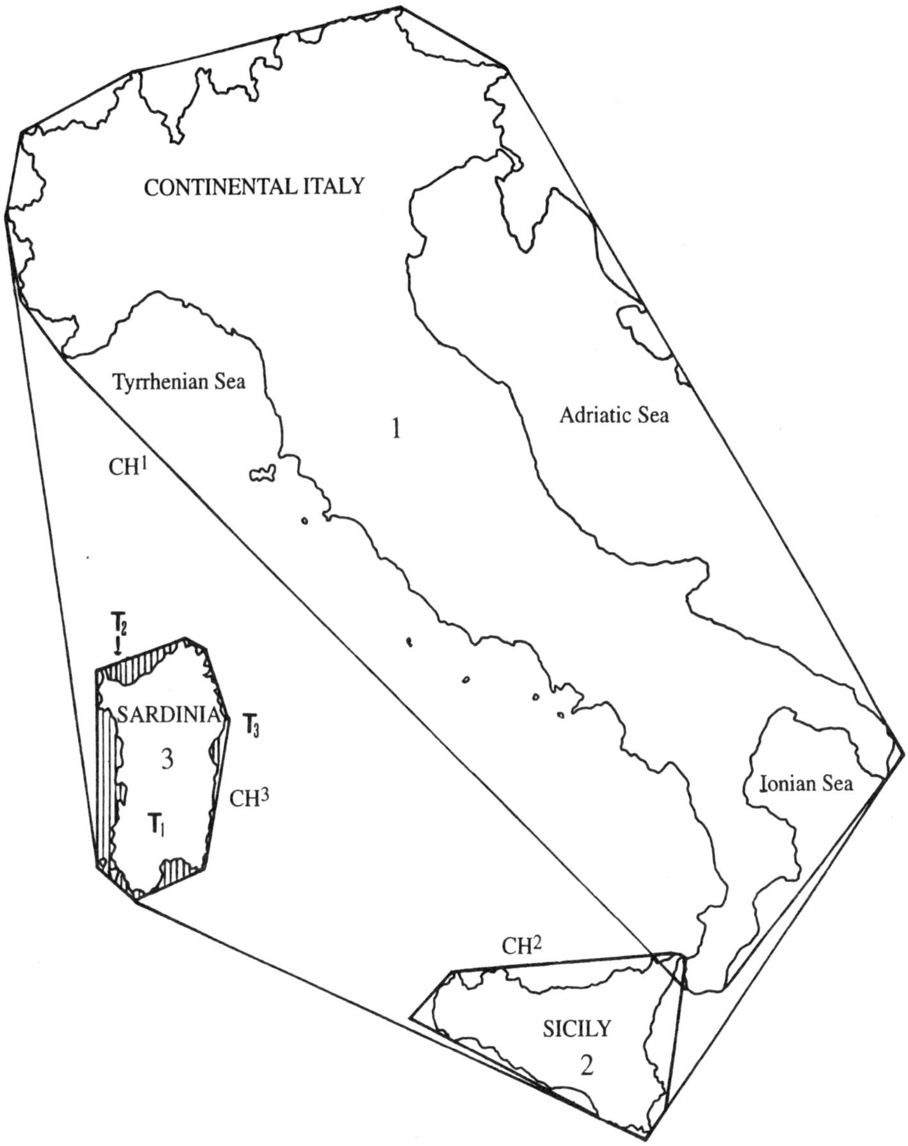

Figure 4.17. Convex hulls around all the fragments of Italian territory (1) Continental Italy (T^1), (2) Sicily (T^2), (3) Sardinia (T^3) (Source: *Statistical Atlas of Italy*, 1988).

and TH; the hatched area is the territory of the match between TC and TH; the territory covered only by horizontal lines can be thought of as being formed by surpluses of TC over TH; and that covered by vertical lines as being formed by the surpluses of TH over TC. Territorial geosets ${}^gM(TC, TH)$, ${}^gS(TC, TH)$, and ${}^gS(TH, TC)$ can be constructed accordingly.

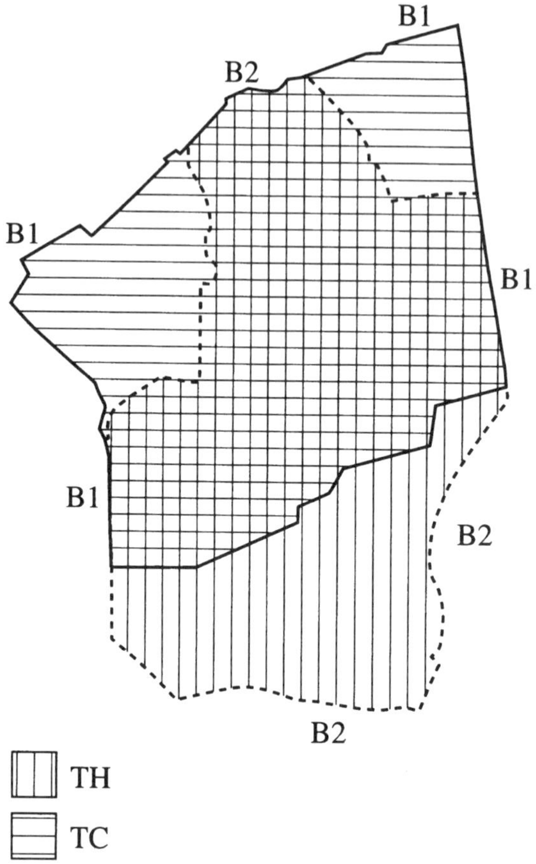

Figure 4.18. Jerusalem. The walls of the Old City dating from two different periods (B1) Suleiman's walls (1537–1541), (B2) Herod's walls (37–4 BCE).

Let us compare the presentation in Figure 4.18 with that of a common Venn diagram such as was shown earlier in Figure 3.1 (Section 3.1). If we assume that in the Venn diagram, circle A encloses territorial elements of TC and circle B those of TH, then both figures would indicate that TC and TH partially overlap. However, the Venn diagram is only symbolic, while maps such as that of Figure 4.18 have a precise geographical meaning, in this case, identifies ${}^{g}TC$, ${}^{g}TH$, ${}^{g}M(TC, TH)$, ${}^{g}S(TC, TH)$, and ${}^{g}S(TH, TC)$ and enables us to calculate the corresponding geostatistical parameters.

3. *Equalizing the number of territorial elements of two geosets:* We saw in Chapter 3 that comparisons between two population geosets ${}^{g}P$ and ${}^{g}U$ or ${}^{g}P$ and ${}^{g}Q$ can sometimes be simplified by assuming that both are formed by an equal integer number of elements. This approach can be extended to territorial geosets. Suppose, for example, that we calculate the minimal distance between ${}^{g}TC$ and ${}^{g}TH$. We denote the respective areas by $A(TC)$ and $A(TH)$ and select any convenient arbitrary integer M. We can assume that M territorial elements are uniformly distributed over $A(TC)$ and M elements are uniformly distributed over $A(TH)$. We calculate the minimal distances by pairing minimally each element of TC with one element of TH.

4. *Territorial geoset of union versus geoset of compound:* To clarify these concepts, we consider a simple example. In a study on the cultivation of a particular vegetable, if we assume that we know the geoset ${}^gW$ of areas of winter crops and the geoset ${}^gZ$ of areas of summer crops, we can determine ${}^gS(W, Z)$, ${}^gS(Z, W)$, and ${}^gM(W, Z)$. If we are interested in studying the territorial distribution of all the cultivated areas irrespective of season, we look at the compound geoset formed by ${}^gS(W, Z)$, ${}^gS(Z, W)$, and twice ${}^gM(W, Z)$, while for the distribution of areas cultivated either once or twice, we consider the union formed by ${}^gS(W, Z)$, ${}^gS(Z, W)$, and ${}^gM(W, Z)$.

4.8. Types of Scales Used to Determine Geosets of Populations and Territories

In the social sciences, a distinction is often made among four types of scales of measurement, which are indicated hierarchically as (1) nominal, (2) ordinal, (3) interval, and (4) ratio. Each type of scale enables us to calculate and use parameters specific to itself in addition to those from hierarchically lower scales. Geosets define locations (X_i, Y_i) of elements i along a special type of two-dimensional interval scale. These scales enable us to calculate parameters of location, dispersion, and shape, as well as those parameters based on such factors as distances, which may provide more information than can be obtained when a territory is treated as a set of regions characterized by nominal or ordinal scales. On the other hand, like zone-dimensional interval scales (used to measure variables such as temperature), location averages, as such, do not lead to the measurement of magnitudes, owing to the fact that territorial locations do not have a true origin (as found in ratio scales).

A distinct advantage of the geosets is that we can link the population geoset to the geoset of the territory over which it is distributed. This makes it possible to introduce refined and/or alternative measurements of the area corresponding to the population (e.g., gross area, net area, and surplus area). Comparison of dispersion measures with areas enables us to express the compactness of the distribution. Other dimensionless measurements are obtained by comparing parameters of population with the analogous parameters of its territory.

Part II
Graphical Presentation of Geographical–Statistical Data

Chapter 5

Statistical Mapping

5.1. Introduction

Previous chapters have included many examples of maps using graphical rational patterns, (GRP), and representing various types of geosets, and from these we have been able to see how GRP maps can help in overcoming the difficulties in presenting long series of data that are detailed by many localities or by irregular territorial units. We found the integration of the geographical approach and of the study of geostatistical parameters to be of value, *inter alia*, for exploratory purposes, for a first orientation in research, for more in-depth analysis, and for comparing geosets and formulating hypotheses concerning their determinants.

We shall proceed in Part II to deal more systematically with the mapping of geostatistical data in general and with graphical presentation by means of GRP in particular. In order to do this, we start with some broad considerations on the state-of-the-art of statistical graphics and mapping and on the evolution of these methods. In our systematic discussion of methods for mapping geographical–statistical data we shall refer mainly to the various types of geosets that were introduced earlier, including, among others:

1. Locations of individual elements of small "populations," (a) unweighted or (b) weighted; the term "population" is to be interpreted in the broad sense as indicated in Section 2.2.
2. Distribution of aggregate locations for large populations; these may be presented as (a) absolute frequencies over territorial units or cells, (b) relative frequencies referred to the total population, and (c) densities per unit area.
3. Systematic comparison of actual, and uniform distributions (${}^gP$ and ${}^gU$) by means of geosubsets derived from them.
4. Simultaneous representation of two geosets ${}^gP$ and ${}^gQ$.
5. Comparison between ${}^gP$ and ${}^gQ$ by means of geosets derived from them.
6. Comparisons between two geosets ${}^gP$ and ${}^gQ$ by means of maps of rates and ratios obtained from the size of P and Q in each cell or territorial unit (such as average income per capita or birthrate).
7. Other types of rates and ratios.

8. A compound geoset and its various components; if these components are distinguished by (a) qualitative attributes (such as marital status or ethnicity), (b) rankings of a variable, (c) different levels of a variable (such as income or age), we obtain a composite presentation of (a) qualitative graphical data, (b) ordered geographical data, and (c) quantitative geographical data. In an analogous way we can present geographical data arranged according to time.

In the foregoing types of presentations geosets are compared through sums, differences, or ratios. We shall also relate to comparisons made by means of the product of standard scores, which underlie the study of ecological correlation.

There are several other types of geographical–statistical data that can also be presented by means of maps, but we could not overburden this book with too much detail. Accordingly, we have only included some brief remarks concerning other important types of geographical–statistical mapping, such as the representation of data on movements (e.g., migrations, travel, shipping) classified by origin and destination or shown along traffic lines or networks.

In its simplest form, a statistical map showing the distribution of a population P over a territory T is an outline of T over which symbols represent the locations of individual elements of P or aggregates of locations in each unit of T.

Territory T is generally not an undifferentiated tract. In the words of Louis Guttman, "if the earth were uniform—well polished, like a billiard ball—there probably would not be any science such as geography."

Much of the interest in the study of a map of population P lies in the possibility of comparing it with maps of other populations or subpopulations over T and/or of "seeing" the characteristics of the distribution of interest over the background of the differential characteristics of various parts of the territory. If the readers of the map of P are well acquainted with these characteristics of T, the map can "speak" to them. Obviously, "acquaintance" with T is often found especially among people living in T. For example, empirical studies have shown that the maps reproduced in this volume from the *Statistical Atlas of Italy* are interpreted rather easily by Italian readers.

However, individuals unacquainted with the physical, political, economic, and social conditions in Italy may encounter difficulty in interpreting the clustering of some phenomena in certain parts of Italy and their partial or complete absence from other parts of the country. This raises the question of how we can convey basic information on the differential characteristics of the various parts of T in addition to the information provided by the map about the distribution of P. This discussion is outside the scope of this book. We shall note only that a simple and very handy method has been found to be well adapted to volumes such as the *Atlas* that include maps of the distribution of many phenomena over the same territory T (such as a country or a city). This method involves adding transparent overlays, in the same format as the maps found in the volume, which give relevant information on the various parts of T. For example, for the study of sets of maps concerning the population of a country, the additional territorial maps may convey indications of, e.g., geographical coordinates, administrative divisions, the location of principal cities, geographical characteristics, divisions by alternative levels, meteorological or climatic information

Plate 1, for example, which was used to produce overlays for the *Atlas*, shows mountains, hills, plains, and principal cities of Italy as well as other geographical indicators that can help in interpreting the maps we gave in Chapter 1 of this book. In the study of sets of maps concerning a city, additional territorial maps might indicate important sites, main streets, administrative and/or historical divisions, distribution of green spaces, and other uninhabited zones, and information concerning principal land use. The degree of detail required depends on the kind of user to whom the information is being addressed and is a function of the scale of the map.

5.2. Revolution in Statistical Graphics and Mapping

5.2.1. Historical Highlights

The history of statistical graphics (including statistical mapping) was explored by Funkhouser as early as 1937 and has been reviewed many times in recent decades (Beniger and Robyn, 1978; Dickinson, 1963; Fitzpatrick, 1962; Haggett, 1966; Bertin, 1973; Feinberg, 1979; Izenman, 1980; Wainer and Thissen, 1981; Feinberg and Franklin, 1975; Tufte, 1983; and Costigan Eaves, 1984).

On the basis of these and other materials, we can highlight some general aspects of the evolution of statistical graphics in general and of statistical mapping in particular:

1. Without going back to attempts in the ancient world, we can date data mapping and the tracing of isolines back to the 18th century. The diagrammatic method developed in analytical geometry at the same time and was later applied to statistics.

2. The 19th century saw the use of various types of cartograms, data curves, bar graphs, histograms, grids, and coordinates, as well as attempts to use stereograms and correlation graphs. The period between 1860 and 1900 was characterized by Funkhouser (1937) as an "age of enthusiasm" for data graphics. Statistical graphs found their way into the presentations of official statistics. Problems of statistical graphics (including the possibility of introducing standards) were discussed at important conferences such as the Third International Statistical Congress of 1857, the Paris Universal Exposition of 1879, and the Jubilee Celebration of the Statistical Society in London in 1885.

3. The interest of professional statisticians in graphical presentation of data continued to increase throughout the first decades of the 20th century and it was in this period that logarithmic diagrams were largely developed. Chapters on graphical methods appeared in many statistical textbooks (Chaddock, 1935; Croxton and Cowden, 1939; Crum *et al.*, 1938; Jerome, 1924; Kelly, 1924; and Secrist, 1920), and several monographs appeared as well (Karsten, 1923; Mudget, 1930; Riggleman, 1926; Rugg, 1925). Some of these books made it clear that errors were often made in the preparation of statistical graphs and that there was a real need for standards.

4. Between the early 1930s and the early 1960s, it appeared that most professional statisticians had lost interest in statistical graphics. Possibly, the focus on inferential problems and on mathematical statistics in those years was responsible for the neglect of methods of data collection and presentation. Besides, statistical graphics seemed to present mainly technical problems, which were unfashionable and of little interest to theory-oriented statisticians. Some introductory textbooks on statistical graphics did appear (e.g., Arkin and Colton, 1936; Brinton, 1939; Pedroni, 1953; and Spear, 1952), but it is probably symptomatic that an initiative by Gini during this period to activate a graphics committee on behalf of the International Statistical Institute did not succeed (Bachi, 1955).

5. Since the early 1960s, there have been dramatic changes in the field of statistical graphics: (a) Progress in automation has led to a revolution in the production of statistical graphics and in cartography (see Section 5.2.2). (b) Important technical developments have occurred in statistical graphics and related fields (see Section 5.2.3). (c) There has been renewed (though still limited) interest in the field on the part of professional statisticians.

The present state of the art is very unbalanced: enormous technical achievements and still greater possibilities for the future fly in the face of insufficient fundamental scientific knowledge and background concerning generally accepted principles of graphical presentation and a lack of proper training of most of the personnel engaged in constructing maps and other statistical graphs. As a consequence, from a scientific viewpoint, a large proportion of the graphs currently being produced are of poor quality and may engender misinterpretation and even misuse (see Section 5.3).

5.2.2. Automation

Automation brought about a revolution in the graphical statistical field and there has been continuous improvement in both hardware and software over recent years. Facilities for many graphical devices, including color mobility and transformations, are commonly found in today's computers, and software packages dedicated to graphical methods, general cartography, geographic information systems (GIS), and statistical mapping are now available. With the continuously decreasing costs and popularization of all these tools, the technical difficulties that hinder the production of statistical graphs and require the skill of specialized draftsmen are gradually disappearing.

These developments in statistical graphics, in general, and statistical mapping, in particular, have many consequences and implications:

1. Graphs can be mass produced today at tremendous speed and at low cost.
2. Automated graphs can attain levels of accuracy and attractiveness exceeding those of handmade graphs.
3. Both the effectiveness and the appearance of automated graphs can be enhanced with color.
4. At each stage of automated production, the operator, by following the

graph on the screen, can substitute and correct its various elements interactively and so experiment quickly and efficiently with alternative graphic solutions.

5. Graphs on the screen that are selected for production can be microfilmed or be reproduced on paper in various ways, e.g., pen plotters, dot matrix printing, laser printing/plotters, and electrostatic plotters.
6. Graphing relevant information can be made even faster, more affordable and more accurate by creating the graph directly from databases in which information related to individual elements or aggregates of the population studied are stored.
7. Many software packages are designed for use by nonexperts and require little more than elementary keyboard skills and a knowledge of computer filing systems.
8. Software packages offer specific menus, making it possible to prepare statistical maps in various ways: e.g., (a) digitized boundaries of territorial units can be shown; and if these are standard political or administrative circumscriptions of the country or city to which the data refer, it is often possible to find those boundaries on computerized media; (b) the area of each territorial unit can be filled with monochrome patterns or with color; (c) alternatively, vector graphics can be used for hatching with lines of different pattern, spacing, or color; (d) vector graphics can also be used to position monochrome or colored symbols such as circles, pie charts, framed or unframed rectangles, divided bars, and line graphs in the center of the territorial units; (e) dots representing frequencies can be distributed over the map; (f) projections of stereograms can be shown and manipulated in different ways.
9. Production of tools for automated statistical graphics and mapping has become a large industry.
10. Apart from offering new graphical facilities, automation has led to an enormous increase in the output of statistical data produced in statistical agencies, government, business, and public institutions, as well as for research, and enhanced the potential for the preparation of such data through growing computer storage capabilities and the further development of databases. Thus statisticians and other users of databases are being increasingly called upon to master large quantities of data quickly, which is an incentive for translating them into automated graphs to select information for use.
11. It seems likely that these developments will lead not only to a vast quantitative increase in graph production, but also to shifts in emphasis with regard to the aims of production.
12. Automation has led to enormous progress in fields such as image processing, pattern recognition and classification, scene analysis, and presentation of images through fractals. While these fields may have some indirect relationships with statistical mapping, a discussion of these links is beyond the scope of this volume, as is the case with the burgeoning area of GIS and related technologies.

5.2.3. Recent Scientific Developments

As we noted earlier, the decades between 1930 and 1960 were a period of stagnation in statistical graphics, but in more recent years new methods have been developed. For example, J. W. Tukey (1983) and Chambers *et al.* (1983) among others at the Bell Laboratories have developed stem and leaf diagrams, box plots, rotograms, quantile-quantile plots, and enhanced scatter plots, and improvements on presentation of data through cumulative sequence polygons or correlation charts and have been used extensively in exploratory research and data analysis. Multivariate analysis has fostered the creation of graphical methods for presentation of basic data and results of analysis, e.g., Chernoff's faces (1973, 1978, 1979); Gabriel's biplots (1971, 1978); and Andrews' glyphs and metroglyphs (1972). However, as these all have little or no bearing on statistical mapping, we will not comment upon them further.

Among new methods applicable to the construction of statistical maps developed since 1960 are:

1. The graphical rational patterns (GRP) (Bachi, 1968), which are discussed in Chapters 6 and 7;
2. The scale of circles proposed by Bertin (1967, 1981, 1983), which we describe in Section 6.7.

There have also been important scientific developments during this period in fields related to statistical graphs such as:

1. *Graphical Semiology*, developed by Bertin, which is research technique dealing with aims, methods, and the nature of processing graphical information, communication, and perception of these statistical data. One of his primary goals is statistical mapping and he has done comparative studies on alternative ways of presenting given sets of data, noting the pitfalls of the common methods. His principal approach for building statistical maps is a scale of distinguishable circles uniformly distributed over the territory (see Section 6.7).
2. Research on graphical perception has long been conducted by psychologists and psychophysicists. There were some applications of statistical graphs to specific problems as early as the 1920s (see, e.g., Schmid, 1983, p. 10). However, in the past two decades or so there has been extensive development in the perception or visual decoding of the information encoded by symbols and color in statistical graphs (see, e.g., Cleveland, 1984, 1985, 1993; Cleveland and McGill, 1984; Wainer, 1980, 1984; Wainer and Thissen, 1981).

This work has helped statisticians to understand, for example, the ways in which the human eye and brain are able to appreciate the quantitative and qualitative information that graphical symbols are purported to convey; the speed at which these operations are performed; systematic errors in perception and their causes; and optical illusions and disturbances created by certain types of symbols and graphs. It might also be of assistance in the selection of symbols to represent given sets of data. From among the findings of Cleveland (1985, p. 254), we can point out that elementary graphical perception tasks (visual decoding) are performed as follows: the best is (a) position along a common scale, followed by (b) position along identical, nonaligned scale, (c) length, (d) angle-slope, (e) area, (f) volume, (g) color hue–color saturation density.

The recent change in attitude on the part of professional statisticians in regard to graphical problems has also manifested itself in several organized fields, e.g., (a) the development of graphic research at the U.S. Bureau of the Census during the period in which it was chaired by Barabba (Barabba and Finkner, 1978, pp. 7–13); (b) the renewed interest of the American Statistical Association in the graphical field, owing partly to some earlier initiatives, which brought about the creation of a Committee in Graphical Statistics and later led to the establishment of the Graphical Section of the ASA; (c) some renewal of interest by the International Statistical Institute culminated in the meeting at Warsaw (see ISI Bulletin, 1975 and Bachi, 1975), where there was also an international exhibition of statistical graphs. Statistical graphics were also the subject of a debate at the ISI session of Tokyo in 1987. It may be noted that some of the scientific research and innovations in graphical symbols during this period were due to professional statisticians.

5.3. The Present State of the Art

Since the earliest appearance of scientific literature on statistical graphics, almost a century ago, authors have lamented frequent errors and misuses and abuses of the maps and other graphs produced. "Amateur" graph makers were often found to ignore or neglect basic principles of graph construction. Some of the literature denounces ways that are adopted in order to "cheat," by means of statistical graphs. A survey of recent studies on statistical graphics does not present a rosier picture of what serious students of this subject think of the level of current statistical–graphical output:

1. C. F. Schmid (1978, p. 76; pp. 1, 111–176, 189), on the basis of the analysis of large numbers of published graphs, reached the conclusion that there is "superabundance of evidence in the form of poorly designed, misleading, and/or dilettantish charts in newspapers, magazines and even learned books and journals.... From observations over the years, one is compelled to conclude that a significantly large proportion of published charts are prepared by persons with little or no knowledge of standards, to say nothing of other aspects of graphic presentation." With regard to statistical maps, his detailed survey shows an impressive list of frequent errors.
2. The book by Tufte (1983) includes a very large number of examples of deceptive graphs in the chapters on graphical integrity and graphical practice. Among Tufte's general conclusions are that, "lying graphics cheapen the graphical art everywhere. Since the lies often show up in news reports,

millions of images are printed." (p.76) "Athough bias and stereotyping are the origin of more than a few graphical distortions, the primary causes of inept graphical work are to be found in the skills, attitudes, and organizational structure prevailing among those who design and edit statistical graphics." (p. 79) "The conditions under which many data graphics are produced—the lack of substantive and quantitative skills of the illustrators, dislike of quantitative evidence,...—guarantee graphic mediocrity. These conditions engender graphics that (1) lie; (2) employ only the simplest designs, often unstandardized..." (p. 87) "Graphical competence demands three quite different skills: the substantive, statistical, and artistic. Yet now most graphical work... is under the direction of but a single expertise–the artistic. Allowing artist-illustrators to control the design and content of statistical graphics is almost like allowing typographers to control the content, style, and editing of prose. Substantive and quantitative expertise must also participate in the design of data graphics, at least if statistical integrity and graphical sophistication are to be achieved." (p. 87).

3. Wainer (1980) studied graph usage in the press and found widespread misrepresentation of data and egregious errors.
4. Cleveland (1984) studied graphs in scientific journals. In one volume of *Science* he found that 30% of the graphs contained errors.
5. Our own research (Bachi, 1978*b*, *c*) with regard to statistical maps prepared in the past in some first-class statistical agencies revealed considerable neglect of important principles,
6. Bertin's book *Semiologic Graphique* (1967) is also full of examples of graphs and in particular of maps that are "bad" owing to methodological errors, moiré ("vibration") of the graphical symbols used, and optical illusions.

The present state-of-the-art of statistical graphics in general, and even more of statistical mapping, thus appears to be paradoxical. On the one hand, automation has resulted in the explosion of output and the enormous technical progress we described in Section 5.3.2. On the other hand, the quality of a large part of the production is low, judged from a statistical point of view.*

This is a matter of very serious concern, because graphs are becoming more and more important in the dissemination of information, research, and decision-making. Mass production of "bad" graphs may lead to wide misinformation to the public, to wrong scientific conclusions, and to misguidance of decision-makers.

The main reasons for this situation are presumably as follows:

1. With the diffusion of software packages for automated preparation of statistical maps (especially on personal computers), the number of individuals preparing statistical maps is continuously growing. Unfortunately, most of these people have received no training in the principles related to the preparation of

*See, e.g., Kadmon (1983), "The easily available and read-made user-friendly" confectionary graphic package programs enable untrained users to produce maps conveying to the recipient a wrong message... [e.g., through use of] a completely hazard [sic] color scale in a map of a continuous quantitative area variable... [or] gradual circles with radius, instead of area, proportional to a thematic variable."

statistical maps and are not even aware of the need to take such principles into account. Some of the software packages available offer an enormous range of options for different graphical solutions, without any guidance as to the implications of each for the accuracy of the information conveyed. More specifically, users are enabled and induced to employ symbols, hatchings, and colors at will, resulting in fancy presentations, without paying attention to the need to adapt a properly selected method to each type of data or to ensure the reliability of the maps produced as reliable images of the data represented.

2. This chaotic situation sometimes even leads to the paradox that a statistical agency may present, in different publications, different maps representing the same set of data graphed for the same purpose. Owing to the use of different keys for colors, symbols, or scales, the messages conveyed to the public by the different maps graphing the same set of data may be widely divergent. Sometimes differences are so great that most readers would not even acknowledge that the maps represent the same data (Bachi, 1978*c*).

3. The arbitrariness and the lack of adherence to scientific criteria found in many of the statistical maps prepared today are in stark contrast to accepted practice in most areas of statistical work, where there are exacting standards for data analysis and real effort is devoted to ensuring reliability of the findings presented to the public. Possibly this difference in attitude toward responsibility or reliability of graphs is related to the retreat of professional statisticians from the graphical field that we alluded to earlier. Despite the recent partial return of some statisticians to the graphical field, this type of statistical work has not regained the prestige that it once had. Presumably, teaching of principles of statistical graphics in secondary schools and universities is very limited, and in-service training by means of internal graphics manuals or through statistical agencies and research centers is practically nonexistent or at best inadequate.

4. It may be added that despite the efforts of many students of the graphical statistical field to define basic statistical principles in graph making, we are still far from any general, international consensus on such principles, apart from those concerning the Cartesian graphical system and related methods.

5.4. Defining Principles and Setting Standards for Statistical Graphs

Each of the authors who laments the "poor quality" of a large percentage of the graphs currently being produced does so either by referring to some list of principles defining a "good" graph or proceeding on the basis of a high level of "graphicacy" acquired through critical familiarity with the graphical field. (The term graphicacy was coined by Balchin and Coleman (1966) to connote an intellectual skill for expressing relationships that cannot be communicated verbally or mathematically.)

Lists of principles are found in many publications. We shall note here just a few of the more recent ones:

1. C. F. Schmid (1983, pp. 14–15) itemizes properties of good graphs under five major categories: (a) *Accuracy*. "The dimensional and other physical aspects of a good chart should reflect the highest degree of accuracy possible... Furthermore, a chart should not be deceptive, distorted, or misleading, or in any way susceptible to wrong interpretation, as a consequence of inaccuracies or careless construction. Also care should be taken not to create optical illusions..." (b) *Simplicity*. "The basic design of a statistical chart should be simple and straightforward, not loaded with irrelevant, superfluous, or trivial symbols and ornamentation. An important design objective is to maximize the information communicated, within the limits of simplicity and conciseness..." (c) *Clarity*. "...the chart can be easily read and understood..." (d) *Appearance*. "...a good chart is one that is designed and constructed to attract and hold attention by reflecting a neat, dignified, and professional appearance..." (e) *Well-Designed Structure*. "...full recognition that all design elements are interdependent... Visual importance and distinction assigned to various elements of a chart should be commensurate with the intellectual significance of the ideas being presented..."

2. According to Tufte (1983), the principles of graphical excellence are: "well-designed presentation of interesting data—a matter of substance, of statistics, and of design; complexities communicated with clarity, precision, and efficiency; it gives to the viewer the greatest number of ideas in the shortest time with the least ink in the smallest space..." (p. 51). Among Tufte's "principles of graphical integrity," we quote the following: "the representation of numbers, as physically measured on the surface of the graph itself, should be directly proportional to the numerical quantities represented; clear, detailed, and thorough labeling should be used to defeat graphical distortion and ambiguity; show data variation, not design variation" (p. 77); "forgo chart junk, including vibration, the grid, and the duck" (p. 121); maximize data ink (pp. 123–137) and data density (p. 168).

3. Cleveland (1985, pp. 100–101) gives a long list of principles of graph construction, focused mainly on the Cartesian method, among which are: (a) Under the item of "clear vision": make the data stand out; avoid superfluity; use visually prominent graphical elements to show the data; do not allow data labels in the data region to interfere with the quantitative data or to clutter the graph; overlapping plotting symbols must be visually distinguishable. (b) Other principles are listed under the items of "clear understanding" and "scales." (c) As "general strategy" the advice is given that a large amount of quantitative information can be packed into a small region and that graphing data should be an iterative, experimental process.

4. Wainer (1984) suggests rules for showing the data accurately and clearly, by indicating through examples 12 ways, "to display data badly."

Presumably, most people who have studied the problems of statistical graphing would agree with many of the above suggestions and with other principles suggested by other authors. However, in order to have an influence on statistical quality of statistical graphs and especially of maps that are actually produced, all this good advice has to be authoritatively systematized and coded into proper rules, possibly at an international level.

The history of attempts at standardization has been described many times (see, among others, C. F. Schmid, 1978; Costigan-Eaves, 1984). This history has very old roots. Proposals to regulate graphic procedures and to provide rules for the purpose of achieving uniformity and comparability were discussed many times in the

International Statistical Congresses held in Europe from 1853 to 1876 and in many of the Congresses of the International Statistical Institute (and especially in the sessions of 1869, 1872, 1876, 1880, 1901, 1908, 1913, 1953, and 1975); see the Bulletin of the International Statistical Institute on the proceedings of these sessions), but none of these attempts led to the practical establishment of general standards.

Considerable progress in this connection was achieved at local levels. In the United States, a "joint committee on Standards for Graphical Presentation" issued a report containing a set of 17 rules for graphical presentation of time series (in the form of arithmetical or semilogarithmic diagrams), which has had considerable practical influence through its repeated publication in many textbooks. Thus report was expanded in subsequent editions in 1936, 1938, and 1960; Some institutions producing statistical graphs have established internal standards for such production.

We are still very far from having a systematic formulation of internationally accepted rules for the graphical presentation of the many types of statistical data available and from achieving many of the aims of this kind of presentation. There are, however, several graphical methods for which there is substantial agreement on the practical suggestions offered by the various authors who have studied statistical graphics. This is particularly true with regard to the graphical methods that are directly or indirectly based on the Cartesian graphical system (see Section 5.6).

5.5. Establishing Principles for Preparing Geostatistical Maps

5.5.1. Basic Considerations

As our central focus in this book is on territorial distributions, we shall only discuss general problems of statistical mapping, taking the following as points of departure:

1. Suggestions for methods to be followed in making statistical maps have been put forward by many scholars. However, these suggestions have not been systematized into a body of rules based on solid scientific principles.
2. The Cartesian system is generally accepted as the best way to represent nonterritorial statistical data. Before describing specific methods for presenting statistical data on maps, we shall discuss: (a) when it is possible to extend the use of the Cartesian method to the presentation of territorial distributions; (b) the difficulties encountered in this (limited) extension; and (c) the main properties of the Cartesian method. We shall see that while in most cases the Cartesian system is not suitable for territorial distributions, its principles can nonetheless be adapted to formulate rules that seem to hold for statistical maps. This, in turn, facilitates critical evaluation of the usual methods of statistical mapping and paves the way for the introduction of the GRP methods used for most of the maps in this book.
3. We have omitted the following topics on the grounds that they are beyond the scope of this volume. (a) mechanics of preparing statistical charts, such as size and proportions of the chart with particular regard to execution,

reproduction, and the medium in which it is to appear; size, location and type of lettering; spacing of various components including margins; formulation and location of scale designation; labels, legends, and wording and placement of the title. These and other problems of techniques of statistical graphics are treated in many different other publications. (b) consideration of ways that render the graphics "visually attractive" and "elegant" (see, e.g., Schmid and Schmid, 1979; Schmid, 1983; Tufte, 1983, Chapter 9).

Problems connected with aesthetics are very important, especially now with the many facilities available for use of color in automated graphics. Color can render the maps very attractive, but it must be utilized in a way that does not introduce biases in the perception of the data. Plate 6 is an example of gain in attractiveness obtained by judicious use of color without damaging accuracy of presentation. This map provides the same information conveyed in black and white by Figures 7.5 and 7.6.

5.5.2. General Properties of the Cartesian Graphical System

The Cartesian graphical method is based on clearly defined and simple principles (which resulted in analytical geometry). If we consider any series of data Y_i (such as frequencies, amounts, or functions) ordered according to values X_i of a variable X (such as income, temperature, or time), each pair (X_i, Y_i) can generally be represented by a point with coordinates proportional to X_i and Y_i. Similarly, for an array of pairs, individual observations indicating the values of two variables (such as income X_i and expenditure Y_i of families i) for each element of the set, a simple solution is to represent each observation by a point i of coordinates (X_i, Y_i), thus obtaining a "scatter plot."

The system is based on the correct location of each X_i along the axis of the abscissas and of Y_i along the axis of the ordinates and on ensuring correspondence between each (X_i, Y_i) and the point i. Correspondence can be ensured by giving abscissas and ordinates proportional to (X_i, Y_i) or to a proper function of X_i and Y_i.

While keeping these general principles of the Cartesian system fixed, the method is very flexible and can be adapted to graphical presentation of many different types of data, for different purposes, and different users. A few elementary examples are: (a) Time series in which X_i are points in time, and distributions according to discrete values of a variable or linearly ordered qualitative attributes can conveniently be represented by segments proportional to Y_i or to a proper function of Y_i; (b) series in which the abscissas correspond to initial and final points of a class or as the central point of a class can be represented by histograms or by frequency polygrams; (c) Correlation tables between two variables of two sets of linearly ordered attributes can be shown by scatter plots.

If properly employed, the Cartesian method makes many important visual perceptions possible: (a) It gives a quick general overview of the series of data presented; (b) it allows comparison of the ordinates (Y_i, Y_j) corresponding to any two abscissas (X_i, Y_j); (c) whenever scales of abscissas and ordinates are properly given, the interested reader can roughly assess the order of magnitude of values corresponding to any abscissa and ordinate; (d) magnifying or shortening the graph changes the accuracy of (c) but, with certain limits, does not alter the results of (a); (e) comparisons are especially evident for ordinates that follow each other; the magnitudes of differences between such ordinates can often be easily evaluated by the Cartesian diagram, which thus adds information to that furnished by the primary data represented.

With some limitations, it is often possible to show more than one series on the same diagram, which sometimes allows one, to determine visually, the difference between points of ordinates corresponding to the same abscissas in the different series. Further, proper transformations of the data, such as expressing abscissas or ordinates or both as logarithms makes it possible to extend visual operations to products and ratios.

One shortcoming of the usual Cartesian presentation is the lack of a valid criterion for linking the scales of abscissas and ordinates. Proposals for linking the scales of ordinates and abscissas in time-series have been suggested over the years (see, e.g., Marshall, 1885; March, 1911; De Pietri Tonelli, 1949; Pedroni, 1965; Cleveland, 1985). By arbitrarily stretching or compressing these scales, the general appearance of any given time series or frequency distribution can be changed, and the message conveyed to the user altered. This danger has often been noted by statisticians (Huff, 1954). However, there are cases in which this shortcoming can be avoided. We note especially two of them:

1. If scatter plots are used for investigating the relationship between two variables X and Y, the values of X and Y are transformed into standard scores $(X_i - \bar{X})/\sigma_X$, $(Y_i - \bar{Y})/\sigma_Y$. Paired observations (X_i, Y_i) will be represented identically by anybody accepting this convention. Differences between two correlation charts will depend only on differences in the data and not on differences in the method of presentation. This is one of the basic principles proposed by Tufte (1983).
2. If we study shapes of unidimensional statistical distributions (probability density functions, empirical frequency distributions, or histograms), irrespective of their locations and dimensions, the scales of abscissas and ordinates can be standardized by having the same length represent σ_X and σ_Y (see Bachi, 1975, 1978*c*; Bachi and Samuel-Cahn, 1976). This simple transformation is sufficient to ensure that a given distribution is presented graphically in a unique way, which greatly facilitates interpretation of both empirical and theoretical distributions and the comparison between them. This system is connected with a battery of parameters describing various aspects of the shapes of the distributions.

5.6. Using the Cartesian System for Statistical Mapping

5.6.1. The Inherent Difficulty in the Cartesian Method

As the Cartesian graphical system is such a good one it would be desirable to be able to use it for statistical mapping. The bidimensionality of the territory obviously, constitutes a basic difficulty, but nevertheless, certain applications are feasible, and we shall deal here with three of them. It should be noted that this bidimensionality is a second-order generalization of the irregular three-dimensional surface of the earth superimposed on the three-dimensional earth spheroid. For any relatively broad geographical region, such as big countries, continents, and, of course, the globe, cartographic projections must be introduced and used judiciously (see, e.g., Robinson, 1953).

5.6.2. Using Dot Maps to Represent Locations of Individual Elements of a Population Distribution

Suppose that we want a graphical representation of the locations of an array of unweighted individual elements i of a population distribution. If we know for each i the coordinates (X_i, Y_i) of its location, we can show this location as a dot on a map. A simple example of this is given in Figure 3.3 (Section 3.2), which illustrates the distribution of hotels in Jerusalem. Such dot maps, which are based on simple principles, sometimes provide a quick overview of the general characteristics of a distribution as well as indications of areas of very high or high density versus areas of low or very low density. If the population is divided into subpopulations, for additional clarity the elements of the different subpopulations can be shown in different colors.

However, the method has serious limitations. While theoretically we conceive of a dot as a dimensionless point, each individual i is actually represented on the map by a small symbol that covers a finite area. If the number of cases is large and/or if many elements are concentrated in small parts of the distributions we may have the following consequences:

1. Symbols may overlap.
2. Sometimes there are patches of such overlapping symbols in which it becomes impossible to distinguish one symbol from another.
3. Comparison of densities in the various parts of the distributions may become difficult or even impossible. In this instance it can be noted that "density" judgments are on the lowest step of the scale of symbols constructed by Cleveland (see Section 5.2.3).

The difficulty of distinguishing different levels of dot densities was described and illustrated in Chapter 3.

The problems in utilizing dot maps are analogous to those encountered in ordinary scatter plots, which are constructed to study relationships between two variables shown, respectively, on the axes of abscissas and ordinates. Symbols may also overlap in scatter plots, and a comparative study of densities in the different parts of the distribution may be difficult.

The solution to these problems suggested by Cleveland and McGill (1982) is to divide the diagram into cells, count the number of observations found in each cell, and place at the center of each cell a symbol called a sunflower ε (see, e.g., Chambers *et al.*, 1983). In this symbol a dot means one observation, a dot with 2,3... means 2,3... observations, and so on.This approach is based on a principle similar to one that we proposed for the graphical presentation of correlation tables by means of GRP (Bachi, 1968). A sunflower can be constructed only when the number of observations falling in the same cell is small. Sunflowers are based on a different symbol for 1 (given by a point) and other numbers (given by lines), and can be used only for a very limited number of observations. On the other hand, GRP are based on one type of symbol only and allows for counting distinctly 1,2,3...,99,100 (or even more) observations per cell.

Thus, the software package we describe later on in Chapter 17 for graphical presentation of population distributions was built on the following principles:

1. Some overlap of points is eliminated by jittering of the type used by Chambers *et al.* (1983, pp. 106–107).
2. If there are strong concentrations of points that would make the analysis of the distribution difficult, the map is constructed using GRP.

There is thus a conceptual passage from dot map to GRP map: For small populations or populations without significant overlaps, a dot map is used. For large populations or populations with significant overlaps, a GRP map is employed.

5.6.3. Extending the Use of Dot Maps to Represent Aggregates of Locations

Frequently the method described in Section 5.6.2 is extended to represent distributions of large populations for which data are available by regional aggregates (such as countries, provinces, or census tracts). Let us indicate the size of the population in each territorial unit j by F_j, and the total size over the entire territory by $F = \Sigma F$. Generally two conventions are adopted: (a) each point is used to represent an arbitrary number N of elements of the population so that the number of dots to be distributed in region J is approximately F_j/N; (b) these dots are distributed evenly over the entire territory of j.

The maps thus obtained have the advantage of conveying the ideas of territorial distribution of frequencies and of density of the distribution. An example of this method of presentation is shown in Figure 5.1, which illustrates the distribution of sugar beet in France. It is seen that here the cross fits this distribution rather well (for the meaning of the cross, see Section 10.7). However, the maps constructed in this way may present the same difficulties as those we discussed in the last section. Moreover, numbers N are often selected arbitrarily. In many atlases with collections of dot maps, no attempt is made to standardize this number, although it must also be a function of the scale of the map. Consequently, comparisons among different maps can become very difficult even when they are of the same scale. Hence the use of dot maps cannot be considered to be a standard solution for graphical presentation of population distributions.

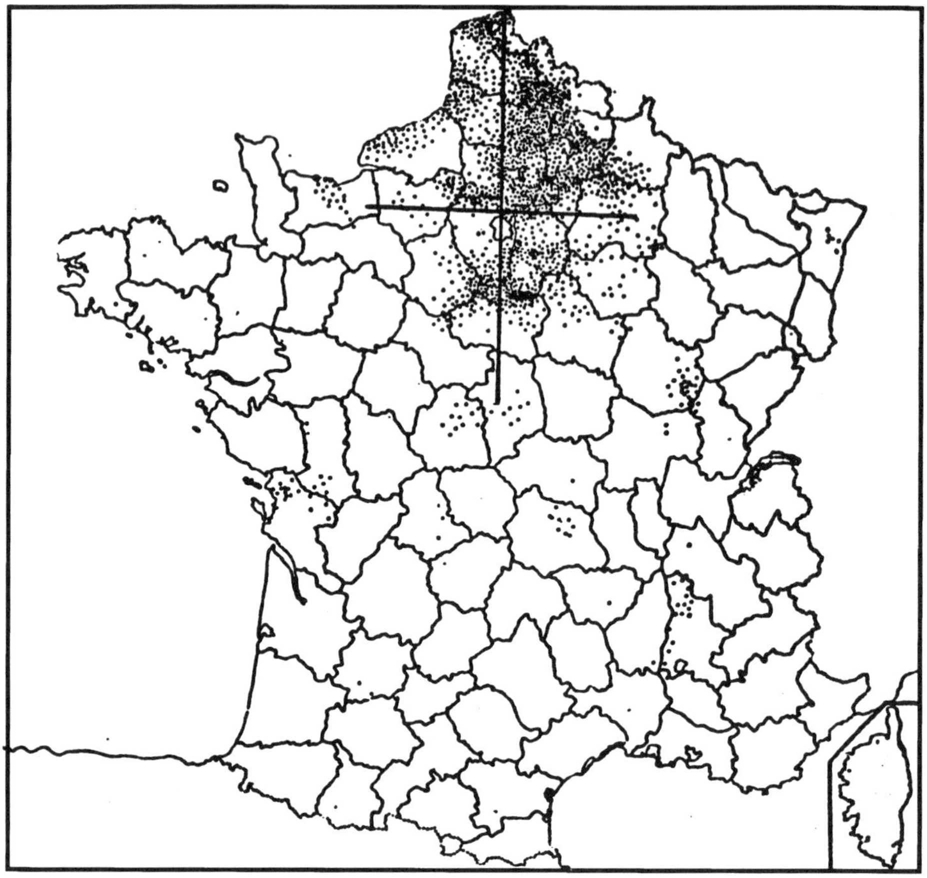

Figure 5.1. France. Industrial sugar beet.

5.6.4. Stereogrammatic Statistical Maps

Another approach to the use of the Cartesian system for statistical mapping is to construct a stereogram over the map and project it on to a plane. The principle upon which this type of presentation is based is as follows: Let us consider any territorial distribution in which a datum V_i corresponds either to a location (X_i, Y_i) of a point i or to a territorial unit (circumscription or cell) i. As two dimensions are required to indicate the distribution of locations i, we can use a third dimension to represent V_i. Values of V_i are supposed to be read along the scale of a third coordinate perpendicular to the scales of X and Y.

If data V_i are referred to territorial circumscriptions or cells i the stereogram can be constructed as a set of prisms having base i and height V_i and can be thought of as a three-dimensional analogue to a histogram. If data V_i refer to discrete points i, the stereogram can show columns proportional to V_i and centered at i, in which case it is analogous to a discrete bar graph.

If the tops or centers of adjacent prisms or columns are connected by lines, tridimensional analogues to frequency polygons can be obtained. If connection is made by smoothing analogues to frequency, curved (3D) surfaces can be obtained. Before automation, the preparation of stereogrammatic maps was not easy.* Sometimes they were actually constructed in three dimensions and then photographed for reproduction on a plane; sometimes they were prepared as plane projections. However, with automation, creating plane projections is no longer difficult.

More recently, sophisticated software packages have included the preparation of stereogrammatic maps, and given any set of data, stereograms representing them can be projected over a plane in any position, displayed on the screen, rotated, reversed, moved in any direction, and colored. Moreover, "dynamic" maps can be generated on the screen by presenting the viewer with a (rapid) succession of images, thus illustrating change over time (see Kadmon, 1975). The same is true of any geographical presentation of statistics.

This flexibility of presentation has advantages in some specific types of analyses, but it does not overcome basic weaknesses in the routine use of stereogrammatic maps for presenting spatial data. We shall now review some of these weaknesses briefly, calling attention to the very thorough analysis made by C. F. Schmid (1983, pp. 152–170) on the basis of many examples:

1. If we keep the set of data V_i, the scales of the location i (X_i, Y_i) and the value of V_i fixed, then solid representing the V_i is uniquely determined by these scales. However its projection on the plane can assume an infinite number of aspects, depending on, e.g., the type of projection, the position of the camera, and the angle at which the projectors strike the plane of the projection. The resulting range of results is bewildering. From practical examples it is seen that a given set of data showing, e.g., the distribution of the population of a country can be presented in forms so different from one another that an unsophisticated reader may have difficulty in even recognizing that they represent the same data. This is in contrast to the rule suggested by Tufte (1983), according to which graphs should show data variation, not design variation.
2. Perception of distances is much more difficult on a map given in perspective form than on ordinary maps.
3. Estimating the values of V_i and comparing values, say, of V_i and V_j corresponding to points or territorial units i and j, may become difficult in stereogrammatic maps and even misleading.
4. High values of V_i on certain parts of the map may hide the values corresponding to other parts of the map.

*The terms "stereogram" and "stereogrammatic map" are somewhat misleading, as both on paper and on the (nearly plane) computer screen they are two-dimensional projections of a three-dimensional body.

5. In consequence of (3) the stereogram may give a distorted overview of the distribution. In consequence of (4) it cannot give correct information concerning details that may be required by the reader. In a true three-dimensional visualization (e.g., in an anaglyph); this objection or drawback can be overcome by interactively changing the viewing point mentioned under (1).
6. Comparison between distributions by means of stereograms is generally very difficult.

5.6.5. Contour Lines

Some of the basic shortcomings of stereogrammatic maps disappear by substituting a map of contour lines for the "mountain" of the stereogram. Representation of contour lines has also been greatly simplified by automation and contour level maps are an excellent research tool whenever we deal with phenomena subject to more or less continuous changes in space. Although there have long been attempts to use contour lines to represent spatial distributions of "populations" this method has not been generally adopted.

To understand the reason for this, we only have to look at the distributions illustrated in this volume, which in many instances show very strong concentrations over very small tracts of territory. The comparative analysis of those concentrations is one of the most important aspects of these distributions. Contour maps would show lines so near to one another that it would be very difficult to evaluate and compare the "heights" of the various concentrations. Further, lines indicating low levels would often emerge as very complex, and it would be difficult for the lay reader to understand what they mean.

However, there is a "true" stereographic method for showing statistics three-dimensionally both on paper and on the computer screen. Known as the anaglyph method it employs the principle of three-dimensional visualization using two stereograms of the same "body" taken from two viewing points separated by a suitable parallax angle and printed (or presented on the screen) in contrasting colors (usually red and blue or red and green). Viewed through blue/red (or green/red) glasses, these are seen as a single truly three-dimensional body. As early as 1971 the Experimental Cartography Unit of the Royal College of Art in London developed an anaglyph method for presenting statistical data on maps. When viewed through red-green glasses, the statistical surface (printed as a green with a red overlay) is seen truly spatially in three dimensions.*

5.6.6. Conclusions

We may conclude that the use of the best graphical–statistical method — the Cartesian system — for the preparation of statistical maps seems to be very limited. Dot maps can be recommended mainly for small populations or samples, and their use for large populations often appears to be of limited utility. Stereograms and their

*_Automatic Cartography and Planning._ Experimental Cartography Unit, Royal College of Art, London, 1971, p. 61 (Anaglyph of ratable land value in London, 1969).

substitutes do not generally constitute a good solution unless viewed interactively on a computer screen.

5.7. Using Non-Cartesian Graphical Methods to Construct Statistical Maps

5.7.1. The Principal Alternative Graphical Systems

Under the circumstances we have just described, statistical maps are generally constructed by means of graphical non-Cartesian systems.

We saw that statistical maps are used to present many types of statistical data for many different purposes. We shall see that this variety requires the use and adaptation of different presentation methods. However, for a preliminary, general discussion, we can roughly distinguish three basic settings according to whether each statistical datum Z_j corresponds (a) to a place (or point) j on the map, (b) to a segment j of a line, or (c) to a territorial unit j, where by "territorial unit" we mean any administrative, political or other circumscription, or a cell of regular geometrical form. In this book we focus primarily on Settings (a) and (c).

Among the symbols used to represent a datum corresponding to a point j [of coordinates (X_j, Y_j)], we shall discuss mainly: simple geometrical figures (such as circles, squares, regular hexagons, and bars), the area of which can vary (in theory) along a continuous scale; and GRP, capable of varying discretely along a very precise scale. Among the symbols used to represent a datum corresponding to a territorial unit, we shall deal with use of repeated simple geometrical figures; repeated GRP; and discontinuous scales of hatchings, shading, or colors used in the preparation of choropleth maps. Examples of applications of these methods are discussed in Sections 6.7 and 6.8, while applications of GRP are described mainly in Chapters 6 and 7.

5.7.2. General Criteria for Statistical Mapping

In Section 5.4 we suggested properties that would be desirable in any system used to construct statistical maps and consideration of these can be of help, in evaluating the various methods described above.

Here we follow the order of presentation of principles as they were given in the evaluation of the Cartesian system in Section 5.5, with some changes. Obviously, these principles must sometimes be modified and made less rigorous, owing to the enormous difficulties encountered in constructing statistical maps.

Let us denote the point or territorial unit to which the datum having a value V_j corresponds by j, and the value (area, density, or intensity) of the symbol used to denote V_j by S_j. For short, we used the word "method," for any technique used to construct statistical maps.

1. The method should be based on clearly defined and simple principles.
2. Each V_j is represented by a graphical symbol(s) falling on the map over point j or territorial unit j.

3. As far as possible, the symbol S_i corresponding to point or region j, should not interfere or overlap with the symbol S_i corresponding to any other point or territorial unit i.
4. The value of symbol S_j should correspond at least roughly to value V_j or to a defined function of V_j.
5. Although the general principles for map-making are inviolate, the method used should enable adaptation to the various types and purposes of graphical presentation (as described in section 5.2).
6. The statistical maps constructed should make it possible to obtain the following visual perceptions: (a) It should give a quick overview of the general characteristics of the distribution of the data represented. (b) It should enable the user to identify, whenever requested, the value of the symbols employed. (c) It should make it possible to evaluate, at least roughly, differences and/or ratios between the values S_i and S_j of any two symbols; this comparison in principle, should not require reference to the key of symbols given in the legend. (d) Magnifying or shortening the graph will change the accuracy of operations (b) and (c) but, within certain limits, should not affect operation (a). (e) If required, there should be the possibility of evaluating the order of magnitude of V_j, represented by the symbol S_j.
7. The perception properties of each method in its different applications should be tested.
8. The graphical method should allow comparison between distributions, which can sometimes be obtained by showing two distributions on the same map. However, this can lead to results that are in conflict with the principle of simplicity described above. Comparison can be obtained by standardizing the method of map-making, representing each distribution as a distinct but comparable map.
9. Comparison can also be obtained by transforming the two distributions compared and obtaining maps representing differences, ratios, and intersections between the distributions.
10. Standardization of methods for linking proper functions of the values to the corresponding graphical symbols and suitable standardization in accordance with the aims of the graphical presentation and the nature of the data shown can eliminate arbitrariness in the graphical presentation.
11. The graphical system should be linked to parameters measuring the primary characteristics of the distributions represented.

It will be seen in Chapters 6 and 7 that the methods usually employed to construct statistical maps fail to comply with some of these criteria, but GRP appear to enable considerable progress in this direction. Thus we describe the GRP system in detail in this volume and many of its uses are compared with other methods. However, even the GRP system in its present form is far from being perfect, and its problems and shortcomings are discussed in Section 6.4.

Chapter 6

Graphical Rational Patterns

6.1. Introduction

In this chapter we deal with graphical rational patterns (GRP), both in general and in particular in connection with simple statistical maps. We begin with the principal types of GRP applications and the GRP scale representing the integers between 1 and 100 by means of squares, and then consider future uses for the system.

GRP were designed to enable graphical representation of important statistical data that cannot be dealt with satisfactorily with ordinary graphical methods. As early as 30 years ago, even before GRP was automated, we proposed possible uses for the GRP system (Bachi, 1968). Ten years later we discussed further possibilities in light of the first steps of its automation (Bachi, 1978c). (As the present volume is devoted primarily to geostatistics, we shall not repeat our discussion of the applications of GRP to nongeographical data.) To complete our overview of the GRP system, we compare its properties and shortcomings to those of the Cartesian system, which is the best existing graphical method.

With regard to the use of GRP for simple maps showing geographical–statistical data, we shall deal mainly with representations of frequencies, quantities, or other values V_j that correspond either (a) to discontinuous dots j on the map, or (b) to continuous territorial units j. To evaluate advantages and possible disadvantages in using GRP for statistical mapping, we compare this system with the other methods in general use, namely scales of symbols to be placed on each distinct dot j; choropleth maps in which each territorial unit is covered by hatching, shading, or color, or a combination of color and either hatching or shading.

6.2. Graphical Rational Pattern Scale Representing Integers between 1 and 100

Figure 6.1 (shown originally as Figure 1.2) shows, among the various uses of GRP, the one that has been used most often in this volume. The GRP are constructed as follows:

1. We indicate any integer number between 1 and 99 by

$$n = 10t + u \quad (t = 0, 1, \ldots, 9;\ u = 1, 2, \ldots, 9)$$

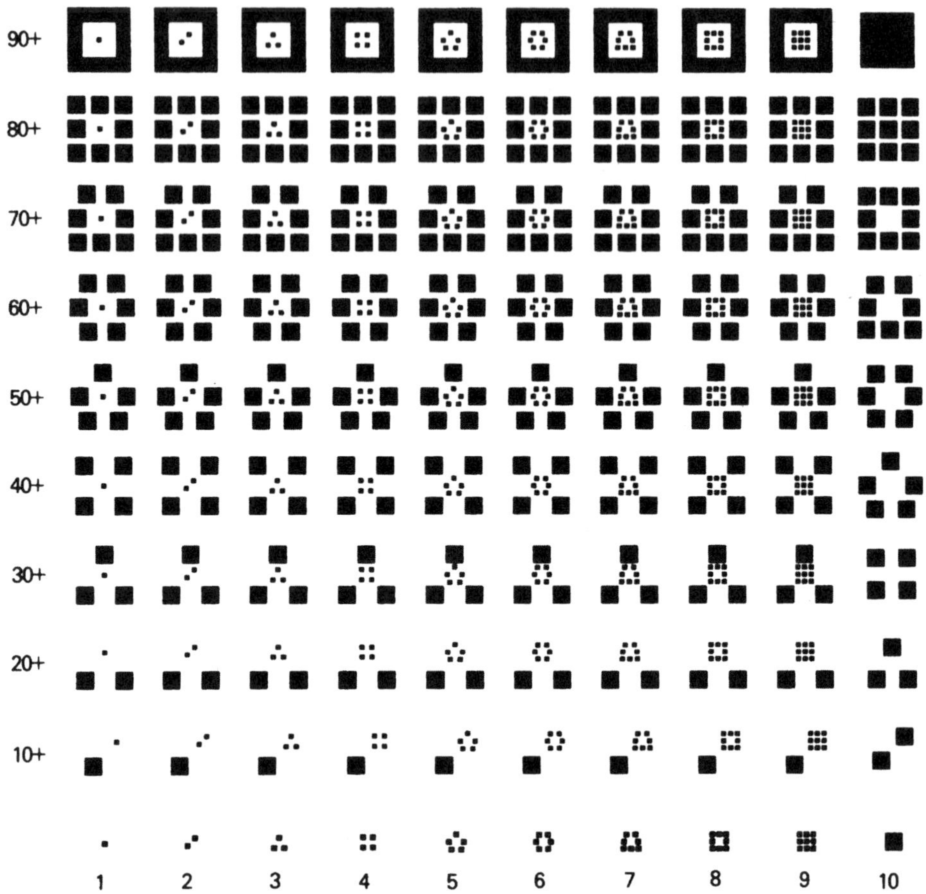

Figure 6.1. Basic scale of a single GRP showing integers up to 100 (Source: *Statistical Atlas of Italy*, 1988).

and the area of a small square by a. Units $u = 1, 2, \ldots, 9$ are represented by $1, 2, \ldots, 9$ squares of area a, arranged in a way that helps to distinguish and read them. Tens $t = 1, 2, \ldots, 9$ are represented by squares of area $10a$, similarly arranged.

2. By combining the symbols of tens and of units within an imaginary 100 frame, we set patterns covering an area $(10a)t + au = na$.
3. An exception to similarity between symbols of tens and of units had to be made for the values between 91 and 99. Here, for obvious graphical reasons a sort of frame covering an area of $9a$ had to be used in place of the symbols formed by 9 squares of area $10a$, which is used to represent 90.
4. 100 is represented by a square of area $100a$ (for higher values, see Section 6.3.3).

We can generalize the principles by which the GRP have been constructed as follows:

1. Any integer number (up to a certain upper limit) is represented graphically by one distinct GRP.
2. A GRP "scale" is formed by a sequence of patterns, each corresponding to an integer.
3. Patterns are drawn in such a way as to require relatively little space. Obviously, this requires a very fine resolution in printing.
4. In the majority of the examples in this book, the pattern itself is formed by clusters of small squares, but the use of square marks is not essential to the method. The total area covered by such marks can be indicated by being inked in.
5. Correspondence between the number n and the pattern representing it is ensured by the fact that the inked-in area in the pattern representing n is n times as large as that of the inked-in area in the pattern representing 1.
6. The fact that the small square marks forming the patterns are clustered in certain ways, which make it comparatively easy to see the numerical value of the pattern quickly, and even to read it whenever the need arises.
7. Readability is based on the use of the decimal system.
8. The efforts that have been made to avoid at least major optical illusions, so that the lay reader can generally arrange the patterns in the right order (e.g., according to increasing magnitude of the numbers represented).
9. We shall describe some of the different uses of GRP for various kinds of statistical maps. However, we can anticipate the following general principle. If we assume that values V_j correspond on the map to individual locations j or to aggregates of locations for territorial units j, and we denote the value corresponding to the unitary square of area a by u, then for $V_j/u = n$ we can represent V_j by a GRP of area na (for the case in which $V_j/u > 100$, see in Section 6.3.3).

We thus have a general solution to the problem of representing any value V_j on the map by a symbol of value approximately proportional to it. This conforms to the criteria (6b) and (6c) suggested in Section 5.7.2 for "good" statistical maps (quick comparability of pairs of symbols; possibility of identifying symbols without resorting to a legend).

6.3. Other Types of Graphical Rational Patterns

6.3.1. Alternative GRP Scale Showing Integers between 10 and 49

The symbols for units and tens shown in Figure 6.1 are well integrated in GRP showing numbers over 50, where the symbols for tens form the outer part of the GRP while those for units are found in the inner part. This is also true to a lesser extent for GRP between 30 and 49, but does not hold for symbols between 11 and

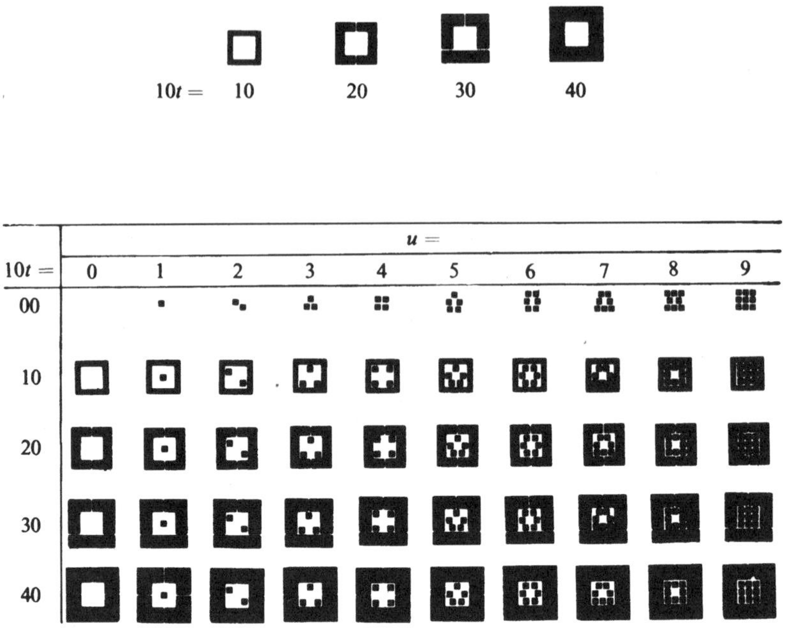

Figure 6.2. Subsidiary scale of a single GRP showing integers between 10 and 49 (Source: Bachi, 1968).

29; in particular, GRP between 11 and 19 are formed by two separate elements, the symbol for tens on the left side and that for units on the right. Sometimes this separation of symbols in the same GRP should be avoided so as not to create misunderstandings in the interpretation of the graph. For this purpose, a subsidiary scale has been designed, as shown in Figure 6.2. Here too, as in the basic scale, the inked-in areas used to represent 10, 20, 30, and 40 are, respectively, $10a$, $20a$, $30a$, and $40a$; however, instead of denoting 10 as a square of $10a$, 20 as two squares of area $10a$ each, and so on, we express 10 as a frame of totally inked-in area $10a$, 20 as an inked-in frame of area $20a$, and so on. To help the reader distinguish quickly among frames representing 10, 20, 30, and 40, we show the frame of 10 as undivided, that of 20 divided into two equal parts, that of 30 divided into three equal parts and that of 40 divided into four equal parts.

6.3.2. Alternative GRP Scale of "Black over White" and "White over Black"

Another alternative scale, shown in Figure 6.3, is constructed by covering 0%, 5%,..., 45% of a given square, respectively, by 0, 1, ..., 9 small square marks of area 5% each; a 10th central pattern is obtained by covering the same area with a chessboard design that is 50% black; patterns from 11 to 20 (55 to 100% of the inked-in area) are obtained by exchanging white for black in patterns 0 to 9. (As the GRP system is based on counting, these transition require "mental readjustment".)

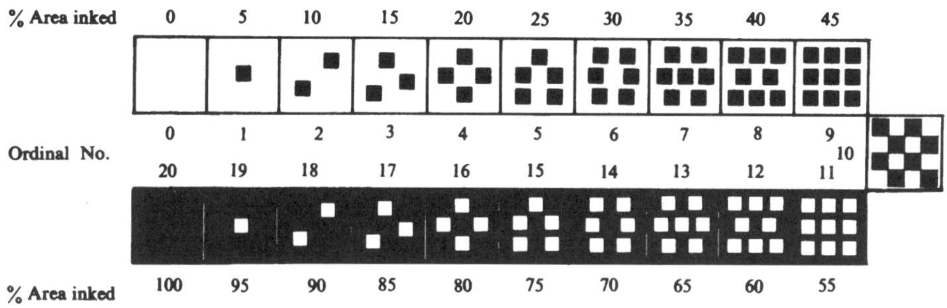

Figure 6.3. Black-over-white and white-over-black scale of a single GRP showing integers between 0 and 20 (Source: Bachi, 1968).

6.3.3. GRP Scales Showing Integers over 100

Theoretically the GRP scale can be extended beyond 100 by introducing simple conventions, consistent with the system explained in Section 6.2. For example, we can use frames showing 100, 200, 300, or 400 which cover, respectively, areas of $100a$, $200a$, $300a$, or $400a$, as shown in Figure 6.4 and place within these frames an appropriate GRP showing hundreds, tens and units, as shown in Figure 6.5.

We have illustrated extensions beyond 499 (Bachi, 1968, pp. 40–41); however they have the shortcoming of taking up too much space. Alternative solutions to the problem can be found by resorting to conventions that have advantages from a visual viewpoint but are not based on a purely scientific criterion. Four of these are as follows:

1. Values between 100 and 200 can be shown by adding a frame indicating conventionally +100 around GRP representing values between 1 and 100.
2. If we wish to add not only +100 but, say, +200, +300, +400, we can use frames as in (1) in four different easily distinguishable hues or colors and indicate increasing values by increasing the intensity of the colors.
3. If color can be used, it is probably even better to show values from 1 to 99 in a "pale" color; e.g., in master printing 100 is denoted by a full square in

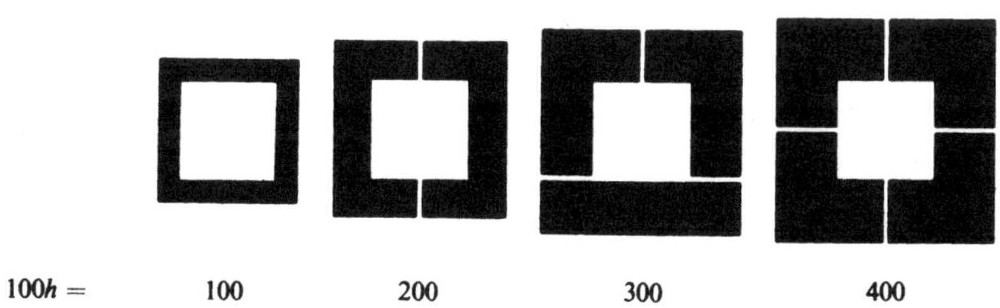

Figure 6.4. GRP representing hundreds: 100, 200, 300, 400 (Source: Bachi, 1968).

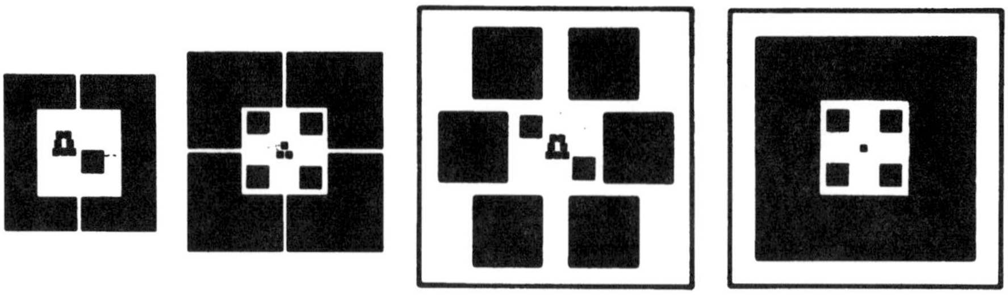

217 443 627 941

Figure 6.5. Examples of GRP representing integers up to 1000 ($n = 100h + 10t + u$) (Source: Bachi, 1968).

that color; values from 101 to 199 are given by GRP in a stronger color superimposed over the full square indicating 100, and so on.

4. If we do not have access to color and we wish to indicate +100, +200, +300, +400 distinctly, we can use different types of black frames.

In this volume we show values between 101 and 199 by a 100 unit of a black GRP bordered by a single thin red frame, between 200 to 299 by a 100 unit of a black GRP bordered by two thin red frames,..., values over 500 by a 100 unit of a black GRP bordered by five red frames. An example of this method, which of course, requires the use of colors, is given in Plate 4 but owing to technical limitations in most of the figures in these pages the red frames are reproduced in black.

6.3.4. Circular GRP

We have proposed various GRP scales based on circular units (Bachi, 1968, pp. 44–45; 1978*c*, pp. 25, 31, 33–36), but we have not used them here.

6.3.5. GRP Representing Positive or Negative Values

Negative integers −10, −9,..., −1, and positive integers +1, +2,..., +10, can be represented by the GRP shown in Figure 6.3. If a scale indicating values −100, −99,..., −1, 0, +1,..., +99, +100 is required, the GRP in Plate 5 can be used.

6.3.6. Repeated Linear GRP

Repeated linear GRP are constructed in order to convey the notion that an integer *n* is to be attributed to an entire line—which can be, e.g., an isoline on a geographical map, a traffic line, or a line showing proportional increase or decrease between two statistical data that it joins. They are obtained simply by repeating a single GRP of the trend described in Sections 6.2 and 6.3.1 on straight or curved lines. A scale of these repeated linear symbols going from 1 to 39 by unit steps is

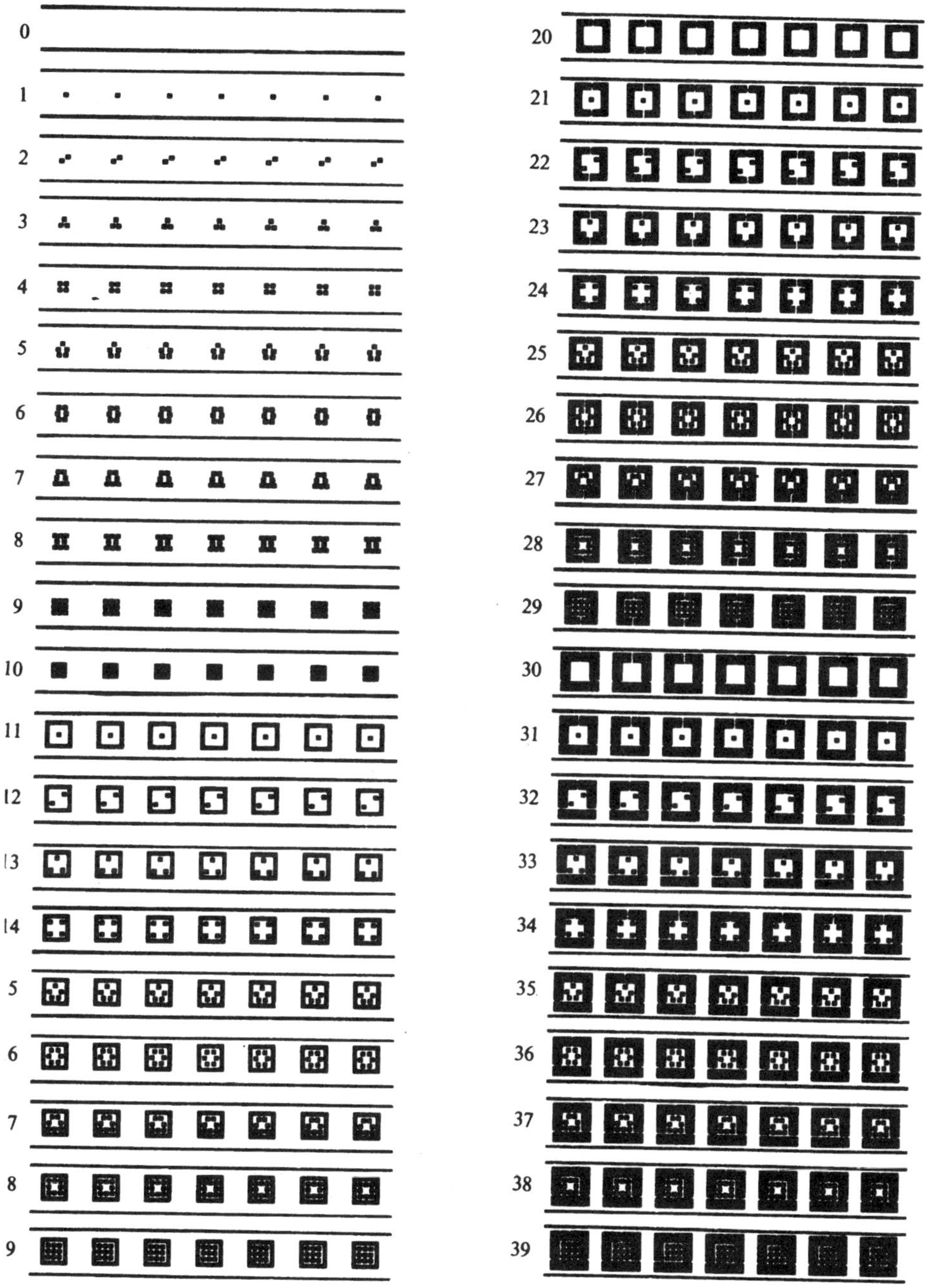

Figure 6.6. Repeated linear GRP of varying sizes representing integers up to 39 (Source: Bachi, 1968).

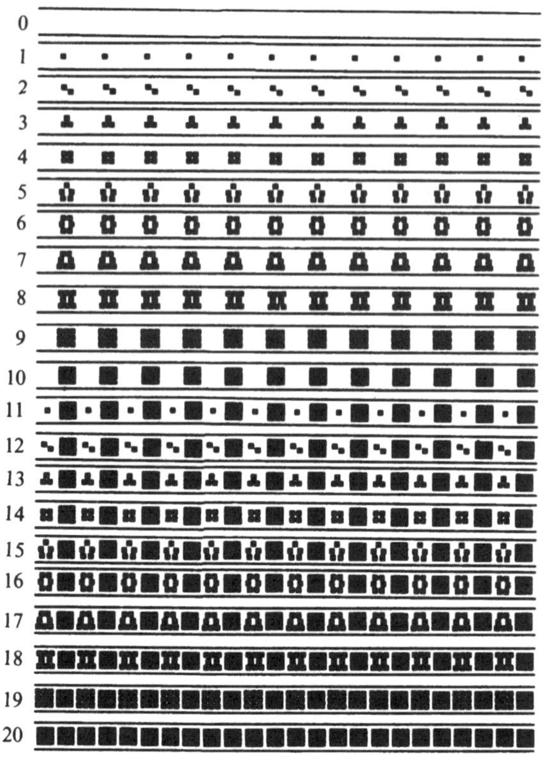

Figure 6.7. Repeated linear GRP representing integers up to 20 (Source: Bachi, 1968).

shown in Figure 6.6. Another scale of repeated linear symbols going from 1 to 20 by unit steps is shown in Figure 6.7. This scale answers the occasional need for keeping the line of GRP to a minimum width. For this purpose, we show values from 1 to 10 by ordinary GRP alternating with empty cells and values from 11 to 20 by replacing the cells with a symbol of 10. For examples of the uses of linear GRP see Bachi (1968).

6.3.7. Repeated Areal GRP

Repeated areal GRP are constructed in order to convey the idea that the integer *n* represented by the GRP is to be attributed to an area (e.g., *n* may represent the degree of population density over a territorial unit). Repeated GRP can be obtained simply by repeating a single GRP over points that are distributed over the area so that each point is equidistant from six neighboring points. Examples are given below. Figure 6.8 shows repeated GRP for $1, 2, \ldots, 10$ or for $10, 20, \ldots, 100$. The use of repeated areal GRP leads to results identical to those obtained from the computer program described in Chapter 17.

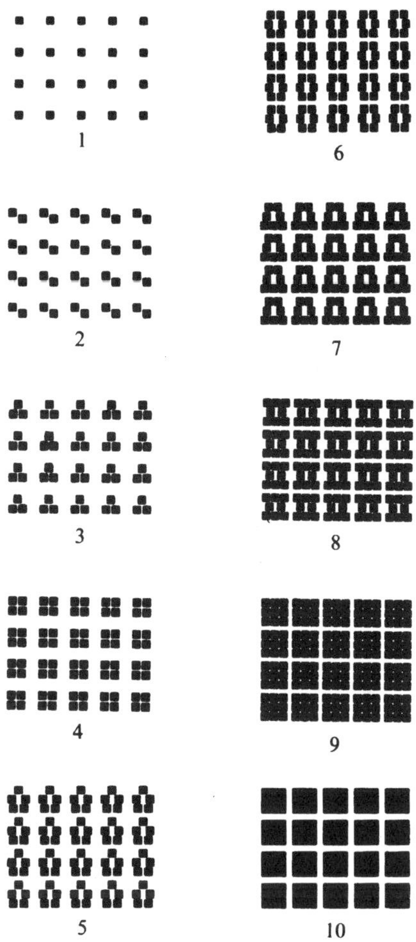

Figure 6.8. Repeated areal GRP representing integers up to 10 (Source: Bachi, 1968).

6.3.8. Multiple GRP

In many fields of graphical representation, including statistical mapping, the need arises to represent jointly two related statistical magnitudes, P_i and Q_i referred to the same element i or to a group of elements i. For instance, we might wish to show a statistical value and its standard error, a rate with its weight, or an average concerning a given region and the population of the region. Sometimes we might wish to represent P_i and Q_i and their product P_iQ_i concurrently: for instance, if P_i is the average income per capita in a region i and Q_i the population of region i, we might want to show the total income in region i, i.e., P_iQ_i.

We offered various solutions to these graphical problems (Bachi, 1968, 1978*c*) and some of them require the use of multiple GRP. Figure 6.9 shows a scale of

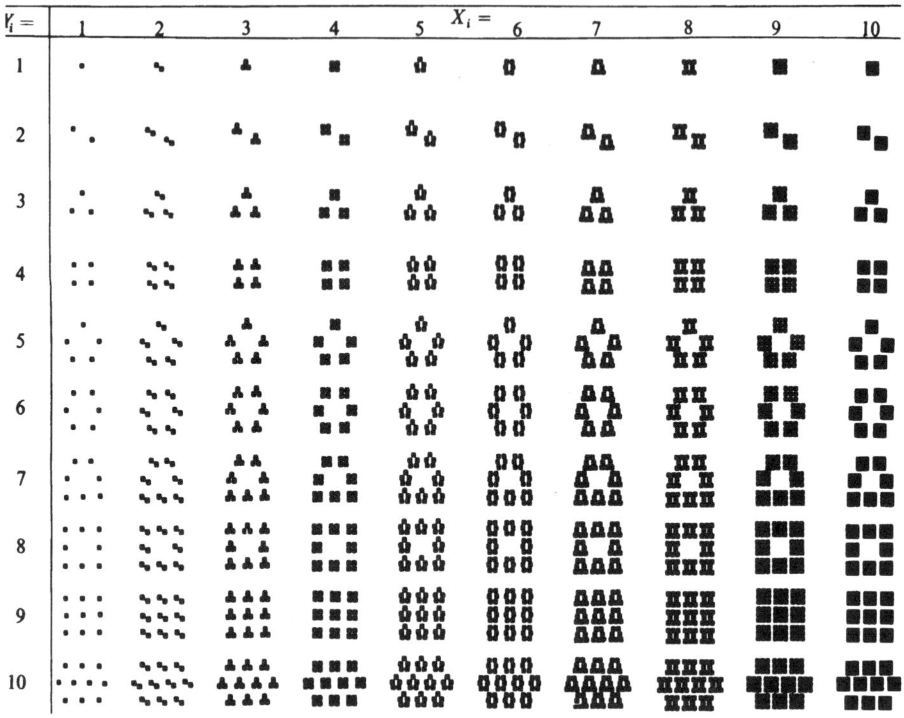

Figure 6.9. Scale of multiple GRP up to $X_i = 10$, $Y_i = 10$ (Source: Bachi, 1968).

multiple GRP indicating values $1 \times 1, 1 \times 2, \ldots, 1 \times 10; 2 \times 1, 2 \times 2, \ldots, 2 \times 10; \ldots; 10 \times 1, 10 \times 2, \ldots, 10 \times 10$. Graphically, this scale produces some crude effects.

6.3.9. Adding Information over GRP

Additional quantitative information can be shown by putting frames around the GRP, over which grades or dots are superimposed. Additional qualitative information can be shown over GRP by using frames of different forms, or styles, or colors. For extensive examples and discussions of these methods see Bachi (1968, 1978*c*).

6.4. Comparing Graphical Rational Patterns and the Cartesian System

Figures 6.10–6.12 enable us to compare GRP showing selected values of integers n to a corresponding simple graphic device—a segment proportional to n on an ordinary Cartesian diagram. Figures 6.10 and 6.11 show, respectively symbols corresponding to units ($u = 1, 2, \ldots, 9$) and to tens ($t = 10, 20, \ldots, 90$). Figure 6.12 shows symbols corresponding to $n = 12, 13$; $n = 42, 43$; $n = 62, 63$; and $n = 82, 83$.

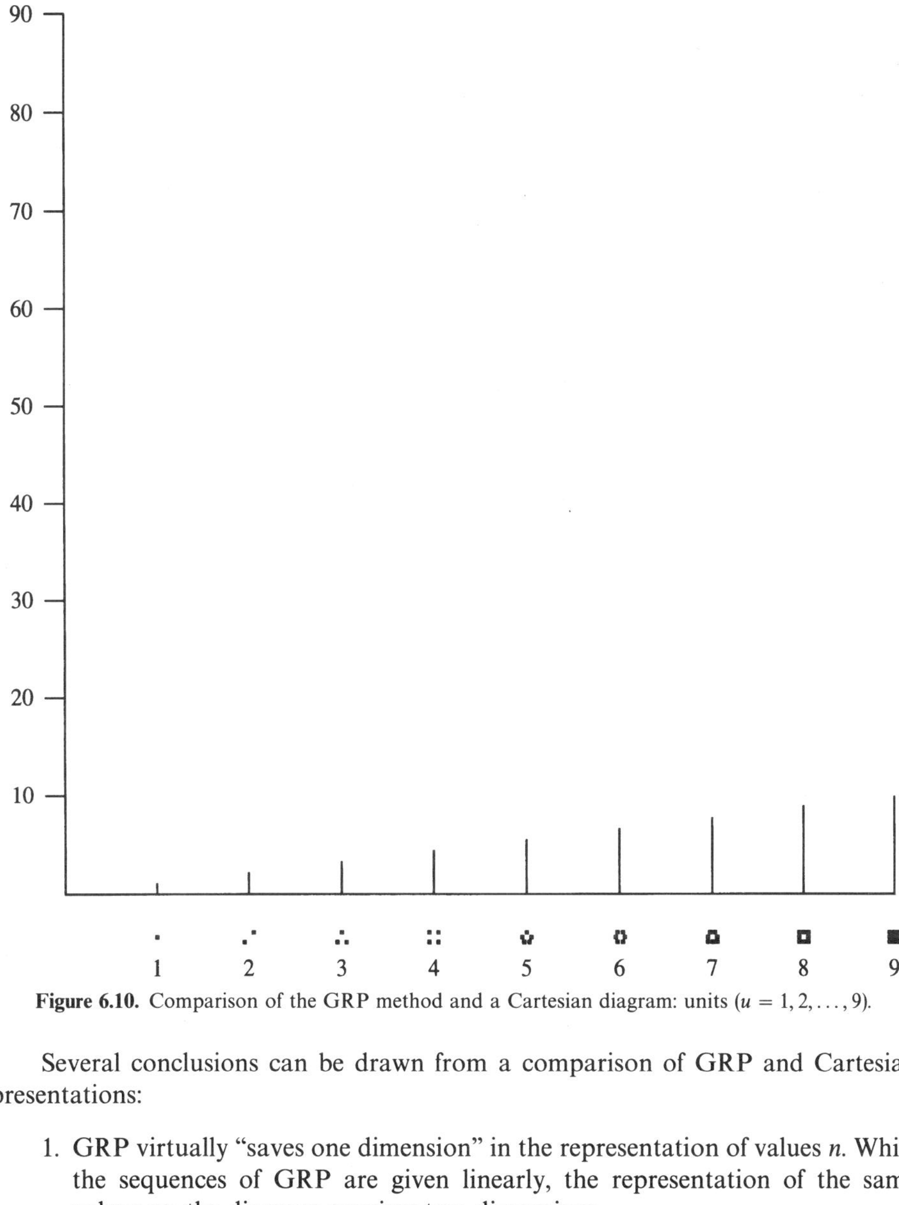

Figure 6.10. Comparison of the GRP method and a Cartesian diagram: units ($u = 1, 2, \ldots, 9$).

Several conclusions can be drawn from a comparison of GRP and Cartesian presentations:

1. GRP virtually "saves one dimension" in the representation of values *n*. While the sequences of GRP are given linearly, the representation of the same values on the diagram requires two dimensions.
2. (a) Perpendicular or horizontal lines over Cartesian diagrams constitute, presumably, the best possible indication with regard to perception (see Cleveland, 1985, p. 254). (b) In Cartesian diagrams it is often legitimate to join end points of subsequent vertical segments (say, j, $j + 1$) and to judge the difference (or in semilogarithmic charts, the ratios) between values V_i and V_{j+1} represented by the segments on the basis of the inclination of the diagrammatic line. Sometimes, without actually joining the end points of the

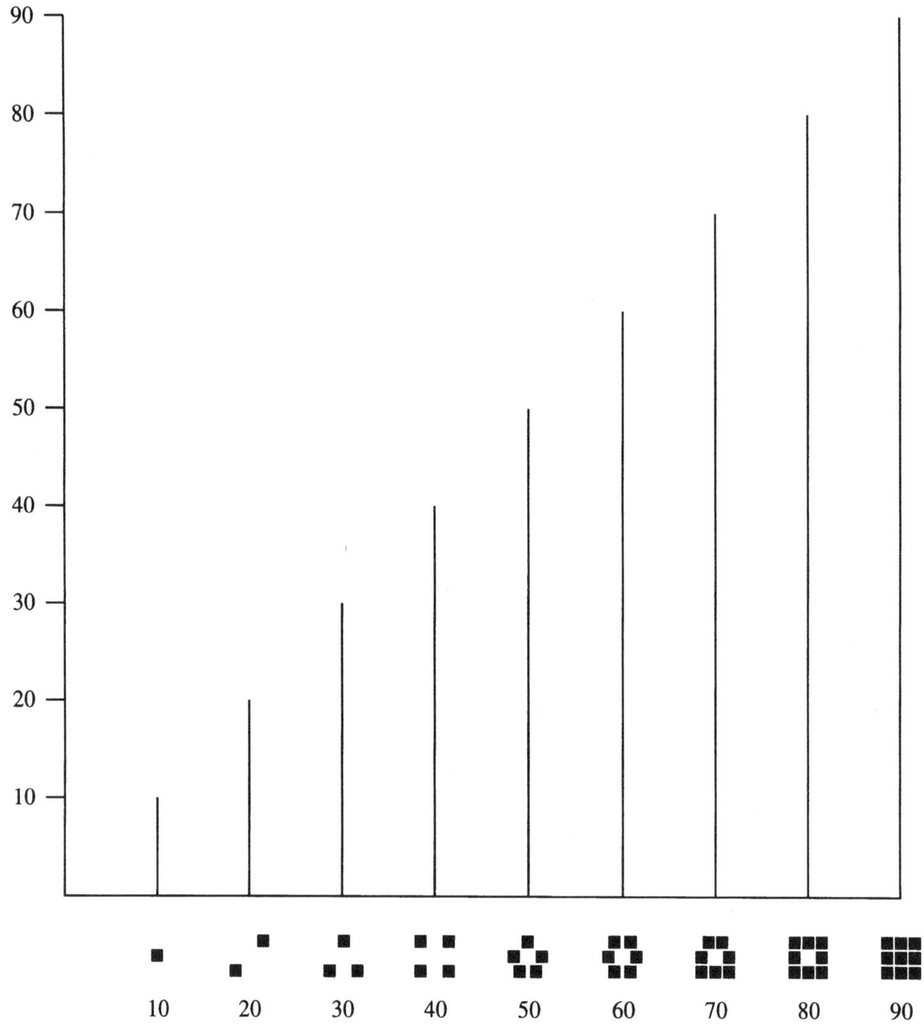

Figure 6.11. Comparison of the GRP method and a Cartesian diagram: tens ($t = 10, 20, \ldots, 90$).

vertical segments, we can nevertheless assume that we do, which allows us to estimate the differences.

3. Obviously the GRP system lacks this important property. However GRP symbols can be easily evaluated and ordered. We know this to be true (albeit not substantiated by systematic experimentation) from polling many readers (including children). Some exercise in the meaning of the GRP alphabet is of assistance here. Readers who understand the GRP method can compare pairs of values as given in Figure 6.12 rather easily. For high values, such as $n = 62, 63$ or $82, 83$, it is easier to estimate the differences between subsequent

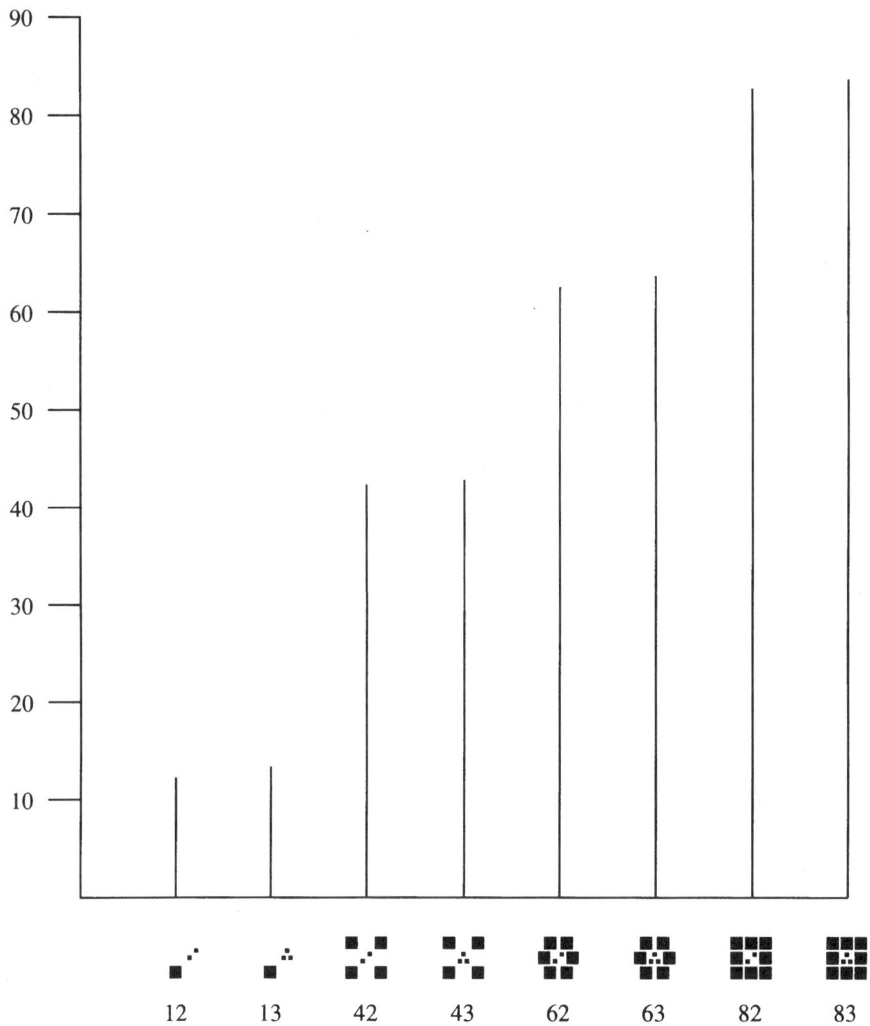

Figure 6.12. Comparison of the GRP method and a Cartesian diagram: n = 12, 13, n = 42, 43, n = 62, 63, n = 82, 83.

values by means of GRP than by vertical segments (because the ratios between these values are very near to 1).

4. In Figure 6.10 and especially in Figure 6.11 the general characteristic of linearity of increases can be more easily understood from the Cartesian presentation than from the GRP system.
5. On the other hand, the comparison of sequences of GRP in Figures 6.10 and 6.11 might suggest equality of ratios such as $2/1 = 20/10$; $3/2 = 30/20$; ..., $9/8 = 90/80$, which is not apparent in the Cartesian representation (if it is not transformed into segments proportional to logarithms).

6. GRP cannot be used for small-scale maps with territorial units smaller than one GRP symbol, e.g., in maps of large areas such as groups of countries, continents, or the globe divided into states. Furthermore, in small GRP symbols the white dividing lines are liable to be visually indistinguishable.

In the graphical presentation of series of data ordered by time, by a variable, or by attributes following each other linearly, it is possible to utilize the Cartesian and GRP methods in tandem, exploiting the advantages of both. Many such examples were included in some of our earlier work (Bachi, 1968, 1978*c*), but they are outside the scope of this book.

6.5. Using Graphical Rational Patterns for Simple Statistical Maps

6.5.1. General Considerations

Having discussed the general characteristics of the GRP system, we now move on to some of its simplest applications to statistical maps, such as the representation of values attributed to individual elements of a population, to individual localities, and to territorial units (Sections 6.5 and 6.6). More sophisticated uses of GRP in statistical maps will be introduced in Chapter 7. The examples of practical statistical maps that were described in Chapters 1, 3, and 4, will be broadened and systematized in this chapter and the next one. This presentation also enables us to generalize the graphic aspects of the study of primary and derived geosets that were introduced in Chapters 3 and 4. This double approach generates some repetition, which is unfortunate but unavoidable.

Using GRP in statistical maps is in line with the geostatistical emphasis and contents of this book. However, we reiterate here that the attempts made in the preautomation era to develop the GRP system had a far more general scope—to develop graphical presentations for many types of statistical data for which the usual methods did not offer satisfactory solutions.

6.5.2. GRP in Maps Representing Values that Correspond to Individual Elements of a Population

We refer here to a geoset indicating the location (X_j, Y_j) and a quantitative value V_j to be assigned to j for individual elements j of a population. For example, j might represent individual industrial firms of a given type and V_j their output during a given period; or j might represent individual hotels in a city and V_j their number of beds. This second example is shown in Figure 6.13 for the hotels of Jerusalem.

The "natural" solution for settings such as these is to show a symbol proportional to V_j, corresponding to location (X_j, Y_j) on the map. A symbol that is often employed here is a circle of area proportional to V_j. However, the use of GRP proportional to V_j is often a better solution for the following reasons among others: The representation of individual values V_j suggests that the map is intended to enable

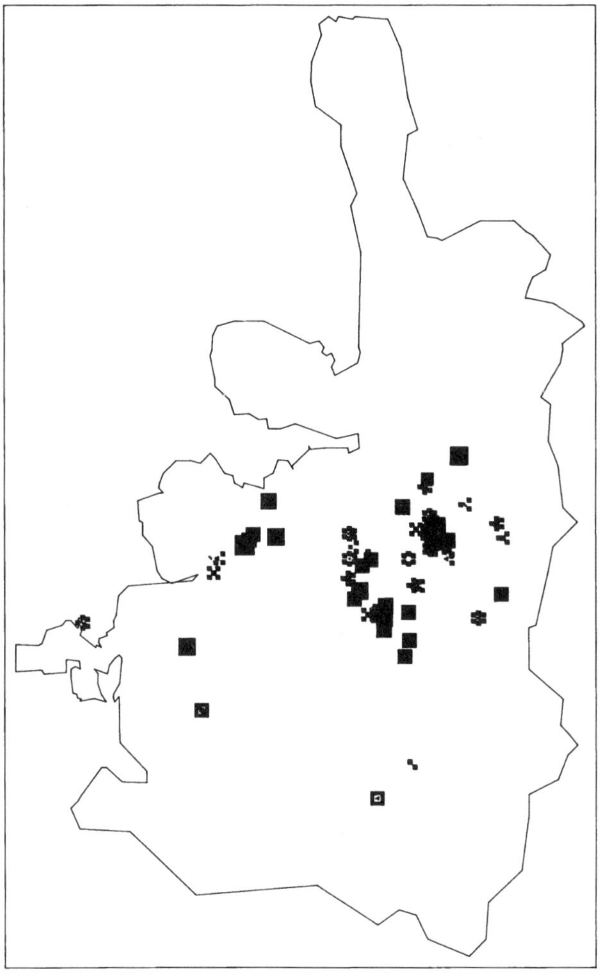

Figure 6.13. Jerusalem. Hotels by the number of beds, 1989: ▪ 14 ▪ 24 ▪ 45 ▪ 61 ■ 118+ (Source of data: Israel Ministry of Tourism).

the reader: (a) to compare the sizes of the various V_j; (b) to evaluate their orders of magnitude.

Concerning (a), let us indicate by a the area of a circle corresponding to the minimal value of the V_j to be represented, and let us also assume that in the scale of the GRP we take S as the area of the unit square (corresponding to a value $S = 1$ in the GRP scale). Under these conditions the GRP scale offers many more distinguishable levels than the scale of the circles. In regard to (b), the symbols have better "readability" in the GRP than in the circles.

In theory, when we keep the size of the map fixed, we can enhance the comparability between circles by enlarging the unit area a used to represent the circles. However, this increases the possibility of overlapping between the circles.

6.5.3. Representing Data Referring to Individual Localities by GRP

We refer here to a geoset of the weights V_j to be attributed to distinct localities j. We assume that we have data V_j classified by localities or places j, for which we do not take account of either territorial boundaries or area. For example, j might represent the individual inhabited places of a country or a region (however defined) and V_j their populations; or j might represent ports of a country and V_j the number of individually departing from those ports during a given period.

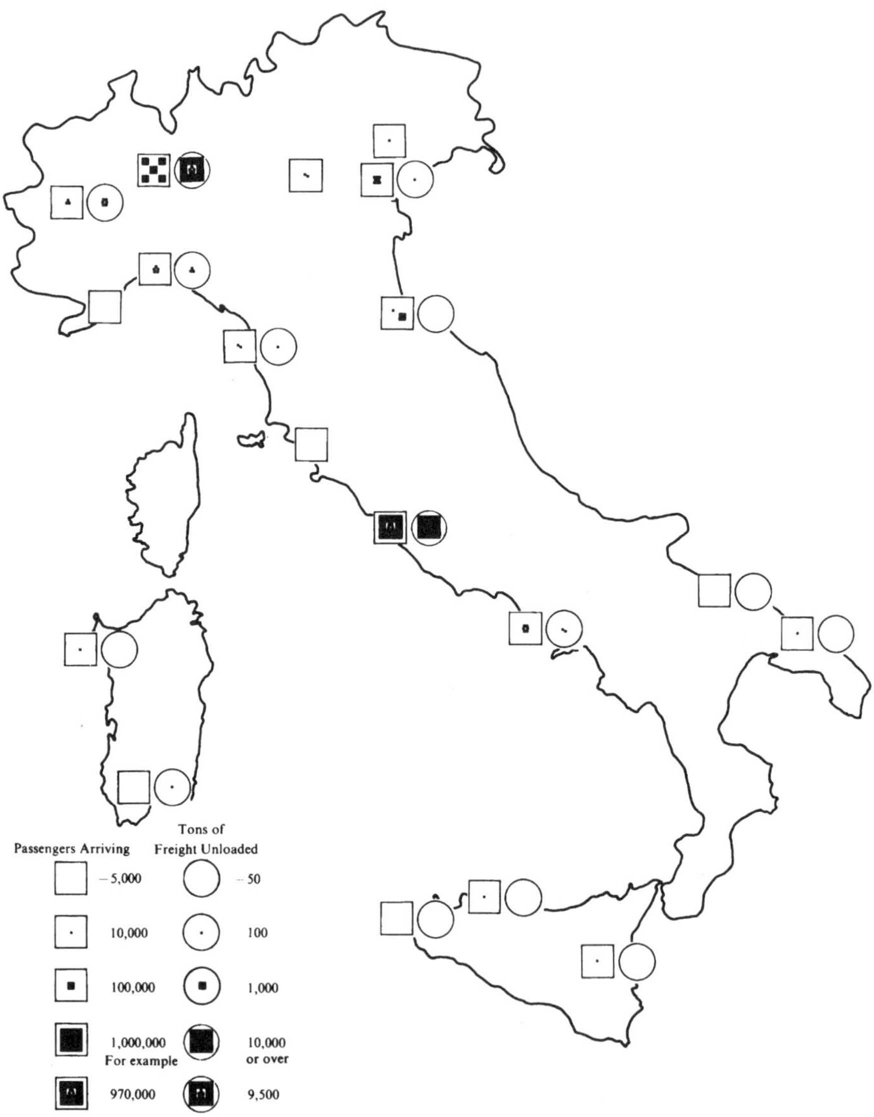

Figure 6.14. Italy. Traffic in the principal airports, 1964: Scale: 10,000 passengers arriving = 1 unit in square GRP; 100 tons of freight unloaded = 1 unit in circular GRP (Source: Bachi, 1968).

As an illustration of this method of presentation, Figure 6.14 shows the distribution of traffic in the principal airports in Italy (1964). Here each airport is given two "weights," represented by GRP, one of which refers to the number of passengers arriving and the other to the tons of freight unloaded. In this example (taken from Bachi, 1968), the key used responds only to criteria of simple expediency: A unit GRP enclosed in a square corresponds to 10,000 passengers and a unit GRP enclosed in a circle corresponds to 100 tons of freight.

To some extent graphical presentation may help us solve a long-standing problem: how to show on the map distinct inhabited places j whose populations (denoted here as V_j) vary from, say, a few hundred to millions of individuals. One way of dealing with this difficulty might be to show "low" values by circles and "high" values by spheres (see, e.g., Schmid, 1983, pp. 147–148). However, distinct perception and comparative evaluation of the volume of the spheres is known to be difficult. The use of the simple GRP scale might give us the possibility of distinct evaluation over a range progressing by individual steps from level 1 to level 100. The use of the various methods suggested in Section 6.3 may help us to extend this range considerably.

6.6. Using Graphical Rational Patterns to Represent Values Corresponding to Territorial Units

6.6.1. Principal Types of Territorial Units

As noted in Chapters 1 and 3, we must often construct maps for which the available data are: (a) the set of frequencies or amounts F_j corresponding to territorial units j into which a territory T is divided; (b) the map of T with the network of the boundaries of territorial units j (from which the area A_j of each unit can be obtained).

Territorial units can be of different types. For example:

1. At an upper hierarchical level, the division of T into units j may have a political meaning: T might represent a continent and j the countries of the continent; T might represent a federation or confederation such as India, the United States, or Switzerland, and j its states or cantons; T might represent a country such as Italy and j those of its regions enjoying some measure of self-government.
2. The division of T into units j may have an administrative meaning. At the lowest hierarchical level, j might represent, e.g., municipalities or *communes* or incorporated areas.
3. At a higher hierarchical level we can consider more important administrative units (such as provinces or *arrondissements*, which in some countries number hundreds or even thousands).
4. The division of T into j might have a purely enumerative aim, such as the determination of ZIP codes or census enumeration districts over a country or the division of a city into census tracts.

5. Units j might have been determined for research purposes and be: "natural" geographical, nodal, or more or less homogeneous regions defined on the basis of socioeconomic, ethnic, or other criteria.
6. Units j might be cells of fixed regular shape (such as rectangles or hexagons) and of fixed area determined for the purpose of analysis and graphical presentation.

6.6.2. Transforming a Detailed Network of Irregular Territorial Units into a Network of Regular Cells

An intrinsic difficulty in presenting the F_j on a map is that we do not know how the frequencies are distributed within each territorial unit j. This is a problem analogous to the one encountered in the analysis of frequency distribution by classes of one-dimensional variables.

In the study of these distributions at an elementary level, we often introduce one of the two following hypotheses: (a) that the frequency or density of each class is evenly distributed within the class interval; (b) that the frequency is attached to a suitably selected actual value of the class. In an analogous manner, in the preparation of the map, we might assure either: (a) that each F_j is distributed evenly (with density F_j/A_j) over the area of unit j; or (b) that each F_j is attached to a suitable central point of j.

However, in constructing the map, things are by far more complex: (a) if we wish to ensure that the map adheres to the criteria listed in Section 5.7.2; and (b) if we wish to discriminate among the various types of territorial units for which the input F_j is given. According to (a), the map should make possible a quick overview of the general characteristics of the distributions, a comparison between frequencies or densities related to various parts of the map that are of interest to the reader, and an evaluation of the order of magnitude of these frequencies or densities, whenever needed.

Let us consider the settings indicated above under (2) and (4), i.e., very numerous and comparatively small administrative units or units formed only for purposes of enumeration. Generally speaking, identification of such individual territorial units over the map is of little interest and is often practically impossible.

Consider, for example, Figure 6.15, which shows the network of more than 30,000 municipalities (*communes*) in France. If we assume that we have data F_j at this level of specification, it would be practically impossible to present them graphically over the network of Figure 6.15 and still keep the map to a reasonable size.

For Italy, despite the fact that the number of *comuni* is smaller (8086), in the preparation of the *Statistical Atlas of Italy*, we also found it practically impossible to use the network of these municipal areas. The reasons for this can be understood from a look at Figure 6.16. Here the communal network is given for 2 out of 95 provinces of the country. We can see that the *comuni* are widely different in size and shape: their areas vary from less than 1 to more than 1500 sq km; and their forms are in many cases, very irregular. Hence, they do not constitute convenient or practical units for the comparative analysis of territorial statistical data.

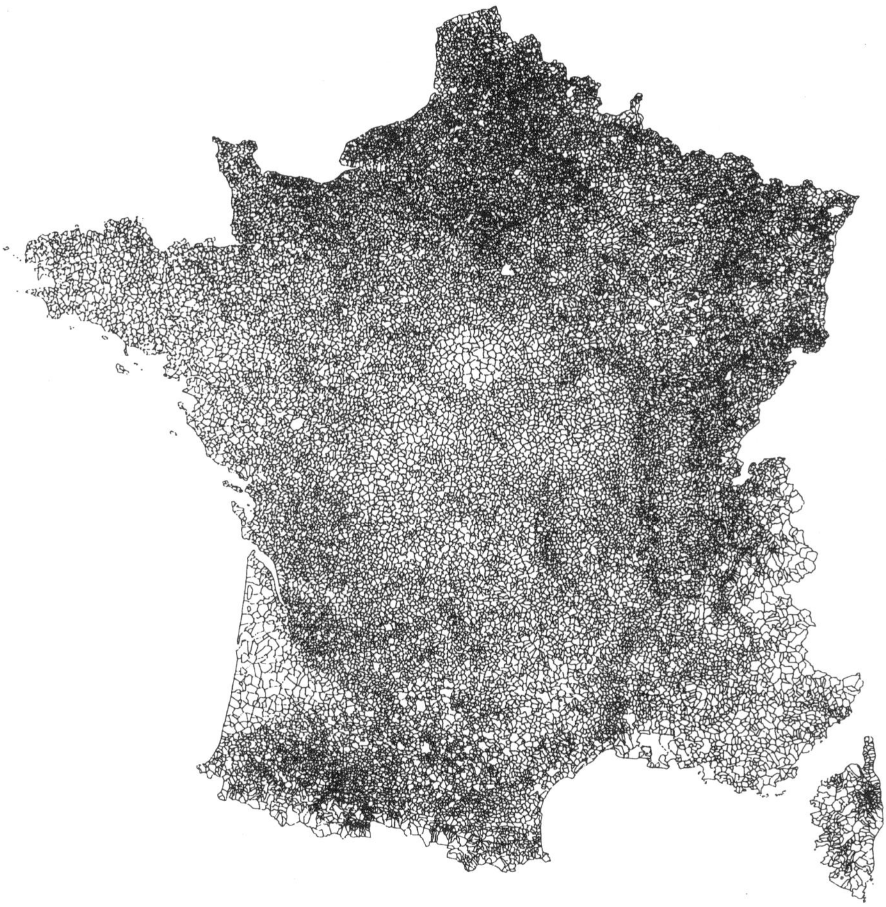

Figure 6.15. France. The network of 30,000 municipal areas (Source: *Le Grand Atlas de La France Rurale*, INRA SCEES sous la direction de André Brun, Jean-Pierre de Monza, France, 1989).

To create a graphical presentation that enables us to carry out simple and direct comparisons among the symbols for frequencies or densities, we must overcome the difficulties created by the variability in area and shape of the territorial units j.

One method sometimes employed for achieving this is to replace the original network of j with a system of units, as noted above under (6), that consists of cells of regular shape and of fixed area, selected so as to facilitate the analysis and at the same time to enable us to use as much as possible of the information furnished by the detailed input data F_j. Most of the applications of this method described in the literature are based on a network of squares formed by a grid, but here we have generally used a network of regular hexagons (see Section 3.2.2). Figure 6.17 and Figure 1.1 (Section 1.2) show examples of these hexagonal networks for the United

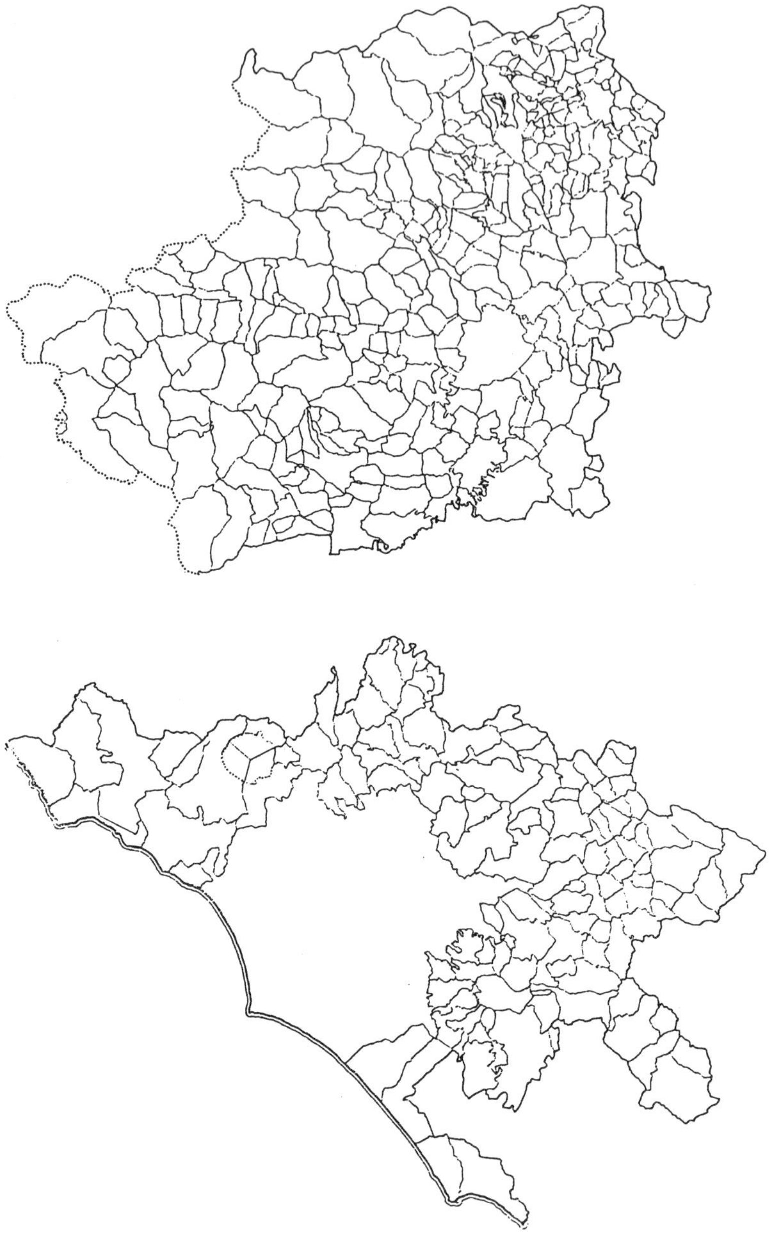

Figure 6.16. Italy. The communal network for the provinces of Torino (above) and Rome (below) (Source: *Statistical Atlas of Italy*, 1988).

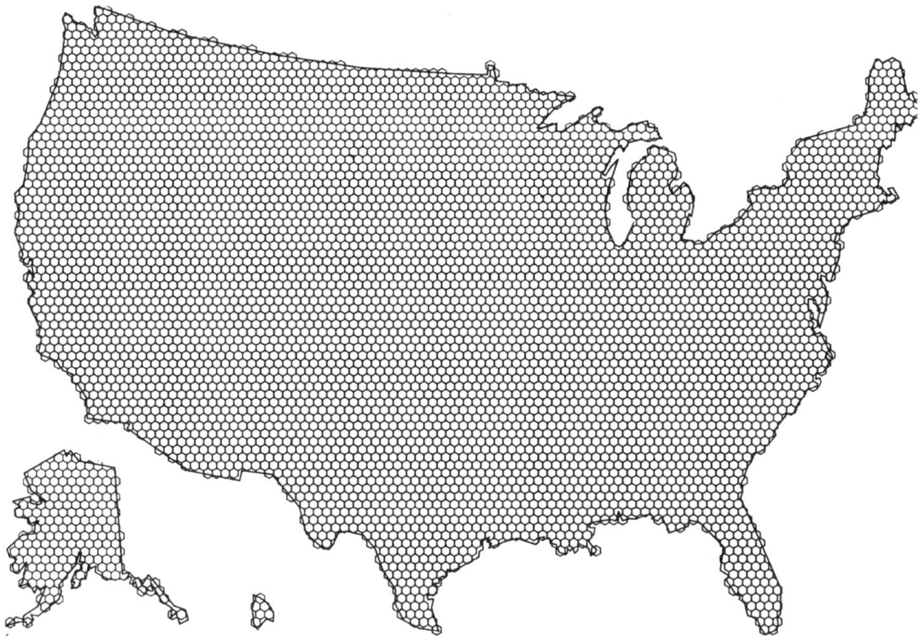

Figure 6.17. Cell network for the territory of the United States.

States (a network of 5000 hexagons) and Italy (a network of 1000 hexagons), respectively.

Similar difficulties are often encountered in constructing maps for cities divided into territorial units such as census tracts. Figure 6.18 shows, for example, the division of the Standard Metropolitan Area of Toronto, Canada, into 608 census tracts. Here, too, the variability of areas and shapes of the tracts is very great, so the same solution for constructing the map is suggested. A network of 1000 hexagons for Toronto is shown in Figure 6.19.

Practical methods for transforming the original input for irregular territorial units into frequencies ${}^{c}F_i$ corresponding to hexagonal cells i are available; in Section 17.8 we discuss the principles of the geostatistical software package. In general terms, whenever the number of units j is very large and these units have areas that do not far exceed that of one hexagonal cell, a sufficiently good approximation can be obtained by assigning to each cell the units j whose centers (however defined) fall within the cell. For units j with a comparatively large areas, a procedure for assigning them to more than one cell is suggested, which is based on the hypothesis of even distribution within unit j. These methods all involve some inaccuracy; there are others that can be suggested that will lead to a better approximation.

The transformation of the original irregularly subdivided territory into a regular network involves: (a) some loss of information (e.g., in the Italian study we replaced

8086 data detailed by municipal area with 1000 data detailed by cells; (b) some inaccuracy owing to a "shift" of density from one cell to another one nearby in cases where the transformation of the original data to the cell network is not very precise. However, the consequences of these "losses" are of little importance compared to the advantages of assigning information to territorial classes of fixed size and shape.

It should be noted here that the use of a regular network also facilitates the calculation of geostatistical parameters, their analysis, their comparison, and their interpretation.

We shall see in Section 6.9 how after transforming the input from the original irregular network into a distribution of frequencies corresponding to regular cells, we can construct "rational maps" by assigning to each cell a symbol (GRP) proportional to its frequency or to one of its functions.

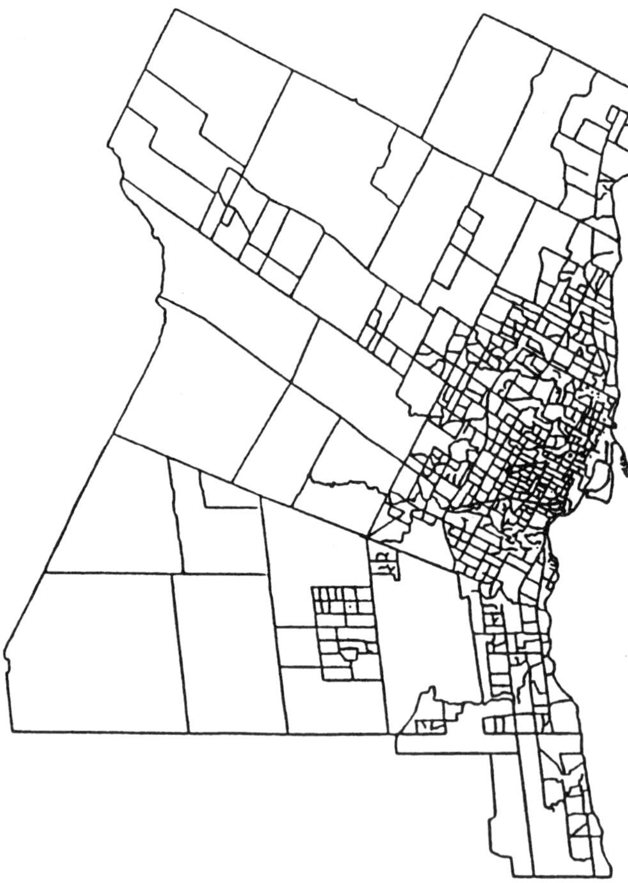

Figure 6.18. Canada. Toronto metropolitan area divided into 608 census tracts, 1981 (Source: Bachi, 1985).

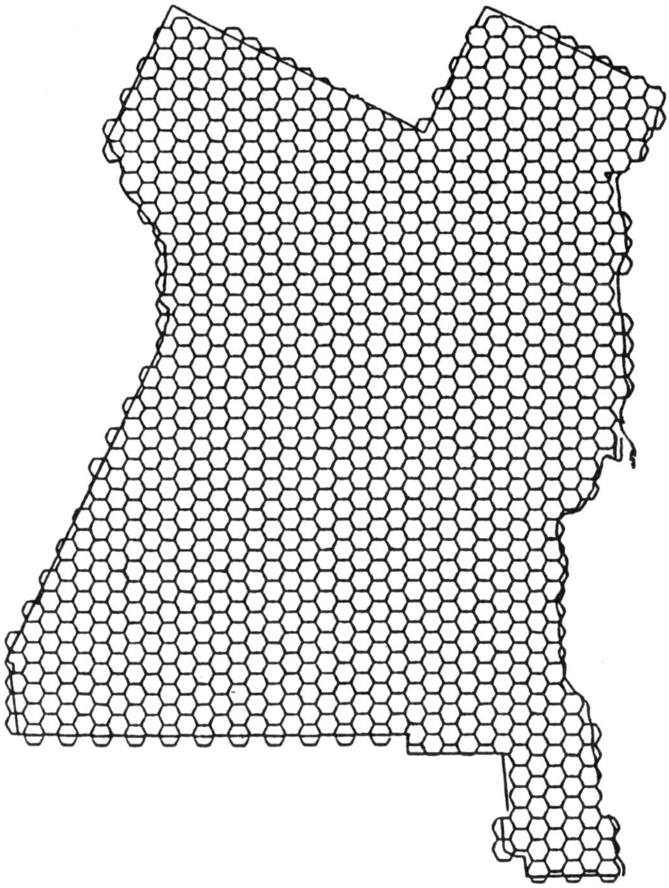

Figure 6.19. Canada. Cell network for the Toronto metropolitan area (Source: Bachi, 1985).

6.6.3. Dealing with a Moderate Number of Meaningful Territorial Units

In the method just discussed the borders between the territorial units j vanish. If these units are of purely local interest or have only an enumerative function this is of no consequence. However, in other cases there might be a reason to distinguish units of types (1), (3), or (5) on the map (see Section 6.6.1), which arises because: (a) these units are meaningful from a political, scientific, or other point of view, or (b) their identification is useful for the interpretation of the information given by the map.

Very often this problem can be solved simply by tracing the boundaries of the territorial units (1), (3), or (5) on the map, which is based on the cell network. Whenever possible, one wants to avoid having these boundaries interfere with the symbols used to represent the cell frequencies, and this result can be obtained, at the cost of some degree of inaccuracy, by making groups of entire cells correspond to each of the meaningful units (1), (3), or (5). An example of this technique is illustrated later in Figure 7.9 (Section 7.3) in which shows the state boundaries of the United States.

If we make use of the method suggested above [letting j denote any one of the meaningful territorial units, ji any one of the cells i included in j, N_j the number of these cells, a the standard area of a hexagonal cell, $A_j\ (=aN_j)$ the approximate area of j, $F_j\ (=\Sigma_i F_{ji})$ the total frequency in j, and F_j/A_j the average density in j], the map we construct, if used properly, can lead to important results:

1. It gives us an overview of the distribution over T.
2. It enables us to compare frequencies or densities in different parts of T that may be of interest to the reader and these parts can be identified with the help of the boundaries of the meaningful territorial units.
3. It allows us to compare and evaluate important characteristics of these territorial units: (a) internal comparison of the F_{ji} for the cells ji of j permits the judgment of variability within j; (b) the number N_j of the cells of j enables us to evaluate, albeit roughly, the area of j; (c) "Visual sum" $F_j = \Sigma_i F_{ji}$ enables us to evaluate, albeit very roughly, the total frequency in j (it is often convenient for the map to show relative frequencies referred to a total such as 10,000, in which case the visual sum represents the proportion of total frequency contributed by the unit j).

Obviously this approach is convenient only when the number of territorial units under consideration is moderate and their meaning is clear to the readers for whom the map has been constructed.

Suppose that instead of showing data by cells ji assigned to each meaningful territorial unit j, we show them only by territorial units j. If we know the total frequency F_j and the number of cells N_j for each j, we might then show the average frequency F_j/N instead of F_{ji} in each of the cells ji belonging to j. Under such conditions, we might still be able to accomplish (3b) and (3c), but (3a) would become impossible and (1) and (2) might be severely impaired.

6.7. Comparing the Graphical Rational Pattern Scale to Other "Regular" Scales of Symbols Used in Statistical Mapping

The principle of representing any datum V_j by a symbol of value approximately proportional to it is often applied in constructing statistical maps. Let us consider the following symbols:

1. The circle, which is frequently used in statistical maps, mainly to show data V_j corresponding to locations of elements j or to centers of regions j.
2. The square, which is used in the same way as the circle as well as in relation to the divisions of territory into square cells by means of a grid.
3. The regular hexagon, which can also be used in the same way as the circle as well as in relation to the division of territory into hexagonal cells.

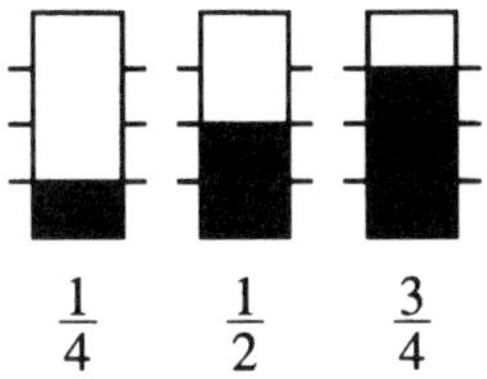

Figure 6.20. Three marks corresponding to 1/4, 1/2, and 3/4 to facilitate one evaluation of V_j.

4. Framed rectangle, proposed by Cleveland (1985, pp. 208–210, 285–289), which is a solid-black rectangle with a fixed base and height, proportional to the value, V_j that it represents on the map. The frame has an area proportional to the highest possible value of the V_j. In Cleveland's proposal, there is a tick mark halfway up the vertical side to facilitate the evaluation of V_j. In Figure 6.20 we have shown three such marks, corresponding to $\frac{1}{4}$, $\frac{1}{2}$, and $\frac{3}{4}$.

Scale of Circles

Figure 6.21 shows a scale of 100 circles with areas proportional to $a, 2a, \ldots, 99a, 100a$, where a is the area of an unitary circle. Let us consider the circles in the lowest line corresponding to units $(1, 2, 3, \ldots)$. The first circles are clearly distinguishable from one another. However, when we go on to consider circles corresponding to 7, 8, 9, and 10, distinguishing between consecutive circles becomes far more difficult. The same can be said for the last column considered from the bottom (10) upward $(20, 30, \ldots)$. In the horizontal lines corresponding to $30+$, $40+, \ldots, 90+$, most circles appear almost equal; for high values, very few circles are distinguishable from one another.

These findings are not unexpected as the possibility of discriminating between two circles is dependent upon the ratios of their areas and not upon the absolute difference in their sizes. Bertin (1967, pp. 368–371) constructed a sequence of circles (*gamme naturelle*), in which one circle should be distinguishable from another. According to him, among the circles shown in Figure 6.21 we should expect to be able to distinguish: (a) between most circles included between 1 and 10, (b) five circles included between 11 and 20; (c) six circles included between 20 and 40; (d) three circles included between 40 and 60; and (e) five circles included between 60 and 100.

Bertin's conclusions are that (a) we can represent values V_i by a circle having an area proportional to V_i; (b) but that such accuracy is of little help, as the human eye can only distinguish between circles included in the natural sequence.

As we consider the Cartesian graphical presentation as the best one available, let us see what the application of (b) to Cartesian axes would mean. This is shown in Figure 6.22, where parallel lines indicate the limits within which we would consider all the included abscissas and ordinates equal. Obviously such an operation

Figure 6.21. Scale of 100 circles.

would degrade many Cartesian graphical presentations (apart from those requiring logarithmical transformations).

Scales of Squares and Hexagons

Figures 6.23 and 6.24 show, respectively, scales of squares and regular hexagons with areas proportional to a, $2a$, $3a, \ldots, 99a$, $100a$. The findings are similar to those found for the scale of circles. Here, too, the ability of the human eye to distinguish among most of the symbols shown is very limited.

Scale of Framed Rectangles

We saw in Section 5.7.2 that one expects from "good" maps that: (a) symbols representing values V_i will correspond to points i or to territorial units i to which they refer, and (b) such symbols will not overlap. To achieve these results, framed rectangles should be of moderate height. Figure 6.25 shows such framed rectangles with inked-in areas proportional to a, $2a$, $3a, \ldots, 99a$, $100a$.

We see that the possibility of distinguishing between pairs of framed rectangles is very limited. Actually we can discriminate only between pairs of rectangles having

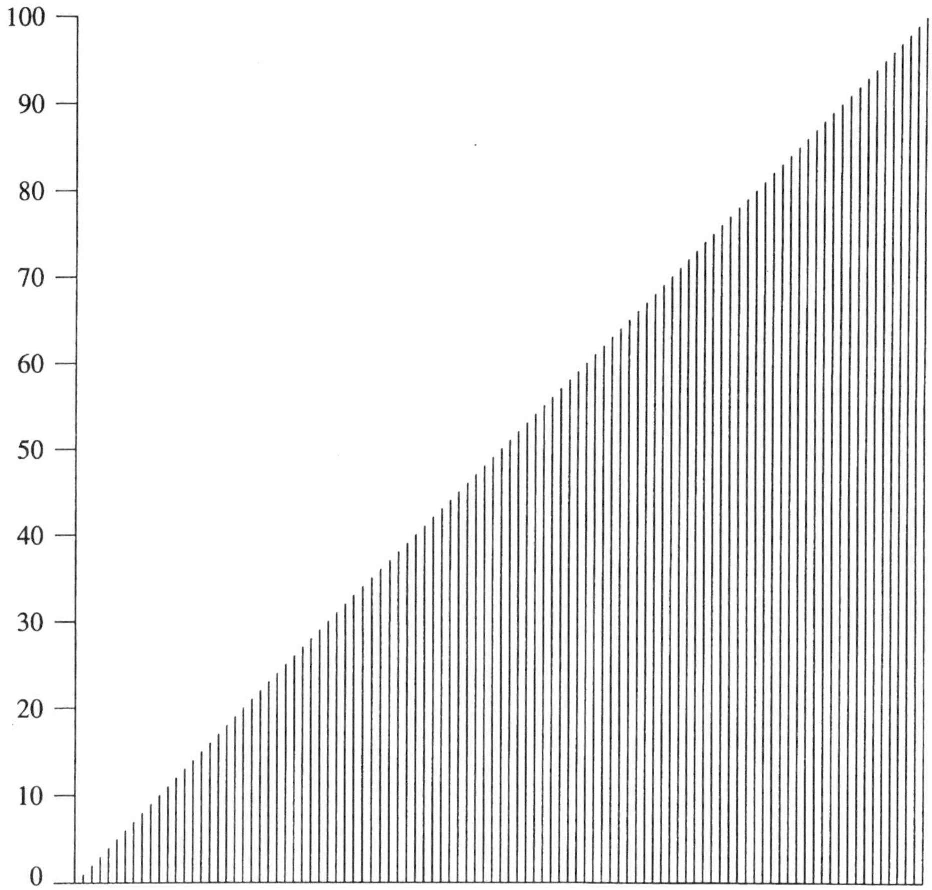

Figure 6.22. Scale of 100 parallel lines.

area ratios that are very different from 1. We can conclude that circles, squares, hexagons, and framed rectangles can theoretically assume an infinite number of sizes; however, in practice, there is only a very limited range of sizes for which direct visual comparison between pairs of symbols is possible. This is contrary to principle (6c) in Section 5.7.2, but GRP appear to correspond to it.

Another aspect of the distinguishability of symbols can be seen by comparing the map based on circles to the corresponding map based on GRP in Figures 6.26a and b, respectively. Suppose we wish to read numerical values given by the maps from the legends. This involves: (a) finding of the symbol on the map, (b) identifying the equivalent symbol in the legend, and (c) reading the numerical value from the legend. Steps (a) and (b) appear to be easier for GRP than for circles. The same can be said for other geometrical symbols. We may thus conclude that GRP also seem to be preferable to other symbols in regard to principle (6c) of Section 5.7.2.

Figure 6.23. Scale of 100 squares.

6.8. Choropleth Maps

The choropleth map has been and is still a very popular type of statistical map, perhaps even the most popular, and a bibliography of examples would include thousands of listings. The choropleth map is purported to represent values V_j related to territorial units j (such as regions and census tracts). The method in use is to cover the territory of each j on the map by a continuous graphical symbol S_j, which can be, e.g.: (a) a step in a scale ranging from white to black through increasing intensities of gray, obtained by shading, hatching, increasingly denser hatched lines, or uniformly distributed dots; (b) a color, such as one hue out of many, or the degree of intensity of a color in a brightness-saturation scale; (c) one of several different texture symbols; or (d) a conventional type of hatching or other pattern.

For a long time, analysts have been warning that thc use of these kinds of

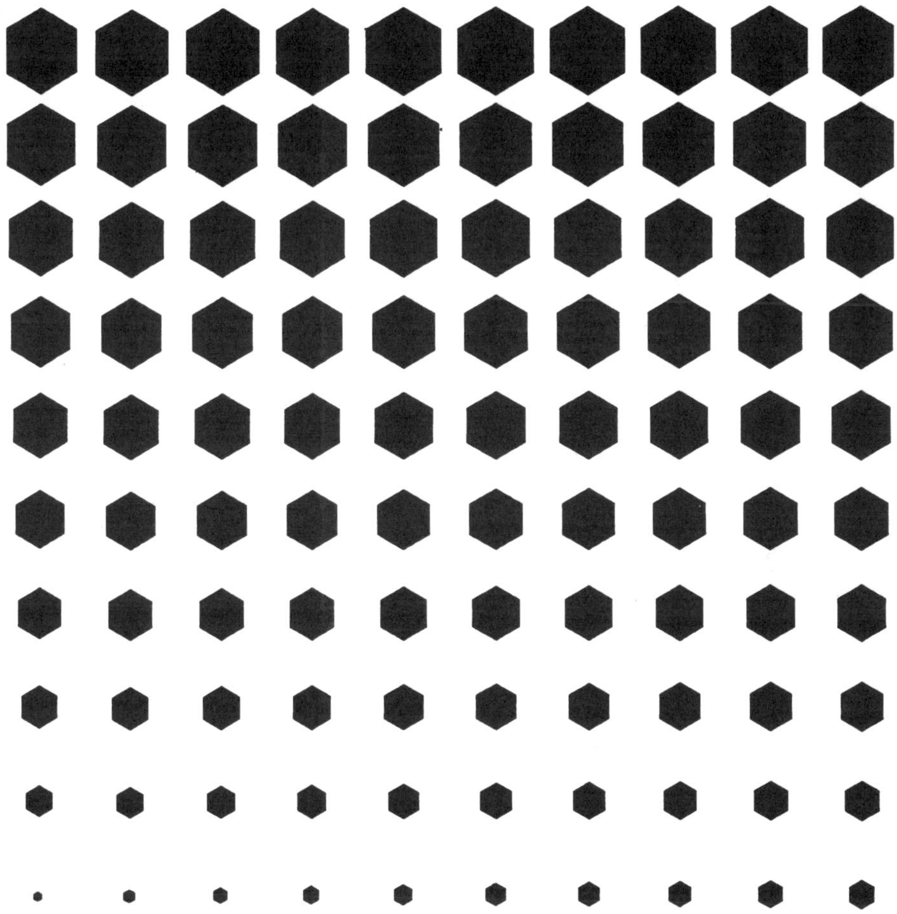

Figure 6.24. Scale of 100 regular hexagons.

symbols requires a study of their complex characteristics and properties and of their adaptation to the types of data shown on the maps as well as to what the presentation is designed to achieve. To find but a few out of many examples of such caveats, see: Bachi, 1968, 1975, 1978*c*; Cleveland, 1985, 1993; Crawford, 1976; Dent, 1993; Kosslyn, 1994; Monkhouse and Wilkinson, 1963; Monmonier, 1991, 1993; Robinson *et al.*, 1984; Schmid, 1983; Smith, W.H., 1924; and Tukey, J.W., 1979.

Constructing choropleth maps used to be difficult for technical reasons. In more recent years many of the problems involved have been solved through automation and there are many software packages that include a wide selection of the necessary symbols. However, as most of these packages unfortunately do not help the user to select the right symbols a large proportion of the choropleth maps produced today are constructed in an arbitrary way and are often biased and even faulty.

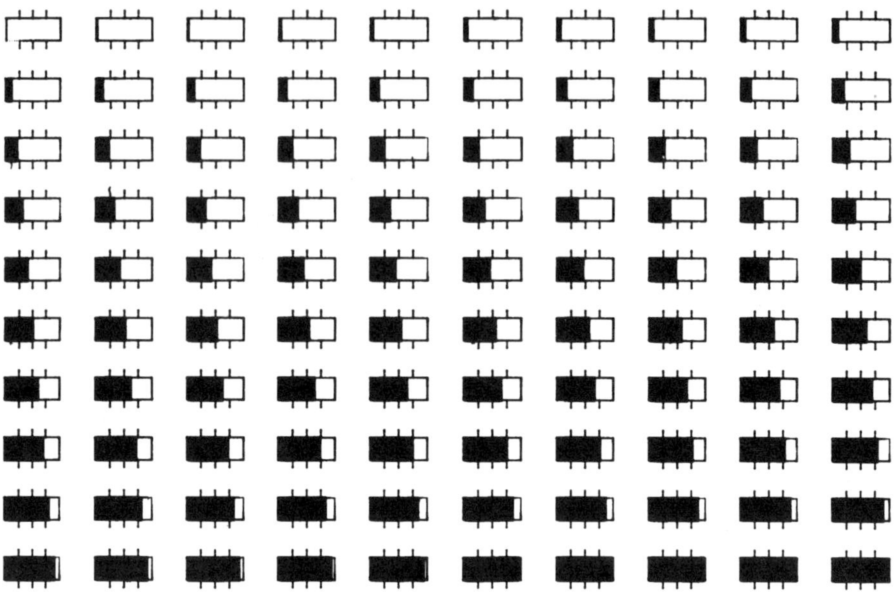

Figure 6.25. Scale of 100 framed rectangles.

The use of choropleth maps can be justified from a technical point of view only under very severe constraints, e.g.:

1. The territorial units referred to should be meaningful in the sense discussed in Section 6.6.3.
2. The selected symbols, S_j, should constitute one step out of a continuous or at least multistep scale based on clear technical criteria.
3. The selected symbols, S_j, should be clearly distinguishable one from the other.
4. The class interval corresponding to each S_j should not be chosen arbitrarily.
5. The mapmaker should take into account the basic fact that the visual impression conveyed by the map with regard to each territorial unit is equal to the product A_jS_j, where A_j is the area of j, which is entirely covered by the graphical symbol S_j. To understand the importance of this property, let us consider the following example: Suppose that we wish to show graphically the absolute size P_j of a population P in each of the territorial units j. The symbol S_j should then be proportional to the average density P_j/A_j of P in j. Unfortunately, in many of today's choropleth maps S_j is selected so as to be approximately proportional to P_j. Consequently territorial units with large areas are strongly emphasized in the map, independently of their actual population density and by changing the network of boundaries while keeping the scale of symbols constant, we can change the aspect of the map, at will.

Figure 6.26. Israel. Total population by 1055 settlements and cities, 1992: (a) map based on circles: · 1011 • 134,700 • 249,800 ● 356,900 ● 556,500; (b) map based on GRP: ▪ 1011 ▪ 10,118 ▪ 19,224 ▪ 50,590 ■ 556,500+ (Source of data: List of Locations, Israel Central Bureau of Statistics, 1993).

6.9. Ensuring Correspondence between the Data and the Graphical Rational Patterns

6.9.1. General Considerations

A basic problem in preparing GRP maps is finding a key to ensure correspondence between the values of the data to be shown and the GRP representing them that will enable users to interpret the numerical value of the GRP at any time. We have studied this issue in detail (Bachi, 1968, pp. 69–74) for many of the series of statistical data, geographic or nongeographic, that can be represented by means of GRP.

We have found that it is possible to base the key on clear and general scientific criteria taking into consideration the different natures of the various sets of data to be represented and the various possible purposes of the graphical format. While still far from being complete or perfect, the system we have proposed seems to enable us to eliminate much of the arbitrariness that affects many of todays statistical graphs. At the same time, it provides for large degree of flexibility in adapting the GRP graphs to a range of needs.

6.9.2. Key to GRP in Simple Maps

One of the main objects of this volume is the analysis of primary and derived geosets, as introduced in Chapter 3. Accordingly, we shall show here how the problem of preparing the key can be solved for simple maps representing primary geosets such as ${}^{g}P = \{X_i, Y_i, w_i^P\}$, where (X_i, Y_i) are the coordinates of each cell i, and w_i^P the proportion of population P found in cell i. The extension of the method to derived geosets follows immediately.

To make this presentation clearer, we take as an example the key used in the preparation of the geoset of the population of Italy in 1987, as shown in Figure 1.3 (Section 1.2). The original input is the distribution over 8086 *communi* of the population P having a total size $F^P = 57{,}290{,}519$. This distribution has been transformed into an equivalent distribution of $M = 10{,}000$ population elements over 1000 hexagonal cells of constant area $a \simeq 301$ sq km. The GRP scale considered is the one generally used in this book, which has steps corresponding to $1, 2, \ldots, 100$ or more.

The relative weight attributed to any cell i is $w_i^P = F_i^P/F^P$, where F_i^P is the size of the population in cell i. The number of population elements in cell i in the equivalent distribution is $M_i \simeq Mw_i^P$, where M_i is the integer nearest to Mw_i. On average, the number of elements of the actual population per unit of the equivalent distribution is $F^P/M = 5729$. The GRP of value S_i corresponding to any cell i may be considered to represent a class of frequencies centered at $M_i(F^P/M)$ and ranging from $(M_i - 0.5)(F^P/M)$ to $(M_i + 0.5)(F^P/M)$.

The GRP shown on the map have a very good degree of approximation: the error in reading the absolute value F_i^P from the GRP $M_i(F^P/M)$ is at most $0.5(F^P/M) < 2865$. The only cells that remain void are those in which $F_i^P < 2865$, which correspond to a density per sq km of fewer than 9.5 inhabitants. Table 6.1

Table 6.1. Key to the Legend of Figure 1.3

Value S_i of GRP	Relative frequencies $100w_i$ of the equivalent distribution included between (2) and (3)			Absolute frequencies in the original geoset included between (4) and (5)			Central value of the class of (4)–(5) $[M_i(F^P/M)]$
(1)	(2)		(3)	(4)		(5)	(6)
—	0	and	0.49	0	and	2,865	1,432
1	0.5	and	1.49	2,866	and	8,594	5,729
2	1.5	and	2.49	8,595	and	14,323	11,458
⋮	⋮		⋮	⋮		⋮	⋮
99	98.5	and	99.49	564,311	and	569,983	567,176
100	99.5	and	100.49	569,984	and	575,770	572,905

shows in detail the relationships between the values S_i of the GRP and the size of the population in the cells i. The same map is thus found to represent in a consistent way: (a) the relative size $w_i^P = F_i^P/F^P$, (b) the absolute sizes, and (c) the densities (explained below). This important result is due to the equality of the areas of the cells:

1. $S_i \simeq M_i$ is the number of the elements of the equivalent population found in i, on the assumption that the total size of the equivalent population is reduced to $M = 10{,}000$.
2. Multiplying $S_i \simeq M_i$ by the constant value $F^P/M = 5729$, we can find the size of the actual population in i. For example, if in a cell i there is a GRP with a value of $S_i = 6$ we can deduce that the size of the actual population in i is about 34,000, because $S_i(F^P/M) = M_i(F^P/M) \simeq F_i^P = 34{,}374$.
3. If the area of each cell is $A \simeq 301$ sq km, then taking F_i^P/A we get the density of the actual population in i. In the numerical example given under (2) the density of the cell with GRP $S_i = 6$ would be (34,374/301), i.e., somc 113 inhabitants per sq km.

Table 6.1, if fully detailed for each of the 100 steps of the GRP, would give a complete explanation of the key used, but it would be cumbersome to attach such a table to each map. In practice, abridged reasons of the key are possible and have been found of use for the different types of data and applications to which they have been adapted. For example:

1. It is possible to select out of the 100 possible steps only those actually used on the map. Figure 6.27 illustrates this approach for Figure 1.3 (Section 1.2). This method was used in the *Statistical Atlas of Italy* in order to facilitate comparisons among the values referred to different cells and it helped the interested reader compare the sizes of the population in cells corresponding to various metropolitan or urban areas. The geostatistical software package that is described in Chapter 17 enables us to obtain the values corresponding to the actual GRP appearing on the map directly.

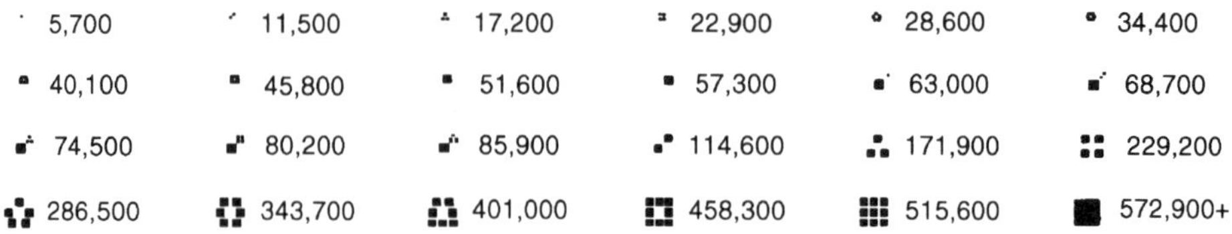

Figure 6.27. Key to Figure 1.3 (1000 individuals) (Modified from *Statistical Atlas of Italy*, 1988).

2. A much simpler approach is to use only the value V_i corresponding to a unitary GRP as the key. For example, in Figure 1.3, we could indicate only that a unit GRP corresponds to 5729. After the reader has become familiar with the 100 GRP ranging between 1 and 100, he/she can easily estimate the absolute value corresponding to any cell. This is the approach we adopted in an earlier work (Bachi, 1968).
3. It has sometimes been found both facilitating and convenient to indicate the values corresponding to the GRP 10 (57,290 in our example) and to GRP 100 (572,900) as well.
4. Another shortcut that was adopted successfully in the *Statistical Atlas of Italy* is the use of abridged scale, e.g., unitary steps between 1 and 15 and steps of 10 afterward (16–25, centered at 20; 26–35 centered at 30, and so on). Figure 6.27 is actually based on this abridged scale, as it was used in the *Atlas*.

Chapter 7

Comparative Maps

7.1. Using Graphical Rational Patterns for Maps of Geosets Derived from Comparisons between Primary Geosets

7.1.1. Selected Examples

We saw in Chapter 3 that many important aspects of geostatistical analysis can be based on geosubsets derived: (a) from the comparison between any primary geoset ${}^{g}P$ and the corresponding uniform distribution over the territory; and (b) from the comparison between two geosets ${}^{g}P$ and ${}^{g}Q$ distributed over the same territory. In Chapters 1 and 3 there were examples of maps representing such geosubsets by means of GRP. For instance: Figure 1.4 (Section 1.2) shows an example of GRP maps of surpluses of the actual distribution over the uniform distribution; and Figures 3.23, 3.24, and 3.25 (Section 3.7) show examples of maps of surpluses of ${}^{g}P$ over ${}^{g}Q$, surpluses of ${}^{g}Q$ over ${}^{g}P$, and matches of ${}^{g}P$ and ${}^{g}Q$, respectively. From a graphical point of view these maps were constructed according to criteria similar to those described in Chapter 6 for GRP maps of simple geosets.

The examples noted above illustrate the way in which GRP maps can help the reader to see quickly both the general macrographic similarities (or dissimilarities) between the geosets compared and the microgeographic characteristics. It often happens that at first sight the primary geosets being compared appear to be almost equal and only by referring to the maps of surpluses do we discover meaningful differences. For example, if we look at Figures 7.1 to 7.4, we see that Figures 7.1 and 7.2 show, respectively, two primary geosets, ${}^{g}P$ and ${}^{g}Q$, which indicate the distributions of children (aged 0–14) and individuals aged 65+ in Rome. Both distributions have been reduced to the same total size (10,000). At first sight, the two geosets appear to be more or less similar and do not seem to suggest that there is any clear and systematic differentiation between the distribution of children and the elderly.

Figures 7.3 and 7.4 show the geosubsets of the surpluses ${}^{g}(SPQ)$ and ${}^{g}(SQP)$, derived from ${}^{g}P$ and ${}^{g}Q$. Both maps are rather lightly marked, as the surpluses constitute only 23.2% of the elements of the equivalent distributions (both reduced

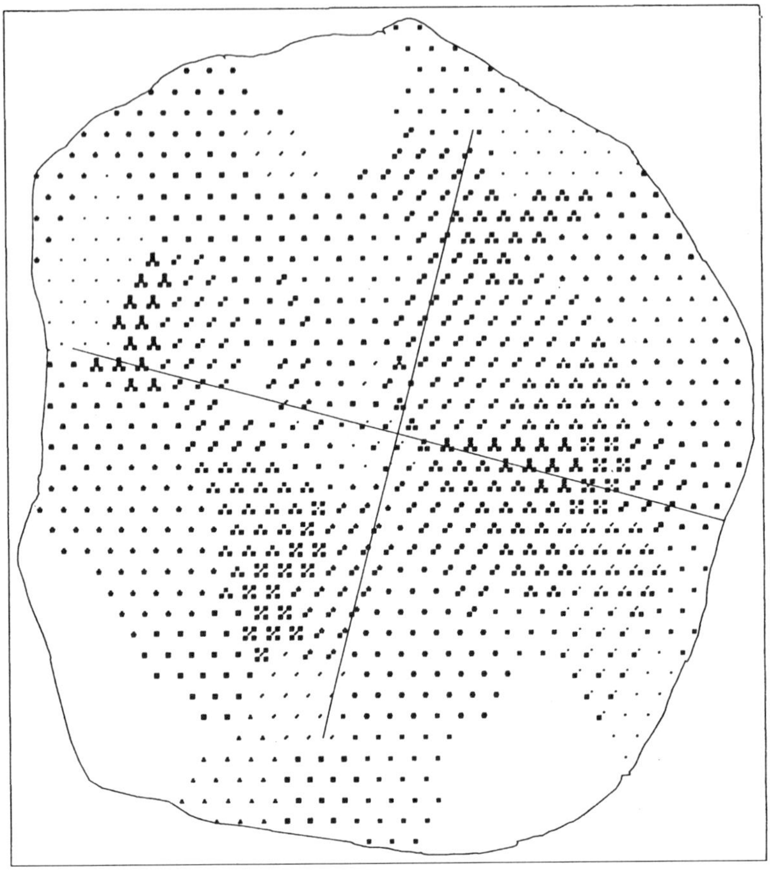

Figure 7.1. Rome. Children aged 0–14, 1981: . 45 ▪ 455 ▪ 910 ▪ 1365 ▪1912 (Source of data: Comune di Roma, Segretariato Generale il Direzione, Ufficio di Statistica e Censimento).

to 10,000). However, the patterns of distribution revealed by Figures 7.3 and 7.4 are very meaningful. They show that the surpluses of children are found mainly in the peripheral zones of Rome, while the surpluses of the elderly are primarily in more central areas, which is a feature common to many cities.

As research interest is focused primarily on these dissimilarities between the distributions, we might find it convenient to make our graphical presentation more efficient by utilizing two different features offered by the geostatistical software package: (a) inflating the geosubsets, and (b) putting the two maps of the surpluses together through transformations on a two-color map.

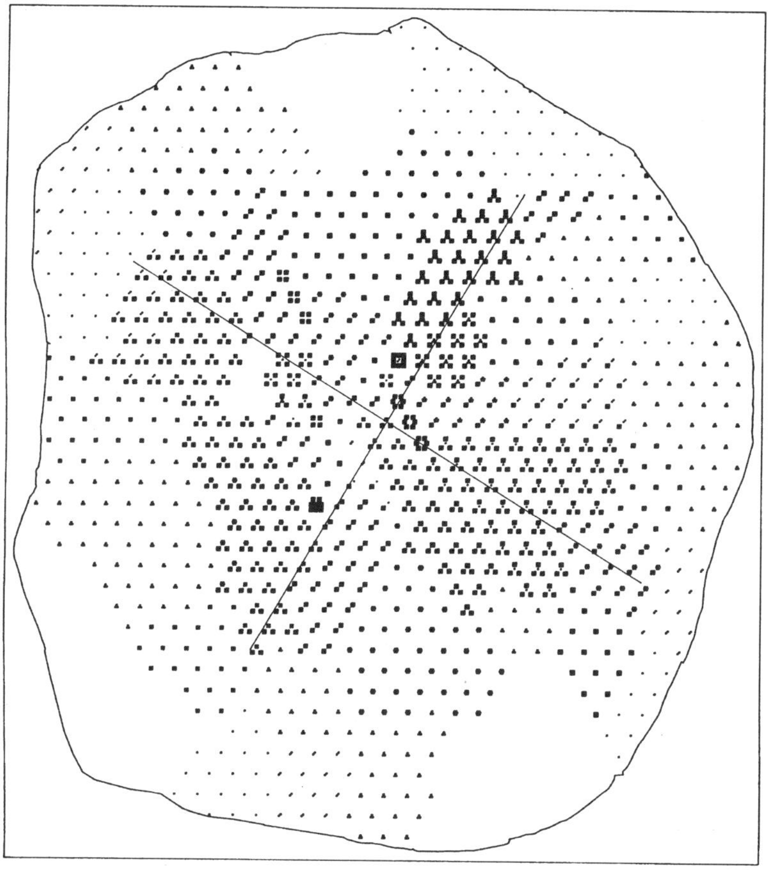

Figure 7.2. Rome. Individuals aged 65+, 1981: · 30 ▪ 297 ▪̇ 595 ∷ 1190 ▣ 2440 (Source of data: Comune di Roma, Segretariato Generale il Direzione, Ufficio di Statistica e Censimento).

7.1.2. Emphasizing the Dissimilarities between Two Primary Geosets by Using Inflated GRP

Figures 7.5 and 7.6 show transformed surpluses ${}^{g}(SPQ)$ and ${}^{g}(SQP)$, respectively, both distributions having been inflated so that the total size of ${}^{g}(SPQ)$ and ${}^{g}(SQP)$ amounts to the standard size of 10,000. Obviously, this simple transformation enables us to "see" the characteristics of the surpluses in a much clearer way than in the noninflated maps of Figures 7.3 and 7.4.

Figure 7.7 shows the map of the matches between children aged 0–14 and individuals aged 65+, also inflated to a total of 10,000. Comparing this map to that

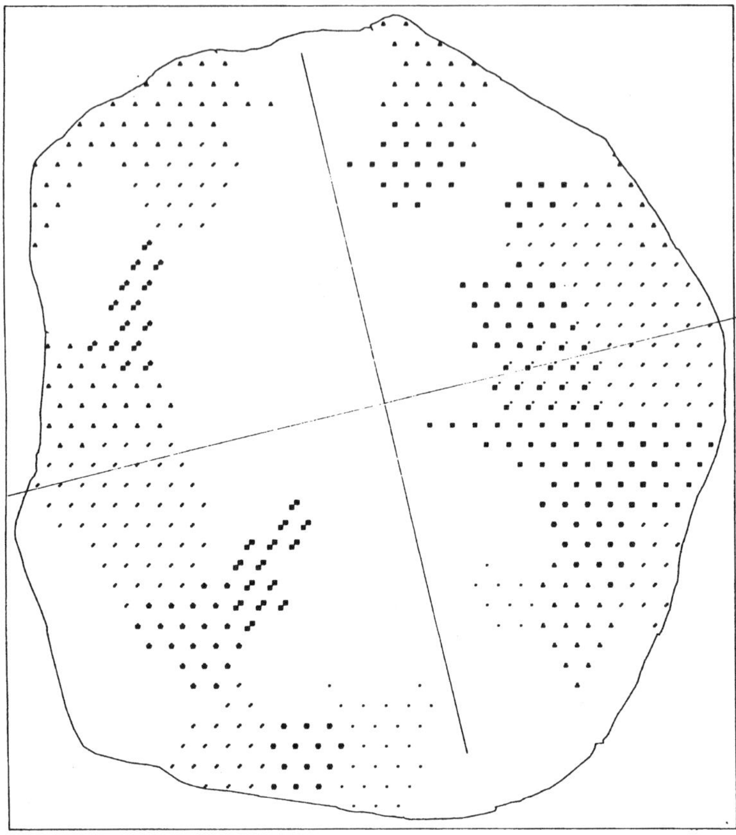

Figure 7.3. Rome. Surplus of children aged 0–14 over individuals aged 65+, 1981: . 45 ▪ 455 ▪ 910 (Source of data: Comune di Roma, Segretariato Generale il Direzione, Ufficio di Statistica e Censimento).

of the total population, as shown in Figure 7.8, we can see that they are similar, apparently indicating that the part common to both the children and the elderly is influenced primarily by the general characteristics of the population distribution.

7.1.3. Presenting Geosubsets of Surpluses on the Same Two-Color Map

The calculation surpluses is done by the computer on the basis of the frequencies of the populations compared (P, Q) in each cell of the territory reduced to the same total standard size (10,000). In each cell in which there are elements of P, Q, or both, we can have a surplus of P over Q, a surplus of Q over P, or no surplus at all. Consequently, surpluses of P over Q cannot be found in cells in which there are surpluses of Q over P (and vice versa), which offers us the possibility of presenting the distributions of P over Q and Q over P together in one map. To keep the two

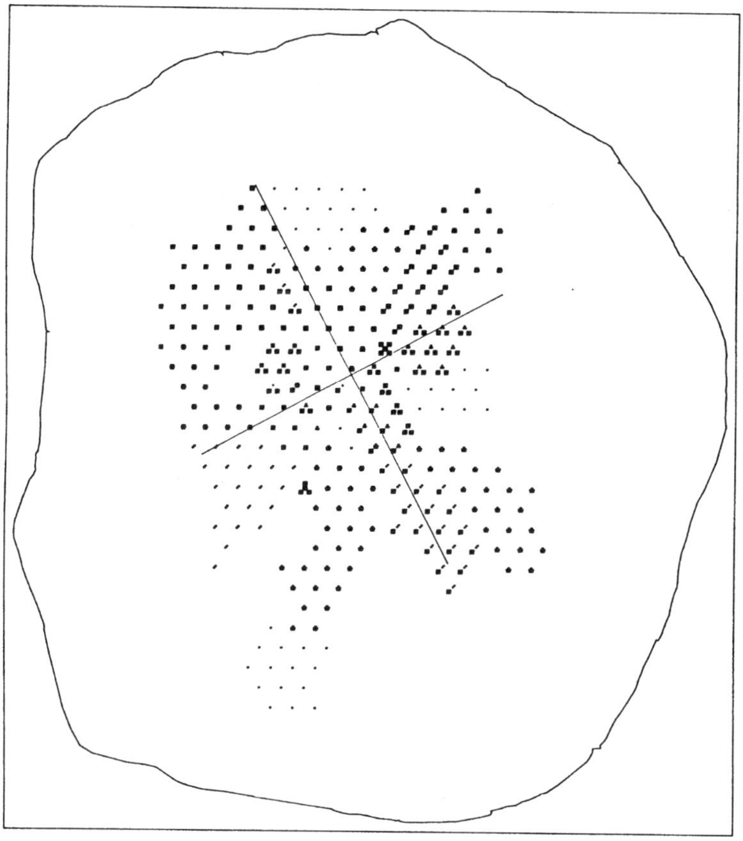

Figure 7.4. Rome. Surplus of individuals aged 65+ over children aged 0–14, 1981: · 30 ▪ 297 ⁘ 565 ⁘ 714 ⁙ 1458 (Source of data: Comune di Roma, Segretariato Generale il Direzione, Ufficio di Statistica e Censimento).

sets of surpluses distinct from one another, we can show them in two different colors. For example, Plate 6 shows: (a) in blue, the surpluses of children over the elderly in Rome (corresponding to Figure 7.5), and (b) in red, the surpluses of the elderly over children (corresponding to Figure 7.6), which enables us to see the primary dissimilarities between ${}^{g}P$ and ${}^{g}Q$ immediately.

Extensive use of these two-color maps in many research projects have proved their utility in facilitating complex comparative studies. For example, in one of our major studies (Bachi *et al.*, 1988) the distribution of Jewish minorities, ${}^{g}P$, was compared systematically to that of the total population ${}^{g}Q$ for many cities in Europe, North America, and Latin America. One of our most successful techniques was the use of two-color maps showing inflated ${}^{g}SPQ$ and ${}^{g}SQP$. Typical characteristics in the dissimilarities between Jews and the total population were discerned as well as similarities among the distributions of the Jewish minorities in most of the cities studied.

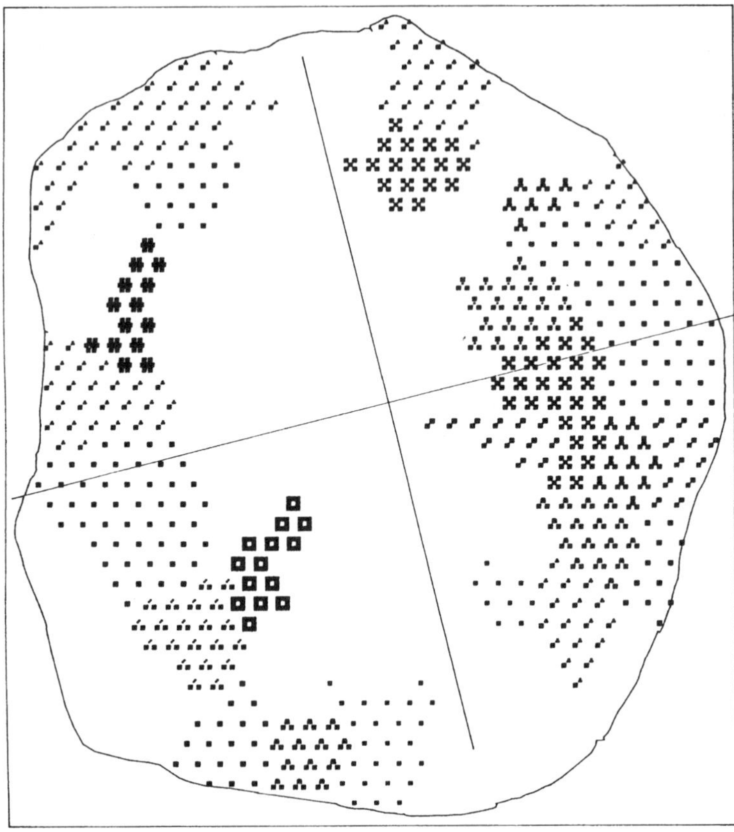

Figure 7.5. Rome. Surplus of children aged 0–14 over individuals aged 65+, 1981 (inflated): ▪ 42 ∴ 136 ◘ 842 (Source of data: Comune di Roma, Segretariato Generale il Direzione, Ufficio di Statistica e Censimento).

7.2. Presenting Two or More Sets of Geographical–Statistical Data on the Same Map

7.2.1. General Considerations

The two-color map of Plate 6 can be thought of as a particular solution of a more general and complex problem: that of presenting—by means of colors—the distribution of two variables or of functions thereof on one map which we shall discuss only in a cursory fashion here.

As has already been noted, in each nonvoid cell i in Plate 6 there is a GRP in one of the two colors showing the geosubset of either surpluses ${}^{g}SPQ$, or surpluses ${}^{g}SPQ$. However, sometimes the primary geosets ${}^{g}P$ and ${}^{g}Q$ or other pairs of variables have to be shown graphically. Accordingly, the values of ${}^{g}P$ and ${}^{g}Q$ have to be shown together in the same cell i, and at the same time have to be kept distinct from one

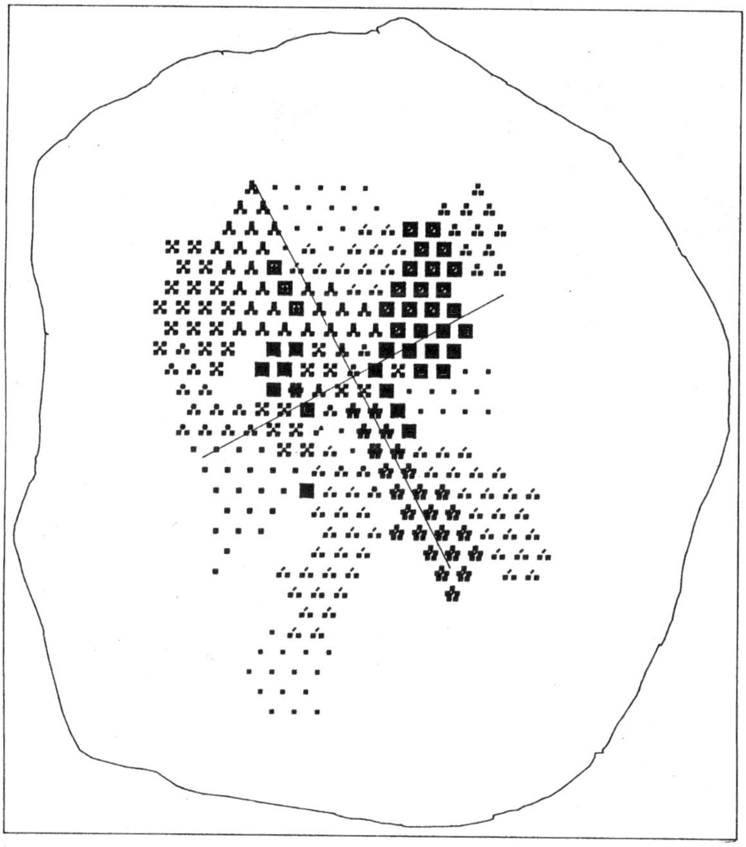

Figure 7.6. Rome. Surplus of individuals aged 65+ over children aged 0–14, 1981 (inflated): . 37 ▪ 357 ⁖ 619 ∷ 1452 ■ 3630 (Source of data: Comune di Roma, Segretariato Generale il Direzione, Ufficio di Statistica e Censimento).

another. An elegant answer to this puzzling requirement was proposed by Barabba and Finkner (1978) and was tested through extensive use in the U.S. Bureau of the Census (1978) (see also, Meyer *et al.*, 1975). The maps prepared were choropleths of census tracts in a selected city, which were purported to represent four levels of two variables (such as median family income A and proportion of high-school graduates B).

Let us denote the four levels of the variable A by A_1, A_2, A_3, and A_4, and the four levels of variable B by B_1, B_2, B_3, and B_4. Levels A_1, A_2, A_3, and A_4, are illustrated by four steps of a chromatic scale, progressing from yellow to pink, red, and dark red. Levels B_1, B_2, B_3, and B_4, are shown by a chromatic scale progressing from yellow to pale blue, blue, and dark blue. There are 16 possible associations of levels (say, A_1, B_1; A_1, B_2;...; A_n, B_n), each being illustrated by the mixture of the two colors corresponding to the associated levels. The maps thus show 1 of the 16 possible mixtures of the two colors over each census tract.

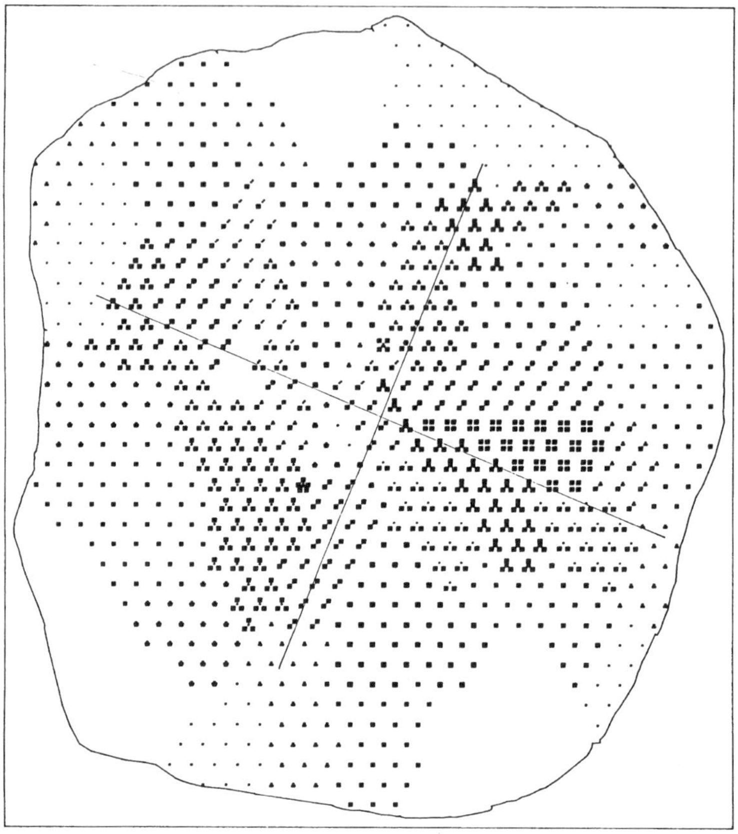

Figure 7.7. Rome. Matched between children aged 0–14 and individuals aged 65+, 1981 (inflated): Number of children aged 0–14: . 35 ▪ 314 ⁘ 598 ∷ 1327 ⁘ 1675; Number of individuals aged 65+: . 23 ▪ 205 ⁘ 456 ∷ 813 ⁘ 1210 (Source of data: Comune di Roma, segretariato generale il direzione, ufficio di statistica e censimento).

Unfortunately, users of these beautiful maps have often been found incapable of distinguishing quickly between these 16 color mixtures and even less capable of associating the mixed colors corresponding to A on one side and to B on the other. (For comments on this method see, e.g., Olson, 1975.)

7.2.2. Two-Color Maps Representing Primary Geosets over Cells of the Same Territory

In a different approach, which seems to achieve much simpler and clearer results, we consider a territory divided into cells i of the same size and shape. Assuming that we want to represent two primary geosets ${}^gP$, ${}^gQ$. We show four GRP in positions 1, 2, 3, and 4, respectively, over each cell i as shown in Plate 7. The GRP representing ${}^gP$ are shown in one color (say, red), while those representing ${}^gQ$ are shown in another color (say, blue). Using the standard GRP scale, we can show 100

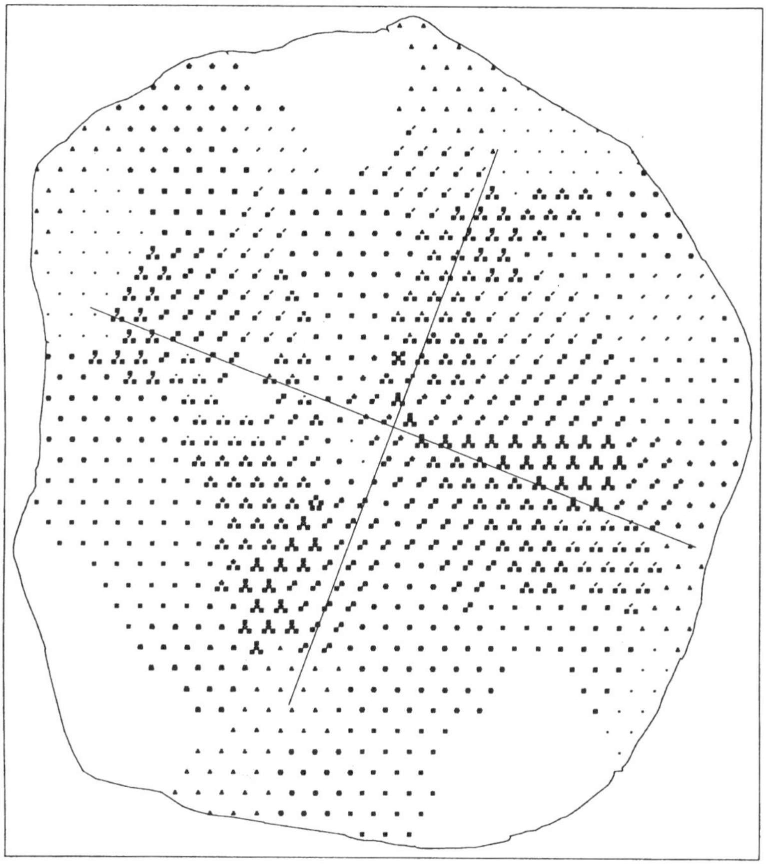

Figure 7.8. Total population of Rome, 1981: . 239 ▪ 2391 ⁘ 4544 ⁙ 11001 ⁘ 11,959 (Source of data: Comune di Roma, segretariato generale il direzione, ufficio di statistica e censimento).

distinct values for ${}^{g}P$ and 100 values for ${}^{g}Q$. To make the comparison clearer we might repeat the red GRP by placing them in positions 1 and 3 and the blue GRP by putting them in positions 2 and 4.

If we wish to adhere to the standard approach of using 1000 cells without changing the size of the GRP, we can multiply the format of the map by 4. Another possibility is to reduce the resolution of the graphical presentation by using, say, a division of the territory into only 300 hexagonal cells.* Two examples of this method are as follows:

1. Plate 9 shows the distribution ${}^{g}P$ of children (aged 0–14) in Rome in red, and the distribution of individuals aged 65+ in blue.
2. Plate 10 shows the distributions of both Asians and Hispanics in the United States in 1980.

*Whatever the solution, thematic resolution could be low so the method cannot be employed in maps of very large areas, groups of countries, continents, the globe.

7.2.3. Maps in Three or Four Colors

In the previous section we divided each cell into four parts (1, 2, 3, and 4), but used only two different colors (one for 1 and 3 and another for 2 and 4). However, there have been attempts to prepare maps with three colors (one, say, for 1 and 3 and two others for 2 and 4, respectively); and maps with four distinct colors for GRP corresponding to parts 1, 2, 3, and 4, respectively, as shown in Plate 8.

Plate 11, which illustrates a four-color map, shows the distribution of dwellings in Rome (1981) by the years they were built (up to 1919, 1920–1945, 1946–1960, 1961–1970). One can see from the map that in later periods there was relatively more construction on the periphery. In this map we divided the territory into only 300 hexagonal cells.

While these approaches seem to be promising, much work remains to be done regarding the colors to be used vis-a-vis the ability of the user to associate and dissociate the GRP shown in these colors.

7.3. Representing Rates and Ratios by Means of Graphical Rational Patterns

7.3.1. General Remarks

The territorial distributions of rates (including ratios) constitute a frequent, and very important theme, of statistical maps. Unfortunately, these graphical presentations are often given on choropleth maps referring to territorial units, without discriminating among the various types of units and the rates represented in them, which can lead to loss of information and biases as well as to misinterpretations.

We shall briefly discuss here several ways to improve the maps of rates: (a) representation of the rates by means of GRP; (b) distinction among the main types of rates; (c) use of proper keys for binding the rates to the GRP representing them; and (d) ways of making it possible for the user to evaluate the importance of each rate represented. We shall use the following notation:

1. Suffix j for a territorial unit, it being understood as desirable to restrict the graphical presentation to meaningful territorial units, as defined in Section 6.6.3.
2. Suffix c for a cell in a network of cells (such as hexagons) of constant area and shape, into which the territory has been subdivided.
3. P and Q for two "populations," understanding this term in its broadest sense (F^P and F^Q are used for the total size of P and Q, respectively, with F_j^P, F_j^Q and F_c^P, F_c^Q being their size in each j or c).
4. A for the total area relevant for the presentation of the rates (A_j for the area of j, $\bar{A} = (A/N^c)$ for the average area of each of the N^c relevant cells).
5. ${}^1R$, ${}^2R$, and ${}^3R$ with suffix j, ji, or c for the main types of rates, which will be explained and discussed below.
6. S_j or S_c for the approximate value of the graphical symbols used (e.g., colors and hatching in choropleth maps and GRP over cells).

7.3.2. Representing ^1R and ^2R by GRP

Conventionally, we denote the rates ^1R, ^2R, and ^3R (with suffix j or c) as follows:

1. $^1R_j = F_j^P/F^P$ and $^1R_c = F_c^P/F^P$, which denotes the proportion of population P found in j or c.
2. $^2R_j = F_j^P/A_j$ and $^2R_c = F_c^P/\bar{A}$, which might denote, e.g., the density of human population per unit of area in j or c, or the average value of agricultural production per unit of area.
3. $^3R_j = F_j^P/F_j^Q$ and $^3R_c = F_c^P/F_c^Q$, which denotes the ratios of the sizes of P and Q in j or c.

With regard to ^1R_c and ^2R_c, we already know that: (a) $^1R_c = F_c^P/F^P$ is equal to the relative weight of the geoset ^gP in any cell c; (b) the GRP map is a satisfactory representation of the geoset ^gP; and (c) the map representing ^gP virtually represents the densities ^2R_c as well.

With regard to $^2R_j = F_j^P/A_j$, if we assume that we use a graphical symbol, roughly proportional to ^2R_j, which covers the entire area of j on the map, then the visual impact of the area of j on the map would be roughly proportional to F_j^P. If the total inked-in area on maps prepared in this way is constant and proportional to F^P, then the maps of densities ^2R_j constitute acceptable presentations of the rates ^1R_j.

However, an error is frequently made in the current choropleth maps of the ^1R_j in that the graphical symbol used has an intensity S_j roughly proportional to ^1R_j. Consequently, if we cover the area of j with this symbol, then the impact of j will be proportional to A_jS_j. For a concrete example of the biases involved here, we consider a map of ^1R_j for the population of the counties of the United States: in the West there are many large counties with comparatively small populations, which are overemphasized, while in the East, many counties with comparatively large populations are inconspicuous as they cover only a small area (see Bachi, 1978*c*).

7.3.3. Representing ^3R by GRP

Let us now go on to consider $^3R_j = F_j^P/F_j^Q$ and $^3R_c = F_c^P/F_c^Q$, eventually multiplied by a convenient coefficient such as 1, 100, 1000, 10,000,... Generally, the direct object of our research is P, while Q is a comparative term that serves as a denominator. In many instances, the object of the calculation of these rates is to net out the influence of Q_j (or Q_c) from P_j (or P_c). The rates might measure, e.g., income *per capita*, birth rate per 1000 population, the proportion of individuals in the labor force who are seeking employment, the proportion of deaths from cancer caused by lung cancer, or the average number of pupils per class in elementary schools.

Most of the current choropleth maps representing ^3R_j use graphical symbols such as colors, grades of shading, or patterns of hatching which are purported to have a value S_j proportional to ^3R_j. This symbol is repeated over all the area of j. Consequently, the visual impression produced by j is proportional to $A_j{}^3R_j = A_j(F_j^P/F_j^Q)$. This is often meaningless and leads to biases such as the following, which were actually found even in maps produced by reputable statistical agencies (see Bachi, 1978*c*): (a) Considering, in a map of the ratios of income *per capita*, two territorial units A and B that have approximately the same ratio, A having a population much larger than B and an area much smaller than B, we see that B, which produces a large share of the total national income might be almost inconspicuous in the map, while the role of A might be unduly emphasized; (b) in a map of percentages of lung cancer over deaths from cancer in general, we might be impressed by a territorial unit j with a large area despite the fact that the proportion of the number of deaths from cancer in j over the national total is very small.

The obvious lesson to be learned from these erroneous presentations is that it is desirable to show the rate ${}^3R_j = F_j^P/F_j^Q$ and a measure of the relative importance of each rate together. Very often the simplest measure of this importance is the value of F_j^Q or of F_j^Q/F^Q.

Let us consider again the example of the study of the income distribution, where ${}^3R_j = F_j^P/F_j^Q$ is the income *per capita*, F_j^Q the size of the population in j, and the product $R_j^3 F_j^Q = F_j^P$ the total income produced in j. GRP may help us to solve the problem of presenting the level of the rates ${}^3R$ together with their importance. Various ways of obtaining this result have been suggested (see, e.g., Bachi, 1968, 1978*c*, and the *Statistical Atlas of Italy*). We shall now consider two of them:

1. Suppose that we wish to represent a rate ${}^3R_j$ over a map in which the territory has been divided into meaningful territorial units j. In the *Atlas*, many maps of birthrates (and, similarly, deathrates and natural increase) per 1000 population, were given on the basis of data for each of the 95 provinces j with excellent results. In each province j the birthrate ${}^3R_j$ was shown by a GRP of value S_j, proportional to a proper function of ${}^3R_j = F_j^P/F_j^Q$. The rate was repeated a number of times, proportional to the size of the population F_j^Q of province j. The visual impression produced by the repeated graphical symbols for province j is proportional to a function of the product ${}^3R_jF_j^Q = (F_j^P/F_j^Q)F_j^Q = F_j^P$, where F_j^P is the absolute number of births in j.

As an example of this approach, Figure 7.9 shows the ratios (%) of Afro-Americans to the total population, in each of the states of the United States in 1987. A proper function of these percentages is given by GRP, corresponding to the ratios (%). Each GRP is repeated a number of times, proportional to the total population of the state. These repeated symbols are distributed evenly over the territory of each state, and thus the total inked-in area of the GRP in each state is proportional to the number of Afro-Americans in that state.

2. Suppose that we have at our disposal values of F_c^P and F_c^Q measuring the size of two populations in each of the cells c of constant area and shape into which the territory has been subdivided. A proper function of ratios ${}^3R_c = F_c^P/F_c^Q$ can be shown by GRP. To indicate the difference of importance of the ratios presented, we can show these ratios in different gradations of a color, the intensity of the shades being selected so as to have some rough relationship to the size of the population ${}^gQ_c$ in each cell c.

This method was used in the *Atlas*, but some further experimentation is needed before we can reach clear-cut conclusions regarding the advantages and disadvantages of this approach.

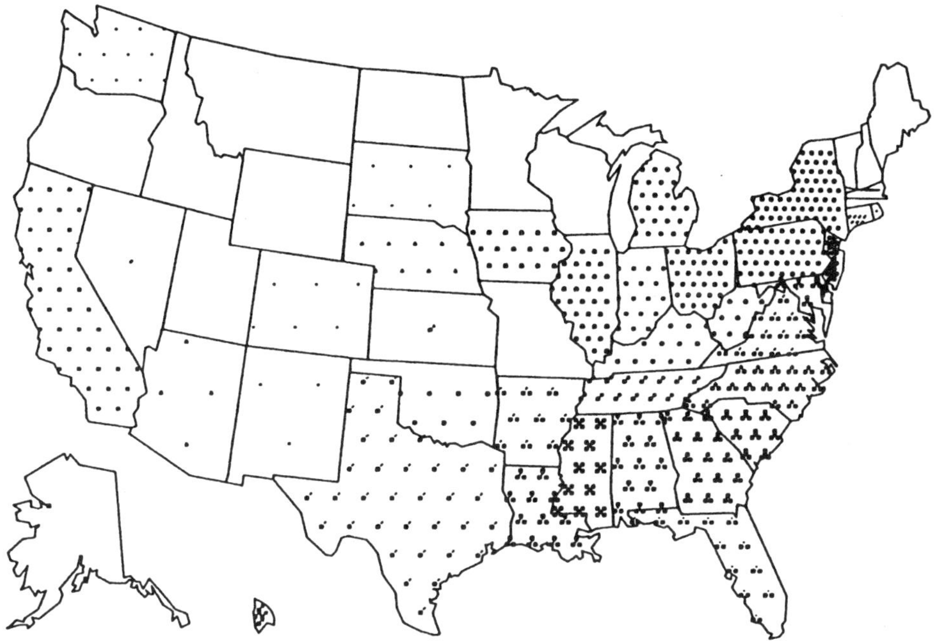

Figure 7.9. United States. Rates per 100 (%) of Afro-Americans by states, 1987. A single GRP pattern represents 250,000 individuals. (Source of data: Strategic Mapping—Counties Data Files, U.S.A.).

7.3.4. Keys for Showing Rates ${}^3R_j$ or ${}^3R_c$ on GRP Maps

The keys that are in current use for relating graphical symbols on choropleth maps to the classes of rates ${}^3R_j$ represented by them are often arbitrary. Having at our disposal a detailed GRP scale, such as the one progressing by unitary steps from 0, to $1, 2, \ldots, 100$, we can select from among various keys in several ways based on principles linking the key to the object of the graphical presentation.

The software for geostatistical programs available today enable us to generate maps based on each of these keys without difficulty. The approach we suggest is to calculate proper functions $f({}^3R_j)$ of the rates and present them graphically.

Let us denote by $\min({}^3R_j)$ and $\max({}^3R_j)$ the minimum and maximum values of ${}^3R_j$. Let us further denote the average of the ${}^3R_j$ by ${}^3\bar{R}$. The computer program has four options for representing these maps:

1. Each GRP gives the ratio as a percent of the geoset, rounded to the nearest integer; e.g., 77.7% is represented by a GRP of 78 units (see, e.g., Figure 7.9).
2. If we wish to eliminate the effects of the order of magnitude of the ratios we can use the quotient of ${}^3R_j/{}^3\bar{R}$ multiplied by 10.
3. Each GRP gives the ratio of any type of rates, e.g. percent rates per 1000, rates per 10,000, or rates per 100,000. The minimum positive rate is shown

by a GRP of 1 unit and the maximum by a GRP of 100 units. This option utilizes the entire range (1 to 100) of the GRP (see, e.g., Plate 12).

4. This option standardizes the rates as follows: the GRP are calculated by using the equation: $h + k[({}^3R_j - {}^3\bar{R})/\sigma_R]$, where σ_R is the standard deviation of ${}^3R_j$ and h and k are properly selected constants, such as, e.g., $k = 5$ and $h = 25$.

Options (3) and (4) eliminate the effects of both the order of magnitude and the variability of the ratios.

Apparently, maps in which the distribution of GRP is close to a "normal" one are easy to understand visually. We have found that the GRP distribution in option 3 is usually close to "normal", and so we strongly recommended it (compare Figure 7.9 with Plate 12).

7.4. Maps of the Products of Standard Scores Used for Calculating Ecological Correlations

A central theme of this book is the comparison between pairs of distributions (e.g., of populations) and of rates over the same territory: many methods are suggested for showing this comparison graphically, analyzing its various aspects by means of detailed geostatistical parameters and by measuring the dissimilarities of territorial distributions synthetically by means of the minimal distance between them.

We have not included the measurement and analysis of ecological correlation in this volume because the coefficient of ecological correlation is not sensitive to geographical characteristics of the distributions. This coefficient measures the association of the levels of the two variables compared within each of the territorial units or cells to which reference is made. In the simple example of the calculation of the correlation between the values V_j^A and V_j^B of two variables over each of the cells j, if we assume that associated pairs of values in each cell are kept constant and that the cells are permuted in all possible ways, then the coefficient of the correlations will remain constant and insensitive to the permutations.

Despite the exclusion of the coefficient of ecological correlation from our study, we shall briefly suggest below a graphical method that gives a distinct map for each of the permutations considered. For the sake of simplicity we will not deal here with the problem of the proper weighting of each cell. Let us denote the ratio between (a) the deviations of V_j^A from the average $\bar{V}^A$ of the values of A and (b) the standard deviation σ_j^A by v_j^A. We obtain

$$v_j^A = (V_j^A - \bar{V}^A)/\sigma_A \tag{7.1}$$

Similarly,

$$v_j^B = (V_j^B - \bar{V}^B)/\sigma_B \tag{7.2}$$

Table 7.1. Correlation Coefficients Obtained by Associating Age Groups with Periods of the Construction of Dwellings in Rome

Period of construction	Age group				
	0–14	15–29	30–44	45–64	65+
Up to 1919	0.1732	0.2575	0.2543	0.3056	0.4740
1920–1945	0.4785	0.5623	0.5600	0.6466	0.8314
1946–1960	0.8461	0.8850	0.8504	0.8940	0.7829
1961–1970	0.8918	0.8189	0.8577	0.7701	0.5337
1971–1981	0.6844	0.6077	0.6391	0.5361	0.3466

The coefficient of correlation r can be expressed as the average of the products $P_j = v_j^A v_j^B$ for each cell j:

$$r = (1/N^c) \sum (v_j^A v_j^B) = \sum P_j / N^c \tag{7.3}$$

where N^c is the number of cells with nonzero values of v_j^A and/or v_j^B. In fact, r is the Pearson correlation coefficient.

Considering each of the N^c cells j and neglecting zero values of v_j^A and v_j^B, we find four possibilities:

(1) v_j^A positive, v_j^B positive, P_j positive
(2) v_j^A positive, v_j^B negative, P_j negative
(3) v_j^A negative, v_j^B positive, P_j negative
(4) v_j^A negative, v_j^B negative, P_j positive

We can construct a map in which the GRP represent the P_j, which are distinguished by four colors denoting (1), (2), (3), and (4), respectively, in a distinct way. By way of illustration Plate 13 shows the comparison between the distribution of individuals aged 45–64 in Rome and the distribution of dwellings built during 1946–1960. In this example the coefficient of ecological correlation is 0.8940, which is the largest among the correlation coefficients obtained by associating age groups with dates of construction of dwellings (see Table 7.1).

7.5. Maps of Habitat and Nonhabitat

We recounted many examples of the division of the territory between habitat H and nonhabitat NH in Chapters 1 and 3 and others will be presented later (mainly in Chapters 8 and 9). H and NH are territorial geosets derived from the comparison between any primary population geoset ${}^gP$ and the uniform distribution ${}^gU$ over the territory. The various examples have made it clear that the geostatistical analysis of ${}^gP$ can be greatly enhanced by the geostatistical parameters of H and NH and by the maps representing H and NH. Strictly speaking, these maps are not "rational" maps in the sense used for GRP, but we mention them here because they can be regarded as a graphical tool coordinated with the GRP maps of primary and derived geosets.

The areas of the hexagons in the maps of H and NH are roughly proportional to those covered by the inhabited parts of each of the N^c cells into which the territory has been divided. The scale of the hexagons is potentially continuous, ranging from a minimum of 0% coverage of the cell to a maximum of 100%. We should, however, take into account what was noted in Section 6.7 concerning the scale of hexagons, i.e., that the human eye has a very limited ability to distinguish the proportion covered exactly. Accordingly, no attempt is made to include a key with the map indicating the percentage of coverage represented by the covered hexagons. While the absence of a key is not justified by the assumption that most users are probably not interested in a micrographic study of the coverage of selected cells, it is true that the location of the inhabited part within each hexagonal cell is conventional and that this imprecision within each hexagonal cell as well as the difficulty in evaluating it are of very little importance from the macrogeographical viewpoint. The primary objective of a map of H and NH is an overview of the general characteristics of H and NH.

Part III

Parameters Measuring Characteristics of Primary and Derived Geosets

Chapter 8

Proportion and Territorial Coverage of Surpluses, Matches, and Densities

8.1. Introduction

Having discussed the various important types of primary and derived geosets and their graphical presentations in Parts I and II, we shall now go on to elucidate the parameters that enable us to measure the primary characteristics of each geoset synthetically and comparatively, namely territorial coverage; density; mean and median location; other indicators of position, including principal direction; absolute territorial dispersion; main aspects of form (concerning the shape of territories and patterns of distribution of population elements), such as oblongity, looseness (or lack of compactness), asymmetry and clustering. In general (apart from a few exceptions, which we shall discuss later), these parameters are obtained from the geosets through the relative weights of the coordinates of corresponding elements or cells. There is thus a clear linkage between a geoset and its map and parameters.

In Chapter 8 we summarize some important aspects of the comparisons of ${}^{g}P$ to ${}^{g}U$ and of ${}^{g}P$ to ${}^{g}Q$. With regard to the first, we calculate the proportion of elements of the population that are in surplus over the uniform distribution; the proportion of the territory corresponding to surplus elements, to other parts of the habitat, and to nonhabitat. With regard to the second, we calculate the proportion of surpluses of P over Q (equal to that of Q over P) and the proportions of matching elements and of territory corresponding to surpluses and matches and the noninhabited remainder. Having refined the measures of territory we go on to refine the usual measures of population density by distinguishing among gross, net, and surplus density.

8.2. Proportions and Territorial Coverage of Surpluses and Matches between ^{g}P and ^{g}U and between ^{g}P and ^{g}Q

Let us go back to the combined analysis of primary geosets ${}^{g}P$ and ${}^{g}U$ of population geosets ${}^{g}SPU$, ${}^{g}SUP$, and ${}^{g}M(P, U)$, and of territorial geosets ${}^{g}H$, ${}^{g}NH$, and ${}^{g}TSPU$ derived from ${}^{g}P$ and ${}^{g}U$. We indicated in Sections 1.3 and 1.4 that these geosets are functionally connected. Accordingly, there are also important relationships among the parameters that summarize their various aspects. We consider here the proportions of the number of surplus and matching elements and the proportion of territory covered by habitat, nonhabitat, and surpluses.

We saw in Sections 3.2.4 and 3.2.5 that these parameters can be measured simply, as follows: Let us indicate the number of elements in the reduced geosets ^gP and ^gU by M (often used here as $M = 10{,}000$); the number of elements ^gSPU and ^gSUP by $M^s = M^{SPU} = M^{SUP}$; the number of elements of ^gP and ^gU that match by $M - M^s$; the area of total territory, habitat, nonhabitat, and territory covered by surplus elements by A^H, A^{NH}, A^{TSPU}, and A^T, respectively; then

(1) $\Delta_P^U = M^s/M =$ Proportion of surplus elements

(2) $1 - \Delta_P^U =$ Proportion of matching elements

(3) $\Delta_P^U = A^{NH}/A^T =$ Proportion of noninhabited area of territory

$$(4)\ \begin{cases} 1 - \Delta_P^U = A^H/A^T = \text{Proportion of inhabited area of territory} \\ A^{TSPU}/A^T = \text{Proportion of area covered by surplus elements} \end{cases}$$

Substituting ^gQ for ^gU in (1) and (2), we can then immediately use these relationships for a comparison of the geosets ^gP and ^gQ distributed over the same territory.

We saw in Section 3.2.4 how the numerical equalities between expressions (1) and (3), and (2) and (4) help to interpret the model of "movements" needed to transform ^gP into ^gU or ^gP into ^gQ. This model is often used in the literature to justify the use of Δ as an index of concentration or segregation. We will now discuss some of the general interpretations of concentration or segregation briefly.

8.3. Links between Δ_P^U and Δ_P^Q and Common Indexes of Concentration and Segregation

In early days, concentration of population over territory was generally measured by using indexes such as those of Lorenz and Gini, originally designed for the study of inequality in income distribution (see, e.g., Wright, 1937; Hoover, 1941, 1948; Duncan and Davis, 1953a,b; Hauser *et al.*, 1956; Duncan, 1957; Duncan *et al.*, 1961), which are based on concentration curves such as the one shown in Figure 8.1. This curve was obtained by (a) arranging the 3109 counties of the conterminous United States according to decreasing order of population density; (b) cumulating them from $j = 1$ to $j = 2$; from $j = 1$ to $j = 3$; ... from $j = 1$ to $j = 3109$; (c) calculating for each of the cumulated percentages X_i of population included in them and cumulated percentages Y_j of their areas. The diagonal shows the proportion of area Y_j^U that would correspond to a cumulative population X_j if there were no concentration $Y_j^U = X_j$. The concentration corresponding to each j is expressed by $Y_j^U - Y_j$. The vertical line in the figure shows the maximum distance $Y_j^U - Y_j$, and its

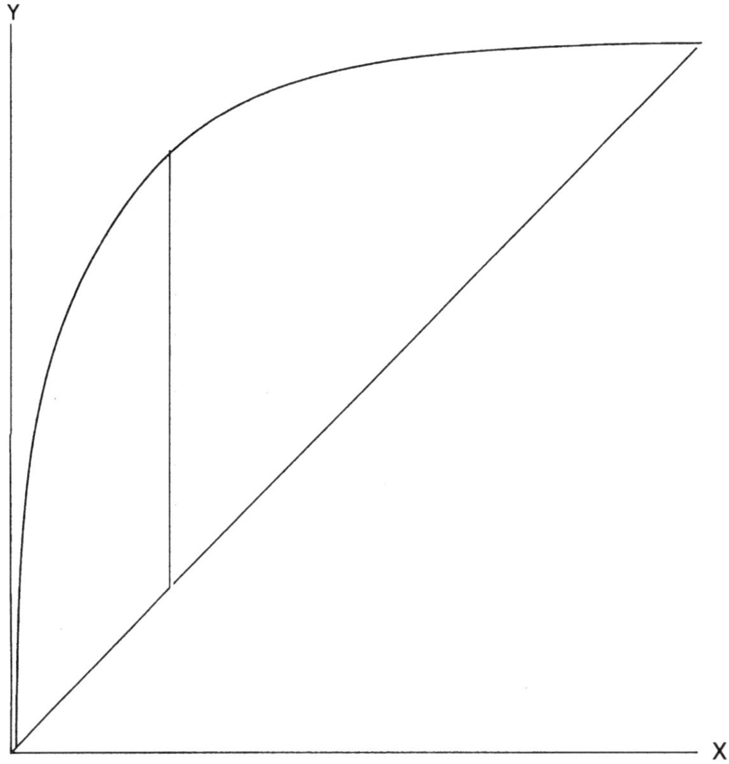

Figure 8.1. United States. Cumulated percent of population (X) by cumulated percent of area (Y) in 3109 counties, 1980 (Source of data: Strategic Mapping—Counties Data Files, U.S.A.).

numerical value is equal to the index of concentration

$$\Delta = \tfrac{1}{2}\sum |p_i - a_i| = 61.73\%$$

where p_i and a_i are, respectively, the proportions of population and area of any county i. This presentation is sometimes interpreted as follows: The object is to determine the population: the proportion of population covered is the "benefit" of the operation and the proportion of the area covered is the "cost." Δ corresponds to minimization of cost/benefit.

The approach used in the study of concentration as a measure of dissimilarities between the distributions of populations and areas has been used extensively to study dissimilarities between the distributions of any two populations or subpopulations. As this kind of research generally concerns comparisons between the distribution of a minority over a majority or a total population, it is often referred to as a measurement of segregation.

In recent decades studies of concentration and especially of segregation have increased multifold, and methodological research has proliferated. Indexes that have been in use for a long time have been subjected to criticism and new measures have been proposed. For instance, Massey and Denton (1988) and Harrison and Weinberg (1992) discussed twenty measures of ethnic segregation and classified these indexes according to their primary target, such as measures of evenness, exposure, centralization, clustering, and spatial proximity. Even earlier White (1986) had distinguished eight groups of such measures (see Section 13.1). Extensive research has been devoted to elucidating relationships among many of these measures, possible algebraic transformation of several of the indexes, properties and defects of each of them, and advantages and difficulties of use.

In this volume the study of territorial concentration is considered within the larger issue of comparing the geosets ${}^{g}P$ of an actual distribution of a population P over a territory T to that of geosets ${}^{g}U$ showing the uniform distribution over T. Similarly, segregation is studied within the broader comparison between the geosets ${}^{g}P$ and ${}^{p}Q$ of two populations. These comparisons have been shown to be complex, to involve many different aspects and to require integrated graphical presentation, analysis of many parameters, measurement of the various characteristics of ${}^{g}P$ as compared to ${}^{g}U$ or ${}^{g}Q$, and, eventually, synthetic measures (discussed in Part IV).

At this point we shall only consider briefly some criticism raised in the literature in regard to Duncan's Δ as a general measure of concentration or segregation and discuss whether these reservations are also valid with respect to Δ_P^U and Δ_P^Q within the narrower, but clearly defined, meanings introduced in this book.

8.4. Interpreting Δ_P^U and Δ_P^Q in the Light of Criticism of Duncan's Δ

8.4.1. Proportion of Area as a Measure of Cost

In the cost/benefit approach it is assumed that the cost of the population enumeration is proportional to the area covered, but this hypothesis is hardly acceptable as a general rule. In many cases we can assume that enumeration costs are linked to the type of distribution of the habitat, e.g., whether it is a solid block or sparred or divided into clusters far apart from one another. Accordingly, we shall take this factor into account in our study. If we assume that we have found a given value of coverage of habitat $(1 - \Delta_P^U) = (A^H/A^T)$, we would then investigate other relevant parameters of habitat, such as measures of territorial dispersion and lack of compactness.

8.4.2. Insensitivity to Geography

The criticism regarding the use of Δ as a measure of concentration can be expressed in a more general way by stating that Δ is insensitive to geography. This characteristic of Δ is well known and it has been found that the same value of Δ may correspond to very different types of distributions of populations over territories (see, e.g., Taeuber and Taeuber, 1965, and especially Appendix C on the measurement of residential segregation). As we showed in Section 3.9, this insensitivity to geography is common to all the parameters based on the study of relative weights of regions or cells, which do not take their locations into account. Most of the measures of concentration and segregation generally employed, as well as the coefficients of ecological correlation, share this insensitivity. Obviously, were we to take any such parameter as the only measure of concentration, segregation, or correlation, we might run into serious risks of misinterpretation. However, these risks are precluded if we consider Δ_P^U and Δ_P^Q to be just proportions of surpluses obtained from well-defined geosets ${}^gSPU$, ${}^gSUP$, ${}^gSPQ$, and ${}^gSQP$. Configurations of these geosets are designed to be shown graphically and their geographical characteristics, related to those of ${}^gP$ and ${}^gQ$, are measured by means of a battery of other parameters.

8.4.3. Controlling the Effects of Aggregation of Territorial Units over Δ_P^U and Δ_P^Q

Duncan *et al.* (1961) made it clear that values of Δ can be heavily influenced by the type of regional network used in their calculation. They found considerable differences in levels and in time trends for Δ as an index of concentration of the population of the United States, depending on whether the basic data came from large geographical units, states, economic subregions, state economic areas, or counties. Concentration values were found to increase with increasing fineness of territorial division. Recent literature has stressed again and again the risk involved in the unwelcome influence of the type of territorial units used; this is also true with respect to other territorial parameters. We have looked at this problem in connection with most of the geostatistical parameters introduced in this book, and the main conclusions we have drawn with respect to Δ_P^U and Δ_P^Q are:

1. If the gross territory T of area A^T is well-defined and if we know the location of each of the elements of P and U or of P and Q, then Δ_P^U and Δ_P^Q can be calculated accurately. Thus they are found to have "true" values.
2. If the populations under study are large and we have to use aggregated data, we can obtain good approximations to true values from aggregations based on very detailed territorial units. With the enormous progress in the preparation of detailed databases this is now often attainable.
3. Other parameters of H and NH, such as centers and measures of dispersion, are much less sensitive than Δ_P^U to the effects of aggregation. Accordingly we can often base the calculations of these parameters and the preparation of maps on the basis of less refined data, while basing calculation of Δ_P^U on more refined data, as suggested in (2).
4. Analogous conclusions can be drawn with regard to Δ_P^Q.

8.5. Calculating Gross, Net, and Surplus Densities of Population

8.5.1. Net versus Gross Density

The measure, analysis, and comparison of densities of populations over territories is considered to be an important aspect of the relationship between population and territory. The refinement in determining the area of a territory also helps to refine the study of density, which is particularly meaningful for human populations, but here we deal with it in general for any population.

We start by considering a simple example, such as was illustrated in Figure 3.3 (Section 3.2), in which we showed the distribution of F^P individual elements of a population (hotels) over a predefined territory T of area A^T. In this example, $F^P = 63$ and $A^T = 108.3$ sq km. The gross density of the population is $F^P/A^T = 0.582$ per sq km. In order to appreciate the meaning of this parameter we can look at Figure 3.5 (Section 3.2) in which the F^P elements are uniformly distributed over the area. Each of these is at the center of one of the F^P cells of area $a = 17.19$ sq. km into which the territory has been divided. As one cell has an area $a = A^T/F^P$, the density over one cell equals the average density over the entire territory: $1/a = F^P/A^T$.

The setting in Figure 3.5 is very different from that in Figure 3.3. Actually the elements of the population are found over a territory of area much smaller than A^T, which we have identified as the habitat of the population. It was shown in Figure 3.8 (Section 3.2) and has an area $A^H = 32.66$ sq km. Knowing the area of the habitat allows us to calculate the net density of the population as the ratio F^P/A^H, which in this example is 1.93 per sq km.

For a more meaningful illustration, let us return to the findings of Section 1.2 for the population of Italy in 1987. The total population in this case is 57,290,519; the total territory of Italy is 301,277 sq km; and the area of the habitat is 160,138 sq km. Thus the gross density is 190.2 per sq km and the net density is 357.8 per sq km.

This way of calculating net density helps to solve a problem we often encounter when comparing the population densities of various countries or regions. In these studies we often find that the usual gross density ratios F^P/A^T have little meaning when large areas of the territory under study are almost devoid of population. These areas may be desert or near desert, or covered by jungle or glaciers, or have climatic conditions that render them almost uninhabitable, but the absence of population might also be due to other reasons that cannot be clearly determined. Eliminating these areas from the study is often difficult. As early as 40 years ago, Duncan (1957) said that "any attempt to refine density figures by basing them on "net" rather than "gross" area encounters a considerable indeterminacy in the notion of "netness." Most presentations of "net" densities are neither clear nor convincing."

In the approach suggested here, we assume that each population element occupies a tract of territory. If the population is "large," the area of these tracts may be "very small." However, in the present context, we are not following the usual convention that the population elements are dimensionless, which seems to circumvent a well-known and difficult conceptual problem. This problem can often also be prevented by, quoting Duncan *et al.* (1961, p. 36):

> The inescapable margin of indeterminacy in density comparisons is a consequence of the mathematical character of the ratio: population/area. The ratio is defined only for non-zero area, and does not necessarily approach a unique limit as the size of a real unit approaches zero. If one reckons that a standing man occupies a square foot of land, then "standing room only" is equivalent to a density of 27,878,400 persons per square mile, a density perhaps locally approximated in ephemeral crowds. Under most circumstances, a standing man is sur-

rounded by at least a small area having zero density. The discontinuity of the distribution of population over space has the following consequence: Taking any arbitrary grid of areal unit boundaries and moving it haphazardly over a territory whose population is not uniformly spaced produces fluctuations in the density of each cell of the grid. By shifting the grid, one could maximize the grid-cell density in the vicinity of a specified point, but the density of any other cell would then vary in an unpredictable manner. Hence, under conditions that may differ rather widely, the densities of any two cells cannot be strictly comparable. It has to be recognized that any set of areal unit boundaries is essentially arbitrary from this standpoint. Consequently, when one computes the density of an areal unit, he obtains a figure which, though well defined, is necessarily relative to areal boundaries and hence fictitious in a sense.

8.5.2. Surplus Density

In the same way in which we calculate net density we can calculate surplus density. Let us indicate the absolute size of the population by F^P, its reduced size by M^P, and the proportion of elements of reduced geoset ${}^gSPU$ to the reduced number M^P of elements of ${}^gP$ (generally taken as $M^P = 10{,}000$) by M^{SPU}/M^P. $F^{SPU} = F^P(M^{SPU}/M^P)$ is the absolute number of surplus elements of the actual population. Referring F^{SPU} to the area A^{TSPU} of the territory corresponding to surplus elements we get the ratio F^{SPU}/A^{TSPU} as the density of surplus elements.

Going back to the example of the population of Italy in 1987 we find: total population = 57,290,519; proportion of surplus elements = 46.85%; absolute size of the surplus population = 57,290,519 × 0.4685 = 26,842,903; area of territory of surpluses of population over uniform distribution = 32,826 sq km; surplus density = 817.7 per sq km. It is evident that surplus density can also be interpreted as density of population in cells in which the density is above the average or, simply, as over average density or density of overpopulated cells.

In the *Statistical Atlas of Italy*, gross, net, and surplus densities were calculated for all the geosets considered. To enable comparative studies of characteristics of densities for many different phenomena we calculated relative net densities and relative surplus densities as follows:

$$\text{Relative net density} = \frac{\text{Net density}}{\text{Gross density}} = \frac{1}{A^H/A^T} = \frac{1}{1 - \Delta_P^U} \tag{8.1}$$

$$\text{Relative surplus density} = \frac{\text{Surplus density}}{\text{Gross density}} = \frac{\Delta_P^U}{A^{TSPU}/A^T} \tag{8.2}$$

These ratios are pure numbers, independent of the absolute sizes of the populations and areas of the phenomena studied. They are functions of Δ_P^U and of the ratio of territory of surpluses to total territory. This analysis has been found to be very useful in many instances.

8.6. Averaging Density Ratios for Cells or Regions

Gross, net, and surplus density ratios were obtained directly by comparing the sizes of relevant populations to areas of corresponding territories. However, we might be interested in determining the density of the population as an average of density ratios for each of the regions or cells of equal area a into which the territory has been divided. Referring back to Section 3.10, let us denote the density of the population in region or cell i by $h_i = P_i/A_i$ and the proportion of area in i by $a_i = A_i/A$. Calculating $\Sigma a_i h_i$ for all the i, we obtain again average density $\bar{h} = P/A$.

Let us consider a simple setting in which the territory T of area A is divided into a number $N^c = F^P$ of cells, all of which have area A/N^c. Then $a_i = 1/N^c$. Let us denote the number of inhabited cells by N^H. The relative weight of any inhabited cell over the entire habitat is $a_i^H = 1/N^H$. Then, taking the weighted average $\Sigma a_i^H h_i$, we obtain P/A^H. Similarly, taking the weighted average of densities of cells having a density greater than $\bar{h}$, we get a value equal to the surplus density.

It is known that Δ_P^U can also be expressed as follows:

$$2\Delta_P^U = \sum |p_i - a_i| = \sum a_i |h_i - \bar{h}|/h \tag{8.3}$$

where $\Sigma a_i |h_i - \bar{h}|$ is the average of absolute deviations of density ratios in each i from their average $\bar{h}$. The average of the densities obtained by taking the proportion a_i of areas in each i as weights corresponds to the usual approach of considering density as a measure of population load carried by a unit of area.

However, we might sometimes prefer to measure the average environmental density at which the population lives, which may be of value when considering the effects of concentration in cities. We assume that people found in i are in an environmental density $h_i = F_i^P/A_i$ and that an average element of the population p lives at environmental density $\bar{h}' = \Sigma p_i h_i$, the average environmental density at which the population lives. This may be of special interest in studies of the effects of urbanization. In this case we can calculate

$$\bar{h}' = \sum p_i h_i = (1/p) \sum (P_i^2/A_i) \tag{8.4}$$

If we follow this approach and wish to calculate an index of concentration, we find the preferred equation to be

$$\Delta_P'^U = \tfrac{1}{2} \sum |(p_i h_i / \sum p_i h_i) - p_i| \tag{8.5}$$

This index compares the proportion of people in each i weighted by their density to the unweighted proportion in i (see Section 3.10). Further, $2\Delta_P'^U$ can be expressed as the ratio of (a) the average of deviations of density ratios from their average to (b) this average:

$$2\Delta_P'^U = \sum p_i |h_i - h'|/\bar{h}' \tag{8.6}$$

Some results obtained with these equations follow:

1. *Population density in statistical regions of Jerusalem (1972)*:

Parameter Weighting:	by area	by population
(a) Average density per sq km	$\bar{h} = 2933$	$\bar{h}' = 14{,}105$
(b) Average deviation of densities	3404	9832
(c) Ratio of (b) to (a)	1.160	0.697
(d) Concentration index	$\Delta_P^U = 0.580$	$\Delta_P'^U = 0.3485$

The ratio $\bar{h}'/\bar{h} = 4.81$ is much larger than its minimal value 1. As expected 2(d) = (c) for both calculations.

2. *Population densities in Israel (1972):* In Israel, which is a country with a considerable proportion of semiarid areas, average density of population per sq km calculated in the usual way, F^P/A, was found to be $\bar{h} = 154$ individuals per sq km; $\bar{h}'$ calculated by dividing the country into 40 natural zones was found to be $\bar{h}' = 1951$. Using rougher divisions we found $\bar{h}' = 1878$ as an average for 14 subdistricts and $\bar{h}' = 1798$ for 6 districts.

8.7. Coverage and Density for Italy

8.7.1. Territorial Coverage of the Habitats for 385 Geosets

In Section 1.5 we gave examples of territorial coverage $1 - \Delta_P^U$ and of the proportions of the surpluses of the actual over the uniform distribution Δ_P^U for several geosets in Italy. The similar but more comprehensive study of $1 - \Delta_P^U$ and Δ_P^U for 385 geosets carried out in the *Statistical Atlas of Italy* made it possible for the analysts of that work to elucidate important aspects of the geosets for which these parameters were calculated. In general, it has been found that territorial coverage varies widely among geosets. Most of the geosets studied cover only modest proportions of the territory of Italy, as shown by the following data for the *percentage of the territory of Italy covered by the habitat*:

Percentage:	3–9.9	10–19.9	20–29.9	30–39.9	40–49.9	50–59.9	60–69.9	70 and over
Number of geosets:	5	23	31	70	97	107	40	12
Total number of geosets: 385								

Half of the geosets studied cover less than 36.1% of the total territory of Italy; three quarters cover less than 53.8%. Low territorial coverage has often been found to be a symptom of the concentration of geosets over tracts of land that have peculiar characteristics, worthy of research. For example, it has been found that many demographic, social, and economic geosets are concentrated mainly in metropolitan, urban, or semiurban areas, or among those areas in subareas with especially strong socioeconomic development, or in zones or spots attractive for national or international tourism. Several geosets relevant to agricultural have been found to have comparatively very small habitats, owing to such factors, as climate, water resources, favorable soil characteristics and proximity to relevant industries or markets.

Sometimes low coverage is accompanied by low territorial dispersion; however, in other cases a different or even opposite setting is found: one typical example is that of the geosets that are concentrated mainly in urban areas that have small territorial coverage and wide territorial dispersion, and consequently are very far from being compact. The complex relationship among lack of compactness, dispersion, and coverage is discussed in Section 11.5.

8.7.2. Evolution of Density and Territorial Coverage of a Habitat over Time

The study of the change of the territorial coverage of a phenomenon over the course of time can be of particular interest. One example is given in Table 8.1, which shows the continuous decrease (with attenuation in the last periods) of the proportion of habitat of the total population of Italy between 1871 and 1987. This population has increased continuously during these 115 years (from 27,300,000 to 57,300,000). At first sight we might expect that its territorial coverage should also have increased. The fact is that it actually decreased might seem paradoxical, but can be explained as follows:

1. The proportion of habitat is measured independently of the actual size of the population. We reduce the population enumerated at each census to a total of $M = 10{,}000$ and compare the geosets thus obtained to the uniform distribution over the territory of 10,000 population elements.
2. As a consequence of growing urbanization as well as other factors, the population tends to become more and more concentrated over the course of time (apart from the last periods). With the increased value of the concentration index Δ_P^U it is obvious that $1 - \Delta_P^U$ should decrease.
3. The increase of population over an unchanged gross territory determines an increase of gross density, as shown by the third column of Table 8.1. The net density, calculated by dividing the total population by the area of the habitat, is obviously much greater than the gross density and has increased at a more rapid rate.
4. The concurrent increases in the population and in Δ_P^U have caused a big increase in the absolute size of the surplus of the actual population over uniformly distributed population. In the meantime, owing mainly to increasing urbanization, the proportion of territory with surpluses has decreased slightly. As a result of these two opposing developments, the density of the surplus population has increased very considerably (see the fourth and fifth columns of Table 8.1).

Table 8.1. Territorial Coverage and Density of the Population of Italy (1871–1987)

Year	Size of population (10^3) (1)	Total territory covered by habitat (%) (2)	Population density[a]		Density of surpluses of population over uniform distribution[a] (%) (5)	Total territory covered by territory of surpluses (%) (6)
			Gross (3)	Net (4)		
1871	27,300	66.3	91	137	234	13.2
1881	28,952	66.0	96	146	252	13.1
1901	32,963	65.1	110	169	299	12.9
1911	35,842	64.2	119	186	339	12.7
1921	39,397	64.1	131	205	394	12.0
1931	41,043(?)	63.1	137	217	432	11.8
1936	42,399	62.5	141	226	466	11.5
1951	47,516	61.1	158	259	556	11.1
1961	50,624	57.9	169	292	687	10.4
1971	54,137	54.1	180	334	793	10.5
1981	56,557	53.3	188	354	819	10.8
1987	57,291	53.4	191	358	818	10.9

[a]Individuals per sq km.

Taking the first point (1) into account, we can explain another seemingly paradoxical finding: a small subpopulation can have a habitat larger than that of the entire population to which it belongs. For example, the percentage of the Italian territory covered by the habitat in 1981 was found to be: 58.47% for widowers, as compared with 53.28% for the entire population.

The average value of Δ_P^U for the 385 geosets analyzed in the *Atlas* was found to be 55.3%. This parameter can be interpreted as indicating concentration, proportion of surpluses of population over uniform distribution, and proportion of nonhabitat territory. The average of the proportions of the territory covered by surpluses has been found to be 11.2%. This value is smaller than (a) the proportion of the habitat and (b) the proportion of the surpluses. Finding (a) is obvious, because the territory covered by the surpluses represents only part of the habitat. Finding (b) has been empircal for all the geosets studied.

It has also been found that there is a negative correlation (Pearson $r = -0.65$) between the proportion of the surpluses over the population and the proportion of territory of surpluses over the total territory. This suggests that higher values of surpluses are associated with more restricted territories and vice versa.

The concept of surplus density may appear to be somewhat complex: it is obtained by taking the ratio between (a) the absolute value of surpluses of population over uniform distribution, and (b) the area occupied by these surpluses. As already noted, this parameter can be interpreted simply as the density of "overpopulated" cells, i.e., the cells with a higher density than the average.

8.7.3. Coverage of Other Subterritories

The proportion of the territory of Italy covered by various types of subterritories was calculated for each of many hundreds of geosets in the second volume of the *Atlas*. Table 8.2 shows the averages, the standard deviations, the first and third quartile, and the nineteeth percentile of these values.

Table 8.2. Percentage of the Territory of Italy Covered by Various Subterritories

Subterritory considered	Average	Standard deviation	First quartile	Third quartile	Nineteenth percentile
From comparisons between ${}^gP$ and ${}^gQ$ (385 geosets)					
Habitats	44.81	14.85	36.08	53.82	62.97
Nonhabitats	55.25	14.83	46.32	63.94	76.21
Territory of surplues of P versus uniform distributions	11.27	22.16	1.79	12.85	14.31
From comparisons between ${}^gP$ and ${}^gQ$ (177 geosets)[a]					
Territory of matching between P and Q	39.41	11.93	30.67	47.66	51.42
Territory of surpluses of P over Q	7.56	5.53	3.54	9.91	14.86
Territory of surpluses of Q over P	11.08	8.28	4.17	15.19	20.48
Void territory	41.95	10.90	36.19	47.66	55.73

[a]In general, P is the geoset constituting the main subject of research, and Q is a more general geoset taken as a term of comparison.

It appears from these parameters that:

1. The coverage is highest for nonhabitats, followed by habitats and void territories (where there are no elements of ${}^gP$ or of ${}^gQ$).
2. However, for these subterritories, too, coverage exceeding 75% of the total territory is rare; for habitats, void territories, and territories of matches, three-quarters of the geosets considered have a coverage that only just exceeds 50%, or is even lower.
3. The territories of surpluses of ${}^gP$ over uniform distribution, of ${}^gP$ over ${}^gQ$, and of ${}^gQ$ over ${}^gP$ have very low coverage. For surpluses of ${}^gP$ over ${}^gQ$, the averages of coverage are less than 8% and only for 10% of the geosets considered do the averages exceed 15%.

8.8. Coverage and Density for the United States

8.8.1. Coverage and Density for the Total Population Geoset

Figure 8.2 shows the geoset of the total population of the conterminous United States in 1980. The territory has been divided into 1000 hexagonal cells and the GRP for each cell has been obtained on the basis of primary data from 3109 counties. Comparing the total population geoset ${}^gP$ to the geoset ${}^gU$ of the uniform distribution of 10,000 elements over the 1000 cells, we obtain: (a) Figure 8.3, which shows the habitat ${}^gH$ (in black) and the nonhabitat ${}^gNH$ (in white); and (b) Figure 8.4, which shows the major clusters of surpluses of the actual population over the uniform distribution.

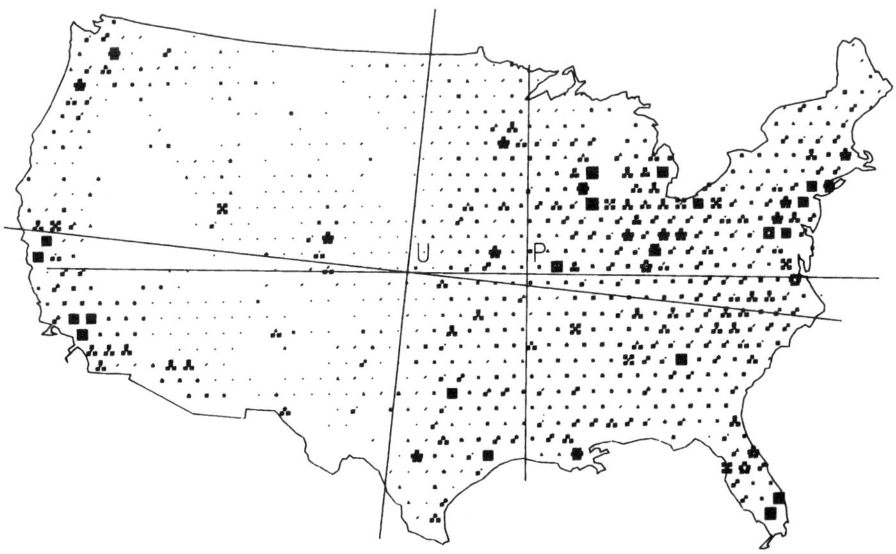

Figure 8.2. Total population of the United States, 1980: . 22540 ▪225,402 ▪450,809 ⁘ 1,137,023 ■ 2,840,098+ (Source of data: Strategic Mapping—Counties Data Files, U.S.A.).

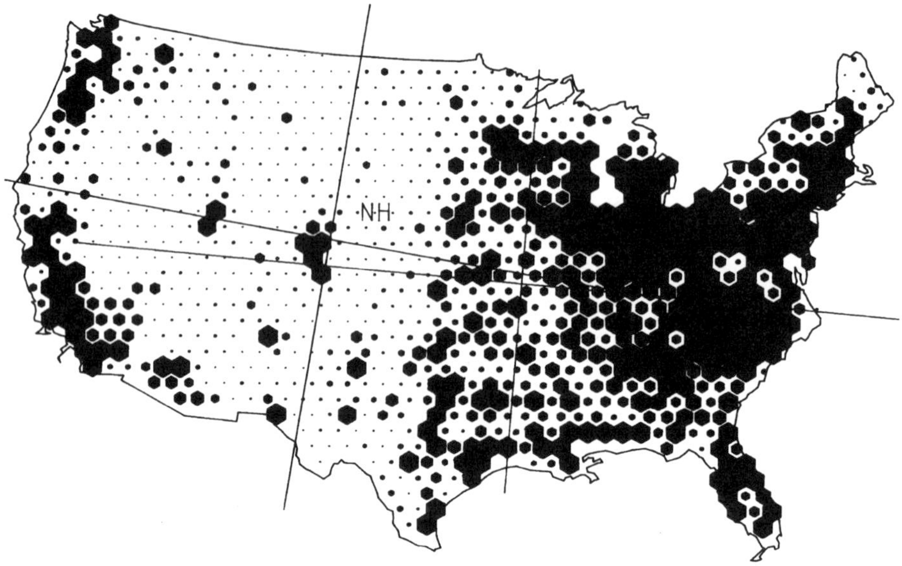

Figure 8.3. United States. Habitat and nonhabitat of total population, 1980: (Source of data: Strategic Mapping—Counties Data Files, U.S.A.).

The nonhabitat covers a wide and rather compact territory (forming the 57.4% of the area of the conterminous United States). The nonhabitat is found mainly over the West–Central and Mountain Census divisions. The compactness of this nonhabitat is attenuated by the existence of a large number of very small nuclei of population and by several larger inhabited tracts corresponding mainly to urban agglomerations. By contrast, the habitat, covering only 42.6% of the territory, is formed mainly by two separate blocks, which correspond roughly to the Atlantic and Eastern–Central Census divisions and to the Pacific division.

The percentage of surpluses of the actual over the uniform distribution is $\Delta_P^U =$ 57.4% of the total population. They are found over a territory covering only the 10.6% of the total area of the conterminous United States. Using a technique described in Section 11.9, we can calculate that 89.4% of the surpluses are found in nine major clusters, as can be seen in Figure 8.4. The largest of these clusters is in the East, the second and third are in California, the fourth and sixth in Florida, the fifth in Louisiana, the seventh and eighth in Texas, and the ninth in Missouri.

As the population of the conterminous United States in 1980 was 225,179,200 we can calculate the following ratios of density (persons per square mile): (a) gross density = 76.56; (b) net density = 178.5; (c) density of surpluses = 413.97; (d) relative net density (b/a) = 2.3; and (e) relative surplus density (c/a) = 5.4.

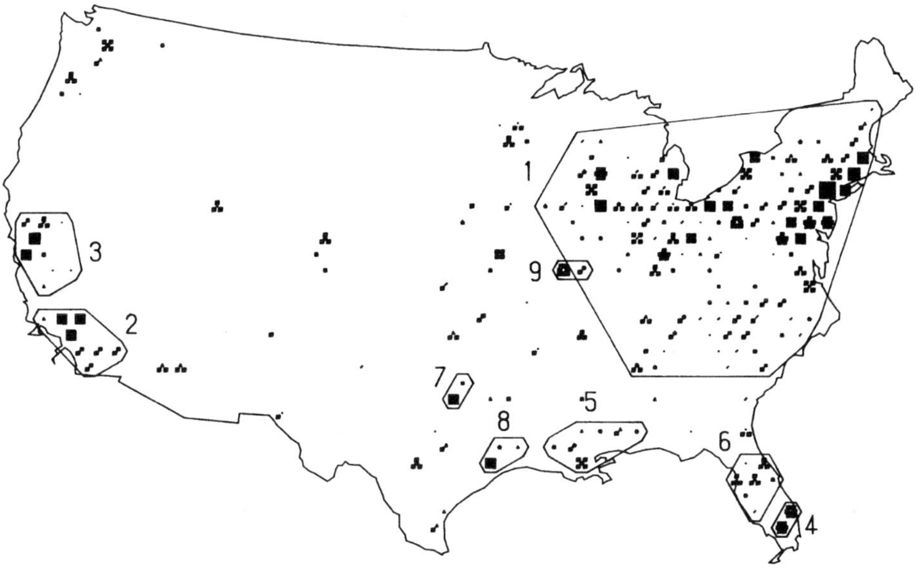

Figure 8.4. United States. Major clusters of surplus of total population, 1980: . 22,540 ▪ 225,402 450,809 1,262,266 ■ 2,276,587+ (Source of data: Strategic Mapping—Counties Data Files, U.S.A.).

8.8.2. Coverage and Density for Whites, Afro-Americans, Asians, and Hispanics

An analysis similar to the one above was done for several subpopulations: Whites, Afro-Americans, Asians (mainly Japanese, Chinese, and Filipinos), and Hispanics (mainly of Mexican, Puerto Rican, and Cuban origin). Maps for Whites are not reproduced here, for they are very similar to those for the total population, as the Whites formed the great majority of the population (83% in 1980). Maps for habitat have also not been reproduced.

The maps given here show, respectively, the detailed distributions by means of GRP and the surpluses of actual over uniform distributions. Figures 8.5 and 8.6 refer to the African-Americans, Figures 8.7 and 8.8 to the Asians, and 8.9 and 8.10 to the Hispanics. The parameters are given in Table 8.3.

1. *Afro-Americans.* The concentration of this subpopulation is very strong ($\Delta_P^U = 73.4\%$). The surpluses are found over tracts of territory that cover only 10.0% of the total area of the conterminous United States. Clusters labeled 1, 4, and 7, extending from the South to the Atlantic division, include 65.3% of these surpluses. Clusters 2, 5, 8, and 9 including mainly East–Central and North–Central metropolitan areas include 24.1%; California clusters 3 and 6, include 7.1% of the surpluses. These nine major clusters contain 96.5% of the surpluses. The remaining 3.5% are shown by isolated GRP.

Figure 8.5. United States. Afro-American population, 1980: . 2649 ▪ 26,480 ▪ 52,980 ⁘ 135,101 ■ 373,516+ (Source of data: Strategic Mapping—Counties Data Files, U.S.A.).

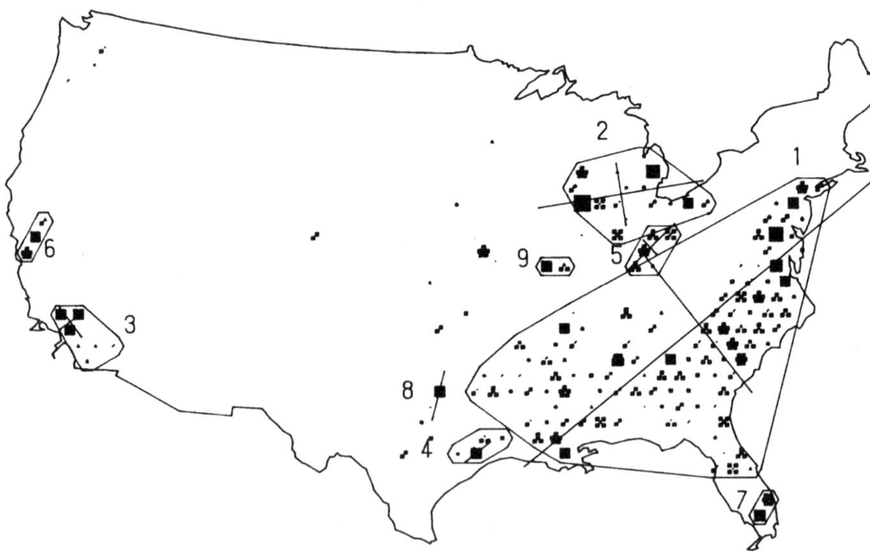

Figure 8.6. United States. Major clusters of surplus of Afro-Americans, 1980: . 3078 ▪30,783 ▪ 61,566 ⁘ 160,071 ■ 307,828+ (Source of data: Strategic Mapping—Counties Data Files, U.S.A.).

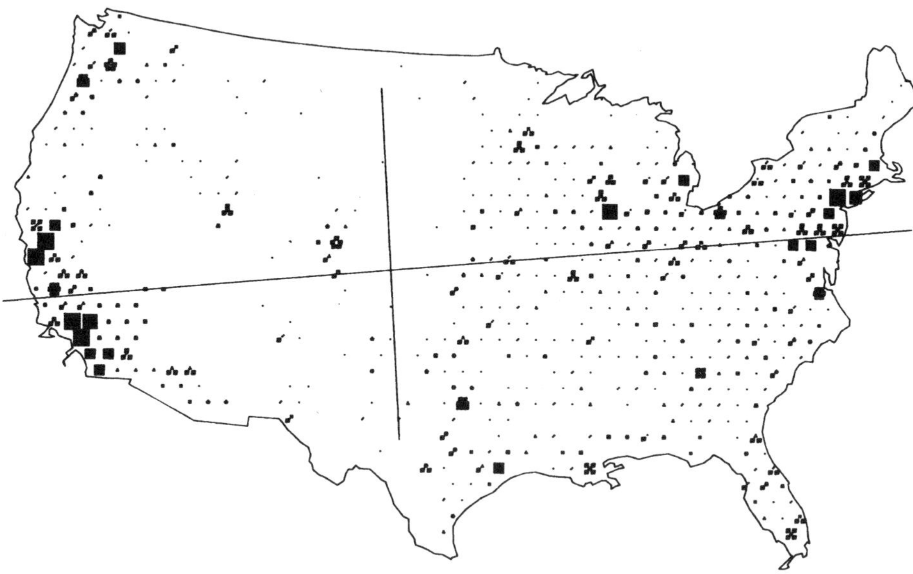

Figure 8.7. United States. Individuals of Asian origin, 1980: . 291 ▪ 2929 ▪ 5824 ▪ 14,560 ■ 29,120+ (Source of data: Strategic Mapping—Counties Data Files, U.S.A.).

Figure 8.8. United States. Major clusters of surplus of Asians, 1980: . 291 ▪ 2912 ▪ 5824 ▪ 16,016 ■ 29,120+ (Source of data: Strategic Mapping—Counties Data Files, U.S.A.).

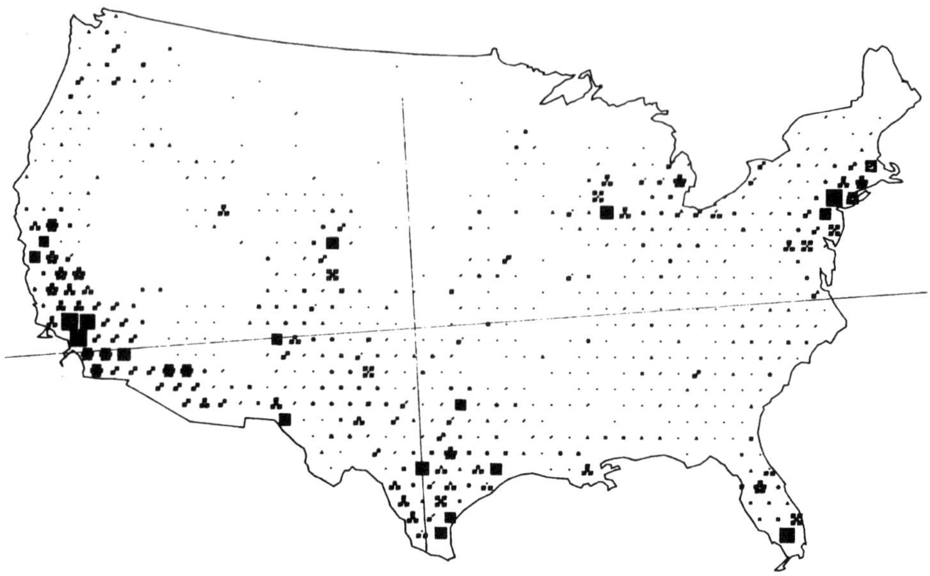

Figure 8.9. United States. Individuals of Hispanic origin, 1980: . 1454 ▪ 14,542 ▪ 29,084 ▪ 74,166 ■ 145,424+ (Source of data: Strategic Mapping—Counties Data Files, U.S.A.).

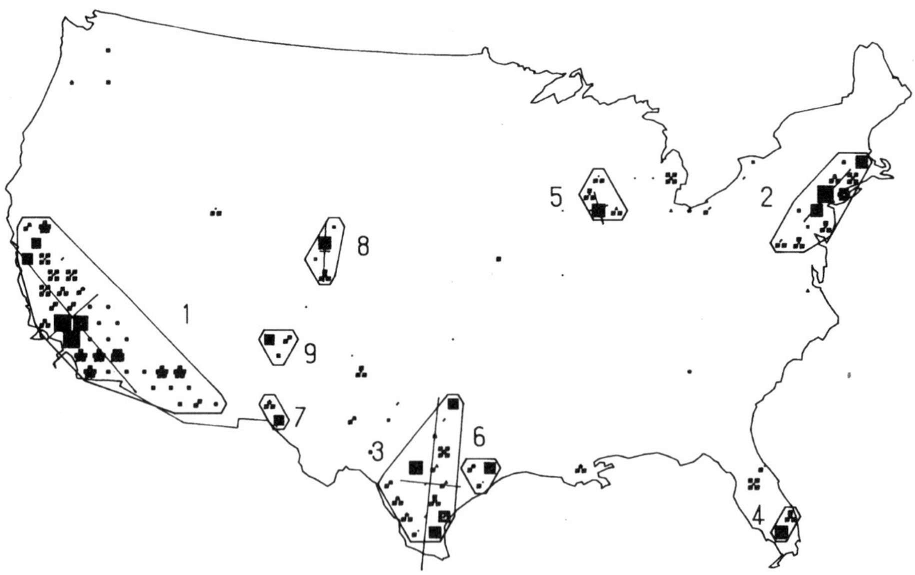

Figure 8.10. United States. Major clusters of surplus of Hispanics, 1980: . 2908 ▪ 29,084 ▪ 110,522 ▪ 133,790 ■ 290,855 (Source of data: Strategic Mapping—Counties Data Files, U.S.A.).

Table 8.3. Territorial Coverage and Density for Population and Subpopulations in the Conterminous United States (1980)

Parameter	Total population	Whites	Afro-Americans	Asians	Hispanics
Percent surpluses of actual over uniform distribution ($100\,\Delta_p^U$)	57.4	56.7	73.4	77.5	75.6
Percent of territory covered by:					
Habitat [$100(1-\Delta_p^U)$]	42.6	43.3	26.6	22.5	24.4
Surpluses	10.6	11.1	10.0	5.7	6.7
Population densities					
Gross	76.56	63.8	9.0	1.0	4.9
Net	178.5	146.5	33.7	4.4	20.1
Surpluses	413.97	324.8	63.6	13.6	55.3
Relative density					
Net (b/a)	2.3	2.3	3.7	4.4	4.1
Surplus (c/a)	5.4	5.09	7.07	13.6	11.29

Table 8.4. Gross Territory Covered by the Habitat of Jewish Population

City	Year	For total population (%)	For Jewish population (%)	Ratio of Jewish population to total population
Rome	1526/7	43.9	11.8	0.269
Rome	1928		19.0	
Rome	1965	63.9	37.6	0.588
Rome	1985	58.5	37.0	0.632
Milan	1965	58.5	31.9	0.545
Turin	1965	66.1	27.7	0.419
Trieste	1875	34.5	12.6	0.365
Trieste	1965	42.2	18.7	0.443
Paris (region)	1975	43.8	24.2	0.553
Brussels	1961	68.6	56.2	0.819
Greater London	1970	75.5	47.9	0.634
Zürich	1970	75.3	50.0	0.662
Vienna	1910	46.1	27.2	0.590
Vienna	1971	59.7	35.5	0.595
Berlin	1933	52.5	28.8	0.545
Warsaw	1882	65.1	53.9	0.828
Warsaw	1913	57.5	40.7	0.705
Warsaw	1938	50.1	18.5	0.369
Lodz	1931	64.5	24.2	0.530
Toronto	1981	34.0	14.9	0.438

2. *Asians.* The concentration of Asians ($\Delta_P^U = 77.5\%$) is even larger than that of the Afro-Americans. These surpluses are found over 5.7% of the total territory and 92.1% of the surpluses are found in six major clusters. The largest (51.67%) is in California, labeled 1. Cluster 2, which includes 26.1% of these surpluses is in the Northeast; cluster 4 is in Washington and includes 4.18%; clusters 3 and 6 are in Michigan and Wisconsin and include 8.1%; and cluster 5 (2.05%) is in Texas.
3. *Hispanics.* The concentration of this subpopulation ($\Delta_P^U = 75.6\%$) is intermediate between that of Asians and Afro-Americans. For the Hispanics, too, the percentage of territory covered by the surpluses is very small (6.7%). Some 61.2% of the surpluses are found in the Southwest in clusters labeled 1, 7, 9, 3, and 6; some 5.7% are in cluster 4 (in Florida); 21.96% are in the Northwest in cluster 2, and 5.7% are in cluster 5 in Wisconsin.

8.9. Coverage and Net and Surplus Densities for Cities

The methods described above for measuring territorial coverage and net and surplus densities for countries can be used for the geostatistical study of cities for many purposes, e.g., for (a) detailed analysis, eventually microgeographic, of the distribution of a population over a city; (b) comparison between the distributions of a population and subpopulations; (c) evaluation of the distribution of a population and subpopulations over the course of time; and (d) comparisons among the geosets of the population in various cities.

Many of these four types of studies were used in a comprehensive report on urban ecology of Jewish minorities in cities of Europe, North America, Latin America, and South Africa (see Bachi *et al.*, 1988). Selected examples are described below. Other, unpublished, applications deal with different fields.

Table 8.4 shows: (a) the territorial coverage $100(1 - \Delta) = 100(A_M/A_T)$ for the total population of the cities considered, (b) the territorial coverage of the Jewish population, and (c) the ratio between (b) and (a).

For the 19 geosets of total population listed in Table 8.4, the first quartile is 43.9%, the median is 58.5%, and the third quartile is 65.1. For the 20 Jewish populations included, the first quartile of the territorial coverage is 18.7%; the median is 28.8%; the third quartile is 37.6%. In all the cities and time periods studied, the habitat of the Jewish population is smaller, and often much smaller, than that of the total population. However, in some cities this tends to become less pronounced over the course of time.

There has been great interest shown in the evolution of territorial coverage and net and surplus densities for cities over long periods of time. For example, for Rome we have reconstructed these parameters for a period that is very important from a historical point of view: 1526–1527, before the conquest by Bourbon troops and for all the censuses taken decennially after Rome was annexed by the Kingdom of Italy in 1871.

In these types of studies, it is often necessary to deal in detail with the problem of properly defining the gross territorial container to which reference is made for the calculation of the habitat in each period. While the gross territorial container for countries is generally easily defined, for cities, we may have alternative definitions, such as municipal area, other local authority area, conurbation, or administrative area netted from rural areas. Parallel use of various definitions is sometimes of help. Sometimes microgeographical analysis is required as a preliminary step.

Chapter 9

Centers of Geosets

9.1. Centers as Averages of Locations

We now go on to consider the centers of geosets, one of which—the mean center—is the oldest and the most popular of all the geostatistical parameters. Summarizing the locations of the elements of a geoset by a center is somewhat analogous to averaging the values X_i of a variable X (such as personal income or family size). We shall discuss the similarities and differences between these operations, denoted, respectively, by centers (of a geoset) and averages (of a one-dimensional variable).

In many statistical textbooks averages are indicated as "measures of location," "central measures," etc., and these appellations can be used effectively for centers as well. Obviously, while an average is expressed by one value such as $\bar{X}$ or the median of X, a center is generally expressed by two coordinates such as latitude and longitude, or $\bar{X}$ and $\bar{Y}$. For a given geoset: the number of centers that can be calculated is infinite (as is the number of the averages); however, only a few of these parameters have meaningful applications. In this book we shall use mainly the mean center (Sections 9.3–9.6) and the median center (Section 9.7), and some substitutes for the elusive modal center (Section 9.9). A few other centers will be mentioned only briefly.

For a given geoset distributed over a territory T, the mean and median centers are uniquely represented by a point on the map of T, as averages can be indicated by a point on the X-axis. However, the numerical expression of the coordinates of the centers may change with a change of scale or a change in the origin of the coordinates.

To reduce practical problems of interpretation of mean and median centers, the following simple rules are often recommended:

1. Take either the center of the territory T (mainly if we are dealing with countries) or of any spot that is easily identifiable as the origin of the coordinates (0,0).
2. Express the coordinates as deviations (in km, miles, etc.) from the origin. Sometimes these deviations are labeled "eccentricities."

3. Whenever convenient, render the eccentricities dimensionless by dividing them through a convenient measure of dispersion of the relevant territory, which enables a comparison of these relative eccentricities for different populations, different countries, or different cities (see, e.g., Sections 9.5 and 9.6).

For the averages, too, one of the most important applications is for comparisons among various populations. However, these are generally meaningful only if they are carried out between pairs of distributions that have more or less the same shape as, for instance, when both distributions tend to Bell-shape, U-shape, or J-shape. In the study of actual geosets we often do not find any tendency to simple patterns of distributions. However, the use of approaches such as (1), (2), and (3) enhances the possibilities for comparisons between centers.

The different nature of the territory T in the study of centers and the X-axis in the study of averages should be taken into account:

1. Territory T of a given geoset is supposed to be clearly defined by its boundaries, while the X-axis can sometimes be regarded as theoretically infinite.
2. Territory T may be fragmented and the habitat of the population distributed over it is often discontinuous. By contrast, the X-axis is generally continuous.
3. Given an X-axis, the location of an average over it has, in general, a very simple and clear meaning. By contrast, the location of the center of a geoset can be interpreted properly only when we take into account the influences exerted over the various parts of the territory and the distribution of the population under survey by many different factors (e.g., geographical, altimetrical, meteorological, historical, ethnic, economic, and social).

In calculating averages, we obtain exact values by working on the list of all the individual values X_i. The same applies to the geosets for which we know the exact locations of each element of the population being studied. For both the averages and the centers, when dealing with large populations, we cannot in practice, avoid using aggregated data, which may introduce some inaccuracy. In the study of a variable X, we often accept the convention that the elements of each class are found at the middle of the class interval. We can introduce similar conventions in the study of geosets, for example we might attribute the coordinates of the center of c to the elements of each cell c: if the data used are based on very refined divisions of the territory and aggregated over a large number of cells of fixed shape and area, the risks of inaccuracy are often not very significant.

One-dimensional variables for which an average is calculated are often obtained by a ratio scale (see Section 4.8) and can be regarded as magnitudes. In contrast, locations of elements of a geoset are based on two-dimensional interval scales. The dot representing their center identifies an average location, but not a magnitude.

Both averages and centers relate to only one aspect of the distribution. We have already seen from the many examples of crosses superimposed on maps that while the location of the center helps in the study of some characteristics of the geoset, other aspects such as dispersion and oblongity must also be taken into account. Many other aspects are discussed in the following chapters. If the need to use a battery of parameters is overlooked there may be misinterpretations for averages and even more for centers.

Averages are often interpreted as indicators of the position of the distribution over the X-axis. The terms position and location are often used indiscriminately in current statistical literature. In this book we prefer to indicate the center as an average of locations. Using the term in an arbitrary way, we consider position as a more complex aspect of the geoset with the average location being only one parameter together with the angle between the principal and conventional axes and the quadrant setting (see Chapter 10).

9.2. Distance-Moments and Potentials

The joint study of centers and parameters of dispersion can be enhanced by applying the concept of moments (first used in physical sciences and afterward widely adopted in statistics to geosets). This approach (under the name of areal moments) was thoroughly investigated by Neft (1966) so we will undertake only a very brief discussion.

Let us indicate the distance between any point i of a population distribution and any point k in the plane by d_{ik}, and the relative weight attached to i by w_i. Then we may denote the sth distance-moment of the distribution around point k as

$$\mu_{s,k} = \Sigma w_i d_{ik}^s \tag{9.1}$$

By integrating rather than summing, we can denote the distance-moment of a territory T of area A around a point k as

$$\mu_{s,k} = (1/A) \iint_T d_k^s dxdy \tag{9.2}$$

where d_k is the distance of any element of T (of area $dxdy$) from k.

Equations (9.1) and (9.2) indicate the value of the sth moment that corresponds to any point k of the plane. They thus define distributions by means of an infinite number of values. The map of such distributions can be constructed with approximations by solving equations (9.1) and (9.2) for selected values of k.

In this volume equations (9.1) and (9.2) will be utilized mainly to determine the points k that correspond to the minimum distance-moments of orders $s = 1$ and $s = 2$. This is, more or less, in line with the practice followed in elementary analysis of one-dimensional distributions. However, some other uses of distance-moments in geostatistical research have been suggested. For instance:

1. Population distributions indicating the number of cases found in each inhabited place are often discontinuous. If we substitute distance-moment distributions for them we can set continuous distributions whose maps may be more easily understood than those of the original distributions.

As we noted above, equations (9.1) and (9.2) indicate the value of the s-moment that corresponds to any point of the plane. They thus define distributions with an infinite number of points. In practice, we work only over points included in T and calculate the moment only for selected values of k and thus obtain values that approximate the true ones. With the help of these values we can construct isoline maps, which show continuous distributions over T corresponding to the distributions of the population studied.

The mean center and the corresponding measures of dispersion (distance-variance d^2 and standard distance d) are obtained by considering $s = 2$; the median center and corresponding measures of dispersion are obtained by referring to $s = 1$. Neft emphasized reference to $s = -1$. In this volume we refer mainly to $s = 2$ and sometimes to $s = 1$ (which can be used together and not necessarily as alternatives). Emphasis on $s = 2$ is related to: (a) the usual consideration of the least-squares principle in statistical research: (b) ease of algebraic manipulation; (c) the relationships among many of the parameters based on $s = 2$; (d) the battery of three parameters defining position (see Chapter 12); (e) the possibility of splitting squared distances into separate squares of deviations along the axes of the abscissas and the ordinates. Squaring may give undue weight to large deviations, in which case use of median center ($s = 1$) may be preferred. For a discussion of the median center see Section 9.7.

The use of $s = -1$ is based on the idea—which is attractive for certain types of research—that "influence" of a population found at point i over territorial point k should be seen as inverse to the distance of k from i. This consideration stands at the basis of population potentials, which have been used extensively since the early days of research in social physics and regional sciences [see, e.g., the works by Stewart, especially 1942, 1948*a*, *b*, 1950*a*, *b*, 1958, Warntz (bibliography in Coffey, ed.), 1988; and Isard *et al*., 1960].

2. Potentials have not been taken into account in this volume for the following reasons: (a) Potentials are generally used for goals that are different from those of the methods discussed here. (b) Were we to use the maps of isolines of potentials only as tools for describing population distributions, we would see that they provide far less information than GRP maps (compare, for instance, the map of potentials of the population of the United States given by Neft (1966, p. 73) to the GRP map in Section 8.8 of this volume. (c) The methods introduced here make the divergence between distributions of population and uniform distribution over territory clear, which potentials are not designed to do; on the assumption of uniform distribution over a convex territory T, the map of potentials would suggest a distribution with a peak in the middle of T and declining isolines in the direction of the periphery.

If we consider a population distributed nonuniformly over T but with high-density peaks toward the center of T, the map of potentials might appear to be quite similar to that derived from the uniform distribution, which could be misleading.

9.3. The Mean Center

9.3.1. Defining the Mean Center

The center most widely employed in geostatistical research is the two-dimensional analogue to the one-dimensional mean. This center may be called the arithmetic mean center or mean center for short, or (whenever there is no danger of misunderstanding) simply the center. It is also often referred to as the center of gravity of a distribution, as it corresponds to the concept used in physics for the center of gravity of a distribution of a mass over an area. The mean center can be found by extending the usual least-squares approach used for one-dimensional distributions to two-dimensional geosets.

Let us determine the point in the plane for which the sum of squares of distances (or the second distance-moment $\mu_{2,k}$) is minimal (k denoting here any point in the plane):

$$\min \mu_{2,k} = \mu_{2,c} \tag{9.3}$$

We find

$$\bar{X}_c = \Sigma w_i X_i = \bar{X} \qquad \bar{Y}_c = \Sigma w_i Y_i = \bar{Y} \tag{9.4}$$

The mean center is thus, simply, the point having as an abscissa the arithmetic average of the abscissas of the elements of the geoset and as an ordinate the average of the ordinates.

9.3.2. Calculating the Mean Center of Population Geosets

In Section 3.1.3 we illustrated the two simple examples of the geosets of hotels taken, respectively, as unweighted or weighted by the number of beds. Table 9.1 shows two examples of geosets formed by the same six locations. In Geoset 1 all the locations have equal weight, while in geoset 2 they have different weights. The center of geoset 1 has coordinates $\bar{X}_1 = 5$, $\bar{Y}_1 = 6$; the center of geoset 2 is $\bar{X}_2 = 6.81$, $\bar{Y}_2 = 8.01$.

Figure 9.1 shows the locations of the elements and of the centers and—by means of GRP—the relative weights attached to the elements of geoset 2. The shift in the north east direction of the center of geoset 2 compared to the center of geoset 1 is due to the relatively strong weight of northern and eastern locations 2 and 3.

In field work, we may find ourselves dealing with various types of population geosets, such as the following:

1. Geosets that list location (weighted or unweighted) for each element of the population. Then the calculation of the center is as shown in Table 9.1;
2. Geosets in which data are aggregated into hexagonal cells, in which case we take the coordinates of the center of each cell as locations.
3. If data are given by regions j, the calculation may follow the steps shown in Table 9.1, provided we can obtain a reasonable estimate of a central place (X_j, Y_j) for each region j, in which we can assume that all of the population of j is concentrated.

Table 9.1. Centers of Population Distributed by Six Locations

Ordinal number of i of the location	Location		Size of population	Relative weight	Products	
	x_i	y_i	N_i	w_i	$w_i x_i$	$w_i y_i$
Geoset 1						
1	7	6	16,666.7	0.167	1.169	1.002
2	9	9	16,666.7	0.167	1.503	1.503
3	5	10	16,666.7	0.167	0.835	1.670
4	3	7	16,666.7	0.167	0.501	1.169
5	2	1	16,666.7	0.167	0.334	0.167
6	4	3	16,666.7	0.167	0.668	0.501
Total			100,000	1.0	5.0	6.0
Geoset 2						
1	7	6	9,000	0.09	0.63	0.54
2	9	9	50,000	0.50	4.5	4.5
3	5	10	20,000	0.20	1.0	2.0
4	3	7	10,000	0.10	0.3	0.7
5	2	1	3,000	0.03	0.06	0.03
6	4	3	8,000	0.08	0.32	0.24
Total			100,000	1.00	6.81	8.01

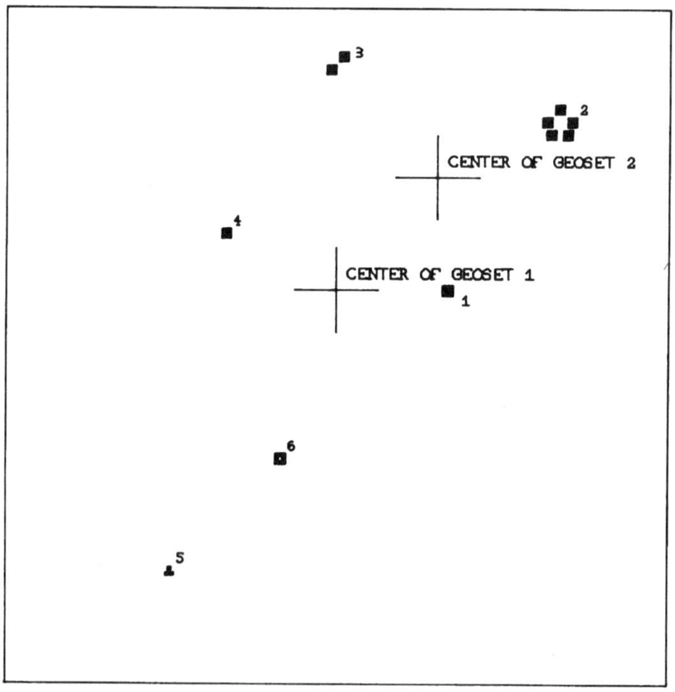

Figure 9.1. Locations of elements and centers of geosets 1 and 2. The GRP indicate the relative weights of the elements of geoset 2.

There are several possible procedures for obtaining the estimates needed for the last possibility:

1. We assume that there is interest in calculating the center of population for a country each year, and that there are annual estimates of the population of each of the regions j into which the territory of the country is divided. These data indicate that in each region j, the relative weight of population is w_j^*. We assume further that data are available for each locality ji within region j from the last census, showing that the relative weight of the population in ji was w_{ji}. Then we can calculate the location (X_j, Y_j) of the center of the population of j at the time of the census. The current estimate of the center $(\bar{X}, \bar{Y})$ for the population of the country can be obtained from
$$\bar{X} = \sum w_j^* X_j \qquad \bar{Y} = \sum w_j^* Y_j \tag{9.5}$$
2. The same procedure can be used if we work on census data for different population groups or on data for phenomena related to population (such as births) and for each group of phenomena. Data w_i are available by regions only, while for the entire population data are available by places. By using equation (9.5) we introduce the assumption that within each region j all groups or phenomena have the same center (X_j, Y_j). This is often not true, but, generally speaking, the error introduced by this assumption is small.
3. If population centers cannot be estimated on the basis of census data by localities, more arbitrary procedures can be employed. For instance, if regions j are comparatively small, we may assume that the population is concentrated in the capital city or the principal city of j, for which coordinates are easily obtained from a map.
4. Alternatively, we can assume that the population of j is evenly distributed over its territory and calculate the center of the territory for each j by methods to be explained in Section 9.3.3.
5. An intermediate situation is often found in which data may be available for each town ji (or for each important town) in j, and for all the rural population (or for the rest of the population) in j, taken together. We can then use the procedure shown in Table 9.1, where for each town we use exact locations (X_{ji}, Y_{ji}), while for the rural population we use the territorial center of j (on the assumption that the rural population is evenly distributed over j).

9.3.3. Calculating the Mean Center of a Territorial Geoset

We saw in Section 4.5 that the mean center of a simply connected territory T divided into infinitesimal elements of area $dxdy$ has the coordinates

$$\bar{X} = \iint_T xdxdy \Big/ \iint_T dxdy \qquad \bar{Y} = \iint_T ydxdy \Big/ \iint_T dxdy \tag{9.6}$$

Practical methods for calculating centers for simply connected territories, fragmented territories, gapped territories, territories with external concavities, and subterritories were discussed earlier in Section 4.6.

9.3.4. Main Properties of the Mean Center

9.3.4.1. Properties of the Lines Passing Through the Center

Figure 9.2 shows again geoset 1 of Figure 9.1 and a line XX, arbitrarily selected, that passes through the center of the geoset.

1. For any line XX passing through the center C we denote the axis orthogonal to XX through the center by YY and take the origin of coordinates $(0, 0)$ at the center C.

Let us denote the distance between any point i and XX by y_i and attach the signs + and −, respectively, to the distances from i to XX for the i's on one side of the line and for the i's on the other. Then the sum of the deviations (or signed distances) from the line XX is zero ($\Sigma y_i = 0$). Or, in other words, the total of absolute deviations on one side of the line equals the total of absolute deviations on the other side.

This property of lines passing through the mean center is analogous to the property of the arithmetic mean of one-dimensional distributions (from which the total of deviations is 0).

2. Let us denote any element of the geoset by i, the mean center of the geoset by c, any point of the plane by p; the coordinates of i, c, p, respectively, by (X_i, Y_i)

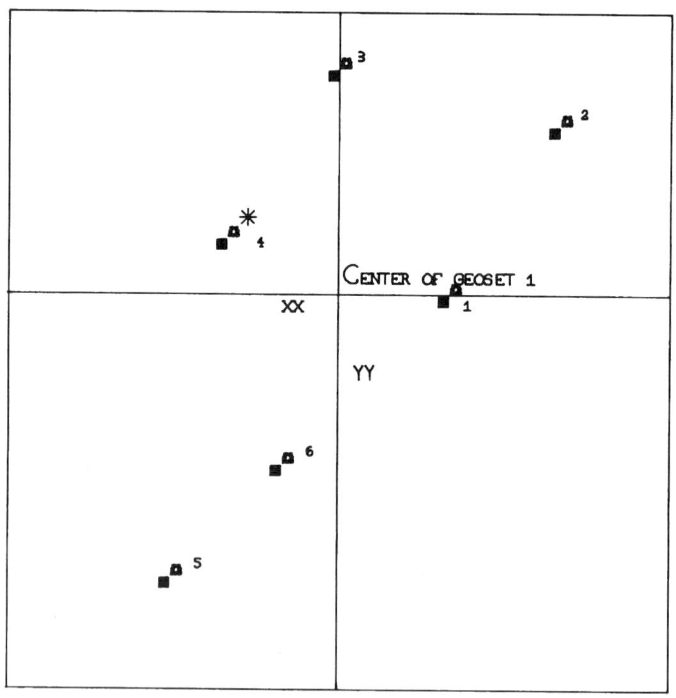

Figure 9.2. Geoset 1 with lines XX and YY passing through the center of the geoset.

$(X_c = \bar{X};\ Y_c = \bar{Y})$, and (X_p, Y_p); and the squared distances between element i and the center and between element i and point p by d_{ic}^2 and d_{ip}^2, respectively. It is well known that $\min \Sigma(X_i - X_p)^2 = \Sigma(X_i - \bar{X})^2$ and $\min \Sigma(Y_i - Y_p)^2 = \Sigma(Y_i - \bar{Y})^2$. Accordingly, $\min d_{ip}^2 = d_{ic}^2$. The mean center is then the point of the plane from which the sum of the squares of the distances is a minimum.

9.3.4.2. Algebraic Manipulations and the Relationship between Centers of Primary and Derived Geosets

The coordinates $(\bar{X}, \bar{Y})$ of the mean center of a population geoset are obtained, as shown above, by working separately on the abscissas and ordinates of the elements of the geosets. We showed in Chapter 3 how geosets derived from the comparison of ${}^gP$ and ${}^gU$ and ${}^gP$ and ${}^gQ$ are related to the primary geosets. These relationships are immediately extended to the mean centers of these parameters. Numerical examples of related centers of primary and derived geosets were given in Table 3.1 (Section 3.2). A simple demonstration of these relationships will be found in Appendix E.

The relationships between the mean centers of primary and derived geosets enable us to construct triangles of centers, which are a special application of the triangles of geosets. These constructions have already been illuminated in Sections 3.2 and 3.3. Very important analogous relationships were found between compound populations and component subpopulations (Section 3.6), which were extended to the compounding of geosets and other settings (Section 3.8). All of these relationships hold for the mean centers of the geosets.

9.3.4.3. Invariance with Transformation of Coordinates

The position of the mean center is invariant to: (1) rotation of the axes, or (2) parallel translation of axes, or (3) any linear transformation of coordinates, as shown below.

1. If we assume that the origin of the coordinates is changed to any point of coordinates (X_0, Y_0), then the new coordinates (X_i', Y_i') for any point i are

$$X_i' = X_0 + X_i \qquad Y_i' = Y_0 + Y_i \tag{9.7}$$

and the coordinates of the mean center expressed in the new coordinates $(\bar{X}_c', \bar{Y}_c')$ are

$$\bar{X}_c' = X_0 + \bar{X} \qquad Y_c' = Y_0 + \bar{Y} \tag{9.8}$$

For instance, let us consider again the example of Figure 9.2 and translate in a parallel way the axes of the coordinates by taking $x_0 = -2$ and $y_0 = 1$. Then the center, as indicated by an asterisk in Figure 9.2 is given by $x' = 3$, $y' = 7$ in the new coordinates, instead of as $x = 5$ and $y = 6$ in the old coordinates. We thus see that the location of the center remains unchanged, while its numerical expression changes.

2. If we assume that the coordinates are rotated by an angle α around the center (Figure 9.3) and consider the coordinates of any point P of the geoset under survey, then for the example in Figure 9.3, P concides with point 2 of Figure 9.2.

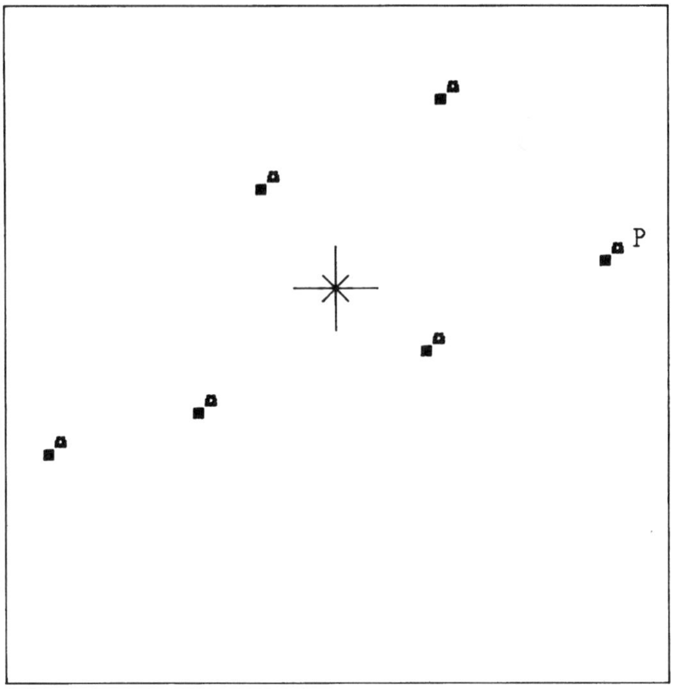

Figure 9.3. Geoset 1 rotated through an angle α equal to 30°.

The transformed points are given by

$$x' = x\cos\alpha + y\sin\alpha \qquad y' = y\cos\alpha - x\sin\alpha \tag{9.9}$$

Then,

$$\bar{x}' = \bar{x}\cos\alpha + \bar{y}\sin\alpha \qquad \bar{y}' = \bar{y}\cos\alpha - \bar{x}\sin\alpha \tag{9.10}$$

where $\alpha = 30°$, $\sin\alpha = 0.5$, and $\cos\alpha = 0.866$. Hence $\bar{x}' = 7.330$ and $\bar{y}' = 2.695$, instead of $\bar{x} = 5$ and $\bar{y} = 6$. The numerical expression of the center has changed again but if we read the point of coordinates $\bar{x}' = 7.330$ and $\bar{y}' = 2.695$ along the new axes, we obviously find the center c again (as shown by an asterisk in Figure 9.3).

3. In general terms, if we perform a linear transformation

$$x' = ax + b \qquad y' = cy + d \tag{9.11}$$

over the coordinates, the expression of the center in the new coordinates becomes:

$$\bar{x}' = a\bar{x} + b \qquad \bar{y}' = c\bar{y} + d \tag{9.12}$$

9.3.4.4. Ease of Calculation

In connection with (1), (2), and (3) above, the separate calculation of $\bar{x}$ and $\bar{y}$ is very easy. The devices that are usually employed for shortening the calculations of the arithmetic average of one-dimensional variables can also be used for the calculation of the mean center. (For special cases, see Section 9.3.2).

9.3.4.5. Sensitivity

Because $\bar{X} = \Sigma w_i X_i$ and $\bar{Y} = \Sigma w_i Y_i$, any change in one of the coordinates or in the weight of any one of the elements *i*, while the others remain unchanged, results in a change in the location of the mean center. This renders the mean center (as the usual arithmetic average) a very sensitive value.

9.3.4.6. Location of Mean Center

The mean center of a territory does not necessarily have to be located inside the territory [see, e.g., Figure 1.11 (Section 1.4)], but it must be located within the convex hull (convex polygon of the minimal area within which the territory is included).

For the mean center of a population, consider, for the sake of simplicity, a setting in which

$$w_i = 1/N \qquad \bar{X} = \Sigma(X_i/N) \qquad \bar{Y} = \Sigma(Y_i/N) \tag{9.13}$$

Every point *i* "contributes" to the center $1/N$ of its X-distance or Y-distance from the origin 0.

In a situation in which most points are near one another and constitute a major cluster while a few other points constitute a minor cluster very distant from the major one, the mean center may fall at a considerable distance from the major cluster.

9.4. Selected Applications of the Mean Centers for Italy

9.4.1. Interpreting Maps and Numerical Values

Let us come back to some examples of mean centers of various geosets distributed over the territory of Italy.

The maps of Figures 1.3 (Section 1.2) and 1.6–1.12 (Section 1.4) show: (a) how the GRP can clarify micrographic and macrographic characteristics of geosets; and (b) how the cross superimposed on the map can help the user toward a general orientation on macrogeographical characteristics. In order to interpret cases of both (a) and (b) correctly, the user should have some preliminary knowledge of the differential aspects (e.g., physical, historical, socioeconomic) of the various parts of the territory (as indicated for Italy in Sections 1.2–1.4). In the interpretation of (b), the mean center (considered together with the principal axes) may play an important role. This is especially true when the mean center appears to be far from the center of the territory for it is then clear that the actual distribution of the geoset differs strongly from the uniform distribution. However, the inverse proposition is not always true; for instance, in Figure 1.3 (Section 1.2), the center of the population is almost coincident with that of the territory, despite the considerable discrepancies between the population and the uniform distributions. It is often useful to express the coordinates as deviations from the center (0, 0) of the territory, measured along two convenient orthogonal axes through the center. In the examples for Italy we followed the method found convenient in the *Statistical Atlas of Italy* to measure the coordinates (X^T, Y^T) of each geoset along the principal axes of the territory T of the country.

The perusal of the mean centers given for hundreds of geosets in the first volume of the *Atlas* demonstrates the utility of the centers, but also reveals examples of two settings that might lead to misunderstandings in the interpretation of the mean center:

1. As the territory of Italy is fragmented and has strong concavities, it happens sometimes that the mean center of a geoset is found outside the territory: an example of a mean center falling in the sea is that given for citrus groves in Figure 1.11 (Section 1.4).
2. Other examples from the *Atlas* show that a geoset can be formed by clusters which are very far from each other and have a mean center that falls in a zone from which the population studied is completely, or almost completely, absent.

Having interpreted the meaning of the mean center of each of the geosets studied, we may want to bring these centers together on one map, as in Figure 1.15 (Section 1.5). In the *Atlas*, approaches (1) and (2) of Section 9.3.2 were used in connection with many hundreds of geosets and found by the analysts to be of great help in their studies. Step (2) was carried out by showing the mean centers of geosets belonging to the same class of phenomena over each of the summary maps. The comparative study of the positions of these mean centers has been successful, having been related to the simultaneous detailed study of each distribution, to the analysis of other parameters, and to expert knowledge in each field. In these summary maps, great distance between the mean centers of any pair of geosets ${}^gP$ and ${}^gQ$ suggests strong divergence between ${}^gP$ and ${}^gQ$, but here too, the reverse is not necessarily true.

Another approach to the summary study of mean centers was used in the second volume of the *Atlas* in which the distribution of the centers of 385 geosets of demographic, economic, and social phenomena along axes X^T and Y^T was studied.

The minor and major axes of the cross for the territory of Italy (as shown by Figure 1.1 in Section 1.2) have, respectively, lengths of $4\sigma_X^T = 660\,\text{km}$ and $4\sigma_Y^T = 1337\,\text{km}$. The dispersion of the 385 mean centers along the minor axis is comparatively small: 80% of the geosets have a value of X^T ranging between -19.6 and $+34.8$ km. By contrast, the dispersion of the mean centers along the axis Y^T is much greater. The first and ninth deciles have, respectively, values of -168.9 and $+204.8$ km. The furthest northwestern center (for rice, see Figure 1.12 in Section 1.4) has a value of $Y^T = +395.0$, while the furthest southeastern center (for citrus, see Figure 1.11 in Section 1.4) has a value of -574.2. A very important operational conclusion has been drawn from this: the location of most geosets along the axis Y^T in Italy has been found to be a very significant parameter and has been utilized very widely in comparative research. In many cases, there was no need to study the value of X^T as well. Thus, the transformation of a two-dimensional parameter into a parameter of one dimension has led to a very considerable saving of research effort.

9.4.2. The Triangle of Five Centers

We have seen that geosets ${}^gSPU$, ${}^gSUP$, and ${}^gM(P, U)$ can be derived from the comparison between a geoset ${}^gP$ and the corresponding geoset ${}^gU$ of the uniform distribution. These geosets are functionally related and so are their mean centers. We have shown through the example of Figure 3.12 (Section 3.2) and Table 3.1 that:

1. The centers of ${}^gM(P, U)$, ${}^gP$, and ${}^gSPU$ are collinear.
2. The centers of ${}^gM(U, P)$, ${}^gU$, and ${}^gSUP$ are collinear.
3. The center of ${}^gH$ is equal to that of ${}^gM(P, U)$.
4. The center of ${}^gNH$ is equal to that of ${}^gSUP$.
5. The center of ${}^gP$ is found on the line between ${}^gH$ and ${}^gSPU$ and cuts this line in parts proportional, respectively, to Δ_p^U and $1 - \Delta_p^U$.
6. The center of ${}^gU$ is found on the line between the centers of ${}^gH$ and ${}^gNH$ and cuts this line into two parts that are proportional to Δ_p^U and $1 - \Delta_p^U$, respectively.
7. The line between the centers of ${}^gP$ and ${}^gU$ is parallel to the line between ${}^gNH$ and ${}^gSPU$.
8. The distance between the centers of ${}^gP$ and ${}^gU$ equals the product of Δ_p^U times the distance between ${}^gSPU$ and ${}^gNH$.

These "laws" are general and they are demonstrated in Appendix E.

In the *Atlas* the triangle of centers was determined for all the geosets considered, and the study of these triangles has been found to be very useful for interpreting and comparing important characteristics of the geosets. We describe two examples in the following paragraphs.

For the first example, Figures 9.4 and 9.5 show the distribution of sheep according to the Italian agricultural census of 1982. Figure 9.4 shows the number of sheep and Figure 9.5 their habitat and nonhabitat. This livestock is concentrated in hilly and mountainous pasture lands on the island of Sardinia and in south-central continental Italy. Figure 9.6 shows the triangle of the centers for the sheep. The center of the habitat is found in the spot indicated by the central point of the cross of Figure 9.5. The centers of ${}^gP$ and ${}^gSPU$ are aligned toward Sardinia. The other line, originating in the center of the habitat, forms an angle $\beta = 92.7°$ with the first line, which indicates that the nonhabitat (white part of Figure 9.5) lies mainly in the northern regions. This line is divided by the center of the territory (shown also in Figure 1.1 in Section 1.2) into two parts proportional, respectively, to $1 - \Delta_p^U = 50.3\%$ and $\Delta_p^U = 49.7\%$.

For the second example, Figure 9.7 shows the triangle of centers for the geoset of Italian citrus groves (illustrated in Figure 1.11 in Section 1.4). Figure 9.8 shows the habitat and nonhabitat of this geoset. We can see here that the triangle of centers degenerates almost to the point of being one line. The lines between centers of ${}^gSPU$,

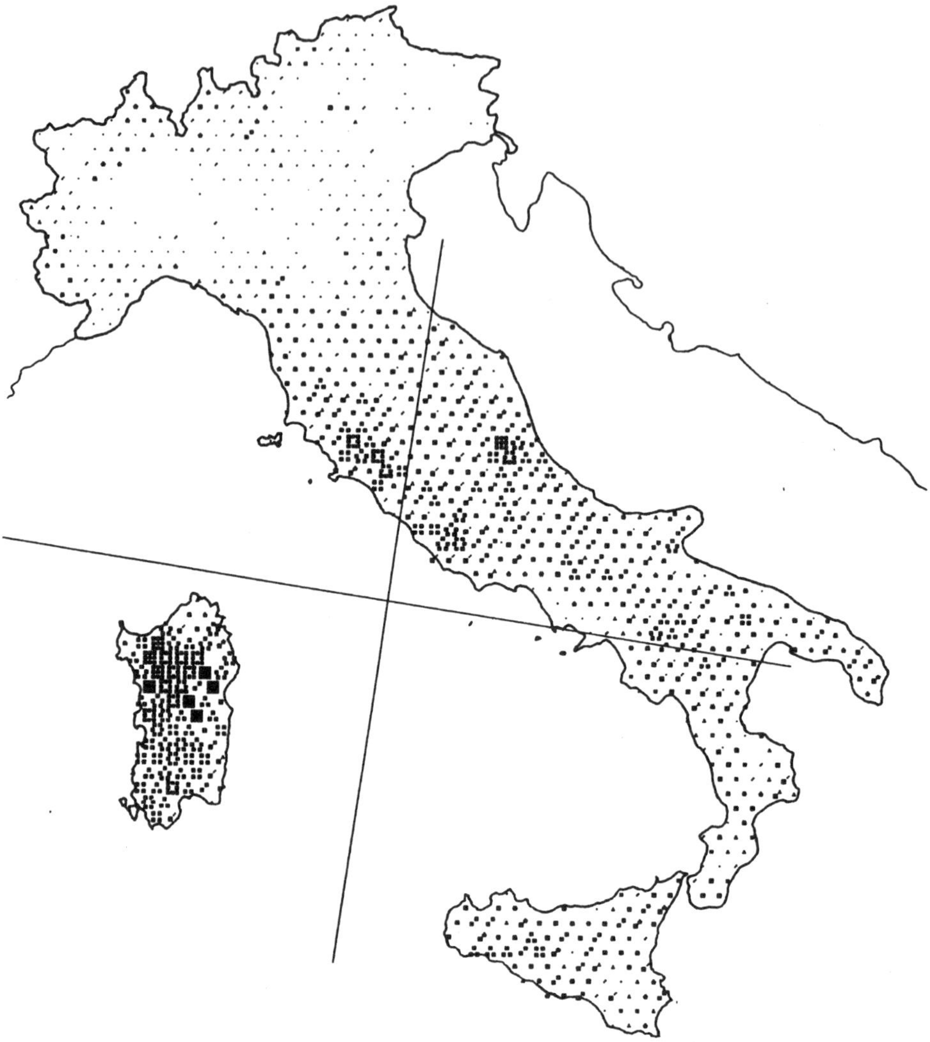

Figure 9.4. Italy: Sheep by number of head, 1982. . 700 ▪ 6700 ▪ 13,500 ⁘ 33,700 ■ 67,500+ (Source: *Statistical Atlas of Italy*, 1988).

${}^{g}P$, and ${}^{g}H$, on the one hand, and centers of ${}^{g}H$, ${}^{g}U$, and ${}^{g}NH$, on the other, form an angle of $\beta = 168.3°$. The interpretation of this setting is simple. Citrus groves are found primarily in the islands and in the south in zones that have suitable soil and climate. The center of ${}^{g}P$ is thus very southern; the center of the surpluses ${}^{g}SPU$ is even more southern, while the habitat is slightly less southern. In contrast, the center of the ${}^{g}NH$ is very far to the north.

Similar alignments of the five centers (${}^{g}SPU$, ${}^{g}P$, ${}^{g}H$, ${}^{g}U$, ${}^{g}NH$) have been found for many geosets in Italy. For 25% of those studied, the angle β is between 163° and 180°; for a further 25% it is between 146° and 163°.

Figure 9.5. Italy. Habitat and nonhabitat of sheep, 1982 (Source: *Statistical Atlas of Italy*, 1992).

For many geosets it has been found also that the distance between the two extremes ${}^{g}SPU$ and ${}^{g}NH$ of the nearly aligned five centers is comparatively large and that the nearly aligned five centers are found very near the major axis X^T of the territory of Italy (oriented roughly in a northwest/southeast direction. These findings suggest that in Italy coherent factors act upon many phenomena in determining a systematic shift of the nonhabitat from the territory and a shift in the opposite direction of the surpluses ${}^{g}SPU$. These factors include the geographical structure of the territory and physical and climatic characteristics as well as historical, social, economic, and other determinants. For some of the phenomena studied, we find the

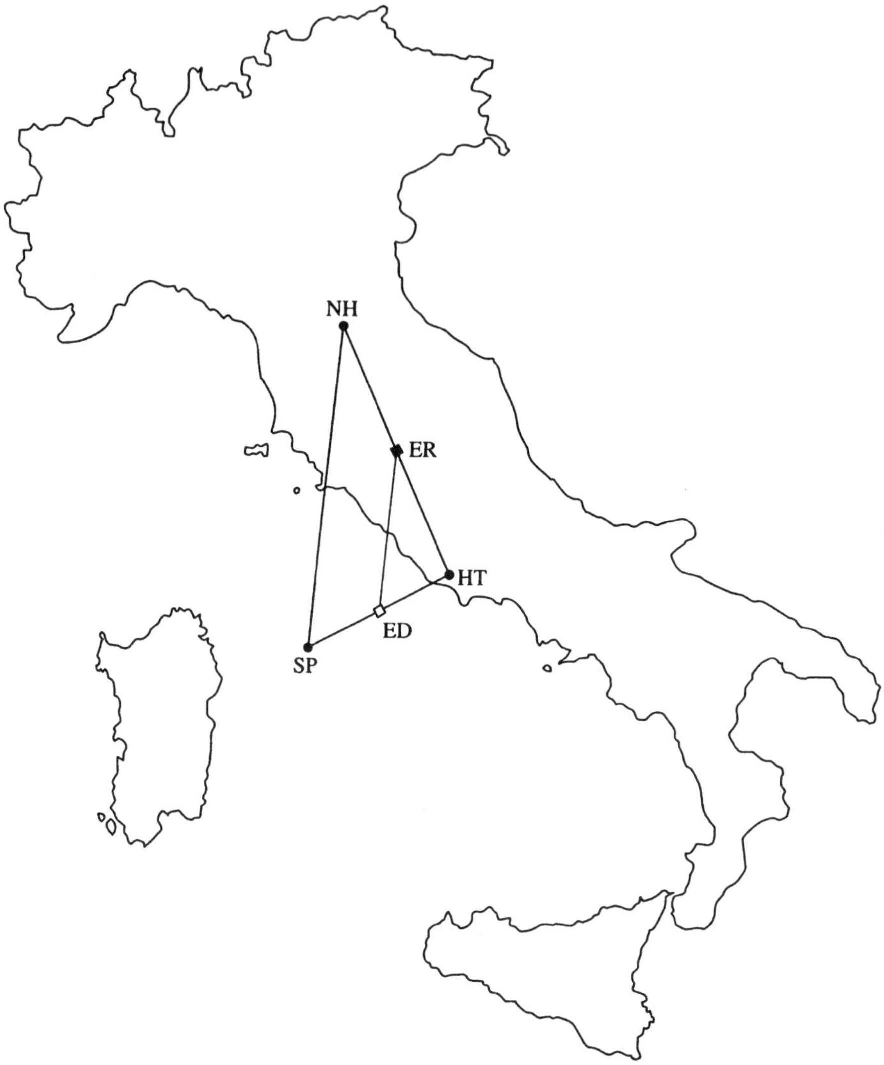

Figure 9.6. Italy. The five centers graph of sheep, 1982: (ED) population, (ER) uniform distribution, (SP) surplus of P over U, (NH) nonhabit, and (HT) habitat (Source: *Statistical Atlas of Italy*, 1992).*

center of ${}^{g}SPU$ in the northwest and the center of ${}^{g}NH$ in the southeast, while for others we find an opposite setting. Systematic study of these alternative settings, of the angle β, and of the distance between the centers of ${}^{g}SPU$ and ${}^{g}NH$ has furnished a very useful background to systematic comparative analysis and to formulation of hypotheses on the existence of meaningful "families" of similar geosets and on their determinants.

*Key to figures 9.6, 9.7, 9.13, 9.14, 9.15, 9.16, and 9.17: (ED) population ${}^{g}P$; (ER) uniform distribution ${}^{g}U$; (SP) surplus of P over U, ${}^{g}SPU$; (NH) nonhabitat ${}^{g}NH$; (HT) habitat ${}^{g}H$.

Figure 9.7. Italy. The five centers graph of citrus groves, 1982: (ED) population, (ER) uniform distribution, (SP) surplus of *P* over *U*, (NH) nonhabitat, and (HT) habitat (Source of data: ISTAT).

The triangle of five centers can be similarly determined in a comparison between any two primary geosets (${}^{g}P, {}^{g}Q$) distributed over the same territory *T*, and geosets ${}^{g}SPQ$, ${}^{g}SQP$, and ${}^{g}M(P, Q)$ derived from the primary geosets.

As an example, we consider the comparison between the distribution ${}^{g}Q$ of illiterates (Figure 9.9) to that of the population ${}^{g}P$ (Figure 1.16 in Section 1.5) and the distribution of surpluses of illiterates over the population ${}^{g}SQP$ (Figure 1.13 in Section 1.4). The centers of these three geosets and the centers of the geosets of ${}^{g}M(P, Q)$ and ${}^{g}SPQ$ are shown in Figure 9.10. Here too, the triangle of centers

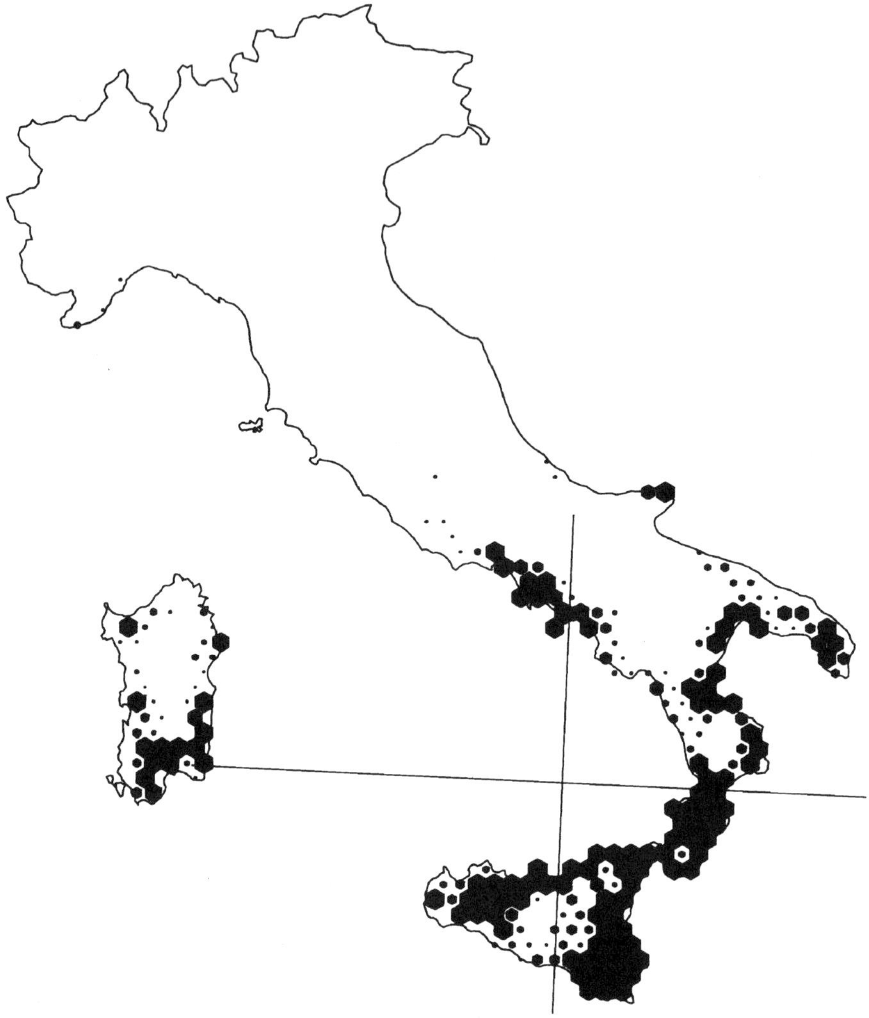

Figure 9.8. Italy. Habitat and nonhabitat of citrus groves, 1982 (Source of data: ISTAT).

actually degenerates into a line. The angle β has a value of 176.7°; the line of centers is very near to the major axis of the territory; the distance between the center of ${}^{g}SQP$ and ${}^{g}SPQ$ is comparatively very large. The small remnants of illiteracy still found in Italy thus evidence a distribution that is clearly related to historical and socioeconomic underdevelopment in some of the areas of the south and of the islands.

The *Atlas* shows the triangle of centers for 177 comparisons between pairs of geosets ${}^{g}P$ and ${}^{g}Q$. The tendency to alignment of the five centers is even more pronounced than for the comparisons between ${}^{g}P$ and ${}^{g}U$. The angle β is between 171° and 180° for 50% of the geosets studied and between 145° and 171° for 25%.

For many of the geosets ${}^{g}P$ and ${}^{g}Q$ the distances between their centers are very

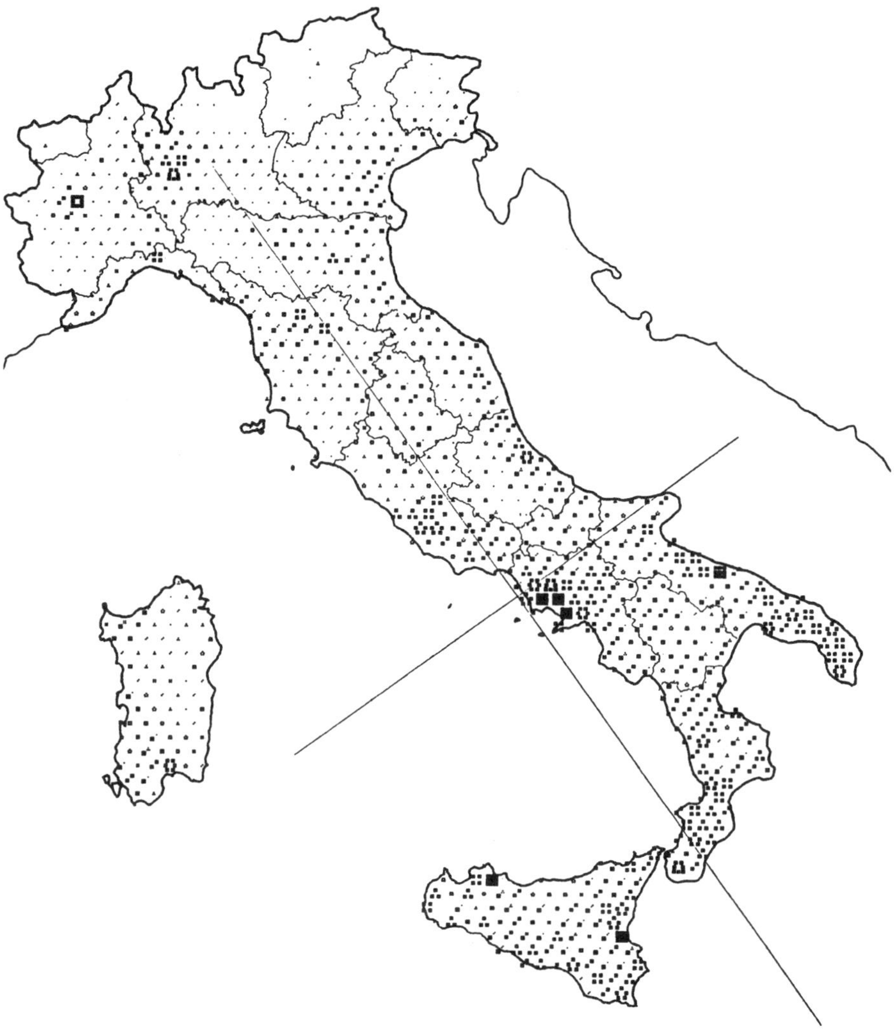

Figure 9.9. Italy. Illiterates, 1981. . 160 ▪1600 ▪▪ 3200 ⁙ 8000 ■ 16100+ (Source: *Statistical Atlas of Italy*, 1988).

marked. Thus, the setting indicated here appears to have a very broad descriptive and analytical value and has also been useful for quick identification of geosets very different from one another and in seeking determinants of peculiar behavior of their five centers. The triangle of centers can be constructed for settings in which we compare a compound geoset to two of its components. (e.g., individuals aged 0–14, 15 and over, and total population). As we discuss a similar theme at length for the triangle of distances (see Section 14.6) we will not say anything more about this theme here. To avoid overburdening this book with too many details, we are also skipping the discussion of the relationships between centers of compound geosets formed by many (positive or negative) components (see Section 3.8).

Figure 9.10. Italy. The five centers graph of illiterates (${}^{g}Q$) and total population (${}^{g}P$), 1981 (Source: *Statistical Atlas of Italy*, 1992).

9.4.3. Mean Centers of Subpopulations Ordered by Levels of a Variable

We now go on to consider an example of the setting introduced in Section 3.8.2. The recruits in the Italian army born in 1961 were divided into ten classes or subpopulations, by height. Data are available for each subpopulation in its geographical distribution (by 20 regions of residence), and geostatistical parameters were calculated for the ten geosets thus obtained. Figure 9.11 shows the mean center of each subpopulation. We see that these centers follow roughly a straight line and that the height decreases more or less regularly from northwest to southeast, which is explained both by historical and anthropological factors and by differences in the socioeconomic development among the varius regions. Similar results were found among army recruits in other years in previous studies (Bachi, 1962*b*) in which linear equations for estimating the centers of the subpopulations were calculated.

To show the differences between the centers of the subpopulation more clearly here, the mean centers for the surpluses of each subpopulation over the population (all the recruits) have been calculated.

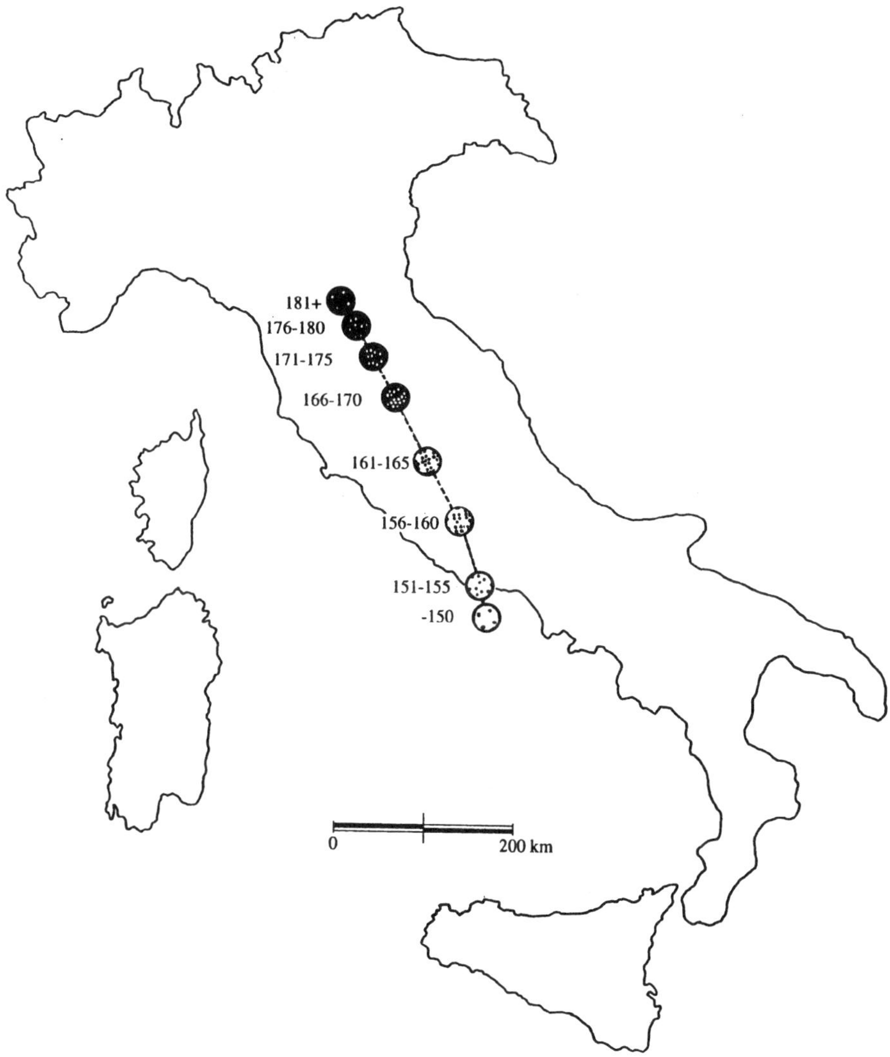

Figure 9.11. Italy. The mean centers of subpopulations of the recruits to the Italian army classified by height.

9.4.4. Mean Centers of Provinces Ordered by the Level of a Variable

In geostatistical studies we often have to deal with data indicating a rate, or a ratio, or an average value of a variable for each of the regions of a country or a city that is relevant to the population of that region. If the number of regions is sufficiently large, it may be useful to divide them into classes and to study the geosets of these classes comparatively. Figure 9.12 taken from unpublished research under-

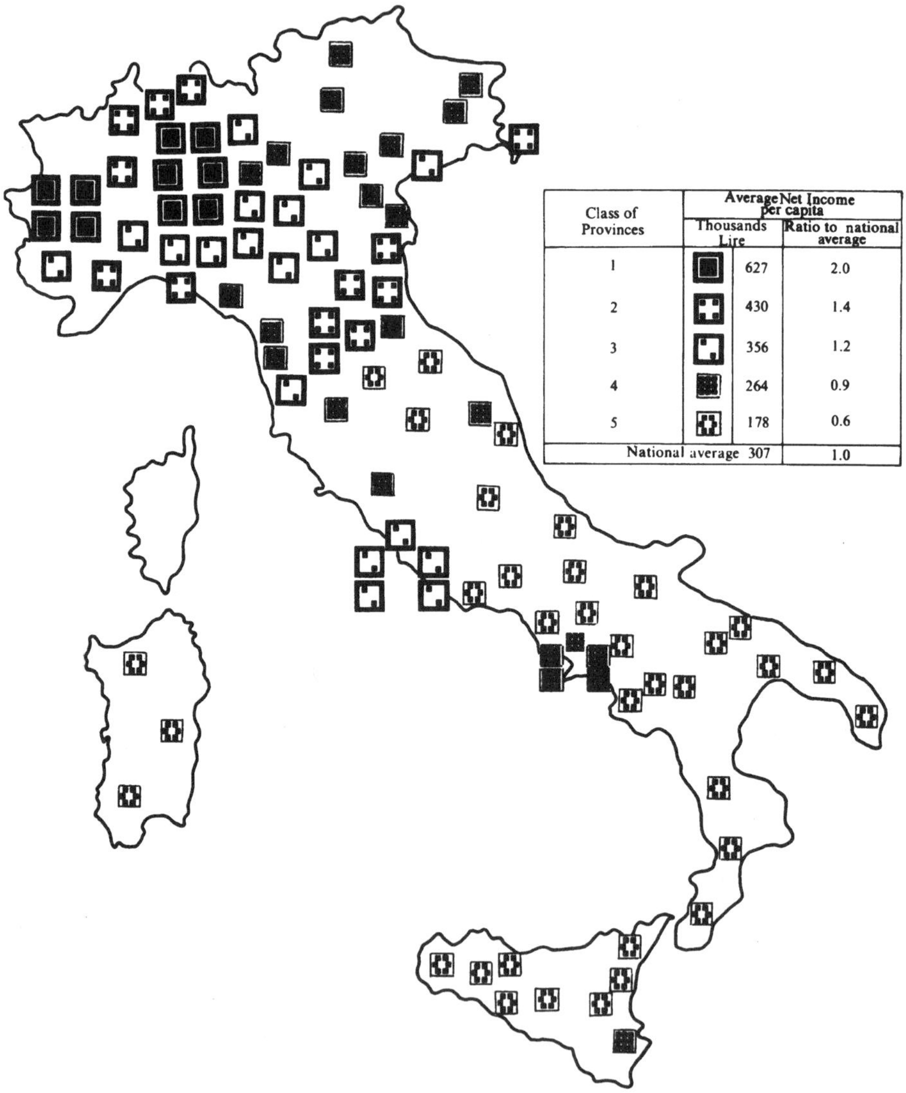

Figure 9.12. Italy (1960–1962). Average income per head in each province. Indexes based on average for Italy (307,000 lira per head) taken as 100 (Source: Bachi, 1968).

taken in 1963–1964 in cooperation with Prof. Guglielmo Tagliacarne (Unione Italian adelle Camere di Commercio, Industria e Agricultura), illustrates this approach. The 92 provinces in existence at the time were classified according to the average level of various economic functions. Figure 9.12 relates to the average income per capita in 1960–1962. The provinces were ranked according to this variable and divided into five quintiles, each of which included about a fifth of the total Italian population. Geostatistical parameters were calculated for the population of each quintile. We can see in Figure 9.12 that the mean centers are roughly aligned typically from northwest (highest average income) to southeast (lowest average income).

9.5. Mean Centers and Their Triangle for the Population of the United States

9.5.1. The Centers of the Total Population

The mean center of the population of the United States is presumably the best-known example of geostatistical parameters: it has been calculated by the U.S. Bureau of the Census at each decennial census; widely publicized in the yearly *Statistical Abstracts of the United States* and other official publications, and it has been discussed extensively in the literature. We see no value to repeating well-known findings here, but we shall give an analysis of the center based on the comparison between the actual and uniform distribution of the population, as introduced in Section 8.8.

Figure 9.13 shows the triangle of five centers (${}^{g}P$, ${}^{g}U$, ${}^{g}H$, ${}^{g}NH$, and ^{g}SPU) of the total population of the United States in 1980. We can evaluate each of these centers by looking at it over the geoset to which it corresponds using Figures 8.2–8.4 (Section 8.8). Figure 9.13 shows that, as expected, the centers of ${}^{g}NH$, ${}^{g}U$, and ${}^{g}H$ are collinear and so are the centers of ${}^{g}H$, ${}^{g}P$, and ${}^{g}SPU$. This suggests the existence of a rather consistent system of determinants working on the five centers. We will now examine these centers with the help of the maps of Section 8.8 that illustrate the detailed distribution of ${}^{g}P$, ${}^{g}H$, and ${}^{g}NH$, and ${}^{g}SPU$.

Centers of ${}^{g}P$ and ${}^{g}U$. Figure 8.2 (Section 8.8) shows the map of ${}^{g}P$. The two crosses superimposed over this map show the principal axes of ${}^{g}P$ and ${}^{g}U$; the central points of these axes correspond, respectively, to the mean centers of ${}^{g}P$ and ${}^{g}U$. The map (as the maps of the United States given elsewhere) have been constructed utilizing projections with respect to distances between points on the surface. To simplify the interpretation of the coordinates, their origin $(0, 0)$ has been taken at the mean center of the territory. We can see from the map that both the principal axes (X'_p, Y'_p and X'_U, Y'_U) of ${}^{g}P$ and ${}^{g}U$ differ only slightly from the conventional axes with which they make, respectively, angles of $+4.40°$ (for ${}^{g}P$) and $-7.95°$ (for ${}^{g}U$). This, together with the widespread habit of referring to cardinal points in popular presentations of geographic locations for the United States, leads us to express the coordinates of X_p and Y_p here as deviations (in miles) from the origin along the axes X and Y.

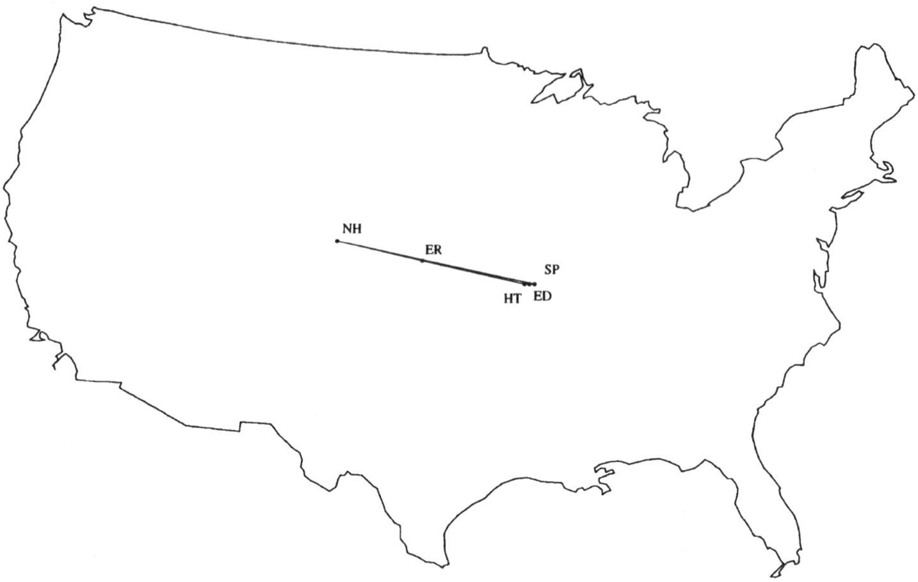

Figure 9.13. United States. The five centers graph of total population, 1980. (ED) population, (ER) uniform distribution, (SP) surplus of P over U, (NH) nonhabitat, and (HT) habitat (Source of data: Strategic Mapping—Counties Data Files, U.S.A.).

Negative values of X_p indicate shifts from (0, 0) to the west, while positive values mean shifts to the east; negative values of Y_p indicate shifts to the south, while positive values mean shifts to the north. The center of ^gP is at a distance of $X_p = +475$ miles to the east of the center of the territory and that of $Y_p = -26$ miles to the south of this center. This location does not imply high population density in its neighborhood. We can interpret it by using, with some slight modifications, the official definition of the mean center given by the Bureau of the Census: The mean center is the point upon which the (conterminous) United States would balance if it were a rigid weightless plane, and the population of each cell c was assumed to have a weight proportional to w_c^p and to exert an influence on the central point proportional to its distance from that point. The location of the mean center is thus the resultant of a very complex interplay of microgeographic and macrogeographic determinants. Consideration of the centers of ^gH, ^gNH, and ^gSPU may help to improve understanding of the effects of some of the macrogeographic determinants.

Centers of ^gH, ^gNH, *and* ^gSPU. Figure 8.3 (Section 8.8) shows ^gH and ^gNH with the crosses of ^gH and ^gNH superimposed. The center of ^gNH falls in an almost uninhabited zone, to the northwest of the center of the territory (^gU). By contrast, the center of the habitat is shifted to the southeast of the center of the territory. The center of the habitat, the center of the population, and the center of supluses ^gSPU are all rather close to one another, near the western border of the largest cluster of surpluses (see Figure 8.4 in Section 8.8). The center of the surpluses is found to the northeast of the two other centers.

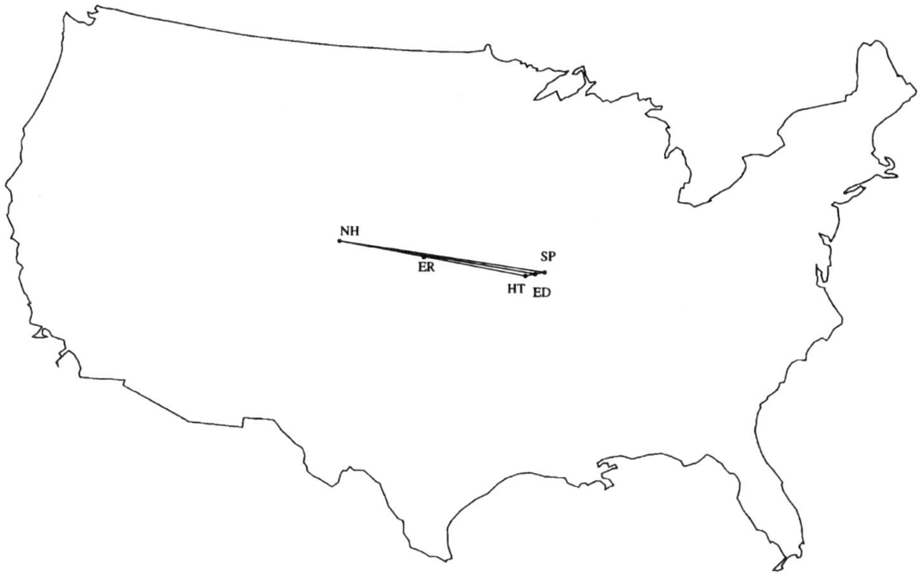

Figure 9.14. United States. The five centers graph of White population, 1980. (ED) population, (ER) uniform distribution, (SP) surplus of P over U, (NH) nonhabitat, and (HT) habitat (Source of data: Strategic Mapping—Counties Data Files, U.S.A.).

Changes in the Center of Population over the Course of Time. It is well known that over the course of time, the center of the total population of the United States has systematically shifted toward the west and, to some extent, toward the south. The line of the centers of the population of the conterminous states for the census years between 1790 and 1980, given in Figure 9.18, confirms that the same tendencies also characterized the centers of these states.

9.5.2. Mean Centers for Subpopulations

We present here an analysis of mean centers for the subpopulations of the United States, whose distributions were introduced in Section 8.8. Detailed maps of their geosets and of the clusters of their surpluses were given in Figures 8.5–8.10. Figures 9.14–9.17 show the triangle of five centers for Whites, Afro-Americans, Asians, and Hispanics, respectively.

Whites. The Whites constituted a large majority of the total population (1870: 84%; 1980: 83%) in the conterminous states during the period considered. Accordingly, their geoset and their triangle of five centers (as shown in Figure 9.14) are very similar to those of the total population illustrated above. Also for the Whites, the mean centers of ${}^{g}SPU$, ${}^{g}P$, ${}^{g}H$, ${}^{g}U$, and ${}^{g}NH$ have a tendency to be aligned from east to west. The historical trend of the center of the White population has been very consistent: the prevailing direction has been from east to west, with a bend to the south, mainly after 1920.

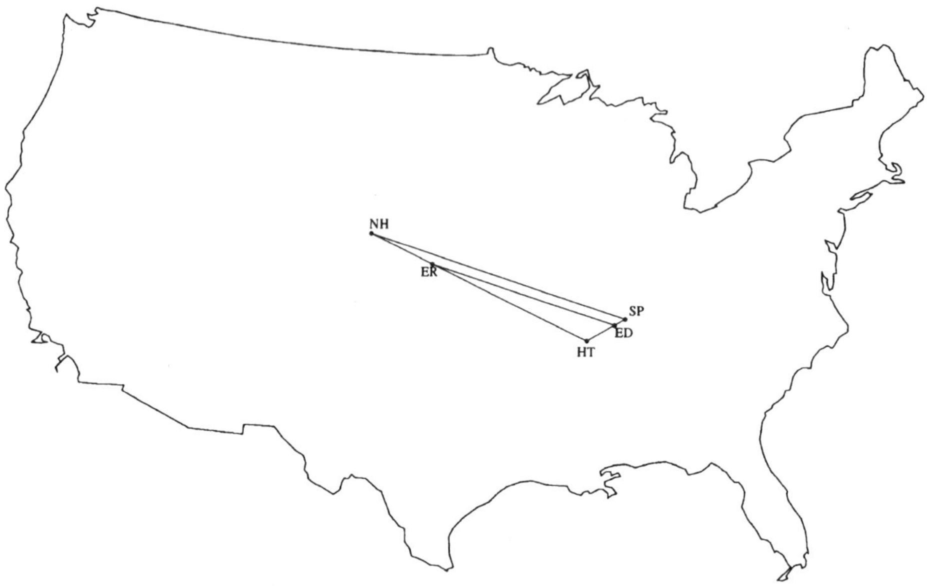

Figure 9.15. United States. The five centers graph of African-American population, 1980. (ED) population, (ER) uniform distribution, (SP) surplus of P over U, (NH) nonhabitat, and (HT) habitat (Source of data: Strategic Mapping—Counties Data Files, U.S.A.).

Afro-Americans. Between 1870 and 1920 the Afro-American population had a comparatively low dispersion (see Bachi, 1962*b*) and was markedly centered in the South. Since then that dispersion has continuously increased, owing primarily to the attraction of urban areas. This has led to a systematic shift northward, which has become very strong in the last decades. The habitat of Afro-Americans is comparatively small (about 25% of the territory in 1980). Its center is still markedly southern. However, the center of the surpluses of the actual population over uniform distribution is more directed to the northeast. The very wide nonhabitat is centered to the west of the center of the territory. According to data for 1970, the center of the Afro-American farm population was even more to the south than the center of the total Afro-American population (see a comparison of the ellipses corresponding to these populations in Section 4.4).

Asians. Between 1870 and 1970, the mean center of the Chinese population has shown a very consistent and strong trend: in the beginning, this center was extremely western (evidently owing to where the Chinese immigrants first settled). Since then it has moved eastward, owing mainly to the attraction of cities of the northeast, which has created an extremely strong bipolarity (for this measure see Bachi, 1962*b*).

The mean center of the Japanese population has also had a strong eastward shift, for similar reasons, but it has become evident only since 1940 and has been less marked. The mean center of the Filipino population is located between the mean center of the Chinese and that of the Japanese (and nearer to the latter).

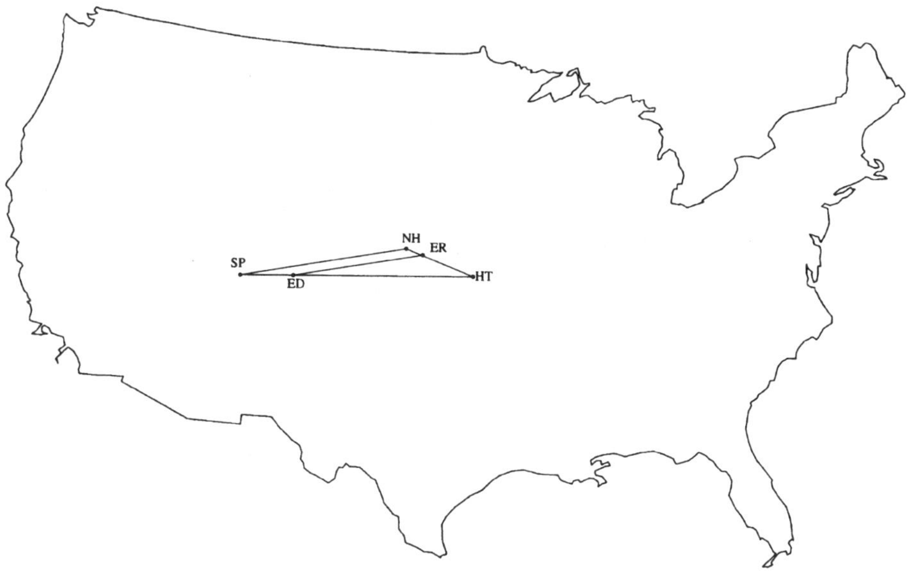

Figure 9.16. United States. The five centers graph of Asians, 1980. (ED) population, (ER) uniform distribution, (SP) surplus of *P* over *U*, (NH) nonhabitat, and (HT) habitat (Source of data: Strategic Mapping—Counties Data Files, U.S.A.).

A comparatively very high proportion of all Asians are found in California and in the large cities of the northeast, while small nuclei of this population are scattered over many other places. Consequently their habitat is comparatively very small and its center falls in the middle of the wide nonhabitat; the five centers are all close to one another.

Hispanics. This population (including Mexicans, Puerto-Ricans, and Cubans among others) is largely concentrated in southern areas, partly bordering Mexico (mainly in California, Texas, and Florida) and in the cities of the northeast. The centers of habitat, population, and surpluses of population over uniform distribution are very near to each other and located not too far from the center of the territory, but with a shift to the south. The center of the nonhabitat is, according to data for the years 1910–1960, more to the north than the center of the territory. The mean center of individuals born in Mexico has been found to have had a southward trend over the course of time.

Overview of Population Centers by Race. Figure 9.19 shows the mean centers of the populations discussed above. Considering the east to west direction we see that the most eastern are the Afro-Americans, followed by the Whites, and the Hispanics; the most western are the Asians. The mean center of all the race subpopulation is influenced by the general characteristics of the distribution of the total population. In order to eliminate this influence, we have calculated the mean centers of the

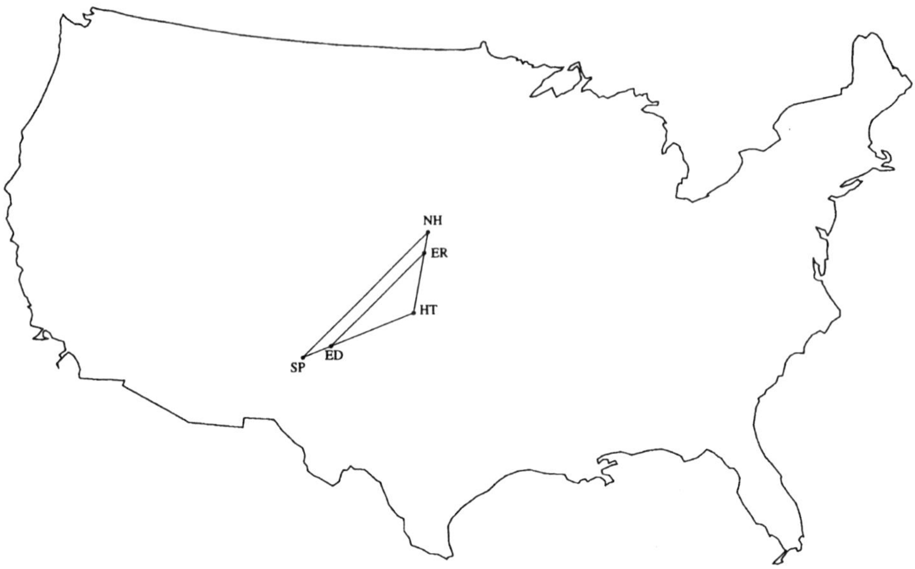

Figure 9.17. United States. The five centers graph of Hispanics, 1980. (ED) population, (ER) uniform distribution, (SP) surplus of P over U, (NH) nonhabitat, and (HT) habitat (Source of data: Strategic Mapping—Counties Data Files, U.S.A.).

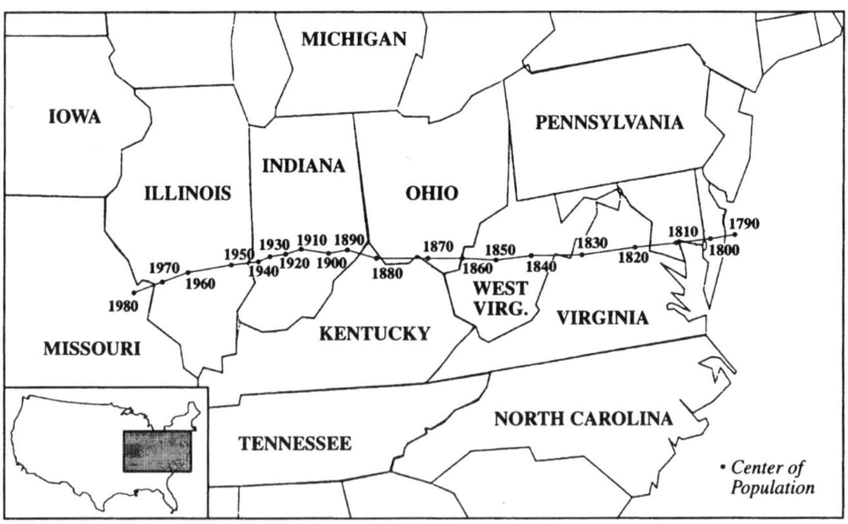

Figure 9.18. United States. Mean centers of total population in the census years between 1790 and 1980 (Source of data: U.S. Bureau of the Census, Graphic Summary of the 1970 Population Census, 1790).

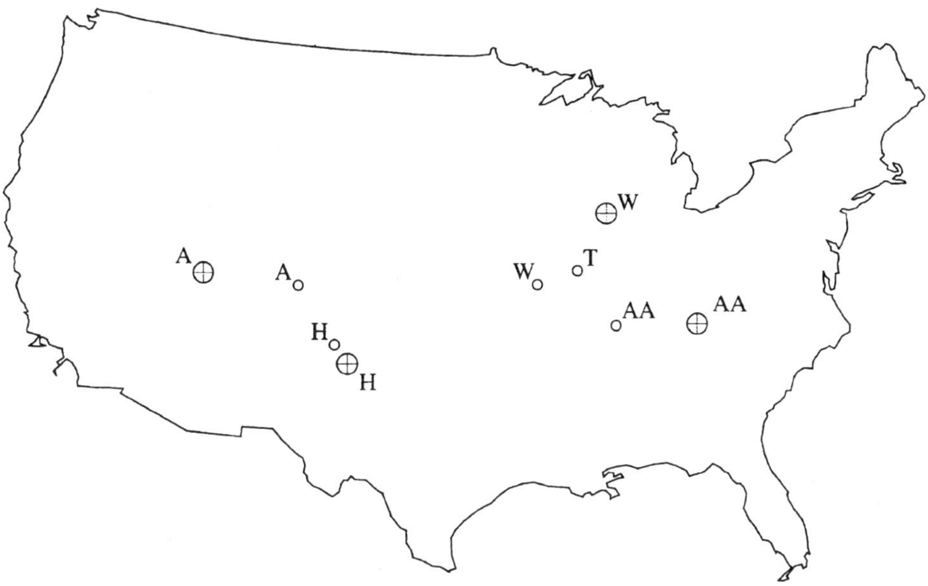

Figure 9.19. United States. Mean centers of Whites, Afro-Americans, Asians, and Hispanics, and mean centers of surplus of these subpopulations over total population, 1980: (○) mean center, (⊕) mean center of surplus, (AA) Afro-Americans, (W) Whites, (H) Hispanics, (A) Asians, (T) total (Source of data: Strategic Mapping—Counties Data Files, U.S.A.).

surpluses of each subpopulation over the population (both reduced to the same standard size of 10,000). These centers are also shown in Figure 9.19. As expected, these centers emphasize the characteristics of the location of each race.

9.6. Comparison of Population Mean Centers and Dispersions for Various Countries

In Section 9.5.1 we discussed the mean center of the population of the United States and its changes over the course of time. Obviously similar analyses can be done, and have actually been done, for other countries. We shall now relate to the question of how such analyses can be made comparable, despite the fact that the territories considered differ in size and shape. For this purpose, we offer as an example comparative tables for four countries: the United States, France, Italy, and Argentina (see Appendix V in the scond volume of the *Statistical Atlas of Italy*).

Several of the parameters introduced below are based on the mean center, which is the principal focus of this chapter. However, to avoid repetition further on we have also included other parameters of absolute and relative dispersion for the four countries considered. The meaning of these parameters and methods for their calculation were introduced in Section 1.5 and are explained in detail in Chapters 10 and 11.

Table 9.2 enables us to obtain a first comparative orientation on the size and some aspects of the shape of the territories of these countries. We can see from Column 1, that the territory of France is almost double that of Italy and those of Argentina and the United States are, respectively, 9 and 26 times larger. For an interpretation of the parameters in Columns 2, 3, and 4 the reader is referred to Section 1.5.6 and to Chapter 11.

The relative scatters of Column 2 suggest that the territories of France, the United States, and Argentina are basically convex, despite the presence of some external concavities. In contrast, the territory of Italy is characterized by a very high relative distance-variance, owing to its fragmentation, to the deep external concavities, and to its oblongity. The territories of the United States and Argentina also have a considerable oblongity, while the territory of France is almost nonoblong. The dispersion of these territories (measured by standard distance d_T or by distance-variance d_T^2) is influenced by both their area and their looseness (lack of compactness owing to relative scatter and oblongity). For a comparative study of geostatistical parameters for a population free of the influence of territorial dispersion, it is convenient to divide by d_T or d_T^2.

Let us denote the standard distance of the population by d_p; the distance between the mean center of the population and the mean center of uniform distribution over the territory T by ${}^cD(P, U^T)$; and the distance between the centers of a population at two different times (1,2) by ${}^cD(P_1, P_2)$.

Table 9.3 shows the following ratios (given in percentages to simplify the readings) between these parameters and d_T, which can be considered as pure numbers, netted from the effects of area and looseness of the territories. They enable us to compare dispersion and distances of the population centers:

1. The ratio d_p/d_T measures the dispersion of the population as compared to that of the territory.
2. The ratio ${}^cD(P, U^T)/d_T$ measures the relative eccentricity of the population over the center of the territory.
3. The ratio ${}^cD(P_1, P_2)/d_T$ measures the relative shift of the center of population, passing from time 1 to time 2.

Table 9.2. Area and Relative Dispersion of the Territories of Italy, France, Argentina, and the United States

Country	Ratio of area to the area of Italy	Parameters of looseness of territory		
		Relative scatter	Relative oblongity	Relative distance-variance
Italy	1.00	2.32	1.25	2.91
France	1.80	1.09	1.01	1.10
Argentina	9.22	1.18	1.59	1.86
United States	25.93	1.10	1.21	1.34

We have also indicated in Table 9.3 index numbers of the size of each population, taking as a basis (= 100) its size in the first period.

Examining the first set of ratios (1) we see that in both France and Italy the ratio of the dispersion of population to dispersion of territory has remained almost constant and that d_p actually almost coincides with d_T. On average we find that in France $d_p/d_T = 101.2\%$ during the period from 1801 to 1962. In Italy the average of ratios d_p/d_T from 1871 to 1987 is 98.6%. Presumably, the identity of behavior of these ratios in France and in Italy is accounted for by the following, determinants: Human settlement in both countries is ancient and continuous and does not show wide and compact nonhabitat areas, while the cities are widely dispersed over the territory. The great urbanization that occurred during the period under survey did lead to increasing concentrations in the big metropolises and other cities. However, the fact that the urban areas are widely dispersed coupled with other compensatory factors resulted in the population dispersion remaining almost unchanged.

The behavior of population dispersion in both the United States and Argentina has been very different owing to the impact of the enormous increase in their populations that resulted from immigration.

At the end of the colonial period in the United States, the population was concentrated largely on the Atlantic seaboard. In the earlier years of independence there was a continuous demographic and economic expansion toward the states and territories that were becoming part of the confederation. At the same time immigrants from abroad were settling mainly on the Atlantic seaboard but also on the Pacific coast and near the borders of Mexico. The internal migrations had a strong impact on the increase of the population dispersion: the standard distance of the origin of internal migrations is found to be generally greater than the standard distance of their destinations.

In consequence of these and other developments, the ratio d_p/d_T continuously increased in the period that was surveyed: in 1880 it was 70.9%; in 1976 it had reached 107.8%, thus showing that the population is more dispersed than the uniform distribution over the territory. This is a clear indication of the great relative weight of the populations of peripheral regions.

In contrast with the continuous increase of d_p/d_T in the United States, in Argentina this ratio decreased continuously between 1869 and 1914, during which time the great increase in population owing largely to immigration from abroad determined a strong and growing concentration of the population in and around Buenos Aires. Between 1914 and 1960, the ratio d_p/d_T had increased somewhat but still remained at a comparatively very low level, the population being far less dispersed than under the hypothesis of uniform distribution over the territory.

Going on, now, to consider the second and third parameters, we find that despite the similarity of other aspects of population distribution in France and Italy, the behavior of mean centers in the two countries is considerably different. The great and growing concentration of population in the Paris metropolitan area and in the wards of the north and the Pas de Calais has led to a slow but continuous shift of the population center toward the northeast. The relative shift of the mean center of the French population between 1801 and 1962 was of 13.1%, while in Italy the center shifted by only 1.4% between 1870 and 1987.

Table 9.3. Ratios of Parameters for Population to Standard Distance of Territory for Italy, France, Argentina, and the United States

Year	Index number of population	Relative standard distance of population, d_p/d_T	Relative eccentricity, ${}^cD(P, U^T)/d_T$	Shift of center of population,[a] ${}^cD(P_1, P_2)/d_T$
Italy				
1871	100	98.1	4.6	1.4
1881	106	99.0	3.3	1.7
1901	121	99.5	2.0	1.3
1911	131	99.4	3.1	2.1
1921	144	99.6	5.1	1.5
1931	150	98.4	6.5	0.9
1936	155	98.1	5.7	2.6
1951	174	98.4	4.5	1.4
1961	185	98.5	4.0	3.4
1971	198	97.8	6.4	1.1
1981	207	97.9	5.5	1.8
1987	210	98.2	4.3	
Average		98.6	4.6	1.7
France (without Corsica)				
1801	100	100.6	15.1	1.3
1831	118	101.0	14.1	0.8
1841	125	101.0	14.6	0.1
1851	131	101.0	14.6	1.9
1861	137	101.3	14.5	0.8
1872	138	101.3	15.2	1.2
1881	143	101.2	16.3	3.1
1901	149	101.3	19.3	2.0
1911	152	101.3	21.2	0.1
1921	143	101.3	21.1	2.7
1931	153	101.6	22.9	0.2
1936	153	102.0	22.7	1.8
1946	148	101.3	22.6	2.3
1954	156	101.3	24.6	1.5
1962	169	101.3	25.2	
Average		101.2	18.9	1.4
Argentina				
1869	100	57.8	46.2	15.0
1895	228	55.9	42.0	7.8
1914	454	54.1	38.8	0.1
1947	915	57.2	39.0	2.9
1960	1152	58.6	36.6	
Average		56.7	40.5	6.5

Table 9.3. (Cont.).

Year	Index number of population	Relative standard distance of population, d_p/d_T	Relative eccentricity, ${}^cD(P, U^T)/d_T$	Shift of center of population,[a] ${}^cD(P_1, P_2)/d_T$
United States (without Alaska and Hawaii)				
1880	100	70.9	92.1	6.3
1890	126	75.4	85.8	1.5
1900	152	77.2	84.4	4.4
1910	183	83.0	79.9	1.7
1920	211	85.5	78.2	2.8
1930	245	89.5	75.5	1.8
1940	263	91.5	73.8	5.1
1950	301	97.2	68.8	5.0
1960	356	104.1	69.9	3.4
1970	403	106.8	67.6	2.6
1976	425	107.8	65.7	
Average			76.5	3.5

[a]From year indicated to previous year.

The systematic shift of the mean center of the French population has determined a continuous increase of the eccentricity of the population over the uniform distribution, as measured by (2), the value of which increased from 14.1% in 1831 to 25.2% in 1962. In contrast with this, in Italy this parameter remained almost unchanged standing at 4.6% in 1871 and 4.3% in 1987.

In the United States, the center of population has shifted continuously toward the west. Between 1790 and 1940 the changes in the center occurred mainly along the 39th parallel. Afterward, there was also some bending toward the south. As the center of territory of the United States is to the west of the center of population, the latter's shift to the west determined a continuous decrease of eccentricity of the population over the center of territory. However, relative eccentricity, as measured by (2), remains much larger in the United States than in the three other countries considered here (1880: 92.1%; 1980: 76.5%).

In Argentina the center of population has had a relative increase mainly toward the south. As the center of territory is to the southwest of the center of population, the shift of the latter has determined a decrease in eccentricity, and ratio (2) has gone from 46.2% in 1869 to 36.6% in 1960.

Obviously, these methods for rendering the distance between the center of the population and the center of the territory in different countries comparable can be applied to the distance between the center of any one geoset from the center of territory. For example, for the 385 geosets studied in the *Statistical Atlas of Italy* it was found that, on the average, they had: (a) a relative distance between the center of the geoset and the center of territory ${}^cD(P, U^T)/d_T = 31.3\%$; and (b) a relative distance between the center of surpluses of the actual over the uniform distribution and the center of the nonhabitat ${}^cD(SPU, NH)/d_T = 53.8\%$.

9.7. The Median Center

9.7.1. Definitions and Properties

Just as we defined the mean center as the point from which the second distance-moment is minimal, we can also define a point from which the first distance-moment is minimal. This point, usually called the median center, is denoted by $\tilde{C}$ and its coordinates by $\tilde{X}$, $\tilde{Y}$. The median center is generally considered to be a two-dimensional analogue of the median of a one-dimensional distribution, as will be discussed in Section 9.8.

Let us indicate any point in the plane by k; any element of the geoset by i with (X_i, Y_i) as its coordinates, and w_i as its relative weight; and the first moment about any k by $\mu_{k,1}$. The minimum value of $\mu_{k,1}$ is found for $k = \tilde{c}$:

$$\min \mu_{k,1} = \min \Sigma w_i \sqrt{(X_i - X_k)^2 + (Y_i - Y_k)^2}$$
$$= \Sigma w_i \sqrt{(X_i - \tilde{X})^2 + (Y_i - \tilde{Y})^2} = \mu_{\tilde{c},1} \qquad (9.14)$$

This implies that the median center is the point at which the entire distribution can be convened with minimal aggregate travel.

The concept of median center was explicated in the early 1930s by Italian statisticians, who distinguished between the "restricted" median center and the "enlarged" median center. The former is one of the points in the actual distribution from which the sum of distances of all points is minimal, while the latter is any point in the plane from which the sum of distances of all points is minimal. The restricted median center is easy to calculate and might have some practical value in planning, e.g., the location of an institution that can only be established in an inhabited location. However, we deal below only with the enlarged median center which is the more important general parameter, and will refer to it simply as the median center.

We note below some properties of the median center that have been found to be of practical value in the applications and in the interpretation of the parameter (see, e.g., Austin, 1959).

1. The median center (and the mean center) is covariant with a rotation or parallel translation of the geoset being considered.
2. The median center (and the mean center) is not necessarily located within the territory of the geoset, but must be located within the convex hull of the geoset.
3. The sum of the cosines of the angles formed by all pairs of lines from the median to the given points is zero, unless one of the points itself is the median.
4. If i is one of the given points and L_i is the line from the median center to i, then i can be moved freely for any distance along L_i without affecting the location of the median center.
5. If $n = 2k$ points are distributed about a point k such that every line through k containing n given points on one side of k also contains n points on the opposite side, then k is the median center.

The median center is thus seen not to be sensitive to changes in the location of the dots as is the mean center, but only to angular relationships of the dots around it.

9.7.2. Method of Calculating the Median Center

Pencil and paper calculation of the median center is by no means an easy job. The relevant calculation for the case in which the distribution is formed by three nonaligned points was discussed as early as the 15th century by the physicist Torricelli and is also found in the more modern literature (see, e.g., Eells, 1930; Flaskämper, 1963). Special cases of the median center of populations distributed in a continuous fashion along a straight or curved road, a curved broken line, or for certain regular territories were discussed by Griffin (1933). However, no single general solution for the calculation was found before the computer era.

Galvani (1934) showed that there is a median center for any distribution of points over a plane, and he proposed an approximate solution to the problem. A computer solution was proposed by Sherf in the early 1960s, which was later improved on by Yam and is routinely employed in Israel.

Another computer solution was offered by Seymour (1965) (see also Seymour, 1970). In this book we have adopted the solution described in Appendix F for geosets and uniform distribution over territories developed by Ever-Hadani, which is practically identical to the one proposed by Seymour.

9.7.3. Uses of the Median Center

The property of the median center as the point in which the elements of the geoset can convene with minimal aggregate travel is often utilized for practical purposes to plan optimal locations for services, businesses, and other establishments. However, the analysis of these problems is outside the scope of this book. We only wish to stress here that in the application of both the mean center and the median center to these problems, the underlying hypothesis is to consider linear distances (simple for median center and squared for the mean center). Actually, the costs of travel may be proportional to other distances as "Manhattan distances," or "city block" distances, "vial distances" (see Flaskämper, 1962), or other measures proportional to the time needed for travel; actual costs of travel may constitute preferable approaches.

Comparison of the median center to the mean center can sometimes be employed in the study of asymmetry, although we do not use this approach here. The median center has been found to be useful for general descriptive and comparative purposes in various geostatistical studies. In some instances joint analysis of the median center and other centers is informative, primarily if it is based on critical consideration of the different properties of these centers (such as, sensitivity of the mean center to the location and to the changes of location of each element of the geoset versus lack of sensitivity of the median center to the location of the outlying elements).

9.7.4. Median Center versus Mean Center of Thousands of Geosets for Italy

In the *Statistical Atlas of Italy*, median centers and mean centers have been calculated for thousands of primary and derived geosets. Obviously, we cannot enter here into the analysis of this huge amount of material, so we shall limit ourselves to noting only some very general findings:

1. For 385 geosets triangles of centers were constructed that show the distances between the mean centers of: (a) habitat and territory; (b) territory and nonhabitat; (c) habitat and population; (d) population and surpluses over uniform distribution; (e) population and territory; (f) surpluses and nonhabitat. As expected, (a), (b), and (c) are collinear as are (c), (d), and (e). The median centers for (a), (b), (c), (d), (e), and (f) were calculated in a parallel way, and it was found that the general behavior of the median centers is similar to that of the mean centers. The maps of the median centers show quasi-triangles, but these do not have the regularities of the mean centers. If we look at the average for all the 385 geosets, we see that the distances (e) are 35% longer for the median centers than for the mean centers, the distances (f) are longer by 30%, and so on.
2. By calculating the mean centers and the median centers for 385 geosets along the major axis of the territory of Italy, we find averages of 29.6 km for the mean centers and of 84.1 km for the median centers. This suggests that median centers tend to be shifted to the northwest as compared to the mean centers.
3. In the calculation mentioned under (2) the distances between extreme values are longer for median centers than for the mean centers. However, in the values obtained in that calculation a very high Pearson coefficient of correlation (98.8%) is found between the mean center and the median center of each geoset.

9.8. Meaningful and Nonmeaningful Median Point

For one-dimensional distributions, the median is generally defined as a value that divides the distribution into two parts, each of which includes half of the elements. The fact that the median is the value minimizing the first moment is generally indicated as a property of the median. In seeking a two-dimensional analogue to the median, the early literature proposed as a median point, the intersection of two orthogonal axes, each of which bisects the population. However, the location of this point can change with changes in the axes. The number of median points can thus be infinite, which detracts from the utility of this parameter. Accordingly, the median point has generally been discarded.

In our opinion, this disparagement of the median point should be somewhat mitigated, as there are pairs of orthogonal axes that are meaningful. For example: (a) conventional axes sometimes refer to meaningful north–south and west–east directions; (b) principal axes are often very meaningful and we can define as the principal median point that point that bisects the distribution and is obtained as the intersection between two orthogonal axes parallel to the principal axes.

The median point can be calculated easily. Many attempts have been made to obtain it for actual geosets and to compare it to the median and the mean centers. However, the discussion of these parameters would require a microgeographic analysis, which cannot be introduced here.

9.9. Modal Centers

In the analysis of one-dimensional distributions the mode is sometimes employed, alone or together with other averages. However, the determination of the mode is possible mainly when an entire population or a large sample are studied, and the hypothesis of an underlying continuous distribution can be accepted. Then, if the distribution of X is supposed to obey some law, indicated by a function $f(X)$, a value of X corresponding to a maximal density of $f(X)$ can often be determined.

In geostatistics analogues to the mode may also be required. However, this calculation is rendered difficult by the complexity of the geostatistical distributions, which are often multimodal or for which the hypothesis of underlying continuous distribution cannot be accepted. Some alternative ways of overcoming this difficulty have been proposed but are not always suitable.

Use of Distance-Moment Distributions. A distance-moment μ_{ks} of any order s can be calculated for any point k of the plane. This enables us to obtain a continuous two-dimensional distribution over the plane, while the original distribution of the dots is discontinuous. In order to calculate a modal center, it is desirable to have a type of moment that is more strongly influenced by nearby points than by distant points, which is a property of moments with negative exponents s:

$$\mu_{ks} = \Sigma(w_i/d_{ik}^s) \tag{9.15}$$

The point that minimizes μ_{ks} (for $s = -1$) is often employed as a modal center and can be denoted as the point with the highest "potential" or as the harmonic mean center. Unfortunately, the criticisms against the use of distance-moments with negative exponents discussed earlier also apply to this measure and we do not think it desirable to use it in its original form.

Partitioning Methods: Centers of Clusters. As the geostatistical distribution is generally a complex one and often multimodal, methods for partitioning it into subdistributions can sometimes be a convenient tool for its analysis. If the distribution is partitioned into subdistributions or major clusters, a subcenter can be calculated for each cluster. Mean centers and/or median centers of the clusters can be adopted as indicators of modal centers.

Chapter 10

Parameters of Dispersion and Oblongity of Geosets

10.1. Measuring the Dispersion of Geosets

10.1.1. Parameters of Dispersion of Populations over Territories versus Measures of Variability of One-Dimensional Variables

We indicated in Section 9.1 that summarizing the locations of the elements of a geoset through a center is an operation that has some similarity (and dissimilarities) to averaging the values X_i of a variable X. Analogously, parameters of dispersion of geosets are somewhat similar to parameters of variability of these values X_i of a variable X. We shall consider these operations, called for short, dispersion (of a geoset) and variability (of a one-dimensional variable).

The basic similarity of the two concepts is shown by the fact that in many statistical textbooks one-dimensional variability is indicated as "dispersion." Variability can be measured using different approaches, which should be selected according to the nature of the data to be summarized and the goals of the study. This is also true for parameters of dispersion. However, as we explain below, some of the basic approaches used or the parameters of dispersion differ from those used in elementary statistics for measuring variability.

The simplest measure of variability is the range (difference between the highest and lowest value of X). The shortcomings of this parameter are well known, but if X is a more or less continuous variable as, for instance, the temperature in a given place on a given day, there may be some justification for the use of range. Owing to the discontinuous character of most geosets and their bidimensional nature, the measure of the "distance range" (Neft, 1966, pp. 65–66, 83, 103–105)—namely the maximal distance between two elements of the geoset—would generally be of very little interest. We have already indicated that the bidimensional parameter analogous to the range is the convex hull of the geoset (see Section 4.2). Taking the area of the convex hull, we obtain a rough measure of the size of the territorial container of the geoset and this measure, if associated with other parameters, may be of some value. However, this parameter cannot be considered as a reliable measure of dispersion of the geoset over the territory.

The measures of variability most widely used are obtained by averaging the deviations between each value X_i of X and a mean of these values. Similarly, the parameters of dispersion most widely employed are based on the averages of distances of the locations of the elements of the geoset from one of its centers. In order to render the measure of variability independent of the selection of a particular mean, Gini introduced the arithmetical or quadratic average of the differences between all pairs of values of the variable. Similarly, we can calculate an average of the distances between the locations of all possible pairs of the elements of the geoset. Using the quadratic average of distance from the mean center, we can easily reduce this parameter to the product of $\sqrt{2}$ and the quadratic average of the distances between the locations of each element and the mean center (see Section 13.2), which simplifies both calculations and concepts.

The study of variability can be based on partition values and differences between them. These parameters can be of considerable use, primarily if the variable X is more or less continuous and if its frequency distribution tends to have some easily recognized regularity. In the study of geosets, extensions of this approach can sometimes be attempted, but, generally speaking, it is not very promising owing to difficulties arising from the bidimensional nature of geosets and the fact that most of them are discontinuous.

Parameters of variability are based on the calculation of the absolute difference between pairs of values, and, similarly, most parameters of dispersion are based on the calculation of Euclidean distances between pairs of locations. However, other approaches might sometimes be preferred. For example: (a) if the territory considered is large, one should take into account the fact that the earth is spherical in the calculation of distances between pairs of points; (b) one might use city-block distances or measures of distances based on time or on the cost of moving from the location of one element to the location of the second element compared with it. However, as a rule, in this book we use only Euclidean distances.

In the calculation of parameters of variability, the squaring of dimensions is often found to be of considerable value both conceptually and in order to utilize the algebraic properties of such parameters. This applies even more to parameters of dispersion, where squaring enables easy linkage of the overall parameters to those along each pair of orthogonal axes. In the study of variability of a variable X, as considered here, we deal with a one-dimensional parameter. We saw earlier that in the study of dispersion we often take into account dispersion along each of two meaningful orthogonal axes, such as major and minor or north–south and east–west axes.

Consideration of dispersions along each of the two principal axes has already led us to pay attention to the oblongity or difference in length between standard deviations of locations of the elements along these axes. In this chapter we shall extend this approach and introduce various measures of oblongity. Separate measurement of overall dispersion and oblongity will produce parameters of dispersion netted from effects of oblongity. These parameters will be found to have important properties (see Sections 10.7.5–10.7.7). Obviously, these approaches are not applicable in the study of variability, where we can obtain dimensionless parameters by comparing parameters of absolute variability with an appropriate average of the values of the X. In the study of dispersion, we cannot compare absolute parameters of dispersion with the numerical value of a center, as this depends on an arbitrarily selected origin. However, we shall see that very meaningful dimensionless parameters can be obtained by comparing the dispersion of the population geosets studied with the dispersion of the uniform distribution over the territory over which the geoset is distributed.

10.1.2. Historical Development of the Parameters of Dispersion

The first attempts to calculate the parameters of territorial dispersion were made in the late 1920s and 1930s by various researchers working independently. We made such an attempt in Italy as a natural development of the flourishing centrographic studies, fostered mainly by Gini (1934).

The calculation of centers was considered as a first approach toward summarizing "territorial series" through a parameter analogous to the mean. We suggested parameters of territorial dispersion as a method that might help in the study of an aspect of territorial series analogous to the variability of one-dimensional variables, and later continued our research on dispersion parameters in Israel (see Bachi, 1958, 1962*b*). Dispersion parameters were used by other authors as part and parcel of the enlarged centrography (see Section 2.3). A parallel development in the study of the dispersion parameter was embraced by social physics (see Section 2.3). An early root of the study of dispersion is found in studies on the behavior of standard deviations of elements along axes through the center (owing mainly to Furfey, 1927). In more recent years studies on dispersion have been extended to the separate analysis of oblongity and of other components.

We shall start by introducing an areal overall measure of dispersion of geosets: the distance-variance d^2. We shall deal, in turn, with calculating dispersion for population geosets on territories and with its algebraic properties (Sections 10.3–10.4). We shall then consider a corresponding linear overall measure of dispersion: the standard distance d (Section 10.5.1). Other parameters of linear dispersion will be introduced in Section 10.5.2. We shall then go on to the detailed definition and analysis of principal axes and of equating axes (Sections 10.7.1–10.7.4), which will pave the way to a better understanding of the meaning of oblongity; alternative parameters of oblongity will be discussed in Section 10.7.7. The simple numerical example introduced in Section 10.8 will be utilized to help the reader to check the equations introduced.

10.2. Average of the Distances between All Possible Pairs of Elements of a Geoset

As a simple introduction to this measure, assume that we are dealing with the geoset of a poulation formed by N elements, $1\,2, \ldots, i, \ldots, j, \ldots, N$, having distinct locations (X_1, Y_1), $(X_2, Y_2), \ldots, (X_i, Y_i), \ldots, (X_j, Y_j), \ldots, (X_N, Y_N)$. We denote the distance between any two elements i and j of the geoset by d_{ij} and assume that each location is given the same weight, $1/N$, in the calculation of a parameter of dispersion. One intuitive approach to the calculation of this parameter is to take the average of the $N(N - 1)/2$ distances calculated over all the possible pairs of elements (Bachi, 1934, 1962*a,b*; Neft, 1966).

Neft called this parameter the "general mean distance deviation." Kellerman (1981*a*, p. 13) termed it the "mean distance." We refer to it as "the average distance between elements," and use the symbol AD, eventually followed by an indication of the geoset to which reference is made:

$$\mathrm{AD} = 2 \sum_{i=1}^{N-1} \sum_{j=i+1}^{N} d_{ij}/N(N-1) \tag{10.1}$$

If we do the calculation "with repetition," by extending it to comparisons of each element to itself, we obtain:

$$\mathrm{ADR} = \sum_{i=1}^{N} \sum_{j=1}^{N} d_{ij}/N^2 \tag{10.2}$$

We might prefer to work on squares of distances in order to obtain a parameter with important algebraic properties. Calling the average of these squares the average of squared distances (denoted by ASD) and working with repetition (ASDR), we obtain

$$\mathrm{ASDR} = \sum_{i=1}^{N} \sum_{j=1}^{N} d_{ij}^2/N^2 \tag{10.3}$$

It can be shown that

$$\mathrm{ASDR} = (2/N) \sum d_{ic}^2 = 2d^2 \tag{10.4}$$

where d_{ic} is the distance between any element i of the geoset and the mean center, and d^2 is the distance-variance (see Section 10.3).

This was first developed by Risser (1937), commenting on our work (Bachi, 1932, 1934). This development is very important as it helps to conceptualize and interpret the distance-variance and reduce the number of calculations needed to reach the ASDR. This reduction may be important for high vaues of N, and it was especially useful in the precomputer era.

In the calculation of ASD, ASDR, d^2, and d, the passage from equations for unweighted to weighted locations is immediate. For example, while d^2 for unweighted elements is given by $(1/N)\Sigma d_{ic}^2$, it is given by $d^2 = \Sigma w_i d_{ic}^2$ for weighted elements.

For ADR [equation (10.2)] no algebraic simplification is available such as that suggested by equation (10.4) for the average of squared distances. However, Neft (1966 p. 59) found empirically that for large populations ADR is very close to $\sqrt{2}\bar{d}$, where $\bar{d}$ is the arithmetic average of the distances of the elements from the mean center (see Section 10.5.2).

10.3. Distance-Variance of a Geoset

10.3.1. Calculation of Distance-Variance

The geostatistic analogous to the variance of a one-dimensional variate is the average of the squares of the distances of the elements i from the center:

$$d^2 = (1/N) \sum d_{ic}^2 \tag{10.5}$$

We term this areal parameter of dispersion distance-variance and denote it by d^2 [see Bachi (1958, 1962*b*) and Neft (1966)]. Obviously $d_{ic}^2 = (X_i - \bar{X})^2 + (Y_i - \bar{Y})^2$ for deviations of elements i measured along the conventional axes X, Y. Let us denote the variances $(1/N)\Sigma(X_i - \bar{X})^2$ and $(1/N)\Sigma(Y_i - \bar{Y})^2$ by σ_X^2 and σ_Y^2, respectively. Then $d^2 = \sigma_X^2 + \sigma_Y^2$, and this equality holds for any pair of variances measured along orthogonal axes through the center. These variances generally change with rotation of the axes, but d^2 does not. These considerations for a geoset formed by the unweighted locations of individual elements of a population can be extended or adapted to other types of geosets.

If each individual element i of the geoset has a relative weight w_i, then $d^2 = \Sigma w_i d_{ic}^2$. As an example, we consider the geoset formed by ten weighted elements shown in Table 10.1, where the coordinates, the distance d_{ic} for the coordinates of the center ($\bar{X} = 168.92$, $\bar{Y} = 132.35$), the squares of the distances d_{ic}^2, and the relative weight w_i are given for each element i. We find accordingly that $d^2 = \Sigma w_i d_{ic}^2 = 14.08$ and $d = 3.75$. A graphical presentation is given in Figure 10.1.

If we consider a population geoset for which information on the locations of the elements is given in an aggregate way, as by regions r (e.g., the *comuni* of Italy or census tracts in cities), where the number of regions or cells is very large and their areas comparatively small, we may neglect within-region dispersion. Then $d^2 = \Sigma w_r d_{rc}^2$, where w_r is the relative weight attributed to region r and d_{rc} is the distance between the mean center of the region (or any other central point of r) and the overall mean center. This also applies to the setting in which the gross territory has been divided into a large number of small cells. However, if the number of regions or cells is comparatively small and their areas comparatively large, within-region or within-cell dispersion cannot be ignored.

Detailed suggestions given in Section 9.3 for the calculation of the mean center for the various types of population geosets can be applied to the calculation of d^2 for these geosets. The distance-variance for the uniform distribution of N elements i over a territory can be calculated by distributing the element i uniformly and by attributing to each of them a weight $1/N$. Then $d^2 = (1/N)\Sigma d_{ic}^2$. The distance-variance of a geoset ${}^gT$ for a territory T can be calculated by the method described in Section 4.5.2, which is based on the calculation of the variances $\sigma_{X,T}^2$ and $\sigma_{Y,T}^2$, by means of the coordinates of points distributed along the boundary of the territory. The distance-variance is obtained as $d_T^2 = \sigma_{X,T}^2 + \sigma_{Y,T}^2$. This method can be applied directly to the calculation of d^2 for territories having regular geometrical figures, where the coordinates taken into consideration are those of the vertices of these figures. A few examples are given in Table 10.2. Calculations of d_T^2 for other types of territories (e.g., fragmented, with external convexities or internal gaps) can be based

Table 10.1. An Example of a Geoset Formed by 10 Weighted Elements[a]

X_i	Y_i	w_i	d_{ic}	d_{ic}^2
173	131	0.10	4.30	18.48
163	132	0.08	5.93	35.16
169	133	0.16	0.65	0.43
172	135	0.06	4.06	16.50
165	137	0.12	6.08	39.96
168	130	0.26	2.53	6.38
170	132	0.16	1.14	1.29
175	133	0.04	6.11	37.39
177	132	0.01	8.09	65.42
172	133	0.02	3.15	9.91

[a] $\bar{X} = 168.92$; $\bar{Y} = 132.35$; $d^2 = 14.08$; and $d = 3.75$.

on the presentation of the geosets of these territories, given in Sections 4.6.1–4.6.4. These calculations will be described in detail in Chapter 11.

Regarding the discussion of distance-moments given in Section 9.2, we see that the second distance-moment around any point k is $\mu_{2,k}$ and that the minimum of this value is found for the second distance-moment around the mean center C, which is equal to the distance-variance: $d^2 = \mu_{2,C}$.

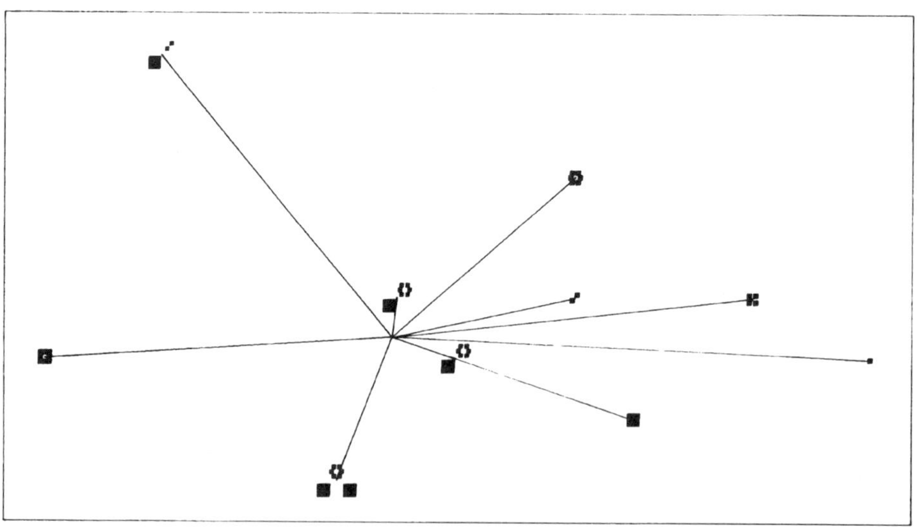

Figure 10.1. Graphical presentation of the geoset of Table 10.1.

Table 10.2. Distance-Variance and Variances for Territories with Regular Shapes

Shape	Dimensions	x-variance σ_x^2	y-variance σ_y^2	Distance-variance d^2	Standard distance
Circle	r = length of radius	$r^2/4$	$r^2/4$	$r^2/2$	$r/\sqrt{2}$
Ellipse	Major and minor axes l (along X-axis) h (along Y-axis)	$l^2/16$	$h^2/16$	$(l^2+h^2)/16$	$\sqrt{l^2+h^2}/4$
Rectangle	Length l (along X-axis); Height h (along Y-axis)	$l^2/12$	$h^2/12$	$(l^2+h^2)/12$	$\sqrt{l^2+h^2}/\sqrt{12}$
Square	Length of side l	$l^2/12$	$l^2/12$	$l^2/6$	$l/\sqrt{6}$
Equilateral triangle	Length of side l	$l^2/24$	$l^2/24$	$l^2/12$	$l/\sqrt{12}$

10.3.2. Properties of Distance-Variance

The distance-variance has many properties, among which we note the following:

1. Distance-variance, as other averages of distances from the center, is independent of the system of coordinates used to indicate the locations of the elements of the geoset.
2. As noted above, d^2 has several equivalent definitions, each of which carries a different intuitive meaning. Therefore, studies based on various approaches can all use d^2 as a convenient parameter.
3. The parameter d^2 can be conveniently manipulated algebraically, as will be seen from many applications described in this volume, and will be easily understood, considering: (a) that $d^2 = \sigma_{x''}^2 + \sigma_{y''}^2$ for any system of orthogonal axes through the mean center (denoted by X'', Y''), and (b) that variances of σ^2 can be manipulated algebraically.
4. The decomposition of d^2 into the sum of two variances, $\sigma_{x''}^2 + \sigma_{y''}^2$, along orthogonal axes contributes to the simplification of the calculation, and enables us to consider d^2 as an overall areal measure of dispersion based on the study of variances along principal axes (see Section 10.7).
5. Similar to the mean center, d^2 is sensitive to any change in weight or location of any element of the geoset. This property is shared by other dispersion parameters and is often found to be quite useful. However, one peculiar and useful property of distance-variance is that we can decompose it by dealing separately with changed and unchanged locations of elements.
6. Similar to the mean center, d^2 is particularly sensitive to the presence of outliers, which is often an undesirable property, but there are ways to compensate for it.

7. In particular, we shall see that the analysis and interpretation of d^2 can be enhanced by the often existing possibility of splitting the geosets into major clusters. The distance-variances of the clusters are algebraically related to the distance-variances of the entire geoset.
8. Distance-variance can be used for inferential purposes (not discussed in this book).
9. As shown in Section 10.7, d^2 can be split into a parameter of oblongity and a parameter of the part of dispersion not affected by oblongity, which is of great help in the analysis of dispersion.
10. Distance-variance is a parameter of areal dispersion. As such, it can reach high absolute values, which may not be easily manageable and/or interpretable. However, this difficulty can be avoided in various ways, such as: (a) working on a linear parameter of dispersion closely connected to d^2, namely square root d, the standard distance (see Section 10.5.1); (b) Comparing d_p^2 for the geoset ${}^gP$ of a population P to the area of the gross or the net territory covered by ${}^gP$, we shall see that this approach leads us to a refined and very meaningful analysis of an important aspect of the geoset ${}^gP$, i.e., its looseness or lack of compactness; this approach is also applicable to the study of noncompactness of territories; (c) analyzing d_p^2/d_T^2 or d_p^2/d_H, i.e., the ratio of the distance-variance of a population geoset to that of its gross or net territory.
11. Other important properties of d^2 emerge from the links that exist between the distance-variance for a compound geoset and the distance-variances of its components.

10.4. Distance-Variances of Components versus Compound Geosets

10.4.1. Distance-Variances within and between Components

In Sections 3.6 and 3.8 we discussed the relationships between the geosets of compound and component populations as well as many extensions of these relationships to other important settings, including the compounding of geosets.

Relationships between these geosets can be immediately translated to corresponding relationships between the distance-variances of the relevant geosets. As a first example, we consider the following setting: Let ${}^gK$ be a compound geoset having many components ${}^gK_j$ $(j = 1, 2, \ldots)$, some of which are positive and some negative. Let us denote the center and distance variance of ${}^gK$ by $\bar{X}_k$, $\bar{Y}_k$ and d_k^2; the corresponding parameters of the components by $\bar{X}_j$, $\bar{Y}_j$ and d_j^2; the proportion of the population of each K_j over that of K by v_j; and a coefficient to be attached to each component according to whether it is positive or negative by $\varepsilon_j = +1$ or $\varepsilon_j = -1$. Then,

$$d_k^2 = \sum \varepsilon_j v_j d_j^2 + \sum \varepsilon_j v_j (\bar{X}_j - \bar{X}_k)^2 + \sum \varepsilon_j v_j (\bar{Y}_j - \bar{Y}_k)^2 \tag{10.6}$$

It might be convenient to designate $\Sigma \varepsilon_j v_j [(\bar{X}_j - \bar{X}_k)^2 + (\bar{Y}_j - \bar{Y}_k)^2] = d_b^2$ as the distance-variance between geosubsets and $\Sigma \varepsilon_j v_j d_j^2 = d_w^2$ as the distance-variance within geosubsets, so that $d^2 = d_b^2 + d_w^2$. From these equations we can easily derive others adapted to the various settings described in Sections 3.6 and 3.8.

10.4.2. Components of Distance-Deviances

It is sometimes convenient to present the relationships discussed above in a somewhat different way. For simplicity we assume that a compound geoset K is formed by N positive elements, each of which has a relative weight $1/N$. The elements are divided among geosubsets having, respectively, $N_1, N_2, \ldots$ elements each. Let us denote the proportion of the elements of geosubset K_j by $v_j = N_j/N$, the coordinates of any element ji of K_j by (X_{ji}, Y_{ji}); and the square of the distance between each ji and the center of the compound geoset ${}^gK$ by $u_{ji}^2 = (\bar{X}_{ji} - \bar{X}_k)^2 + (\bar{Y}_{ji} - \bar{Y}_k)^2$. We may designate as distance-deviance the average of the u_{ji}^2. Distance-deviance is thus $(1/N)\,\Sigma_j \Sigma_i u_{ji}^2 = d_k^2$, and we can calculate the relative contribution to distance-deviance of each geosubset, its eccentricity, and its dispersion within as follows:

	Proportional contribution to distance deviance owing to: Contribution of K_j	Contributions of all K_j
Weighted square distance between the center of K_j and center of K (eccentricity)	$v_j{}^cD^2(K_j, K)/d^2$	$\Sigma\, v_j{}^cD^2(K_j, K)/d^2$
Weighted distance-variance within K_j (dispersion within)	$v_j d_j^2/d^2$	$\Sigma\, v_j d_j^2/d^2$
Total	$v_j[{}^cD^2(K_j, K) + d_j^2]/d^2$	$\Sigma\, v_j[{}^cD^2(K_j, K) + d_j^2]/d^2 = 1$

We now go on to consider a setting that is very important in both theory and in practice in which a compound geoset ${}^gK$ is formed by two positive component geosubsets ${}^gH$ and ${}^gJ$. Let us denote the proportion of the number of elements of ${}^gH$ and ${}^gJ$ to the number of elements of ${}^gK$ by v_H and v_J, respectively, so that $v_H + v_J = 1$; the centers and distance variances of ${}^gH$, ${}^gJ$, and ${}^gK$ by $(\bar{X}_H, \bar{Y}_H; d_H^2)$, $(\bar{X}_J, \bar{Y}_J; d_J^2)$, and $(\bar{X}_K, \bar{Y}_K; d_K^2)$, respectively; and the distance between the centers of ${}^gH$ and ${}^gJ$ by ${}^cD(H, J)$. Then,

$$d_K^2 = v_H d_H^2 + v_J d_J^2 + v_H v_J{}^cD^2(H, J) \tag{10.7}$$

If $v_H = v_J = \frac{1}{2}$, this reduces to

$$d_K^2 = \tfrac{1}{2}(d_H^2 + d_J^2) + \tfrac{1}{4}{}^cD^2(H, J) \tag{10.8}$$

If $d_H^2 = d_J^2 = d^2$ and the centers of H, J are coincident, then

$$d_K^2 = d^2 \tag{10.9}$$

It was found useful to derive some further equations from those above for the average of the squares of distances between all possible pairs formed by one element of ${}^gH$ and one element of ${}^gJ$. Geosubsets ${}^gH$ and ${}^gJ$ are, again, assumed to have the same relative weight, $v_H = v_J = \frac{1}{2}$. Let us denote by ASDR, the average of the squares of the distances (with repetition) between all possible pairs of elements of a geoset. Then ASDR(K) is obtained by summing $\frac{1}{4}$ASDR(H), $\frac{1}{4}$ASDR(J), and half of the sum of the squared distances between each element of H associated in all possible ways with one element of J. This leads to

$$\text{ASDR}(K)^2 = \tfrac{1}{2}[\text{ASDR}(H) + \text{ASDR}(J)] + \tfrac{1}{2}{}^{c}D^2(H, J) \tag{10.10}$$

If the geosets H and J are concentric or are transformed so as to have the same center, this reduces to ASDR(K) $= \frac{1}{2}$[ASDR(H) + ASDR(J)]. Important applications of these findings are presented in Section 15.2.2.

10.5. Linear Measures of Dispersion

10.5.1. The Standard Distance

The approach followed in the calculation of the distance-variance as an areal measure of the dispersion of a geoset is based on the squared distances of the location of each element from the mean location. We might stay with this approach particularly if we prefer a linear measure of dispersion expressed in the same dimension as the distances of the elements from the center. One of the tools for achieving this is the square root d of the distance-variance d^2, which has the advantage (over other averages of distances introduced in Section 10.2) of being a function of d^2 so that it is possible to obtain d^2 through an easy transformation, thus enabling us to utilize the various properties of d. We proposed the term standard distance in 1958, discussed it at length in 1962, and used it extensively in subsequent publications as did other authors. In 1958, Stewart and Warntz introduced an analogous parameter, which they termed the "dynamical radius" (Neft, 1966).

10.5.2. Other Averages of Distances of the Elements from a Center

Many measures of dispersion could be obtained by taking different types of averages and different centers (see examples in Neft, 1966). However, we believe that only a few dispersion measures should be considered in practice, for the same reasons for which only a few parameters of location were considered. We discuss some of them below.

Arithmetic Average of Distances from the Mean Center. We again denote the distance between the location of an element i and the mean center by d_{ic}, and the relative weight to be attached to i by w_i. The arithmetic average $\bar{d}_c$ of the distances from the center is then

$$\bar{d}_c = \sum w_i d_{ic} \tag{10.11}$$

This parameter might be preferred over the standard distance $d = \sqrt{(\Sigma w_i d_{ic}^2)}$ or be needed as a supplementary parameter of dispersion, for instance: (a) whenever the influence of a few long distances over d is a disrupting factor; (b) in cases in which we are interested in distances as measures of the cost of travel for reaching the center and there is no justification for thinking that such costs are linked to the squares of distances (this consideration may be important in planning connected with location issues, which are, however, beyond the scope of this volume). An interesting relationship between $\bar{d}_c^2$ and d^2 can be indicated as follows: Let $\sigma^2(d_{ic})$ be the variance of the length of the distances from the mean center. Then

$$\sigma^2(d_{ic}) = \sum w_i(d_{ic} - \bar{d}_c)^2 = d^2 - \bar{d}_c^2 \tag{10.12}$$

Writing $d^2 = \bar{d}_c^2 + \sigma^2(d_{ic})$, we see that distance-variance can be split into two components: the arithmetic average of the distances and the variability of such distances [measured by the variance $\sigma^2(d_{ic})$]. Only when all the i are at the same distance from the center, being located on a circle of radius $\bar{d}_c$, and only then, do we have $d = d_{ic} = \bar{d}_c$. In all other cases $d > \bar{d}_c$.

Arithmetic Average and Median of Distances of the Elements from the Median Center. When we calculate the median center (see Section 9.7), we might want to obtain a parameter of dispersion by taking a mean of the distances d_{im} between the location of each element i and the median center. Using the average of these distances, $\bar{d}_m = \Sigma w_i d_{im}$, has the advantage that we can utilize the property that the first distance-moment of a geoset is minimal when calculated around the median center. Consequently, $\bar{d}_m \leqslant \bar{d}_c \leqslant d$. However, the median of distances from the median center can also be of some use, particularly if it is taken as a member of a set of partition values.

Average of Distances from a Selected Central Point. In many types of studies it may be useful to calculate a mean of the distances d_{ik} between each element i and a central point that is selected based on the nature of the geoset and the goal of the specific research. For instance, in urban studies, we might select a point that represents, e.g., the business center, a central market, the municipality, or central historical site. While this type of selection may be perfectly justified, the analytical advantages of the parameters of dispersion may be lost. In this instance, it might sometimes be of interest to relate the average of the d_{ik}^2 to d^2:

$$\sum w_i d_{ik}^2 = d^2 + (\bar{X}_k - \bar{X}_c)^2 + (\bar{Y}_k - \bar{Y}_c)^2 \tag{10.13}$$

10.6. The Covariance between the Coordinates of the Elements of the Geoset

In the study of dispersion, we often have to use the covariance between the coordinates of the locations (X_i, Y_i) of the elements i of the geoset. Formally, the equations for this covariance are equal to those employed in the calculation of the covariance between any two variables (X, Y) in ordinary statistical work. For example, we can write:

1. $\mathrm{cov}(X, Y) = \Sigma\, w_i[(X_i - \bar{X})(Y_i - \bar{Y})]$
2. $\mathrm{cov}(X, Y) = \Sigma\, w_i[X_i Y_i - \bar{X}\bar{Y}]$
3. For a compound geoset K formed by components K_j, we denote the covariance between the coordinates of the elements ji of K_j by the expression $\mathrm{cov}_j = \Sigma\, w_{ji}[(X_{ji} - \bar{X}_j)(Y_{ji} - \bar{Y}_j)]/w_j$; the covariance within the components by $\mathrm{cov}(w) = \Sigma \varepsilon_j w_j \mathrm{cov}_j$; the covariance between components by $\mathrm{cov}(b)$; and the mean center of the compound geoset by $(\bar{X}_c, \bar{Y}_c)$; $\mathrm{cov}(b) = \Sigma\, \varepsilon_j w_j[(\bar{X}_j - \bar{X}_c)(\bar{Y}_j - \bar{Y}_c)]$. For the compound geoset, $\mathrm{cov}(X, Y) = \mathrm{cov}(w) + \mathrm{cov}(b)$.

However, the meaning of covariances in geostatistics differs from that in the study of usual variables X, Y whose meaning is well defined. In geostatistics, we are primarily interested in seeing how covariance changes with the rotation of the axes.

10.7. The Principal Axes

10.7.1. Definitions of the Principal Axes

We noted in Section 10.3.1 that $d^2 = \sigma_{x''}^2 + \sigma_{y''}^2$ for any system of orthogonal axes (X'', Y''). In many geostatistical studies we are interested in considering separately and then comparing the variances $\sigma_{x''}^2$ and $\sigma_{y''}^2$ or the standard deviations $\sigma_{x''}$ and $\sigma_{y''}$ corresponding to each axis. However, as the number of pairs of such orthogonal axes (X'', Y'') through the mean center is infinite we study the laws of variation of $\sigma_{x''}$ and $\sigma_{y''}$ with the rotation of the axes around the center and/or select for consideration axes that have a particular meaning. Sometimes conventional axes oriented in north–south or west–east directions or vice versa are found to be of practical interest. An example of this approach was given with regard to the population and subpopulations of the United States (see Section 9.5). However, in general, our interest is usually focused on the pair of principal axes, many examples of which have been given in the maps of geosets shown in previous chapters.

Such axes can be defined in various ways, all of which yield the same solution. As we shall see later, these axes have important mathematical properties that render them particularly suitable for geostatistical research.

In the following, we shall use three definitions of principal axes. Let us consider all the systems of orthogonal axes (X'', Y'') through the center of the geoset:

1. We can select as the major axis the line nearest to the elements of the geoset, in the sense of the least squares, namely the line from which the total of the squares of the distances of the elements of the geoset is minimal. This line indicates the main linear orientation of the distribution. This axis and the one that is perpendicular to it through the center (minor axis) constitute the principal axes.

2. We can select the system of axes along which one of the standard deviations of the elements is maximal while the other is minimal. It turns out that the former coincides with the major axis and the latter with the minor axis.
3. As we consider the standard deviations as tools for measuring the dimensions of the geosets, it may be desirable to use axes along which the coordinates are uncorrelated. It can be shown that these axes coincide with the principal axes as defined in (1) and (2).

10.7.2. Identifying the Principal Axes

To identify the principal axes, we shall use the third of the above definitions. To introduce our problem, we refer first to Figure 10.2 in which there is a point P that has as coordinates along the conventional axes (X, Y), the deviations $x = OB$ and $y = BP$ from the origin (0, 0). Two arbitrary orthogonal axes through origin 0 are drawn. We indicate by X'' the rotated axis forming an angle α ($<45°$) with axis X and by Y'' the axis forming an angle α with the conventional axis Y. In our example, the rotation is counterclockwise.

The coordinates of P along the rotated axes can also be expressed as deviations (x'', y'') from the origin 0.

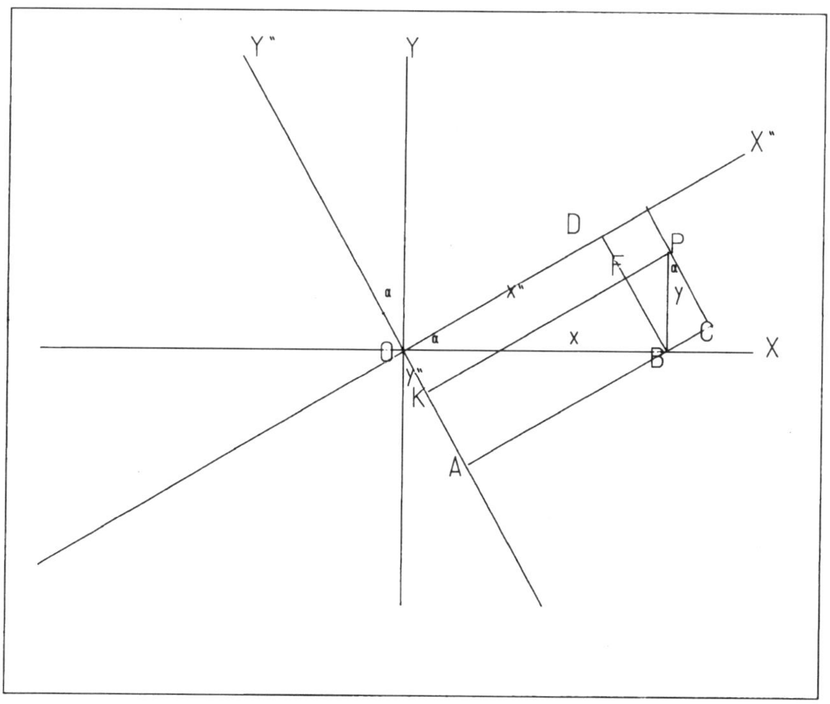

Figure 10.2. Rotation of axes.

We see that $x'' = AC = AB + BC$ and $y'' = OK = OA - AK$ from which we obtain the usual equations of notation:

$$x'' = x \cos \alpha + y \sin \alpha \tag{10.14}$$

$$y'' = y \cos \alpha - x \sin \alpha \tag{10.15}$$

If we assume that we know the coordinates (x_i, y_i) and (x_i'', y_i'') for all the elements i of a geoset, and we take the mean center as the origin of the coordinates, we can then obtain the variances and the covariances of the geoset along the rotated axes as functions of variances and covariances along the conventional axes, as follows [denoting the coordinates of any element i of the geoset along the conventional and the rotated axes by (x_i, y_i) and (x_i'', y_i'') respectively and the relative weight of the element i by w_i]:

$$\sigma_x^2 = \sum w_i x_i^2 \qquad \sigma_{x''}^2 = \sum w_i x_i''^2 \tag{10.16a}$$

$$\sigma_y^2 = \sum w_i y_i^2 \qquad \sigma_{y''}^2 = \sum w_i y_i''^2 \tag{10.16b}$$

$$\operatorname{cov}(x, y) = \sum w_i x_i y_i \qquad \operatorname{cov}(x'', y'') = \sum w_i x_i'' y_i'' \tag{10.16c}$$

$$\sigma_{x''}^2 = \sigma_x^2 \cos^2\alpha + \sigma_y^2 \sin^2\alpha + 2\operatorname{cov}(x, y) \sin \alpha \cos \alpha \tag{10.17}$$

$$\sigma_{x''}^2 = \sigma_x^2 \cos^2\alpha + \sigma_y^2 \sin^2\alpha + \operatorname{cov}(x, y) \sin 2\alpha \tag{10.18a}$$

$$\sigma_{y''}^2 = \sigma_x^2 \sin^2\alpha + \sigma_y^2 \cos^2\alpha - \operatorname{cov}(x, y) \sin 2\alpha \tag{10.18b}$$

$$\operatorname{cov}(x'', y'') = \operatorname{cov}(x, y) \cos 2\alpha - \tfrac{1}{2}(\sigma_x^2 - \sigma_y^2) \sin 2\alpha \tag{10.19}$$

From these equations we see that both the variances and the covariance $\operatorname{cov}(x'', y'')$ between the coordinates of the elements may change with the rotation of the axes (namely with changes of α). However,

$$\sigma_{x''}^2 + \sigma_{y''}^2 = \sigma_x^2 + \sigma_y^2 = d^2 \tag{10.20}$$

which confirms that the distance variance d^2 is invariant under the rotation of the axes, and

$$\sigma_{x''}^2 - \sigma_{y''}^2 = (\sigma_x^2 - \sigma_y^2) \cos 2\alpha + 2\operatorname{cov}(x, y) \sin 2\alpha \tag{10.21}$$

Let us denote the principal axes by (X', Y'), the angle formed by each of these axes with the corresponding conventional axis (X, Y) by α', and the coordinates of elements i along the principal axes by (x_i', y_i'). To identify the principal axes we must find the specific angle α'. In this search we consider as principal those rotated axes along which the correlation between the coordinates (x_i, y_i) of the elements i of the geoset is zero, and, consequently, the covarience is also zero.

Taking $\operatorname{cov}(x_i', y_i') = 0$, we obtain from equation (10.19)

$$\tfrac{1}{2}(\sigma_x^2 - \sigma_y^2) \sin 2\alpha' = \operatorname{cov}(x, y) \cos 2\alpha' \tag{10.22a}$$

$$\tfrac{1}{2}(\sigma_x^2 - \sigma_y^2) \tan 2\alpha' = \operatorname{cov}(x, y) \tag{10.22b}$$

and, consequently,

$$\tan 2\alpha' = 2\,\mathrm{cov}(x, y)/(\sigma_x^2 - \sigma_y^2) \tag{10.23}$$

$$\alpha' = \tfrac{1}{2} \arctan[2\,\mathrm{cov}(x, y)/(\sigma_x^2 - \sigma_y^2)] \tag{10.24}$$

It can be shown that if the other definitions of the principal axes are adopted, the same angle α' is obtained.

Having the value of angle α', we can arrive at the standard deviations along the principal axes by using the equations of rotation (10.18a) and (10.18b)

$$\sigma_{x'}^2 = \sigma_x^2 \cos^2\alpha' + \sigma_y^2 \sin^2\alpha' + \mathrm{cov}(x, y) \sin^2\alpha' \tag{10.25a}$$

$$\sigma_{y'}^2 = \sigma_x^2 \sin^2\alpha' + \sigma_y^2 \cos^2\alpha' - \mathrm{cov}(x, y) \sin^2\alpha' \tag{10.25b}$$

Table 10.3 details a practical example of the calculation of σ_x^2 and σ_y^2 for the geoset

Table 10.3. Calculations for Principal Axes of the Population of Italy (1987)[a]

Parameters for conventional axes

Standard deviations:	$\sigma_x = 228.92$	$\sigma_y = 285.68$
Variances:	$\sigma_x^2 = 52{,}404.37$	$\sigma_y^2 = 81{,}613.06$
Distance variance and covariance:	$d^2 = 134{,}017.4$; $d = 366.08$; $\mathrm{cov}(x, y) = -46{,}157.44$	

Calculation of angle α' between principal and conventional axes

$\sigma_x^2 - \sigma_y^2 = -29{,}208.70$	$\arctan(+3.161) = +72.440°$
$2\,\mathrm{cov}(x, y)/(\sigma_x^2 - \sigma_y^2) = +3.161$	$\alpha' = +36.22°$
$\sigma_x^2 + \sigma_y^2 = d^2 = 134{,}017.4$	

Calculation of variances and standard deviations along principal axes

As $\alpha' = +36.22°$ and $2\alpha = 72.44°$

$\sin^2\alpha' = 0.3492 \quad \cos^2\alpha' = 0.6508$

$\sin 2\alpha' = 0.9534 \quad \cos 2\alpha' = 0.3017$

Consequently, we obtain:

$$\sigma_{x'}^2 = \sigma_x^2 \cos^2\alpha' + \sigma_y^2 \sin^2\alpha' + \mathrm{cov}(x, y) \sin 2\alpha'$$
$$= 34{,}106.42 + 28{,}496.7 - 44{,}007.18 = 18{,}595.94$$

$$\sigma_{x'} = 136.37$$

$$\sigma_{y'}^2 = \sigma_x^2 \sin^2\alpha' + \sigma_y^2 \cos^2\alpha' - \mathrm{cov}(x, y) \sin 2\alpha'$$
$$= 18{,}297.95 + 53{,}116.36 + 44{,}007.2 = 115{,}421.5$$

$$\sigma_{y'} = 339.74$$

Checking

$$\mathrm{cov}(x', y') = \mathrm{cov}(x, y) \cos 2\alpha - \tfrac{1}{2}(\sigma_x^2 - \sigma_y^2) \sin 2\alpha$$
$$= 13{,}924 - 13{,}924 = 0$$

[a] The parameters are taken from the second volume of the *Statistical Atlas of Italy* (linear parameters are in km, areal parameter are in sq km).

of the population of Italy (1987) that is shown in Figure 1.3 (Section 1.2). The approach discussed here for determining an angle α' formed between the principal axes and conventional axes was first suggested in the geostatistical context by Linders (1933).

If one wishes to obtain α' on the basis of an equation that makes reference to $\tan\alpha'$ instead of $\tan 2\alpha'$, then we have:

$$\tan\alpha' = [-(\sigma_x^2 - \sigma_y^2) + \sqrt{(\sigma_x^2 - \sigma_y^2)^2 + 4\operatorname{cov}^2(x, y)}]/2\operatorname{cov}(x, y) \qquad (10.26)$$

(see Ebdon, 1977, p. 114; Kellerman, 1981*a*, p. 20). Kellerman pointed out that this approach is more cumbersome than the one recommended above.

10.7.3. The Major Axis as an Indicator of the Principal Orientation of the Geoset

We have noted repeatedly that the major axis is the line that has the minimal sum of squared distances from the elements of the geoset. As we are dealing with two-dimensional distributions, it may seem strange to "summarize" it by means of a line. It often happens, however, that in empirical research the average of the squared distances d^2 for the major axis has an absolute value that can be considered "rather small" when judged by proper criteria. In this case the major axis may acquire considerable interest in the study of the geoset, for it can be considered as an indicator of the latter's principal orientation. We saw the example of the orientation of the territory of Italy, whose approximate direction is northwest–southeast, in Section 10.7.2. This was found to be so meaningful in the analysis of many geosets for Italy that we have chosen the principal axes of the territory as the system of axes for the analysis of the mean centers of these geosets. Other examples of combined studies of the rotation angles of principal axes and of the values of d^2 will be given in Section 10.8.

10.7.4. Identifying Equating Axes

It may sometimes be of interest to consider the axes that form an angle of 45° with the principal axes. To facilitate the identification of these axes, we rotate the principal axes so that they coincide with the conventional axes. We then get $X' = X$ and $Y' = Y$ and $\operatorname{cov}(x, y) = 0$. To identify the variances along the axes forming an angle $\alpha'' = 45°$ with the axes $X' = X$ and $Y' = Y$, we can use equations (10.18a) and (10.18b). As $\sin^2 45° = \cos^2 45° = \frac{1}{2}$, the variances along the axes forming an angle of 45° with the principal axes are equal to $\frac{1}{2}d^2$. We call these the "equating axes."

Let us denote the variance along the minor axis m by σ_m^2 and the variance along the major axis M by σ_M^2. If we assume that $M = X$ and $m = Y$ and consider an axis through the center that is rotated counterclockwise so as to form an angle α'' with axis $M = X$, then the variance along this axis is $\sigma_M^2 = d^2 - \sigma_m^2$ for $\alpha'' = 0$; it decreases to $\frac{1}{2}d^2$ for $\alpha'' = 45°$; it further decreases to $\sigma_m^2 = d^2 - d_M^2$ for $\alpha'' = 90°$; it increases up to $\frac{1}{2}d^2$ for $\alpha'' = 135°$; and again reaches its maximum value of σ_M^2 for $\alpha'' = 180°$.

10.7.5. Oblong and Nonoblong Geosets

In the preceding sections on principal axes, we assumed that we were dealing with geosets for which there are orthogonal axes through the center along which the standard deviations differ. This condition is generally fulfilled by the geosets one deals with in empirical geostatistical research. The identification of the major and minor axes has been based on equation (10.24) or (10.25).

We identified the angle α'' between the principal and the orthogonal axes, under the assumption that the covariance between the coordinates of the elements is 0 along the principal axes. It is, however, necessary to add an important caveat. In geosets for which we find that $\text{cov}(x_i'', y_i'') = 0$ and $\sigma_{x''} = \sigma_{y''}$ along one system of orthogonal axes (X'', Y''), equations (10.24) and (10.25) become equal to (0/0), and the principal axes cannot be identified because the standard deviations along all axes through the center are equal. To envisage this special setting, one should think of a geoset whose elements are evenly distributed over a square or a circle. Such geosets, which may have a theoretical interest, can be termed nonoblong. In our discussion of principal axes we deal only with geosets for which we can distinguish a major axis M and a minor axis m, and as they have different lengths, we can call them oblong.

It must be stressed that the study of oblongity is based on principal axes only. We saw above that all oblong geosets have: (a) a position (the equating axes) along which standard deviations are equal; (b) a position (principal axes) along which the standard deviations have maximal difference, and (c) an infinite number of intermediate positions.

In early geostatistical studies, attention was focused on changes of standard deviations $\sigma_{\alpha''}$ with a rotation of axes around the center (changing values of angle α''): (see the often quoted controversy between Lefever and Furfey). We do not discuss this problem here because it seems rather irrelevant to the description of the characteristics of actual geosets, for which purpose we concentrate on α', $\sigma_{x'}$, and $\sigma_{y'}$. The meaning of α' as an indicator of the orientation of the geoset was discussed in Section 10.7.2 In the following, we shall focus on $\sigma_{x'}$ and $\sigma_{y'}$ as tools for enhancing measures of dispersion and oblongity.

10.7.6. Scatter

Figure 1.3 (Section 1.2) shows the values of the standard deviations $\sigma_{x'}$ and $\sigma_{y'}$ along the principal axes for the geoset of the population of Italy in 1987 (for numerical values, see Table 10.3). To simplify the presentation, we have put principal axis X' over the conventional axis of the abscissas X and Y' over axis Y. The diagonal gives the standard distance d, which was introduced above as a linear overall parameter of dispersion. The distance variance $d^2 = \sigma_{x'}^2 + \sigma_y^2$ is obviously the corresponding areal measure of dispersion. However, it may seem appropriate to introduce another areal measure of dispersion by considering the area $\sigma_{x'}\sigma_{y'}$ of the rectangle that has the standard deviations $\sigma_{x'}$ and $\sigma_{y'}$ along the principal axes as its sides. To study the relationships between $\sigma_{x'}$, and $\sigma_{y'}$, and d^2, we refer to Figure 10.3. Here, the squares A and B have an area proportional to $\sigma_{x'}^2$ and $\sigma_{y'}^2$, respectively, so that $A + B = d^2$; square C has an area proportional to $(\sigma_{y'} - \sigma_{x'})^2$; the rectangles

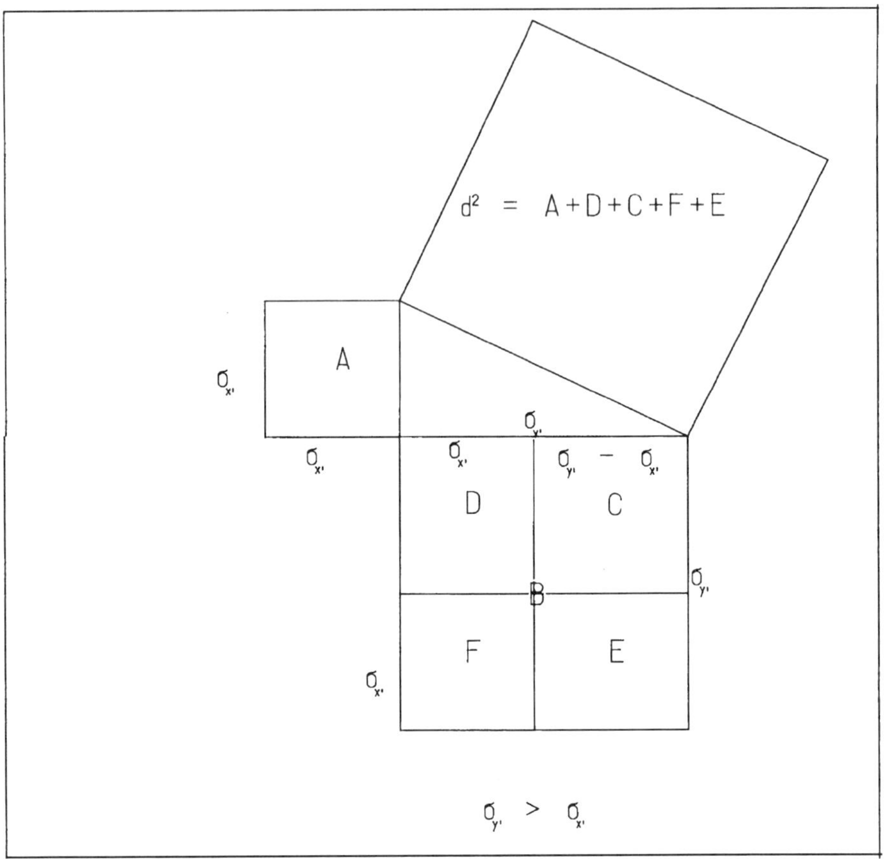

Figure 10.3. Areal measure of dispersion.

formed by D + F and by E + F both have an area proportional to $\sigma_{x'}\sigma_{y'}$ and squares F and A are equal in area. We see graphically that

$$d^2 = 2\sigma_{x'}\sigma_{y'} + (\sigma_{x'} - \sigma_{y'})^2 \tag{10.27}$$

We thus can decompose the overall areal measure of dispersion d^2 into two components: an areal measure of the oblongity $(\sigma_{x'} - \sigma_{y'})^2$ denoted by Ω^2, and the part of the areal dispersion that is not due to oblongity, $2\sigma_{x'}\sigma_{y'}$.

We call, arbitrarily, the parameter $s^2 = 2\sigma_{x'}\sigma_{y'}$ scatter (other appellatives have been used in previous publications), and note that the doubling of $\sigma_{x'}\sigma_{y'}$ in the calculation of the scatter is justified by the development shown in equation (10.26). We thus obtain: $d^2 = s^2 + \Omega^2$.

Next we go on to assume that $2\sigma_{x'}\sigma_{y'}$ is kept constant and there is no oblongity, i.e., $\sigma_{x'} = \sigma_{y'} = \sigma$ and, consequently, $\sigma_{x'}\sigma_{y'} = \sigma^2$. Figure 10.3 shows what would happen in this setting: $d^2 = 2\sigma^2$ would correspond to a scatter equal to $2\sigma^2$, i.e., to two squares each of which has an area proportional to $\sigma^2 = d^2/2$. This is the setting that would be found if we compared the length of the two variances along the equating axes: $\sigma_e^2 = d^2/2$.

Oblongity $(\sigma_{x''} - \sigma_{y''})^2$ along any pair of orthogonal axes (X'', Y'') is maximized by referring to the principal axes (X', Y'). Accordingly, $2\sigma_{x''}\sigma_{y''}$ is minimized by referring to (X', Y'). We see again that reference to the principal axes enables us to obtain parameters of dispersion and oblongity having a clear meaning, which no other pair of axes (X'', Y'') would provide. Later we shall see other very important properties of $2\sigma_{x'}\sigma_{y'}$.

We now assume that we know the standard deviations σ_x and σ_y along any system of orthogonal axes through the center and $\text{cov}(x, y)$ between the coordinates of the elements along these axes. Let us denote the coefficient of correlation between these coordinates by $r = \text{cov}(x, y)/\sigma_x\sigma_y$. It can be shown that

$$2\sigma_{x'}\sigma_{y'} = s^2 = 2\sigma_x\sigma_y\sqrt{1 - r^2} \tag{10.28a}$$

and that

$$\sigma_{x'}^2\sigma_{y'}^2 = \sigma_x^2\sigma_y^2 - \text{cov}^2(x, y) \tag{10.28b}$$

We know that for a given geoset, both $\sigma_x\sigma_y$ and $\text{cov}(x, y)$ may change if we calculate them along different pairs of orthogonal axes. However, $2\sigma_{x'}\sigma_{y'}$ is invariant as it is calculated along uniquely defined principal axes.

We see from equation (10.27) that $2\sigma_x\sigma_y\sqrt{(1 - r^2)}$ is also invariant. We also see that equality between $2\sigma_x\sigma_y$ and $2\sigma_{x'}\sigma_{y'}$ is reached only when $r = 0$ and consequently when $\text{cov}(x, y) = 0$ as well. This happens only if the arbitrary axes considered coincide with the principal axes.

We also see that apart from this case $2\sigma_{x'}\sigma_{y'} < 2\sigma_x\sigma_y$, which confirms that for a given geoset the scatter along the principal axes is a minimum in comparison to $2\sigma_x\sigma_y$ along other pairs of axes.

10.7.7. Measuring Oblongity

Oblongity and the impact of oblongity on dispersion can be measured in various ways, depending on the nature of the geosets considered and the aims of the research. We indicate below some alternative approaches. We refer in general to oblong geosets and take as a numerical example the geoset of the population of Italy (1987). The parameters used in these calculations are (in km):

$$\sigma_{x'} = 136.37 \qquad \sigma_{x'}^2 = 18{,}596.78$$

$$\sigma_{y'} = 339.74 \qquad \sigma_{y'}^2 = 115{,}423.30$$

$$d^2 = 134{,}020 \qquad d = 366.09$$

$$s^2 = 2\sigma_{x'}\sigma_{y'} = 92{,}661$$

In the specific example considered:

$$\sigma_{y'} = \sigma_M \qquad \sigma_{x'} = \sigma_m$$

1. The simplest way to measure oblongity is through the ratio "relative linear oblongity" denoted by ${}^{L}b$:

$$ {}^{L}b = (\sigma_M - \sigma_m)/\sigma_m = 1 - (\sigma_m/\sigma_M) \tag{10.29} $$

In earlier publications oblongity was denoted by o, but we have used b in order to avoid confusion with zero.

Parameter ${}^{L}b$ can vary between "a little" above 0 and "a little" under 1. Strictly speaking, zero is reached when the geoset is nonoblong, and, as we noted in Section 10.7.5, in a case such as this we cannot distinguish between σ_M and σ_m as we cannot identify the principal axes. However, for geosets with very small oblongity, this parameter is near 0. At the other extreme, 1 can be reached only when the geoset degenerates into a linear distribution and then, too, we cannot distinguish between σ_M and σ_m. In our example ${}^{L}b = 0.5986$, which suggests that the two principal axes have very different lengths. This is confirmed by a look at Figure 1.3 (Section 1.2).

2. Going on to consider squared measures, we can use the approach introduced in Section 10.7.6, where we denote the scatter by s^2 and the square of the difference between the standard deviations along the principal axes by Ω^2:

$$ \Omega^2 = (\sigma_M - \sigma_m)^2 \tag{10.30} $$

This parameter can be termed the "absolute squared oblongity." We have already indicated that $d^2 = s^2 + \Omega^2$, which offers a neat decomposition of distance-variance between effects of oblongity, Ω^2, and dispersion netted from the influence of oblongity, $s^2 = d^2 - \Omega^2$. In our example we have

$$ \Omega^2 = (339.74 - 136.37)^2 = 41{,}359 $$

$$ s^2 = 92{,}661 $$

$$ d^2 = 134{,}020 $$

3. Using this approach we see that the percentage of distance-variance owing to scatter, s^2/d^2, is 69.14%, while that owing to squared oblongity, Ω^2/d^2, is 30.86%.

4. Under (2) we decomposed d^2 into two summable components $s^2 + \Omega^2$. In many cases, for important calculations, it is desirable to compare d^2 to s^2 by means of a ratio ${}^{r}b^2 = d^2/s^2$, which is called the relative oblongity, and can be interpreted in various ways:

1. This parameter measures the relative increase in areal dispersion owing to oblongity in passing from s^2 to d^2, so obviously, if there is no oblongity, it reaches its minimal value of 1. If we write $d^2 = s^{2\,r}b^2$, we see that d^2 can be split into two multiplicative factors measuring the influence of scatter and oblongity, respectively.
2. The calculation of this parameter can also be rationalized as follows: We can measure relative oblongity either through the ratio σ_M/σ_m or through the inverse ratio σ_m/σ_M. If we suppose that there is no particular reason to prefer one ratio over the other, we take an average:

$$ {}^{r}b^2 = \tfrac{1}{2}(\sigma_M/\sigma_m + \sigma_m/\sigma_M) = d^2/s^2 \tag{10.31} $$

In our example, ${}^{r}b^2 = \frac{1}{2}(2.4913 + 0.4014) = 1.4464$. If there is no oblongity, then ${}^{r}b^2 = 1$ and $d^2 = s^2$. We should remember when interpreting ${}^{r}b^2$, that this parameter is not very sensitive. A few examples follow, showing values of ${}^{r}b^2$ that correspond to selected values of ${}^{L}b$:

${}^{L}b$	0.1	0.2	0.3	0.4	0.5	0.6	0.7	0.8	0.9
${}^{r}b^2$	1.0056	1.025	1.064	1.133	1.25	1.45	1.82	2.60	5.05

3. In Chapter 11 we discuss "relative parameters of dispersion," which are based on the comparison of d^2 and s^2 to the area A of the geoset considered: the relative scatter is ${}^{r}s^2$ measured by $2\pi s^2/A$ and the relative distance-variance ${}^{r}d^2$ by $2\pi d^2/A$. The ratio ${}^{r}d^2/{}^{r}s^2$ measuring effects of oblongity and called here relative oblongity equals d^2/s^2. We have accordingly denoted $d^2/s^2 = {}^{r}d^2/{}^{r}s^2$ as the relative oblongity.

5. In some cases it may be necessary to indicate whether σ_M coincides with $\sigma_{x'}$ or with $\sigma_{y'}$. This can easily be done in approach (1) by substituting $\sigma_{x'}$ or $\sigma_{y'}$ for σ_M and $\sigma_{y'}$ or $\sigma_{x'}$ for σ_m in the equations above. Also, the ratio $(\sigma_{x'} - \sigma_{y'})/(\sigma_{x'} + \sigma_{y'})$ has been found to be of use in this type of study.

10.8. Selected Parameters of Dispersion and Oblongity

10.8.1. Types of Examples Considered

The parameters reviewed above, especially standard distance and distance-variance, have been employed rather widely in studies of the dispersion of territories and subterritories, populations and subpopulations, and other phenomena, for many different countries, regions, and cities. For some relevant bibliography the reader is referred to the review of cartographic and geostatistical literature cited in Chapter 2.

To fully understand the meaning of the parameters of dispersion and oblongity, one must analyze them: (a) by using the maps of the geosets to which they refer, and (b) by studying them as part and parcel of the entire battery of parameters describing the geosets. This approach has been followed with regard to a few geosets selected among the thousands of primary and derived geosets that are analyzed in the *Statistical Atlas of Italy* (see Sections 10.9.3–10.9.5). We shall add some other examples of the same approach for the United States (see Section 11.7) and for selected cities (see Section 11.8). Owing to limitations of space we cannot enlarge this detailed presentation. Many other examples will be presented as averages or other summaries of the values of parameters of dispersions and oblongity that have been obtained from research in which large numbers of geosets have been studied. These collections of values enable us to draw some general conclusions as to the behavior of each of the parameters considered.

10.8.2. Parameters of Dispersion and Oblongity for Selected Population Geosets

In Section 1.5.4 we discussed examples of parameters of dispersion and oblongity for ten primary and derived population geosets (for Italy) for which maps are given in this volume. Let us denote these geosets generically by Z and the uniform distribution over the territory of Italy by U. Table 1.1 (Section 1.5) showed (a) the ratios d_Z^2/d_U^2 between the distance-variances of Z and U; (b) the ratios s_Z^2/s_U^2 between the scatters of Z and U as parameters of dispersion. It also showed ${}^r b_Z^2 = {}^r d_Z^2 / {}^r s_Z^2$ as a parameter of relative oblongity. Obviously, $d_Z^2/d_U^2 = (s_Z^2/s_U^2)({}^r b_Z^2 / {}^r b_U^2)$. Perusing the detailed maps for each of the geosets, we have seen that these parameters of dispersion and oblongity "make sense": they have values that appear

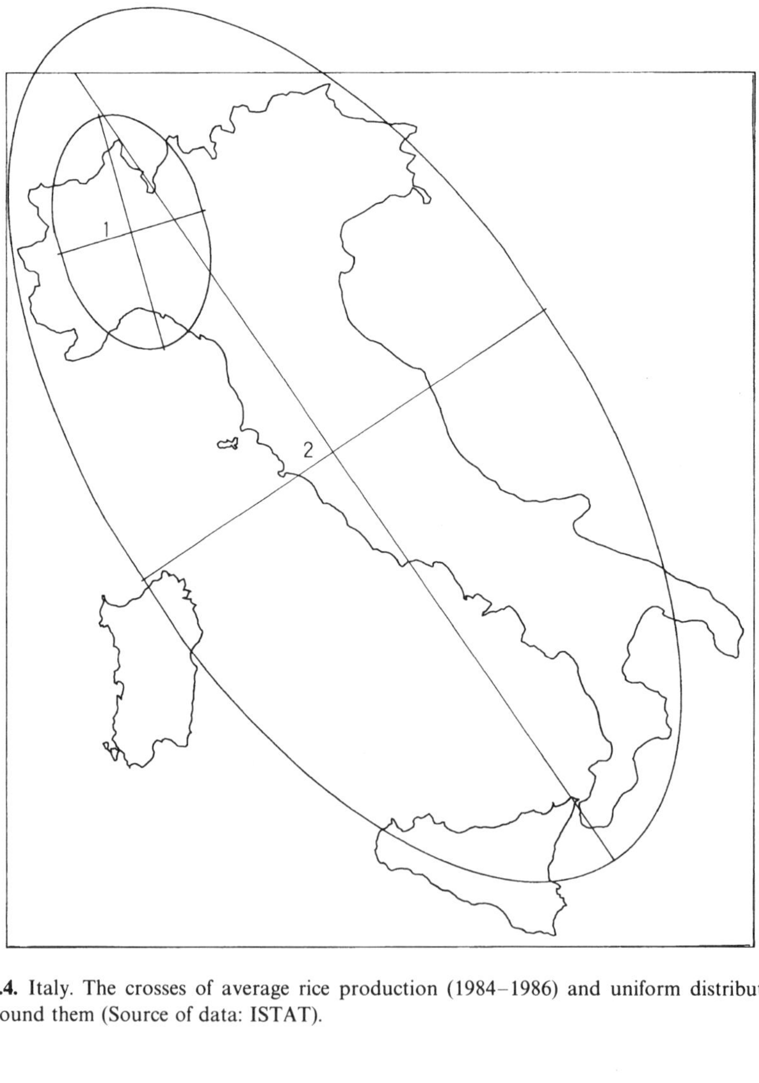

Figure 10.4. Italy. The crosses of average rice production (1984–1986) and uniform distribution with ellipses around them (Source of data: ISTAT).

to correspond well to what might be expected from what we know of the maps and the substantive characteristics of the geosets. The comparison between the maps of each geoset Z and the parameters of dispersion and oblongity can be made easier by using a pair of ellipses, as illustrated in Figure 10.4, which refers to the geoset of rice (shown in Figure 1.12 in Section 1.4) as compared to the uniform distribution U over the territory of Italy (as shown in Figure 1.1 in Section 1.2). Of all the geosets considered, rice is the one with the smallest distance-variance.

Figure 10.4 shows two ellipses labeled (1) and (2), and Figure 10.5 shows two circles, also labeled (1) and (2). Ellipse (1) has been constructed around the cross of Figure 1.12 (Section 1.4). Its area is $A_1 = 2\pi s_{\text{rice}}^2$, as the principal semiaxes have lengths proportional to $2\sigma_M$ and $2\sigma_m$. Ellipse (2) has been constructed around the cross of Figure 1.1 (Section 1.2). Its area $A_2 = 2\pi s_U^2$.

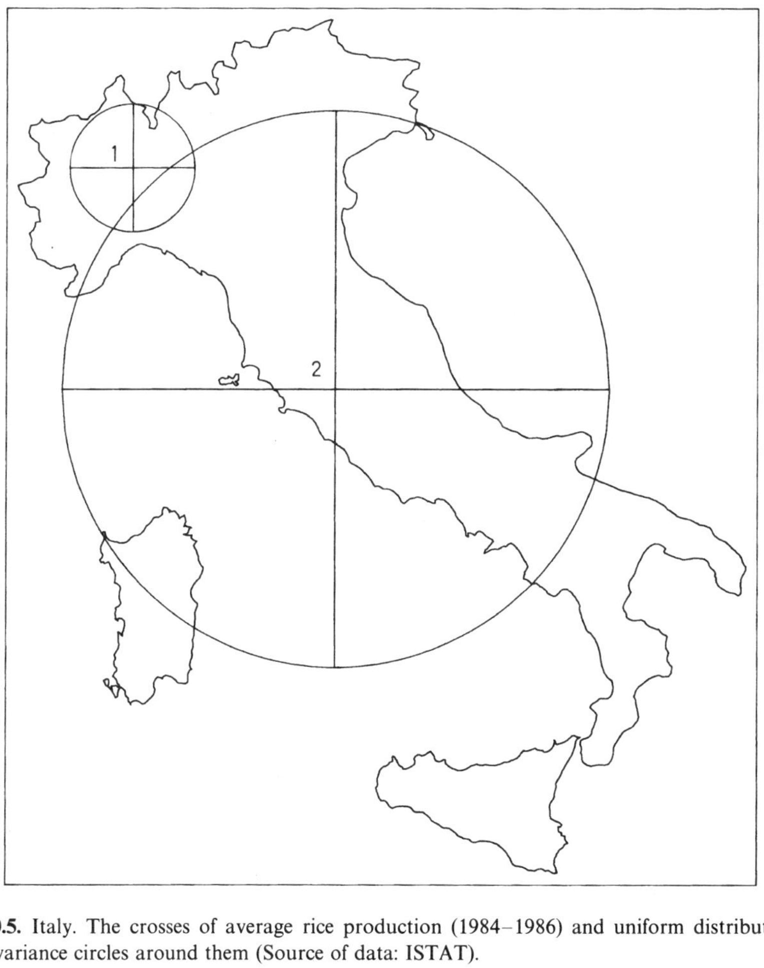

Figure 10.5. Italy. The crosses of average rice production (1984–1986) and uniform distribution with distance-variance circles around them (Source of data: ISTAT).

The ratio between A_1 and A_2 is equal to the ratio s_1^2/s_2^2, which has the very low value of 9.4%. The oblongity of the rice appears to be somewhat smaller than that of the uniform distribution (${}^rb^2 = 1.13$ for rice as compared to 1.26 for the uniform distribution).

Circles (1) and (2) in Figure 10.5 represent, respectively, the distance-variance of rice and of the uniform distribution. For each of these circles, the radius, $r = 2\sqrt{(s_\pi^2 + \sigma_r^2)} = 2d$, can be easily identified as shown in the first quadrant of ellipse (2): a radius is drawn there making an angle of $+45°$ with the axis of the abscissas and an angle of $-45°$ with the axis of the ordinates. The coordinates of the end point of this radius are $x = \sqrt{2}d$ and $y = \sqrt{2}d$ and thus its length is $r = 2d$. The oblongity is measured by comparing the area of circle (1) to that of ellipse (1) for the rice, and the area of circle (2) to that of ellipse (2) for the territory.

10.9. Decomposition of Distance-Deviance of Compound Geosets

10.9.1. General Equations

We saw in Section 10.4 that the distance-deviance d_K^2 of a compound geoset K formed by component geosubsets K_j can be decomposed into: (a) the distance-variance between components, $d_b^2 = \Sigma\, \varepsilon_j v_j {}^cD^2(K_j, K)$, and (b) the distance-variance within components, $d_w^2 = \Sigma\, \varepsilon_j v_j d_j^2$.

The contribution of each component K_j to d_K^2, $\varepsilon_j v_j[{}^cD^2(K_j, K) + d_j^2]$, can be calculated separately. The analysis of d_K^2 can be enhanced if we take into account the individual impacts of the scatter and of the oblongity over the distance-deviance, which can be expressed by

$$d_K^2 = s_K^2 {}^rb_K^2 \qquad d_K^2 = s_K^2 + \Omega_K^2$$

$$d_j^2 = s_j^2 b_j^2 \qquad d_j^2 = s_j^2 + \Omega_j^2$$

The additive equation $d_j^2 = s_j^2 + \Omega_j^2$ enables us to decompose the distance-variance within components into two parts: $\Sigma\, \varepsilon_j v_j s_j^2$, which measures the impact of the scatters of the components, and $\Sigma \varepsilon_j v_j \Omega_j^2$, which measures the impact of the oblongities of the components. Taking together $\Sigma\, \varepsilon_j v_j s_j^2 + \Sigma\, \varepsilon_j v_j \Omega_j^2 + \Sigma\, \varepsilon_j v_j ({}^cD^2(K_j, K)$, we again obtain d_K^2. We thus have a complete analysis of the contribution of the compound geoset K to the distance-deviance d_K^2: (a) by each component K_j; (b) by each determinant (scatter, oblongity, and eccentricity) of the distance-variance of each component; and (c) by each determinant for all the components taken together. There is no functional relationship between the decomposition of d_K^2 into

$$\sum \varepsilon_j v_j [d_j^2 + {}^cD^2(K_j, K)]$$

and the direct decomposition of d_K^2 into $s_K^2 + \Omega_K^2$ or $s_K^2 {}^rb_K^2$.

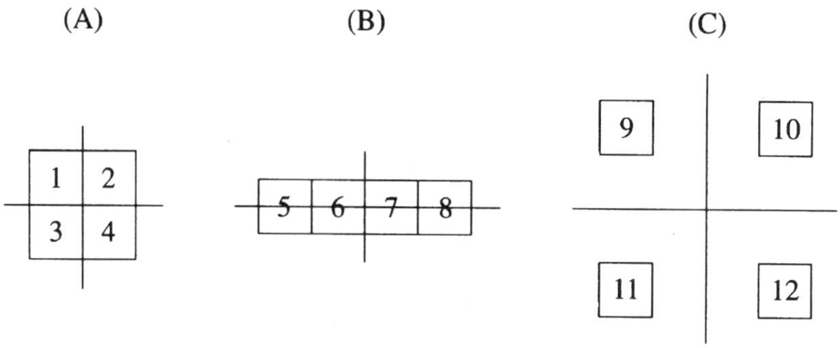

Figure 10.6. Theoretical example of decomposition.

10.9.2. Simple Territorial Models of Decomposition of Distance-Deviance

Before describing some applications of the important types of decomposition of distance-deviances, we will introduce some theoretical examples of decomposition of territorial models. To simplify the presentation, we shall deal here with territories (labeled A, B, C) that have equal areas, each of them being formed by four small squares whose sides are of length 1, so that the total area of each territory is always equal to 4 and each square has a relative weight of 0.25 (see Figure 10.6). The squares are numbered from 1 to 12. Territories and subterritories are formed by sets of these squares, as shown below, and their parameters are given in Table 10.4.

- $A = \{1, 2, 3, 4\}$ is a square—an example of a simply connected, convex, and nonoblong territory. From direct calculation we find that $d_A^2 = s_A^2 = 0.667$ and that there is no oblongity ($\Omega_A^2 = 0; {}^r b_A^2 = 1$).
- $B = \{5, 6, 7, 8\}$ is a rectangle also (as A) is a simply connected and convex territory; however (unlike A) it is oblong. From direct calculation we find that $d_B^2 = 1.417$, $s_B^2 = 0.667$ (as for A); $\Omega_B^2 = 0.75$; ${}^r b_B^2 = 2.125$.
- $C = \{9, 10, 11, 12\}$—a territory (unlike A and B) not simply connected, but fragmented, the four cells being separated from one another. From direct calculation we find $d_C^2 = s_C^2 = 4.667$, so there is no oblongity ($\Omega_C^2 = 0; {}^r b_C^2 = 1$). However, each of the four elements has a squared eccentricity of 4.5. This is the determinant of the higher distance-variance of C as compared to that of A. *Ceteris paribus*, increasing eccentricity leads to an increase in distance-variance.

Table 10.4. Parameters of Territories and Subterritories

K	A_K	d_K^2	s_K^2	Ω_K^2	${}^r b_K^2$
Each square $\{1\}\{2\}, \ldots, \{12\}$	1	0.167	0.167	0	1
$A = \{1, 2, 3, 4\}$	4	0.667	0.667	0	1
$B = \{5, 6, 7, 8\}$	4	1.417	0.667	0.750	2.125
$C = \{9, 10, 11, 12\}$	4	4.667	4.667	0	1
$D = \{1, 2\}$ or $E = \{3, 4\}$	2	0.416	0.333	0.083	1.250

Let us consider A as a compound territory formed by two components labeled, respectively, D and E. Here each component is formed by two cells, each of which has relative weight 0.5.

- $D=\{1,2\}$; $E=\{3,4\}$. Both components have $d^2=0.416$ of which $\Omega^2=0.083$, $s^2=0.333$, and squared distance 0.25 between their centers and the center of the compound territory A. These values also indicate the contribution of the scatters, the oblongities, and the eccentricities of D and E to the total distance-deviance, $d_A^2=0.667$, of A. Had we divided A into other subsets, such as $F=\{1\}$, $G=\{2,3,4\}$ or $H=\{1,4\}$, $I=\{2,3\}$, we would have found different values for the total contribution of the scatters, oblongities, and eccentricities of F, G or H, I. However, their total would always have been equal to $d_A^2=0.667$. We have seen that this is the value found for d_A^2 in the direct analysis of A; however, in this analysis $s^2=d^2$: oblongity and eccentricity do not play any role.

10.9.3. Decomposition of an Actual Fragmented Territory

The approach suggested above for model C can be used whenever we deal with an actual territory formed by well-defined, separated fragments. Let us consider the territory of Italy, T, formed by three main fragments: T_1 = Continental Italy, T_2 = Sicily, T_3 = Sardinia, as shown in Figure 4.17 (Section 4.6). Using equation (10.32) for T and $T_j = T_1, T_2, T_3$ and denoting the areas by A_T, A_1, A_2, and A_3, respectively, we find that the proportions of the area of each fragment to the total are A_j/A_T, denoted by a_j, and we obtain

$$d_T^2 = \sum \varepsilon_j(s_j^2 + \Omega_j^2 + \varepsilon_j^2) \tag{10.32}$$

We might also wish to measure the contribution of the scatter, the oblongity, and the eccentricity of each fragment T_j to the distance-deviance of T, and then the total contribution of each fragment and of each determinant. Tables 10.5 and 10.6 show how this calculation can be carried out.

It is found that Sicily, which covers only 8.5% of the territory of Italy, contributes 21.9% of the distance-deviance d_T^2; Continental Italy, which covers

Table 10.5. Basic Parameters for the Decomposition of the Distance-Deviance of the Territory of Italy

		Ratio to d_T^2 of the			
Fragments	Percentage of total area $a_j=A_j/A_T$ (1)	Scatter (s_j^2/d_T^2) (2)	Absolute squared oblongity (Ω_j^2/d_T^2) (3)	Squared eccentricity (ε_j^2/d_T^2) (4)	Total [(2)+(3)+(4)] (5)
T_1 Continental Italy	83.6	0.463	0.310	0.051	0.824
T_2 Sicily	8.5	0.035	0.012	2.529	2.57
T_3 Sardinia	7.9	0.025	0.013	1.127	1.1

Table 10.6. Percentage Contribution of Each Fragment and Each Determinant to the Distance-Deviance of the Territory

Fragments	Percentage of total area $a_j = A_j/A_T$	Weighted determinants of distance-deviance: Scatter $a_j(s_j^2/d_T^2)$	Oblongity $a_j(\Omega_j^2/d_T^2)$	Squared eccentricity $a_j(\varepsilon_j^2/d_T^2)$	Total $a_j(d_j^2+\varepsilon_j^2)/d_T^2$
T_1 Continental Italy	83.6	38.7	25.9	4.3	68.9
T_2 Sicily	8.5	0.3	0.1	21.5	21.9
T_3 Sardinia	7.9	0.2	0.1	8.9	9.2
T Italy	100.0	39.2	26.1	34.7	100.0

83.6% of the territory, has a share of only 68.9% of d_T^2; the more important determinant of this setting is the strong eccentricity of Sicily, which contributes 21.5% to d_T^2. Both the eccentricity and the oblongity are very important determinants, being, respectively, 34.7% and 26.1%; scatter is only 39.2% of the magnitude it would be if the territory of Italy were circular.

10.9.4. Decomposition of Distance-Deviance of a Territory into that of the Habitat and the Nonhabitat

The method described above can be extended to other important settings, as shown by the many applications in the *Statistical Atlas of Italy*. Some examples are given in Table 10.7.

Table 10.7. Contribution of Habitat and Nonhabitat of Population and Citrus Groves to the Distance-Deviance of the Territory of Italy

	Percentage of total area $a_j = A_j/A_T$	Ratio to d_T^2 of the: Scatter $a_j(s_j^2/d_T^2)$	Absolute squared oblongity $a_j(\Omega_j^2/d_T^2)$	Squared eccentricity $a_j(\varepsilon_j^2/d_T^2)$	Total $a_j(d_j^2+\varepsilon_j^2)/d_T^2$
Population					
Habitat	53.3	39.6	14.3	0.1	54.0
Nonhabitat	46.8	38.8	7.1	0.1	46.0
Total	100.0	75.4	21.4	0.2	100.0
Citrus Groves					
Habitat	14.7	7.3	0.5	21.4	29.1
Nonhabitat	85.3	55.8	11.4	3.7	70.9
Total	100.0	63.1	11.9	25.1	100.0

1. The upper half of Table 10.7 is based on the decomposition of the territory T of Italy into habitat and nonhabitat with regard to the population. The table shows the relative contribution of each subterritory and of each determinant to the distance-deviance of T. We see that the percentage of distance-deviance contributed by the habitat of the population is almost identical to that of the area covered by the habitat; as the centers of the habitat and of the nonhabitat of the population are very near to each other and to the center of the territory, the contribution of the eccentricity to d_T^2 is very small, while the contribution of the scatter is dominant.
2. The lower half of Table 10.7 shows a similar calculation with respect to the habitat and the nonhabitat of citrus groves. Here the results are very different from those found for the population. The percentage of d_T^2 contributed by the habitat is almost twice that of the territory covered by the habitat. As the center of the habitat is very far from the center of the territory, the contribution of the eccentricity of the habitat is very large; as the nonhabitat is very big, the contribution of its scatter and oblongity are important.

10.9.5. Decomposition of Distance-Deviance of Primary Geosets into Determinants of Derived Geosets

There is no need to discuss simple applications of the decomposition of d_p^2 of a population divided into subpopulations as they are similar to those described above for d_T^2 of a territory T divided into subterritories. However, we wish to introduce methods of decomposition that are applicable when we compare a population geoset ${}^gP$ to (1) the uniform distribution ${}^gU$, or (2) to the geoset ${}^gQ$ of another population.

Let us take as an example of (1) the comparison of the geoset of the population of Italy in 1987 to the uniform distribution, taking into account two derived geosets: (a) the geoset ${}^gM(P, U)$ of the matches, which indicates the part of ${}^gP$ that is common to ${}^gU$; and (b) the geoset of the surpluses of ${}^gP$ over ${}^gU$ (see Figure 1.4 in Section 1.2). Table 10.8 shows the decomposition of the distance-deviance of ${}^gP$ into matches and surpluses with the contributions of scatter, oblongity, and eccentricity.

Table 10.8. Percentage Contributions to Distance-Deviance d_P^2 of the Population of Italy of Matches and Surpluses versus the Uniform Distribution

		Ratio to d_P^2 of the			
	Percentage out of total population, v_j	Scatter $v_j(s_j^2/d_P^2)$	Absolute squared oblongity $v_j(\Omega_j^2/d_P^2)$	Squared eccentricity $v_j(\varepsilon_j^2/d_P^2)$	Total $v_j(d_j^2+\varepsilon_j^2)/d_P^2$
Matches ${}^gM(P, U)$	53.3	41.1	14.8	0.1	56.0
Surpluses of population over uniform ${}^gSPU$	46.8	27.5	16.4	0.1	44.0
Total	100.0	68.6	31.2	0.2	100.0

Table 10.9. Percentage Contributions to Distance-Deviance d_P^2 of Citrus Groves in Italy of Matches and Surpluses versus the Uniform Distribution

		Ratio to d_P^2 of the			
	Percentage out of total population, v_j	Scatter $v_j(s_j^2/d_P^2)$	Absolute squared oblongity $v_j(\Omega_j^2/d_P^2)$	Squared eccentricity $v_j(\varepsilon_j^2/d_P^2)$	Total $v_j(d_j^2+\varepsilon_j^2)/d_P^2$
Matches ${}^gM(P, U)$	14.7	29.0	1.8	4.3	35.1
Surpluses ${}^gSPU$	85.3	62.7	1.5	0.7	64.9
Total	100.0	91.7	3.3	5.0	100.0

We see here that the shares of the deviance of the population owing to ${}^gM(P, U)$ and ${}^gSPU$ are almost identical to the properties of elements of ${}^gM(P, U)$ and ${}^gSPU$ over those of ${}^gP$. As the center of population and the center of territory are very near to each other, the contribution of eccentricity is almost nil. As both the ${}^gP$ and ${}^gU$ are very oblong, the contribution of oblongity is strong.

Table 10.9 shows another example of (1) taken from the comparison of citrus groves to uniform distribution. Here the overwhelming share is that of the scatter, especially of the surpluses ${}^gSPU$.

Chapter 11

Parameters of Form

Shapes of Territories and Patterns of Populations

11.1. Introduction

In this chapter we shall discuss parameters of the main aspects of the "form" of a geoset. We use this term in a somewhat arbitrary way to indicate both the shape of a territory and the pattern of distribution of a population geoset. Our interest in a territory is mainly (but not exclusively) focused on its function as a container of a population. Parameters of shape of a territory measure patterns of the uniform distributions and at the same time serve as a basis for the comparison of patterns of distributions of actual population geosets over the territory. In speaking of these patterns of distribution, we refer mainly to macrogeographical characteristics.

As data on uniform distributions are easier to deal with than those for actual population distributions, we shall start here by discussing aspects of the shape of territories, which in itself is a complex concept. We shall focus principally on ways to study and measure one very important aspect of this feature, namely, lack of compactness (or looseness) of territories. We shall see that it can be expressed by means of an overall parameter known as the relative distance-variance, which can be conveniently split into relative scatter and relative oblongity (see Sections 11.2 and 11.3).

To facilitate the interpretation of relative scatter of actual territories, we study the behavior of this parameter for various simplified models, such as regular geometrical figures, other simply connected figures (see Section 11.4), territories with internal and/or external concavities, and fragmented territories (see Section 11.5).

We shall then examine three selected aspects of patterns of population geosets and the parameters that quantify them:

1. Looseness of the distribution of populations and subpopulations over the territory: We shall see that concepts and equations for measuring looseness of territories and subterritories can be extended, with certain modifications, to the study of lack of compactness of primary and derived geosets.
2. In the analysis of a geoset of the surpluses of an actual population distribution over the uniform distribution, we might be interested in deter-

mining whether, how, and how much the surpluses are clustered. We can identify blocks of contiguous cells with surpluses, concentrate on the major clusters, analyze the relevant parameters for each of them, and give overall parameters of clustering (see Section 11.9).

3. Asymmetry is another aspect of form that can be studied in a similar way for territories and population geosets. However, for technical reasons, the discussion of asymmetry is postponed (see Section 11.11).

Obviously, there is some arbitrariness in our selection of the aspect of form of territories and population geosets to be measured. However, experience has shown that the methods suggested enable us to obtain summary geostatistical parameters that are useful for the description and interpretation of territories and population distributions. Examples are given in this chapter for selected geosets for Italy, the United States, and several individual cities.

11.2. Shape and Parameters of Looseness of Territories

11.2.1. Shape of Territories

The shape of territories has been defined by Boots and Lamoureux (1972) as a set of properties possessed by a closed figure of at least two dimensions, which has a planar representation and precise boundaries. The need to express characteristics of the shape of a figure quantitatively, has been felt in many scientific areas, including geography (both physical and human), geology, mineralogy, biological sciences, psychology, and scene analysis, as well as in applications such as districting. There is in consequence an extensive and complex literature, but we will limit ourselves here to a cursory mention of methods described by various authors, that seem to be pertinent to the specific problems of measuring characteristics of the shape of territories in geographical–statistical research.

For instance, it has been proposed to utilize for this purpose a ratio between two geometrical parameters of a figure of a territory *T*, such as (a) the perimeter and the area of *T*; (b) the perimeter of *T* and that of a circle of equal area; (c) the area of *T* and that of a circle with the same perimeter; (d) the largest axis of *T* and its area (e); the longer and the shorter axes of *T*; or (f) the area of *T* and the area of the smallest circle enclosing *T*. All of these alternatives have rightly been criticized (see, e.g., Blair and Biss, 1967) for being based on geometrical parameters that can change significantly even if the basic characteristics of the shape remain constant. For example, the perimeter can change owing largely to more-or-less deep zigzag peripheral indentations in *T*; the smaller and larger axes of *T* can be seriously affected by the accidental position of only two points and so on. Better results were obtained by methods using (a) the ratio between a function calculated based on all the elements of *T* (the distance-variance); and (b) a measure of variability of the radials drawn from the center to the periphery of a simply connected territory, as proposed by Boyce and Clark (1964) (see also Clark and Gaile, 1973). However, all of these measures try to summarize shape with only one parameter and owing to the complexity of shapes, these approaches are clearly inadequate (see Lee and Sallee, 1970).

Other authors have adopted far more complicated methods, such as the study of the distribution of the distances between all the possible pairs of elements of T (see, e.g., Shimony, 1963, Luu-Mau-Thanh, 1962, Taylor, 1971) or certain functions of the distances between pairs of selected vertices of the polygon substituted for T (Bunge, 1962). While interesting from an analytical viewpoint, the results obtained in these ways are far from satisfactory in terms of discriminating among the different types of territories. Attempts have also been made to expand into a Fourier series the lengths of radials drawn from a center to points on the outline of T and calculated as a function of an angle made with an arbitrary line through the center. While in special cases this method can be of use in constructing a model of T, it does not help to visualize the principal characteristics of T quickly.

Maceachren (1985) reviewed many of the measures of the shape of territories and emphasized the fact that it has several aspects, each requiring different measures. Among these aspects, compactness is particularly important. We shall, accordingly, devote a great deal of attention in this chapter to lack of compactness (called for short, looseness) and to ways of measuring it and its determinants separately. In the study of looseness of territories we find it convenient to distinguish among convex territories, territories with external or internal concavities, and fragmented territories.

Analogous parameters of looseness of population geosets will be introduced. Measures of the effects of the clustering of populations will be developed that are analogous to the parameters of the fragmentation of territories.

11.2.2. Parameters of Looseness of Territories

In the endeavor to measure the degree of compactness of a territory T, different authors have proposed: (a) basing the measure of compactness of territory T on the comparison between the dispersion and the area of T; (b) evaluating the actual degree of compactness of T by comparing it to that of a territory having the most compact shape, namely the circle (see Bachi, 1958, 1962*a*, *b*, 1973, 1989*a*, *b*, 1993; Blair and Biss, 1967; Maceachren, 1985; Massam and Goodchild, 1971; Whittington *et al.*, 1972).

If we take for (a) the ratio of the distance-variance d_T^2 to the area A_T of T, we can calculate d_T^2/A_T as a measure of dispersion per unit of area. Since for the circle d_T^2 is $r^2/2$ (where r is the radius) and the area of the circle is πr^2, the ratio between d^2 and A is $\frac{1}{2}\pi$. For the parameter of interest we can take either

$$\text{(i)}\ [(d^2/A_T)/\tfrac{1}{2}\pi] = 2\pi d_T^2/A_T \qquad \text{or} \qquad \text{(ii) Its reciprocal, } A_T/2\pi d_T^2$$

Parameter (i) varies between 1 (for circular territories) and ∞, while (ii) varies between 1 (for a circular territory) and 0. As (i) is more sensitive than (ii), it may be preferred, because sensitivity is a very desirable property in this context.

However, it should be noted that when A_T is a constant the ratio $2\pi d_T^2/A_T$ increases with increasing dispersion (measured by d_T^2). Thus this ratio does not give a direct measure of compactness, but measures rather the "departure from compactness" or "lack of compactness" or "noncircularity" or "looseness" of T. We can call $2\pi d_T^2/A_T$ the relative distance-variance (denoted by ${}^r d_T^2$) and consider ${}^r d_T^2$ a measure of the looseness of T; the minimum value ${}^r d_T^2 = 1$ is found for a circular T. For any T having a shape other than circular, ${}^r d_T^2 > 1$.

Maceachren compared (a) ${}^r d^2$ and similar measures to other parameters suggested in the literature, such as (b) perimeter–area measurement; (c) parameters of circles drawn within or around the figure; or (d) comparison to a standard shape. According to Maceachren, the "most accurate" indexes of compactness are those of type (a).

In order to give some examples of relative distance-variances, we list below the values of ${}^r d^2$ for selected territories (see Figure 11.1): Spain and Portugal, 1.09; France without Corsica, 1.10; Ireland, 1.25; the Arabian Peninsula, 1.26; Australia, 1.29; India, 1.33; conterminous United States, 1.34; South America, 1.36; Turkey, 1.63; Calebes (Indonesia), 2.44; Great Britain, 2.18; Italy, 2.88; Pakistan (before the division between West Pakistan and Bangladesh), 4.67; and Chile, 11.68.

The enormous differences in the looseness of the territories considered correspond to what could be expected intuitively: e.g., the Iberian Peninsula and France are very compact and Ireland, Australia, and the Arabian Peninsula are fairly compact. On the other hand, Turkey is quite oblong and Chile is extremely oblong. High values of relative distance-variances are also found for territories divided into fragments, such as Italy or Pakistan (before the division). Celebes is neither fragmented nor considerably elongated, but has strong external concavities.

Thus it is evident that while ${}^r d^2$ is a sensitive general parameter of looseness, it may be influenced by various determinants. For a more in-depth analysis, it is desirable to isolate and measure the effects of each of these determinants. A very important first step in this direction is based on the split of ${}^r d^2$ into the separate effects of scatter and oblongity (see Section 11.3). Further steps include measuring looseness of convex territories and the effects of fragmentation and concavities.

Before we go on to a detailed analysis of the effects of various determinants over ${}^r d^2$, it is important to stress the fact that ${}^r d^2$ is independent of two other important aspects of the territory:

1. The parameter ${}^r d^2$ is not affected by the position of the territory in the plane. If T is transferred by rigid motion, through parallel translation, rotation, or reflection its ${}^r d_T^2$ does not change.
2. We consider a territory T with area A_T, distance-variance d_T^2, and relative distance variance ${}^r d_T^2$ and, for the sake of simplicity, assume that we take the mean center as origin of the coordinates and $d_{ic}^2 = (x_i^2 + y_i^2)$ as the squared distances of each element i from the mean center. We suppose, further, that we multiply both the x_i and the y_i by any real and positive coefficient k, and we indicate the transformed territory by T' and, its area and distance-variance by $A_{T'}$ and $d_{T'}^2$ respectively. We then obviously obtain: $A_{T'} = k^2 A_T$ and $d_{T'}^2 = k^2 d_T^2$. Consequently, ${}^r d_{T'}^2 = {}^r d_T^2$.

Territories T and T' can be considered as similar, because they differ only in the extent of their areas. We thus have found that similar territories have equal relative

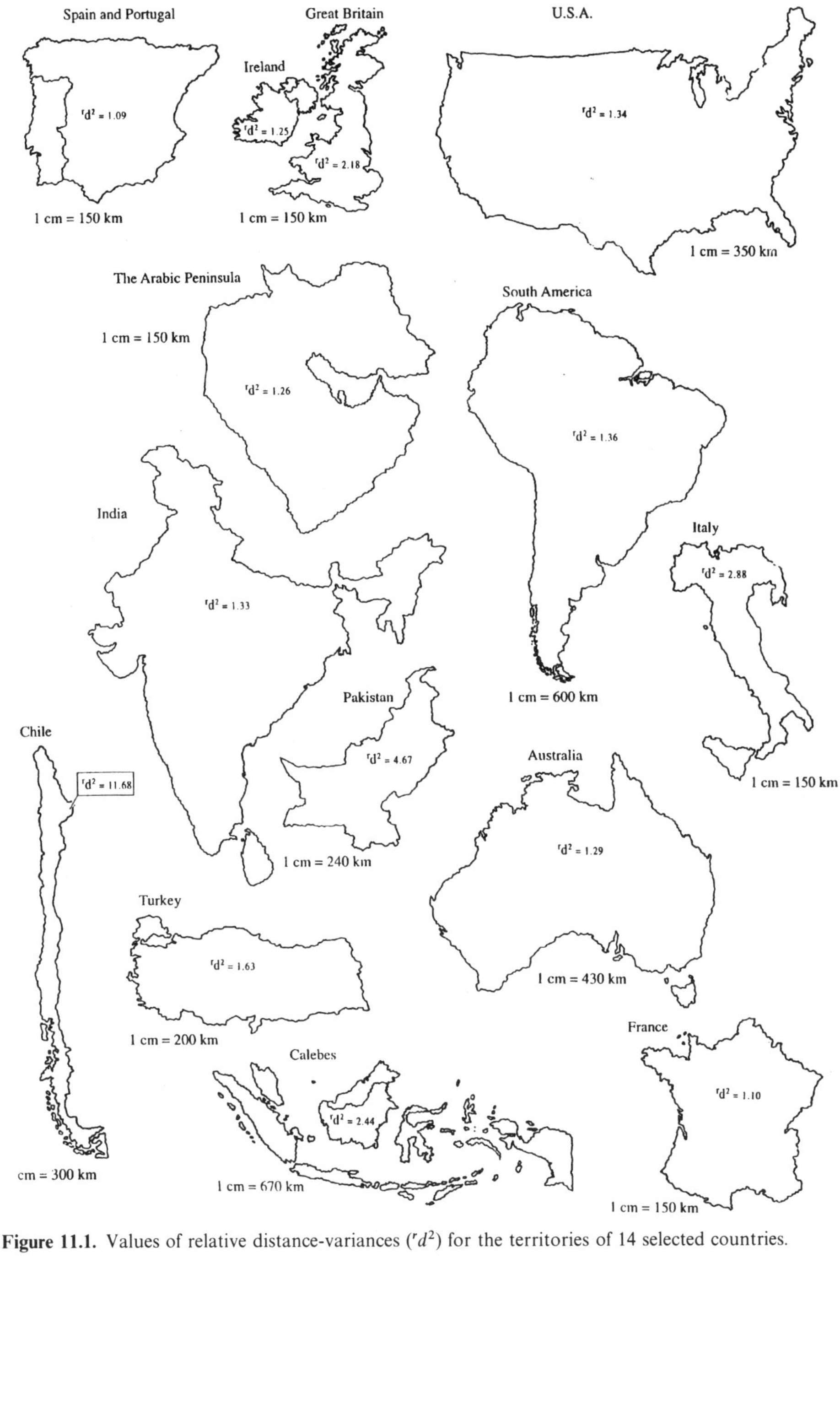

Figure 11.1. Values of relative distance-variances ($^rd^2$) for the territories of 14 selected countries.

distance-variance. Equality of ${}^rd^2$ is a necessary but not a sufficient condition for the similarity of territories and territories with equal ${}^rd^2$ may be dissimilar.

11.3. Relative Scatter of a Territory: General Considerations

The initial approach to the analysis of relative distance-variance is based on the split of distance-variance into scatter and relative oblongity (see Bachi, 1973, 1989*a*, *b*). For any territory $d_T^2 = s_T^{2\,r}b_T^2$ we can write

$$ {}^rd_T^2 = 2\pi d_T^2/A_T = (2\pi s_T^2/A_T)^r b_T^2 \tag{11.1}$$

and thus isolate the effects of oblongity (measured by ${}^rb_T^2$) from those of other factors, as measured by $2\pi s_T^2/A_T$. We call this parameter as the relative scatter,

$$ {}^rs^2 = 2\pi s_T^2/A_T = 4\pi(\sigma_{M,T}\sigma_{m,T})/A_T \tag{11.2}$$

which can be interpreted as follows: Let us denote the scatter of T per unit of area by s_T^2/A_T. Prof. Gad Nathan has shown that the minimal value of s_T^2/A_T is reached when T has the shape of an ellipse (see Appendix G). In this case, $\min(s_T^2/A_T) = \pi/2$. We now compare the scatter per unit of area to its minimum and obtain the relative scatter:

$$ {}^rs^2 = (s_T^2/A_T)\colon(\pi/2) = 2\pi s_T^2/A_T = 4\pi(\sigma_{M,T}\sigma_{m,T})/A_T \tag{11.3}$$

This parameter measures the component of looseness that does not depend on oblongity. Obviously,

$$ {}^rd_T^2 = {}^rs_T^{2\,r}b_T^2 \tag{11.4}$$

For figures of territories that can each be obtained from one another by affine transformation, relative scatter is constant. This can be shown by considering two affinely equivalent figures of territories T_1 and T_2. The coordinates along the conventional axes of the deviations from the mean center of the corresponding elements i in T_1 and T_2 are denoted by

$$ x_{2i} = hx_{1i} \quad \text{and} \quad y_{2i} = ky_{1i} \tag{11.5}$$

where h and k are positive real coefficients (constant for all the elements). Accordingly:

$$ \sigma_{x,2}^2 = h^2\sigma_{x,1}^2 \qquad \sigma_{y,2}^2 = k^2\sigma_{y,1}^2 \qquad \operatorname{cov}(x_2, y_2) = hk\operatorname{cov}(x_1, y_1) \tag{11.6}$$

For the areas of T_2 and T_1, we find

$$ A_2 = hkA_1 \tag{11.7}$$

From equations (10.28a) and (10.28b),

$$ \sigma_{x,2}^2\sigma_{y,2}^2/A_2^2 = [\sigma_{x,2}^2\sigma_{y,2}^2 - \operatorname{cov}^2(x_2,y_2)]/A_2^2 = [\sigma_{x,1}^2\sigma_{y,1}^2 - \operatorname{cov}^2(x_1,y_1)]/A_1^2 \tag{11.8}$$

and thus ${}^r s_2^2 = {}^r s_1^2$ (see Bachi and Samuel-Cahn, 1976). We can define affinal equivalent territories as territories having the same structure, so relative scatter can be considered a measure of structural looseness: for ellipses ${}^r s^2 = 1$ while for any nonelliptical territory ${}^r s^2 > 1$.

For a clearer understanding of the effects of affine transformation, let us consider the special case of diagonal transformation of a nonoblong territory T_1, where ${}^r b^2(T_1) = 1$ and ${}^r d^2(T_1) = {}^r s^2(T_1)$. Rotating T_1 around its center, to obtain a territory T_2, will not affect either the area or the relative scatter $[A(T_2) = A(T_1)$; ${}^r s^2(T_2) = {}^r s^2(T_1)]$. However, if we induce oblongity, thus obtaining a territory $T_3[{}^r b^2(T_3) \neq 1]$ over the rotated territory, we find that the affinely equivalent territory T_3 has ${}^r d^2(T_3) \neq {}^r d^2(T_1)$ owing to the effects of oblongity $[{}^r d^2(T_3) = {}^r s^2(T_1){}^r b^2(T_3)$ as compared to ${}^r d^2(T_1) = {}^r s^2(T_1)]$.

There are simple equations for calculating s_T^2 and A_T for simple geometrical figures (see Section 11.4.1) and we can thus obtain ${}^r s_T^2$ for them as well; as we can induce ${}^r b_T^2$ at will, we can obtain complete a priori knowledge of ${}^r d_T^2$ and its factors. This approach can be extended to other "regular" territories whose shape can be determined mathematically (see Section 11.4.2), but for "irregular" territories (as most actual territories are) such a priori information is not available. However, we can calculate ${}^r s_T^2$, ${}^r b_T^2$, and ${}^r d_T^2$ on the basis of coordinates of the elements of T or of the map of T. We will discuss below some general rules that may help in interpreting these parameters (see Section 11.5). However, it should be stressed that while territories that have the same structure have the same ${}^r s_T^2$, the reverse is not necessarily true, i.e., territories with the same ${}^r s_T^2$ may have different structures.

11.4. Relative Scatter of Regular Geometrical Figures

11.4.1. Relative Scatter of Regular Polygons

The relative scatter for regular (equilateral and equiangular) polygons with n sides can be obtained very easily. Standard deviations, σ, along all the axes through the center are equal. Relative oblongity is measured by ${}^r b^2 = 1$ and, consequently, $d^2 = s^2 = 2\sigma^2$. Let us denote the length of a side of the polygon by L and the number of sides of the polygon by n. It can be shown that

$$s^2 = d^2 = (L^2/24)[1 + 3\cot^2(180°/n)] \tag{11.9}$$

The area is $A = (n/4)L^2 \cot(180°/n)$, and thus the scatter per unit of area is

$$[\tan(180°/n) + 3\cot(180°/n)]/12n$$

Accordingly, the relative scatter ${}^r s^2 = {}^r d^2$ of the polygon is

$$\pi[\tan(180°/n) + 3\cot(180°/n)]/3n = \pi\{[\tan(180°/n)/3] + \cot(180°/n)\}/n$$

We thus arrive at the important result that the relative scatter of regular polygons depends only on the number of sides n. From Table 11.1, which shows the relative scatter for selected polygons, we can see that:

1. The relative scatter for regular polygons decreases when the number of sides

Table 11.1. Relative Scatter for Selected Regular Polygons

Polygon	Number of sides	Relative scatter
Equilateral triangle	3	1.209275
Square	4	1.047198
Pentagon	5	1.016975
Hexagon	6	1.007701
Heptagon	7	1.003890
Octagon	8	1.002272
Nonagon	9	1.001414
Decagon	10	1.000912
Dodecagon	12	1.000447
Eicosagon	20	1.000193
	100	1.000053
Circle	∞	1.000000

increases, reaching its lowest limit (${}^{r}s^2 = 1$) when the number of sides is infinite (a circle).

2. The decrease of the relative scatter for a polygon with n sides to a polygon with $(n + 1)$ sides is large in going from an equilateral triangle (${}^{r}s^2 = 1.2092$) to a square (${}^{r}s^2 = 1.0472$), but becomes gradually smaller as the number of sides increases. For $n = 7$ the value of ${}^{r}s^2$ is already very near to 1.
3. All regular polygons with even n have a relative scatter limited to a very narrow range, $1.000 \leqslant {}^{r}s^2 \leqslant 1.0472$.

11.4.2. Relative Scatter of Figures Affinely Equivalent to Regular Polygons

Figure 11.2a shows an equilateral triangle with one of its sides parallel to the conventional axis of the abscissas. A horizontal line L_1 parallel to this axis is drawn through the center. A vertical line V_1 perpendicular to L_1 is also drawn through the center. Figure 11.2b shows a triangle T_2 that has been obtained by rotating T_1 around the center by an arbitrary angle $\alpha = 28°$. Lines L_1 and V_1 have also been transformed into L_2 and V_2 by a rotation of 28°. For both T_1 and T_2 we obtain ${}^{r}d^2 = {}^{r}s^{2r}b^2 = {}^{r}s^2 = 1.2093$ (as there is no oblongity). In Figure 11.2c the length of axis L_2 has been doubled ($L_3 = 2L_2$) while the length of V_2 has been halved ($V_3 = 0.5V_2$) resulting in triangle T_3.

This operation (diagonal transformation) has left the area of the triangle unchanged ($A_3 = A_2 = A_1$), but axes L_3 and V_3 have become principal axes of T_3. Their relative oblongity is measured by ${}^{r}b^2 = 0.5[(0.5/2) + (2/0.5)] = 2.125$. So ${}^{r}d^2 = 1.2093 \times 2.125 = 2.5698$.

Triangle T_3 is thus described by α, ${}^{r}b^2$, and the fixed value of ${}^{r}s^2 = 1.2093$, which is true of all triangles. Similarly, by knowing α, ${}^{r}b^2$, and ${}^{r}s^2 = 1.00$ we can describe any ellipse, while any parallelogram can be characterized by α, ${}^{r}b^2$, and ${}^{r}s^2 = 1.0472$, and so on.

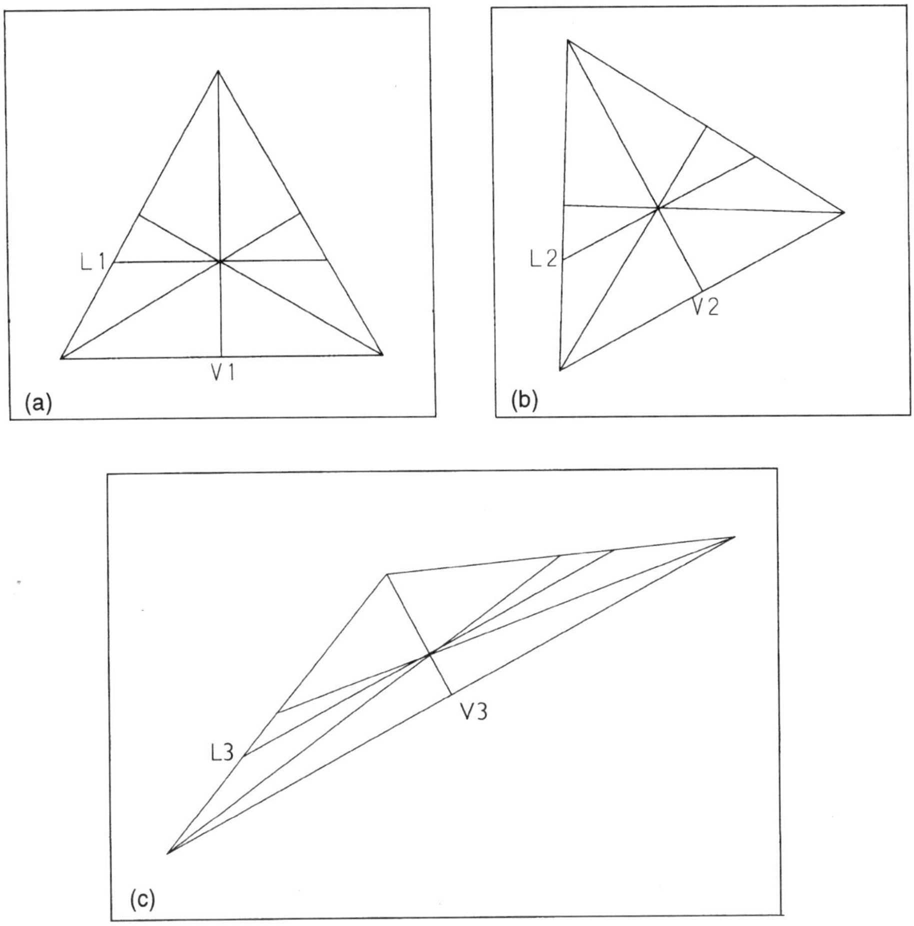

Figure 11.2. Relative scatter of figures affinely equivalent to regular polygons: (a) Triangle T_1; (b) Triangle T_2; (c) Triangle T_3.

11.4.3. Relative Scatter of Isosceles Trapezoids

Because the gap between the relative scatter of triangles and parallelograms is large, it is convenient to obtain ${}^r s^2$ for intermediate figures. Let us consider an isosceles nonoblong trapezoid of fixed area conventionally chosen as equal to 1. Let us indicate the length of the shorter and the longer basis by b and B, respectively, and the ratio between these lengths by $k = b/B$. It can be shown that the relative scatter for any isosceles trapezoid is given by

$$ {}^r s^2 = \frac{2\pi\sqrt{(1 + k^2)(1 + 4k + k^2)}}{3\sqrt{3}\,(1 + k)^2} \tag{11.10} $$

Table 11.2. Relative Scatter for Selected Isosceles Trapezoids

k $(=b/B)$	Relative scatter	$k(=b/B)$	Relative scatter
0.0	1.20928		
0.1	1.1926	0.6	1.0681
0.2	1.1616	0.7	1.0578
0.3	1.1304	0.8	1.0515
0.4	1.1039	0.9	1.0482
0.5	1.0832	1.0	1.0472

Table 11.2 gives the values of the relative scatter for selected values of k. We see that the highest value of ${}^{r}s^2$ $(2\pi/3\sqrt{3} = 1.2092)$ is obtained for $k = 0$, when the trapezoid degenerates into a triangle. The lowest value corresponds to $k = 1$, when the trapezoid becomes a parallelogram ${}^{r}s^2 = \pi/3 = 1.0472$. Here, too, we can fix α and ${}^{r}b^2$ arbitrarily and thus obtain the value of ${}^{r}d^2$ for any figure obtainable by affine transformation of isosceles trapezoids.

11.4.4. Relative Scatter of Other Simply Connected Figures

The examples of ${}^{r}d^2$, ${}^{r}s^2$, and ${}^{r}b^2$ given above relate to elementary geometrical figures that are not only simply connected, but also convex. There have been other applications to simply connected figures—convex and not convex, symmetrical and asymmetrical (Bachi and Samuel-Cahn, 1976)—but we shall not discuss them here as they relate to figures that do not have the same nature as territories, but rather represent theoretical statistical distributions, having values of a variable along the X-axis, and a frequency or probability distribution along the Y-axis. The possibility of combining two of these distributions—above and below the X-axis—to obtain models having some utility in geostatistics has been explored, but the approach has not been sufficiently developed to warrant presenting conclusions.

11.5. Parameters of Looseness for Territories with Concavities: Examples for Italy

11.5.1. Parameters of Looseness for Territories with Internal Concavities

We will go on now to study the behavior of parameters of looseness for territories with concavities. We have already presented examples of various types of territories with internal concavities, such as the Sea of Galilee in northern Israel, in Figure 6.26a (Section 6.7), with external concavities such as the United States in Figure 6.17 (Section 6.6), and fragmented such as Italy in Figure 1.1 (Section 1.2).

Here we shall approach this research by comparing the looseness of the territory studied to both the looseness of the concavities and the looseness of the territory if those concavities are eliminated. Further on we shall discuss measures of the impact of fragmentation.

We begin by considering the setting in which there are concavities or void areas such as lakes or deserts inside a well-defined territory. Hundreds of examples of this setting were studied for the *Statistical Atlas of Italy* and were found to be very meaningful. In the *Atlas*, the territory of Italy was divided into the habitat and the nonhabitat of many phenomena (see Figures 9.5 and 9.8 in Section 9.4). Examples of habitat and nonhabitat of cities are given in Figures 3.8 (Section 3.2) and 11.3 (on the basis of data on individual or aggregated locations). The setting discussed here can be broadened by indicating that the area of the territory is exhaustively subdivided into two classes of zones.

In the examples given in the *Atlas*, the zones of the habitat have been represented in black, while those of the nonhabitat have been left white. Adopting the symbols used in the *Atlas*, we denote the territory under study by T; the habitat and the nonhabitat by H and NH; the proportion of the area of the nonhabitat (internal concavities) over the area of T by $\Delta_{NH}^{T} = A_{NH}/A_T$ (shortened here as Δ); the relative distance-variances by ${}^{r}d_T^2$, ${}^{r}d_H^2$, and ${}^{r}d_{NH}^2$; the relative squared oblongities by ${}^{r}b_T^2$, ${}^{r}b_H^2$, and ${}^{r}b_{NH}^2$; the relative scatters by ${}^{r}s_T^2$, ${}^{r}s_H^2$, and ${}^{r}s_{NH}^2$; and the squared distance between the centers of any pair of territories by ${}^{c}D^2$.

We know that

$$d_T^2 = (1 - \Delta)[d_H^2 + {}^{c}D^2(H, T)] + \Delta[d_{NH}^2 + {}^{c}D^2(NH, T)] \qquad (11.11)$$

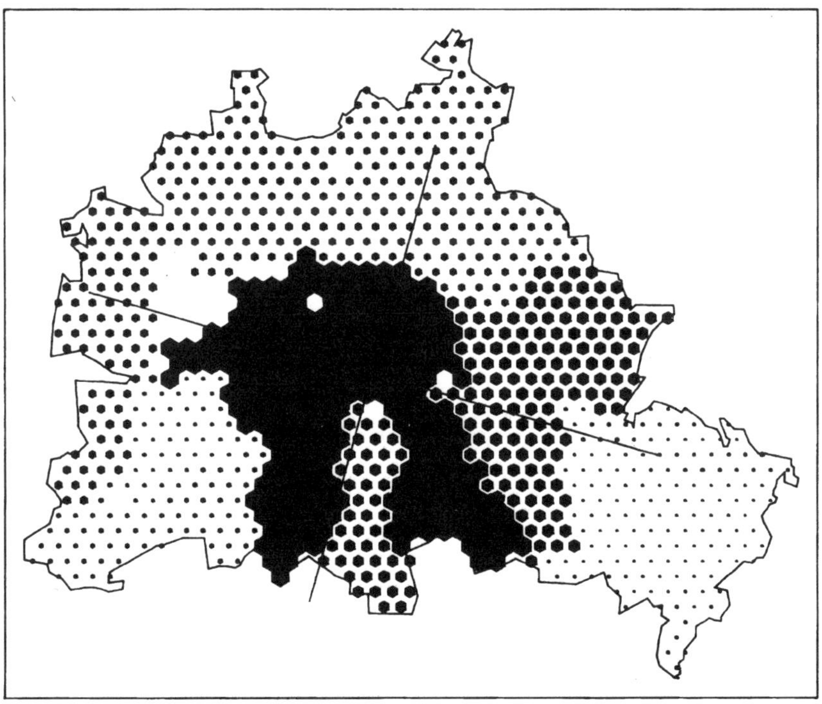

Figure 11.3. Berlin. Habitat and nonhabitat of total population, 1933.

Table 11.3. Relative Distance-Variance and Its Determinants for the Habitats of Population, Sheep, and Citrus in Italy

	Total territory	Population 1987	Citrus	Sheep
Relative distance-variances of:				
(a) Habitat		5.50	10.42	4.54
(b) Total territory	2.91			
(c) Nonhabitat		6.14	2.69	5.31
Total area (sq km)	301,278			
Percentage of area of nonhabitat		46.8	85.3	49.7
Distance between centers of nonhabitat and total territory (km)		18.6	77.5	147.7
Relative scatter for habitat		4.05	9.81	4.17
Relative square oblongity for habitat		1.36	1.06	1.09
Relative scatter for total territory	2.31			
Relative oblongity for total territory	1.26			
Relative scatter for nonhabitat		5.19	2.23	3.85
Relative square oblongity for nonhabitat		1.18	1.20	1.38

and it can be shown that

$$(1-\Delta)[{}^{c}D^{2}(H,T)+\Delta[{}^{c}D^{2}(NH,T)]=[\Delta/(1-\Delta)^{3}]{}^{c}D^{2}(NH,T)/A_{T} \quad (11.12)$$

and

$${}^{r}d_{H}^{2}={}^{r}d_{T}^{2}/(1-\Delta)^{2}-[\Delta/(1-\Delta)]{}^{r}d_{NH}^{2}-[\Delta/(1-\Delta)^{3}]2\pi{}^{c}D^{2}(NH,T)]/A_{T} \quad (11.13)$$

Thus, we see that the relative distance-variance of H is determined by the relative distance-variance of T (which is constant for the habitats for the 385 geosets in the *Atlas*), by the proportion of area of the NH, and by the distance between the centers of NH and of T, standardized through division by area A_T.

Ceteris paribus,* the relative distance-variance of the habitat increases: (a) with increasing proportion of the area of the nonhabitat, (b) with decreasing looseness of the nonhabitat, and (c) with decreasing distance between the mean centers of NH and T. Table 11.3 shows the relative distance-variance of the habitats of the geosets for population, sheep, and citrus in Italy, which are shown in Figures 1.5 (Section 1.2) and 9.5 and 9.8 (Section 9.4). The table also gives the values of ${}^{r}d_{NH}^{2}$, Δ, and ${}^{c}D^{2}(NH,T)$, with which we can obtain the actual value of ${}^{r}d_{H}^{2}$ indirectly using equation (11.13). If we assume that ${}^{r}d_{T}^{2}$ is fixed, we can standardize d_{H}^{2} by dividing it by d_{T}^{2}. We then obtain

$${}^{r}d_{H}^{2}/{}^{r}d_{T}^{2}=({}^{r}s_{H}^{2}/{}^{r}s_{T}^{2})({}^{r}b_{H}^{2}/{}^{r}b_{T}^{2}) \quad (11.14)$$

where ${}^{r}s_{H}^{2}/{}^{r}s_{T}^{2}=(s_{H}^{2}/s_{T}^{2}):(A_{H}/A_{T})$.

*Here and in Sections 11.5.2 and 11.e5.3 *ceteris paribus* may not be feasible in practice.

11.5.2. Analysis of Factors of Looseness of Habitats and Nonhabitats for 385 Geosets of Italy

The ratio d_H^2/d_T^2 has been found to depend directly on an oblongity factor ${}^r b_H^2/{}^r b_T^2$ and a dispersion factor s_H^2/s_T^2 and inversely on the territorial coverage of H, namely A_H/A_T.

In the analysis of the looseness of the habitat of many phenomena over the territory of Italy, it has been found of interest to eliminate the impact of the looseness of the total territory, s_T^2, which is constant, from each ${}^r s_H^2$. Calculating the ratio of these parameters, we obtain

$$ {}^r s_H^2/{}^r s_T^2 = (s_H^2/s_T^2):(A_H/A_T) \tag{11.15} $$

where s_H^2/s_T^2 is the ratio of the scatter of the habitat to the scatter of the entire territory and A_H/A_T is the territorial coverage of the habitat. Similarly, for the nonhabitat we have

$$ {}^r s_{NH}^2/{}^r s_T^2 = (s_{NH}^2/s_T^2):(A_{NH}/A_T) \tag{11.16} $$

This split has been found to be of considerable help in applied research. Table 11.4. gives examples for the territorial geosets listed in Table 11.3.

Ranking the values of ${}^r s_H^2/{}^r s_T^2$, we observe that they are countergraduated with respect to the rankings of the territorial coverage. This finding, based on a few cases, suggests that smallness of territorial coverage may play a very important role in raising the relative scatter.

Table 11.5 shows averages and measures of variability of the looseness parameters for the habitats and nonhabitats of the 385 geosets analyzed in the *Atlas*. We can see that:

1. The relative distance-variances and the relative scatters have great variability and appear to be very sensitive parameters. This is true for the nonhabitats and even more so for the habitats.
2. Relative distance variance and relative scatter often reach very high values, exceeding by far those of the territory of Italy.

Table 11.4. Ratios of Relative Scatters of Habitats and Nonhabitats to Relative Scatter of the Total Territory in Italy

	Population 1987	Citrus	Sheep
Ratio between scatters of habitat to total territory	1.757	4.247	1.805
Ratio between scatters of habitat and total territory	0.936	0.625	0.910
Territorial coverage of habitat	0.532	0.147	0.503
Ratio between relative scatters of nonhabitat and total territory	2.247	0.965	1.667
Ratio between scatters of nonhabitat and total territory	1.048	0.824	0.828
Territorial coverage of nonhabitat	0.468	0.853	0.497

Table 11.5. Averages and Variability of Parameters of Looseness for Habitats and Nonhabitats in Italy

	Arithmetic average	Standard deviation	Coefficient of variation[a]	1st quartile	3rd quartile	90th decile
Habitats						
${}^r d_H^2$	6.53	3.07	0.471	5.13	7.06	9.19
${}^r s_H^2$	4.92	2.33	0.474	3.79	5.26	6.79
${}^r b_H^2$	1.33	0.08	0.060	1.31	1.38	1.40
${}^r d_H^2/{}^r d_t^2$	2.244	1.06	0.472	1.76	2.43	3.16
${}^r s_H^2/{}^r s_T^2$	2.130	1.008	0.474	1.641	2.277	2.939
${}^r b_H^2/{}^r b_T^2$	1.056	0.064	0.060	1.039	1.095	1.111
s_H^2/s_T^2	0.851	0.155	0.182	0.798	0.940	0.991
A_H/A_T	0.448	0.149	0.332	0.361	0.538	0.630
Nonhabitats						
${}^r d_{NH}^2$	5.56	1.84	0.330	4.41	6.23	7.57
${}^r s_{NH}^2$	4.65	1.55	0.332	3.66	5.27	6.55
${}^r b_{NH}^2$	1.20	0.060	0.050	1.18	1.22	1.25
${}^r d_{NH}^2/{}^r d_T^2$	1.911	0.632	0.331	1.516	2.141	2.601
${}^r s_{NH}^2/{}^r s_T^2$	2.013	0.671	0.333	1.584	2.281	2.836
${}^r b_{NH}^2/{}^r b_T^2$	0.952	0.048	0.050	0.936	0.968	0.992
s_{NH}^2/s_T^2	1.027	0.105	0.102	1.005	1.087	1.113
A_{NH}/A_T	0.552	0.148	0.268	0.463	0.639	0.762

[a]Ratio of standard deviation to average.

These findings can be explained as indicated under (3), (4), and (5) with reference to the factors of relative looseness noted previously:

3. Oblongity may have some influence in many distributions considered separately. However, dealing with broad collections of geosets, we find that the average of the relative oblongity of the habitats only slightly exceeds that of the total territory, while for the nonhabitat the relative oblongity is generally slightly lower than that of the territory. We can conclude that both for habitats and nonhabitats the role played by relative oblongity is a very minor one. Under these circumstances, the average values of relative distance-variances are found to be very near to those of the relative scatter, and we can concentrate on the relative scatter.
4. As already noted, the values of the relative scatter are comparatively very high for both habitats and nonhabitats.
5. We have seen above that in dealing with any individual geoset, the ratio of its relative scatter of H (or NH) to that of the territory can be transformed into the ratio between a dispersion factor s_H^2/s_T^2 and the territorial coverage A_H/A_T.

Dealing with our collection of 385 geosets, we find that, on the average, the ratio s_H^2/s_T^2 has a value of 85% and for the majority of the geosets it ranges between 80 and 99%. For the nonhabitats, most of the scatters only slightly exceed the scatter of the territory. Territorial coverage shows a very different picture: for the habitats it averages only 45% and is strongly variable, while for the nonhabitats it averages 55% and exceeds 76% for only 10% of the geosets. The leading factor of the high values of relative scatter is thus the generally low territorial coverage of habitats and nonhabitats. This is confirmed by the high inverse correlations between ${}^{r}s_{H}^{2}$ and A_H/A_T and between ${}^{r}s_{NH}^{2}$ and A_{NH}/A_T.

11.5.3. Extension of the Findings Concerning Looseness to Other Subterritories

We have just looked at the looseness of two subterritories, H and NH, into which a territory T is divided. We can extend the methods suggested to the setting in which a territory T is subdivided into several or even many subterritories T_Z. The subdivision can be along political, administrative, socioeconomic, geographical-physical, meteorological, or other lines. It is understood that each T_Z can be defined over a map of T.

There were extensive applications of this approach in the *Statistical Atlas of Italy*, where by comparing pairs of geosets ${}^{g}P$ and ${}^{g}Q$, exhausive divisions of the territory T of Italy in the following four parts have been obtained (as explained in Section 3.3.4): (a) subterritory $TM(P, Q)$ over which the matches between ${}^{g}P$ and ${}^{g}Q$ are found; (b) subterritories $TSPQ$ and (c) $TSQP$ corresponding, respectively, to the surpluses of ${}^{g}P$ over ${}^{g}Q$ and of ${}^{g}Q$ over ${}^{g}P$; and (d) void subterritory $TV(P, Q)$ in which there are no elements of ${}^{g}P$ or ${}^{g}Q$. Table 11.6 shows averages of relative distance-variances calculated over all these subterritories, which are analyzed in the *Atlas*. The territory $TSPU$, corresponding to the surpluses of ${}^{g}P$ over the uniform distribution ${}^{g}U$, is also taken into account.

Table 11.6. Average and Variability of Relative Distance-Variances for Subterritories Obtained by Comparisons between Pairs of Geosets for Italy[a]

Subterritory corresponding to	Average	Standard deviation	Coefficient of variation	1st quartile	3rd quartile	90th decile
Matches						
$TM(P, Q)$	8.09	4.83	0.596	5.97	8.53	10.89
Surpluses of P over Q						
$TSPQ$	52.40	50.96	0.972	20.31	67.65	110.51
Surpluses of Q over P						
$TSQP$	40.80	44.67	1.095	16.50	48.12	87.55
Void						
$TV(P, Q)$	7.56	3.63	0.480	6.23	7.52	9.99
Surpluses of P over U						
$TSPU$	21.32	8.08	0.379	16.68	25.06	28.43

[a]For $TSPU$ the parameters are given for 385 territorial geosets. For other territories the parameters are for 177 territorial geosets.

Table 11.7. Averages of Factors of Relative Distance-Variances for Subterritories of Table 11.6

Subterritories corresponding to	Ratios of parameters for the subterritory to corresponding parameters for the territory of Italy				
	Relative distance-variance	Relative scatter	Relative oblongity	Scatters	Area (territory covered)
Matches					
TM(*P*, *Q*)	2.78	2.58	1.08	0.89	0.394
Surpluses of *P* over *Q*					
TSPQ	18.01	17.77	1.00	0.80	0.076
Surpluses of *Q* over *P*					
TSQP	14.02	13.81	1.03	0.83	0.111
Void					
TV(*P*, *Q*)	2.60	2.77	0.94	1.05	0.420
Surpluses of *P* over *U*					
TSPU	7.33	6.60	1.10	0.72	0.113

We can see from this table that relative distance-variances of the subterritories have a very strong variability and can reach very high values and that this parameter is especially sensitive for territories corresponding to surpluses (*TSPQ*, *TSQP*, *TSPU*). In order to explain these findings, Table 11.7 shows the averages of the ratios of the looseness parameters for the subterritories, standardized by comparison to the corresponding parameters for the total territory of Italy. It appears that the impact of oblongity is very low. Consequently, the ratios for relative distances-variances are almost identical to the ratios for relative scatters. We can thus concentrate on the ratios for relative scatters, where the dominant factor is the extremely variable territorial coverage (see detailed data in Section 8.7).

These findings are in line with those for habitats and nonhabitats discussed in Section 11.5.2.

11.5.4. Parameters of Looseness for Territories with External Concavities

We consider here a simply connected territory T_1, which is not convex, as it has external concavities, and we wish to measure the impact of these concavities on its looseness. To do so, we must define the concavities and construct a model that includes both T_1 and the concavities.

In Section 11.5.1, a somewhat similar problem was solved simply by defining the concavities as gaps internal to the contour of the total territory. The contour of the territory was also clearly defined and included—according to the setting under consideration, e.g., habitat and nonhabitat, water area and land area, or desert area and nondesert area.

In the study of external concavities we do not have an a priori definition of the concavities so that a convention for this definition is required. An acceptable solution seems to be that of drawing around T_1 the convex hull, i.e., the convex figure with minimal area that includes T_1. For example, Figure 4.17 in Section 4.6 (taken from the second volume of the *Statistical Atlas of Italy*, p. 62) shows the convex hulls drawn around each of the simply connected major fragments of the territory of Italy (Continental Italy, Sicily, Sardinia). We focus now on Sardinia and on the convex hull (defined in Section 4.2) drawn around it.*

Let us look at the convex hull around Sardinia as shown in Figure 4.17. The territory of the island, which is labeled T_1 and left blank on the map, constitutes the main object of our study here. The external concavities included between the convex hull and T_1 are marked by vertical lines. We label them as T_2, despite the fact that they are actually in the sea. The convex hull T_3 includes both T_1 and T_2. The parameters of T_1, T_2, and T_3 are denoted by the usual symbols, followed by 1, 2, or 3. The proportion of the area of the concavities, T_2, over that of the territory, T_1, is denoted by h. It can be shown that

$$^r d_1^2 = (1 + h)^{2r} d_3^2 - h^{2r} d_2^2 - [(h/(1 + h)][2\pi^c D^2(\tfrac{1}{2})/A_1] \qquad (11.17)$$

where

$$^r d_3^2 = {}^r s_3^{2r} b_3^2 \qquad \text{and} \qquad {}^r d_2^2 = {}^r s_2^{2r} b_2^2 \qquad (11.18)$$

Equations (11.17) and (11.18) can be useful in constructing theoretical models and in interpreting the empirical values of $^r d_1^2$ and $^r s_1^2$. We see that *ceteris paribus* the relative distance-variance and the relative scatter of a territory with external concavities increase: (a) with an increase in the proportion h of the area of the concavities over the area of the territory; (b) with a decrease in the looseness of the concavities; or (c) with a decrease in the standardized distance between the center of the concavities and the center of the territory.

Table 11.8 shows the parameters for Sardinia that are relevant for equations (11.17) and (11.18), as well as the parameters for Sicily and Continental Italy. We see that, owing to the elongated shape of the island, the oblongity of T_1, measured by $^r d_1^2$, is rather high. The same applies to T_2 and T_3. As $^r b_1^2$, $^r b_2^2$, and $^r b_3^2$ are rather similar, the main determinant of the enormous differentiation between the relative distance-variance of T_2 and that of T_1 and T_3 is the structural looseness, measured by $^r s_2^2$ as compared to $^r s_1^2$ and $^r s_3^2$. As might be expected, the structural looseness of the convex hull ($^r s_3^2 = 1.02$) is very low and quite near to that of a regular pentagon. On the other hand, the structural looseness of the geoset of the external concavities (considered together) is very high ($^r s_2^2 = 5.94$). This is explained by the facts that: (a) the concavities are very dispersed, forming a sort of fringe around large sections of the coast of Sardinia; and (b) they cover a rather small percentage of the area (only 29.3%) of the island.

Under the impact of these concavities the relative scatter of Sardinia is 1.08 as compared to 1.02 of the convex hull. We know that convex figures have an $^r s^2$ of less than 1.2093, but this does not imply that every figure with $^r s^2 < 1.2093$ is convex.

*Had more accurate work been requested, the problem might have arisen as to how to deal with small islands that belong to Sardinia and have been neglected in its convex hull. A possible approach to the solution of this problem is as follows: rank these small islands by area, excluding those that are so small as to be negligible; check for each of the others the effect they might possibly have on the convex hull and exclude those that may be considered "outliers," for they would unduly enlarge the area of the hull by including considerable tracts of sea within it.

Table 11.8. Relative Distance-Variance and Its Determinants for the Territories of Continental Italy, Sicily, and Sardinia

	Continental Italy	Sicily	Sardinia
Relative distance variances for the:			
Territory considered (land) (T_1)	2.69	1.53	1.38
Concavities (water) (T_2)	2.62	5.84	7.04
Convex hull (land and water) (T_3)	1.40	1.26	1.22
Area of territory considered (sq km)	251,479	25,709	24,090
Proportion of area concavities to area of the territory considered (%)	97.22	34.10	29.28
Distance between the center of the territory considered and the center of its concavities (km)	15.28	40.62	31.05
Relative scatter for the territory T_1	1.61	1.28	1.08
Relative oblongity for T_1	1.67	1.20	1.27
Relative scatter for the concavities T_2	2.29	5.37	5.94
Relative oblongity for T_2	1.14	1.09	1.19
Relative scatter for the convex hull T_3	1.04	1.11	1.02
Relative oblongity for T_3	1.34	1.11	1.20

It implies only that external concavities, if they exist, have limited influence and that the territory can like Sardinia be considered rather compact.

For Sicily, too, the oblongities of T_1, T_2, and T_3 are rather similar. The main factor in the large differentiation between the relative distance-variances of T_2 as compared to T_1 and T_3 is structural, as measured by ${}^r s_2^2$, as compared to ${}^r s_1^2$ and ${}^r s_3^2$.

The relative scatter of the concavities around Sicily is very high (owing to the fact that they have large dispersion and cover only a small area). However, the impact of the concavities over the relative scatter of the island is rather small, as shown by the comparison between ${}^r s_3^2 = 1.11$ and ${}^r s_1^2 = 1.28$. The territory of Sicily can also be considered as rather compact as relative scatter ${}^r s_3^2 = 1.28$ would exceed only slightly that of a triangle.

Considering the convex hull around the territory of continental Italy, we find a value for relative scatter ${}^r s_3^2 = 1.04$, which is not far from that of a parallelogrm, and a very strong oblongity ${}^r b_3^2 = 1.34$. The proportion h of the area of the concavities over the area of the territory is very high, which has a strong potential impact for raising the standard looseness of the territory, ${}^r s_1^2$. Strong oblongity contributes to increasing ${}^r d_1^2$. On the other hand, the high relative scatter of the concavities works in the opposite direction. The final result is a relative distance-variance for Continental Italy (${}^r d_1^2 = 2.69$) higher by far than those found for Sardinia and Sicily.

11.5.5. Parameters of Looseness for Fragmented Territories

Let us consider a territory T formed by clearly defined, separated fragments, and let us take again as an example the territory of Italy, T, formed by three major fragments: Continental Italy, Sicily, and Sardinia. Considering T as a compound

Table 11.9. Relative Distance-Variance and Its Determinants for the Total Fragmented Territory of Italy

	Total territory of Italy (land)	Gap between fragments (water)	Convex hull (land and water)
Relative distance-variance	2.91	1.66	1.11
Relative scatter	2.32	1.62	1.02
Relative oblongity	1.25	1.02	1.09
Percentage of area of convex hull	40.6	59.4	100
Distance between the center of the interfragmentary gap for the center of T (km)		86.7	

geoset and the three fragments as components, we can calculate geostatistical-parameters for T (see Section 10.4). Table 11.9 shows the distance-variance, the scatter, and the oblongity for T calculated in this way.

In order to see the impact of fragmentation over T, we can draw the convex hull CH around T. The water area included between CH and T can be considered as the gap between fragments, GF, resulting from the fragmentation. The calculations are similar to those for the comparison of each fragment to its concavities and convex hull. The relevant parameter obtained for T, its CH, and GF are given in Table 11.9. If we denote the proportion of the area of the interfragmentary gap to the area of territory T by h it can be shown that

$$^rd_T^2 = (1-h)^2 d_{CH}^2 - h^{2r} d_{GF}^2 - [h/(1+h)]2\pi^c D^2(T, GF)/A_T \qquad (11.19)$$

where $^rd_T^2 = {}^rs_T^2 {}^rb_T^2$, $^rd_{CH}^2 = {}^rs_{CH}^2 {}^rb_{CH}^2$, and $^rd_{GF}^2 = {}^rs_{GF}^2 {}^rb_{GF}^2$.

Ceteris paribus the relative distance-variance and the relative scatter of a fragmented territory increase: (a) with an increase in the proportion h of the area of the interfragmentary gap over the area of the territory; (b) with a decrease in the looseness of the interfragmentary gap; or (c) with a decrease in the standardized distance between the center of the interfragmentary gap and the center of T.

Let us compare the parameters of looseness and its determinants for the total fragmented territory of Italy, as given in Table 11.8 to the corresponding parameters for Continental Italy, as given in the first column of Table 11.9. We see that in the former the proportion of water area to land area is higher; the relative scatter of water area is smaller (1.62 instead of 2.29); the standardized squared distance between the centers of the water area and land areas is higher; and the oblongity of land area is smaller (1.25 instead of 1.67). The relative scatter of the land area increases from 1.61 for Continental Italy to 2.32 for the total territory of Italy. However oblongity acts strongly in the opposite direction. The final result is that the relative distance-variance for the total territory of Italy (2.91) is not much greater than that of Continental Italy (2.69).

11.5.6. Parameters of Fragmentation Insensitive to the Location of the Fragments

Sometimes we are interested in measuring the fragmentation of a territory T, independently of the location of the fragments. For this purpose, we can adapt a parameter that was introduced by Gini (see Gini, 1939 and earlier papers) for measuring the variability (in Italian: *mutabilita*) of a distribution of qualitative values or attributes. Let us denote the well-defined fragments ($j = 1, 2, \dots, f$) into which T is divided by T_j. We assume that each fragment T_j can be divided into a network of N_j^c hexagonal cells of equal, very small but finite areas. We denote the total number of cells of T by $\Sigma N_j^c = N_T^c$ and the proportion of the area of T_j over the total area of T by a_j, which is roughly proportional to N_j^c/N_T^c.

We then form all the possible N_j^2 pairs of cells with repetition, namely by comparing each cell to itself and to all the other cells. Pairs belonging to the same fragments are denoted by difference 0, while those belonging to different fragments are denoted by difference 1.

Fragmentation can be measured by calculating the proportion of pairs of elements belonging to different fragments. This, in turn, is obtained by calculating

$$Fr = 1 - \sum a_j^2 \tag{11.20}$$

We give below a few theoretical examples of Fr, for a fragmented territory T, divided in different ways:

Divisions into two important fragments:

(1) $a_1 = 0.95;\quad a_2 = 0.05;\qquad Fr = 0.095$
(2) $a_1 = 0.90;\quad a_2 = 0.10;\qquad Fr = 0.18$
(3) $a_1 = 0.80;\quad a_2 = 0.20;\qquad Fr = 0.32$
(4) $a_1 = 0.60;\quad a_2 = 0.40;\qquad Fr = 0.48$
(5) $a_1 = 0.50;\quad a_2 = 0.50;\qquad Fr = 0.50$

Divisions into three important fragments:

(6) $a_1 = a_2 = 0.4;\quad a_3 = 0.2;\qquad Fr = 0.76$
(7) $a_1 = a_2 = a_3 = \frac{1}{3}\qquad Fr = 0.89$

Divisions into many fragments:

(8) $a_1 = a_2 = \cdots = a_{10} = 0.10;\qquad Fr = 0.9$
(9) $a_1 = a_2 = \cdots = a_{20} = 0.05;\qquad Fr = 0.95$
(10) $a_1 = a_2 = \cdots = a_{100} = 0.01;\qquad Fr = 0.99$

11.6. Parameters of Looseness for Primary and Derived Population Geosets

11.6.1. Looseness of Primary Population Geosets

The concepts and equations of the parameters of looseness of territories and subterritories can be extrapolated to measure the looseness of primary and derived population geosets, and extensive applications in the second volume of the *Statistical Atlas of Italy* confirm the feasibility and practical utility of this approach.

We consider any primary population geoset ${}^{g}P$. We can calculate its parameters of absolute dispersion (distance-variance d_P^2; scatter s_P^2). Moreover, we assume that we know the area A_H of the net territory (or habitat) over which the population geoset is actually distributed. This enables us to calculate the distance-variance d_P^2/A_H and scatter s_P^2/A_H of ${}^{g}P$ per unit of area of net territory, which is justified by the fact that d_P^2 and s_P^2 measure territorial dispersion of ${}^{g}P$ and can thus be compared to the area of the net territory over which P is found.

In the analysis of the relative dispersion of territories, we calculate the dispersion and scatter of territory per unit of area, d_T^2/A_T and s_T^2/A_T, for any territory T. To make the interpretation easier, we multiply these ratios by a fixed coefficient 2π; and so obtain ${}^{r}d_T^2 = 2\pi d_T^2/A_T$ and ${}^{r}s_T^2 = 2\pi s_T^2/A_T$, where ${}^{r}s_T^2 = 1$ for elliptical territories and ${}^{r}s_T^2 > 1$ for all others.

We can also use 2π to facilitate the reading and interpretation of the scatter of population per unit of area, ${}^{r}s_P^2$, which is called the relative scatter of population. Similarly, ${}^{r}d_P^2 = 2\pi d_P^2/A_H$ is the relative distance-variance of the population. In the interpretation of ${}^{r}s_P^2$ and of ${}^{r}d_P^2 = b_P^2 \cdot {}^{r}s_P^2$, we must be aware of the following: Let us write ${}^{r}s_P^2 = (s_P^2/s_H^2) - {}^{r}s_H^2$. If the population geoset ${}^{g}P$ is uniformly distributed over a net territory (habitat) having the shape of an ellipse, then ${}^{r}s_H^2 = 1$ and $s_P^2/s_H^2 = 1$ and thus ${}^{r}s_P^2 = 1$ as well. However, ${}^{r}s_P^2 = 1$ is not necessarily a minimal value for ${}^{r}s_P^2$. If we assume that the net territory is an ellipse but that the population is not uniformly distributed over it, then we can have ${}^{r}s_P^2 < 1$ and even have values approaching 0.

In order to make the interpretation of ${}^{r}s_P^2$ easier, it is convenient to follow a procedure similar to the one we employed earlier in this chapter (Section 11.5.2) for the analysis of the habitat. We start by eliminating the impact of the relative distance-variance of the gross predetermined territory T over which ${}^{g}P$ is distributed from the relative distance-variance of the population geoset ${}^{g}P$. We then obtain

$$ {}^{r}d_P^2/{}^{r}d_T^2 = {}^{r}b_P^2({}^{r}s_P^2/{}^{r}s_T^2)/(A_H/A_T) \tag{11.21} $$

Thus we see that, *ceteris paribus*, the relative distance-variance of ${}^{g}P$ increases with increasing oblongity and scatter of ${}^{g}P$ and with decreasing net territorial coverage A_H/A_T.

We used this approach earlier in the study of the relative distance-variance of several population geosets for Italy (see Section 1.5.6), where the findings could be judged to correspond well to what might be expected on the basis of relevant maps of the geosets considered and knowledge of their characteristics.

Table 11.10. Averages and Variability of Parameters of Looseness for 385 Primary Population Geosets for Italy[a]

Parameters[a]	Average	Standard deviation	Coefficient of variation[b]	1st quartile	3rd quartile	90th decile
${}^{r}d_P^2$	3.24	0.558	4.55	6.12	7.23	5.82
${}^{r}s_P^2$	4.20	2.33	0.567	3.27	4.55	5.64
${}^{r}b_P^2$	0.16	0.116	1.31	1.45	1.49	1.39
Ratio of ${}^{r}d_P^2$ to ${}^{r}d_T^2$	2.00	1.11	0.555	1.56	2.10	2.48
Ratio of ${}^{r}s_P^2$ to ${}^{r}s_T^2$	1.82	1.01	0.555	1.42	1.97	2.44
Ratio of ${}^{r}b_P^2$ to ${}^{r}b_T^2$	1.10	0.13	0.118	1.04	1.15	1.18
s_P^2/s_T^2	0.75	0.20	0.276	0.64	0.87	0.98
A_H/A_T	0.149	0.332	0.361	0.538	0.630	0.448

[a]Symbols are as follows: ${}^{r}d^2$ = relative distance-variance; ${}^{r}s^2$ = relative scatter; ${}^{r}b^2$ = relative oblongity; s^2 = scatter; P = population geoset; T = territory of Italy.
[b]Ratio of standard deviation to average.

11.6.2. Analysis of Factors of Looseness for 385 Primary Population Geosets for Italy

We can obtain a wider overview of the behavior of the parameters of looseness of population geosets by considering together all of the 385 geosets studied in the *Statistical Atlas of Italy* at the same time. Table 11.10 shows averages and measures of variability of the looseness parameters for these geosets. In presenting the main conclusions that stem from these values, we can follow the same line as in Section 11.5.2 with regard to parameters of looseness of habitats. This is justified by the similarity of the problems studied as well as by the high correlations found between parameters of looseness of each population geoset and those of the corresponding habitat.

We see from Table 11.10 that:

1. The relative distance-variance and the relative scatter of primary population geosets often reach very high values, exceeding by far those of the territory of Italy. This coupled with their great variability suggests that these parameters are very sensitive.
2. Considering the entire collection of 385 geosets, oblongity is found to play only a minor role in differentiating the looseness of the geosets. Accordingly, we can focus our attention on the relative scatters.
3. If we standardize the relative scatters of each population geoset by dividing by the relative scatter of the territory of Italy (which is a constant), we can present the result as a ratio between the standardized scatter of the population geoset and its territorial coverage. This simple split does not hold when we consider averages (as in Table 11.10). However, from this table we know that both the coverage and the standardized scatter are rather sensitive parameters.

Table 11.11. Average and Variability of Relative-Distance Variance for Derived Population Subsets for Italy[a]

Derived subsets	Average	Standard deviation	Coefficient of variation[b]	1st quartile	3rd quartile	90th decile
Surpluses of ^gP over uniform distribution	19.20	7.95	0.415	14.18	23.15	25.89
Matches of ^gP and ^gQ	7.55	4.90	0.649	5.62	8.03	9.91
Surpluses of ^gP over ^gQ	46.96	46.02	0.98	17.18	68.18	98.59
Surpluses of ^gQ over ^gP	37.10	39.20	1.057	15.12	46.48	79.52

[a]For surpluses of ^gP over uniform distribution the parameters are given for 385 primary geosets ^gP. Other parameters refer to 177 primary geosets ^gP and to 177 primary geosets ^gQ.
[b]Ratio of standard deviation to average.

11.6.3. Analysis of Factors of Looseness for Derived Population Geosets

The above methods were used in the *Statistical Atlas of Italy* to study looseness of population geosets derived from the comparison of a primary population geoset ^gP to the uniform distribution ^gU and to another primary population geoset ^gQ. Table 11.11 shows averages and measures of variability of relative distance-variances for such derived population geosets. The most striking findings are the very high values this parameter reaches, as well as its enormous variability. Table 11.12 enables us to analyze the factors of the very high looseness of derived population geosets: territorial coverage appears here as a dominant factor.

11.7. Parameters of Looseness for Primary and Derived Geosets for the United States

Table 11.13 shows parameters of looseness for the territory and the population of the conterminous United States for some subpopulations and some derived geosubsets, which have been calculated on the basis of data detailed by 3109 counties. Columns (1), (2), and (3) give the relative distance-variance, the relative scatter, and the relative oblongity. Apart from some minor inaccuracy owing to rounding off of the parameters, we find that (1) = (2) × (3).

Table 11.14 shows in columns (4), (5), and (6), respectively, the ratios of the parameters of Table 11.13 to the corresponding parameter for the territory of the conterminous United States. Obviously here, too, apart from minor inaccuracies, we find (4) = (5) × (6). Column (7) gives the ratios between the scatter of the primary or derived population geoset considered and the scatter of the territory, and column (8) the territorial coverage of the geoset (i.e., the proportion of its territory to the total territory of the conterminous United States). We find, apart from minor

Table 11.12. Average of Factors of Relative Distance-Variances for the Population Subsets of Table 11.11

Derived subset	Ratios of parameters for the subset to corresponding parameters for the territory of Italy: Relative distance-variance	Relative scatter	Scatters	Area (territorial cover)
Surpluses of ${}^gP$ over ${}^gU$	6.60	5.72	0.63	0.113
Matches of ${}^gP$ and ${}^gQ$	2.59	2.29	0.80	0.394
Surpluses of ${}^gP$ over ${}^gQ$	12.75	11.93	0.69	0.076
Surpluses of ${}^gQ$ over ${}^gP$	16.14	14.96	0.74	0.111

inaccuracies, that (5) = (7):(8), i.e., that the relative scatter of each geoset, standardized by the relative scatter of the United States, is explained by the standardized scatter divided by the territorial coverage.

Tables 11.13 and 11.14 indicate that:

1. The territory of the conterminous United States is structurally very compact, having a relative scatter of only 1.096. Its relative distance-variance (1.375) is largely due to its strong oblongity (1.254).
2. In many of the geosets considered, oblongity appears to be a factor of considerable importance in determining the relative distance-variance and its variability.
3. Despite this fact, the relative scatters appear as the dominant factor of relative distance-variance of the majority of the geosets and of its great variability.
4. The relative scatter of all the geosets considered is higher than the relative scatter of the territory. The ratios between these two parameters, shown by column (5) in Table 11.14, range between 1 and 2 for four geosets; between 2 and 3 for five geosets, between 4 and 5 for three geosets, and between 5 and 10 for five geosets, and have very high values for four geosets.
5. Very high looseness is found for Asians and Hispanics and for surpluses of actual geosets over uniform distribution and for the corresponding territory. In all these cases the territorial coverage is very low and appears to be the dominant factor in the strong looseness.
6. The looseness is much smaller for nonhabitats than for habitats.

11.8. Parameters of Looseness for Cities

11.8.1. An Almost Circular City: Rome

We now go on to some examples of parameters of looseness for Rome [within the limits of the Grande Raccordo Anulare, as shown by Figures 7.8 (Section 7.1) and 14.7 and 14.11 (Section 14.2)]. The relative scatter of this territory is 1.01 and the relative oblongity is 1.01; taking both factors together, we obtain a relative

Table 11.13. Relative Distance-Variance, Relative Scatter, and Relative Oblongity for the Territory, Population, and Subpopulations of the Conterminous United States

Geoset	Relative distance-variance (1)	Relative scatter (2)	Relative oblongity (3)
Territory	1.375	1.096	1.254
Total population 1980	3.465	2.483	1.395
Total population 1989	3.352	2.426	1.382
Whites 1980	3.655	2.561	1.427
Afro-Americans 1980	4.036	3.086	1.308
Asians 1980	8.857	4.903	1.796
Hispanics 1980	6.406	4.384	1.461
Geosubsets obtained by comparing a population geoset to the uniform distribution			
Total population 1980			
Habitat	3.027	2.332	1.298
Nonhabitat	1.815	1.647	1.102
Surpluses of population	15.192	10.183	1.492
Territory of these surpluses	12.655	9.273	1.365
Afro-Americans			
Habitat	3.889	2.973	1.308
Nonhabitat	1.591	1.332	1.195
Surpluses of population	10.560	7.896	1.312
Territory of these surpluses	8.804	6.805	1.293
Asians			
Habitat	7.472	5.164	1.447
Nonhabitat	1.607	1.344	1.196
Surpluses of population	33.156	17.950	1.847
Territory of these surpluses	36.967	24.059	1.537
Hispanics			
Habitat	5.858	4.422	1.325
Nonhabitat	1.757	1.347	1.290
Surpluses of population	23.040	14.791	1.558
Territory of these surpluses	22.451	16.452	1.365
Subsets obtained by comparing the population of 1989 to the population of 1980			
Matches between 1980 and 1989	4.122	2.862	1.44
Surpluses of 1989 over 1980	109.80	61.37	1.789
Surpluses of 1980 over 1989	31.85	23.727	1.342

Table 11.14. Ratios of the Parameters of Looseness of Table 11.13 to the Corresponding Parameters for the Territory

	Ratios of relative distance-variance for the geoset to relative distance-variance of territory $({}^rd_P^2/{}^rd_T^2)$	Ratios of relative scatter of geosets to relative scatter of territory $({}^rs_P^2/{}^rs_T^2)$	Ratio of relative oblongity of geoset to relative oblongity of territory $({}^rb_P^2/{}^rb_T^2)$	Ratio of scatter of geoset to scatter of territory (s_P^2/s_T^2)	Territorial coverage of geoset (A_H/A_T)
	(4)	(5)	(6)	(7)	(8)
Total population 1989	2.80	2.44	1.16	1.045	0.430
Total population 1980	2.66	2.31	1.16	0.999	0.433
Afro-Americans 1980	2.94	2.80	1.04	0.750	0.268
Asians 1980	8.24	4.96	1.68	1.125	0.227
Hispanics 1980	6.49	5.20	1.23	1.293	0.249
Geosubsets derived from comparison of primary geosets to uniform distribution (1980)					
Total population 1980					
Habitat	2.30	2.21	1.04	0.956	0.433
Nonhabitat	1.30	1.47	0.88	0.831	0.567
Surpluses of population	11.48	8.98	1.28	0.967	0.108
Territory of these surpluses	9.81	9.15	1.08	0.986	0.108
Afro-Americans					
Habitat	2.78	2.73	1.01	0.730	0.267
Nonhabitat	1.13	1.21	0.94	0.887	0.733
Surpluses of population	7.61	7.12	1.07	0.739	0.1025
Territory of these surpluses	5.80	6.22	0.93	0.675	0.1025

Asians					
Habitat	5.73	4.95	1.16	1.124	0.227
Nonhabitat	1.14	1.20	0.96	0.929	0.773
Surpluses of population	33.60	17.56	1.91	1.028	0.059
Territory of these surpluses	28.52	22.03	1.29	1.289	0.059
Hispanics					
Habitat	4.68	4.38	1.07	1.090	0.249
Nonhabitat	1.23	1.22	1.01	0.912	0.751
Surpluses of population	25.58	19.88	1.29	1.343	0.068
Territory of these surpluses	20.87	18.02	1.16	1.217	0.068
Geosets derived from the comparison of 1989 to 1980					
Matches between 1980 and 1989	2.99	2.61	1.15	1.04	0.77
Surpluses of 1989 over 1980	79.68	55.89	1.43	1.03	0.37
Surpluses of 1980 over 1989	23.11	21.61	1.07	0.38	0.63

Table 11.15. Parameters of Looseness for Rome

	Population 1981	Dwellings built up to 1919	Dwellings built between 1971 and 1981
Relative scatter	1.06	0.59	1.28
Relative oblongity	1.01	1.02	1.00
Relative distance-variance	1.07	0.61	1.28
Ratio of scatter of geoset to scatter of territory	0.62	0.13	0.87
Territorial coverage	0.59	0.22	0.69

distance-variance of 1.02, which suggests a shape that is not far from circular, and this conclusion is in good agreement with the map of the territory.

Table 11.15 lists some parameters of looseness for: (a) the population (1981); (b) dwellings built up to 1919; and (c) dwellings built between 1971 and 1981. As might be expected by looking at the corresponding maps: relative oblongity plays a very minor role; both scatter and territorial coverage increase, moving from (b) to (a) and (c); and looseness as measured by relative distance-variance also increases, moving from (b) to (a) and (c). The relative distance-variance of dwellings built up to 1919 gives us an example of a value of this parameter, which is well below 1.

The results in Table 11.15 are easily explained by the fact that old dwellings tend to form an internal compact block while new ones are largely distributed over peripheral zones. However, when we consider the distribution of other phenomena over the territory of Rome we may find very different values of looseness.

11.8.2. Other Cities*

Table 11.16 gives examples of parameters of looseness (absolute scatter in sq km and relative oblongity) for total populations and Jewish minorities in selected cities (adapted from Bachi *et al.*, 1988).

The data in Table 11.16 relate a broad selection of European cities and Toronto, Canada, mostly throughout the 19th and 20th centuries. The Jewish communities studied range from quite large, as in the case of early 20th-century Vienna, Berlin, Warsaw, and Lodz, or late 20th-century London, to fairly small, as in the case of the cities in Italy and Switzerland. Regulations concerning the right of Jews to settle freely in the selected cities varied considerably over time and place from extremely restrictive to rather liberal.

Important differences appear in the metric measure of scatter for different populations over time and space, but without exception, the scatter measure of Jews was significantly smaller than that of the total population, which in turn was smaller than that for the total urban habitat. On the other hand, with the main exception of 16th-century Rome, no clear pattern of differentiation appears from the measures of relative oblongity for Jewish and total populations.

*This section was reconstructed by Prof. Sergio DellaPergola.

Table 11.16. Parameters of Looseness for Total and Jewish Populations in Selected Cities (adapted from Bachi *et al.*, 1988)

		Scatter, $2\sigma_M\sigma_m$ (sq km)			Relative oblongity	
City[a]	Year	Total habitat	Total population	Jewish population	Total population	Jewish population
Rome	1526/27	1.4	0.9	0.2	1.010	1.145
	1965	28.3	24.9	12.1	1.022	1.016
Milan	1965	23.0	17.2	7.9	1.013	1.119
Vienna	1910	47.1	28.1	14.8	1.033	1.045
	1971	58.9	40.0	19.1	1.005	1.005
Berlin	1933	185.1	114.9	60.0	1.058	1.024
Zurich	1970	14.0	11.7	7.7	1.002	1.000
Greater London	1970	253.0	204.9	175.8	1.014	1.024
Warsaw	1839	3.6	2.8	2.5	1.033	1.050
	1897	4.2	3.5[b]	2.5	1.013	1.029
	1938	14.9	10.1	2.8	1.000	1.002
Lodz	1931	7.8	6.5	3.6	1.161	1.153
Toronto (Metro)	1981	487.4	291.5	62.5	1.078	1.075

[a] As territorially defined at given date.
[b] Catholics.

11.9. Determination and Analysis of Clusters of Surpluses of Population Geosets over the Uniform Distribution

11.9.1. Identifying Clusters in a Simple Setting

Clustering is discussed in the statistical and statistical–geographical literature in many contexts. Here we shall deal with this concept in a specific way that enables us to acquire a more in-depth understanding of territorial concentration and helps to describe and analyze important aspects of the form of the distributions of populations over territories.

We begin by coming back to the simple example of the geoset of hotels in Jerusalem (see Sections 3.1 and 3.2), where we have a small population geoset (shown in Figure 3.3) formed by $F^P = 63$ elements, for each of which we know the exact location over a well-defined gross territory T (the municipality of Jerusalem). The description of the geoset can be enhanced by comparing it to the theoretical uniform distribution over the territory, a comparison that is made easier by dividing the territory T into $N^c = F^P = 63$ hexagonal cells. Comparing the actual number of elements of the population in each cell to the number 1, expected in the uniform distribution, we are able to identify the cells in which the actual number of elements exceeds 1 and to form the geoset of surpluses, ${}^gSPU$, as shown in Figure 3.10 (Section 3.2). We noted in Chapter 3 that the geoset under survey is very far from being uniformly distributed, as there are 44 surpluses out of 63 elements. We thus have a very high value for the concentration ratio $[\Delta_P^U = 44{:}63 = 69.8\%]$.

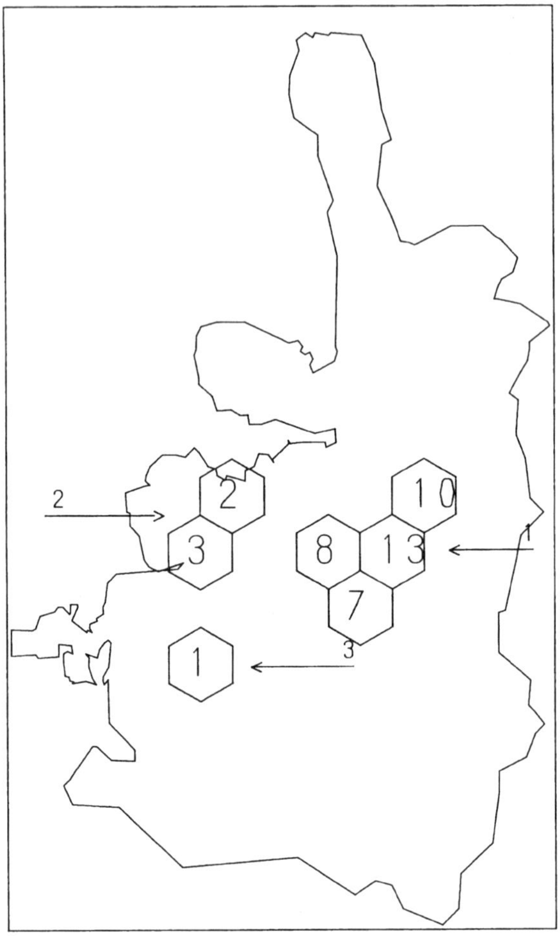

Figure 11.4. Jerusalem. Clusters of surplus of hotels over uniform distribution, 1989 (Source of data: Israel Ministry of Tourism).

We base the study of clusters of surpluses of the actual over the uniform distribution on the consideration of the contiguity of cells with surpluses (two hexagonal cells having a side in common are contiguous). Looking at Figure 3.10, we can neatly distinguish three clusters of cells with surpluses. To help in their identification, they have been reproduced in Figure 11.4: Cluster 1 is formed by four contiguous cells having 10, 13, 8, and 7 surpluses, respectively; Cluster 2 is formed by two contiguous cells having 3 and 2 surpluses, respectively; Cluster 3 is formed by an isolated cell with 1 surplus.

In our example, clusters are thus simply connected tracts of territory in which there are surpluses of actual over uniform distribution. The fact that one cluster includes 86.4% of the surpluses and that an additional 11.4% are found in another cluster suggests that the geoset under survey is strongly clustered.

We can use the clusters in order to simplify both the microgeographical presentation through the map and the macrogeographical description through geostatistical parameters. We assume that, for our purposes, we can neglect the study of density variations among cells of the same cluster. As shown in Figure 11.5, we can illustrate the geoset on a "summary map," by showing its general contour, the contours of clusters 1, 2, and 3, and the number of surplus elements in each cluster (38, 5, 1, respectively) (see Bachi, 1981). We shall thus be able (as we explain below) to calculate geostatistical parameters describing the main characteristics of each cluster separately, the impact of each of them on the distribution of the surpluses, and the overall impact of clustering.

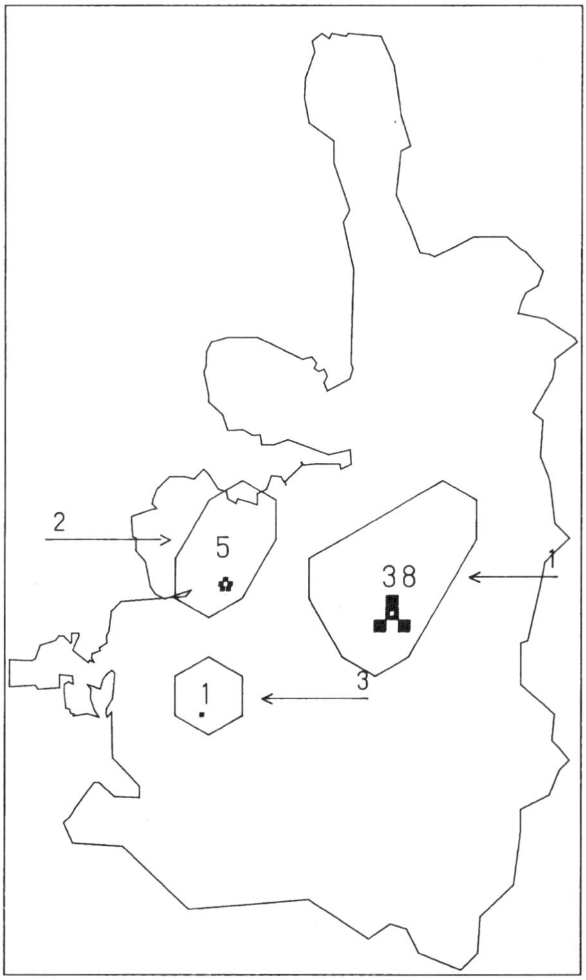

Figure 1.5. Jerusalem. "Summary map" of the clusters of hotels, 1989 (Source of data: Israel Ministry of Tourism).

Looking at Figures 11.4 and 11.5, we can intuitively surmise that clustering is maximal when the surpluses are found in one cell only; it is nonexistent when the actual distribution coincides with the uniform distribution and consequently, there are no surpluses of actual over uniform distribution.

11.9.2. Extending the Study of Clusters to Large Populations

Generally we deal with large populations for which the input of geostatistical information is given in aggregate form. We saw in Chapter 3 how this information can be presented, albeit with some loss of precision, by giving the distribution of the geoset over a network of cells for each of which we know the location of the center and the relative weight. Thus, knowing the primary geoset ${}^{g}P$ and uniform geoset ${}^{g}U$, we derive from them the geoset of the surpluses ${}^{g}SPU$ of the actual over the uniform distribution. The identification of clusters of surpluses then becomes easy, as will be seen from the examples given in Sections 11.9.3 and 11.9.4.

The determination of clusters is conceived as a practical operation, used largely for descriptive purposes, that can lead to various solutions depending on the aims of the particular study and the degree of precision that we want to attain. The software for geostatistical analysis (see Chapter 17) enables us to select the more desirable solutions among those available for each study. For example:

1. In order to reduce the amount of graphical and analytical work, we can decide to consider only a limited number of major clusters, which might be defined as those clusters that have a proportion of surpluses exceeding a prefixed limit, and so taken in a cumulative way include a considerable majority of the surpluses.
2. If some clusters are seen to be relatively close to each other on the map, we can reduce the amount of analytical work by introducing the notion of quasi-cluster. Let us denote by k the distance between the centers of two contiguous cells. We can decide that two surplus cells having centers at distance hk form a quasi-cluster (h being a proper multiplier larger than 1).
3. The definition of the territory of a cluster may be simply the set of contiguous cells. However, we might also want to introduce the convex hull that can be traced around the contiguous cells of the surplus. Comparative consideration of these two territories may be of interest in certain cases in order to enhance the study of the form of the cluster. We might also want to take account of the ellipse traced around the cross of the principal axes of the cluster.
4. If we define the territory of a major cluster by means of its convex hull, we might also wish to include within the cluster isolated cells or minor clusters having surpluses of actual over uniform distribution that are found within the convex hull.
5. If the main aim of the study of clusters is to help in the microgeographical analysis, it might be convenient not to overly restrict the number of clusters considered, at least in the initial phases of the research. This may help to render the determination of clusters very significant in some settings, such as the following: (a) when the different parts of the territory have different climatic, socioeconomic, ethnic, rural-urban-metropolitan or other character-

istics, and some of the clusters are found to fall over territorial sections having characteristics that may explain the clustering of the geoset; (b) when the population to which each of the clusters refers is formed by meaningful subpopulations (as, e.g., in the study of the geoset of orchards, we may find that each cluster corresponds to a given type of fruit tree).

The study of clustering has been discussed above for the surplus geoset ${}^{g}SPU$ derived from the comparison of any population geoset ${}^{g}P$ to the geoset of the uniform distribution ${}^{g}U$, but this method can be extended to the study of clusters in geosets of surpluses ${}^{g}SPQ$ and ${}^{g}SQP$ derived from the comparision of any two geosets ${}^{g}P$ and ${}^{g}Q$ distributed over the same territory.

11.9.3. Examples of Clusters: Surpluses of the Population of the United States over the Uniform Distribution

To give an example of the clusters of surpluses of an actual geoset as compared to the uniform distribution, we turn again to Figure 8.4 (Section 8.8), and the surpluses of the total population of the conterminous United States. We found that 57.4% of the total population of the United States were in surpluses over the uniform distribution of surpluses. Out of these surpluses 89.4% were found in nine major clusters. Surpluses covered only 10.6% of the total territory of the conterminous United States.

The nine major clusters were shown in detail in Figure 8.4. Each cluster can be considered as a derived geoset and we can apply to it all the geostatistical parameters that we have described up to this point. However, to avoid redundancy, we will consider here only some of the more important ones: Parameters that describe position, coverage, dispersion, and looseness for the three of the more important clusters are given by Table 11.17. Interested readers can check the correspondence between the parameters and the detailed distribution as shown in Figure 8.4.

Table 11.17. Selected Geostatistical Parameters for Three Major Clusters of Surpluses of Actual Population over Uniform Distribution for the United States (1980)

Parameters	Cluster 1	Cluster 2	Cluster 3
Percentage of the surpluses in the cluster	62.9	8.5	4.7
Eccentricity, X[a]	+886.0	−1117.6	−1250.7
Eccentricity, Y[a]	+90.2	−235.6	36.8
Standard distance[b]	318.1	57.0	48.1
Relative distance-variance	1.85	0.87	0.55
Relative scatter	1.42	0.70	0.44
Relative oblongity	1.30	1.24	1.24
Ratio of scatter of population to scatter of territory of cluster	69.9	35.3	31.1

[a]Distance (in miles) of the center of the cluster from the center of territory: $-X$ denotes distance to the east; $+X$: to the west; $-Y$: to the south; $+Y$ to the north.

[b]To be compared to standard distance of uniform distribution over the territory 861 miles; total population 905 miles.

Cluster 1 is the only one that has a considerable territorial coverage. All of the clusters are characterized by small dispersion (see parameters 6 and 7). Clusters 2, 4, 5, and 6 are also characterized by very small loseness and so can be considered very compact.

11.9.4. Overall Parameters of Clustering and Its Impact

The parameters just illustrated help us in the microgeographic study of each of the major clusters considered. We might also be interested in obtaining the overall parameters of clustering and its effects, calculated over all the major clusters taken together. These parameters may be required for the quick characterization of a geoset or for comparisons among many geosets, and may constitute a separate domain of study or an addition to the detailed descriptive parameters of each of the major geosets.

We list below using numbers 1 through 11 several of these overall parameters, taking as examples their values for the total population of the United States, *TP*, in 1980 and for three subpopulations: Afro-Americans, *AA*; Asians, *A*; and Hispanics, *H*. Numbers and a short description of the major surplus clusters for *AA*, *A*, and *H* were given in Section 8.8.2 and the figures therein.

(1), (2). In order to evaluate the importance of major clusters, we start by considering (1) the percentage of the elements of the population considered that are in surplus with respect to the uniform distribution (*T*: 57.3%; *AA*: 73.3%; *A*: 77.5%; *H*: 75.5%). Within these surpluses, we calculate (2) the percentage of surplus elements included in the major clusters (*T*: 87.3%; *AA*: 96.5%; *A*: 92%; *H*: 96.3%).

(3), (4). In parallel, we can evaluate (3) the percentage of the total territory of the United States that corresponds to surpluses (*T*: 10.6%; *AA*: 10%; *A*: 5.7%; *H*: 6.7%). Then, we calculate (4) the percentage of all the convex hulls of the major clusters out of the total territory: (*T*: 33%; *AA*: 32%; *A*: 9%; *H*: 13%).

(5). An important characteristic of clustering is that the density of the population over clusters tends to be comparatively large. We are interested in the evaluation of this density by means of dimensionless parameters that enable us to make comparisons among different populations. To this end we can calculate (5) the ratio of density surplus elements to the gross density of population (*T*: 5.4%; *AA*: 7.1%; *A*: 13.6%; *H*: 11.3%).

(6), (7), (8). To evaluate the impact of clusters over the dispersion of the population, we can proceed as follows: We start by calculating the distance-variance of all the major clusters of surpluses taken together. To render this parameter dimensionless, we present it as (6) the ratio of the distance-variance of surpluses in clusters to the distance-variance of total population (*T*: 96%; *AA*: 99.8%; *A*: 98.8%; *H*: 99.8%). We then distinguish (7) the percentage of distance-variance of clusters of surpluses found within the clusters (*T*: 9.5%; *AA*: 22.8%; *A*: 1.3%; *H*: 1.3%) from (8) the distance-variance between the clusters (*T*: 90.5; *AA*: 77.2%; *A*: 98.7%; *H*: 98.7%).

(9), (10). The parameters of clustering discussed above make an abstraction of the position of the clusters over the territory. They are, however, sensitive to the geography of the surplus cells that constitute each major cluster and (in 6, 7, 8) to their reciprocal positions. Sometimes, we are interested in obtaining indexes of clustering even more abstract and insensitive to geography. Since clustering is, in a certain sense, an opposite process to fragmentation we can follow an approach analogous (but opposite) to that suggested in Section 11.5.4 for obtaining overall indexes of fragmentation. Let us denote the major clusters by $j = 1, 2, \ldots, m$; and the proportion of surpluses in each cluster over the total population by p_j. Then we can take $\Sigma\, p_j^2$ as (9) index of population clustering (*T*: 0.41; *AA*: 0.4; *A*: 0.34; *H*: 0.23). This index measures the allocation weights of the cells among the major clusters, bounded by 1. The index is a minimum when all the major clusters contain an equal number of the weights of the cells. Similarly, let us denote the number of habitat cells forming cluster j by a_j. Then we can take $\Sigma\, a_j^2/M^2$, with $M = 1000$ as: (10) index of areal clustering (*T*: 0.306; *AA*: 0.033; *A*: 0.075; *H*: 0.003). This index is also bounded by 1 and its minimum value is obtained when all the major clusters have an equal number of cells.

(11). The number of major clusters considered: *T*: 9; *AA*: 9; *A*: 6; *H*: 9.

11.10. Elementary Methods for Partitioning a Geoset

The determination of clusters, as described just above has the great advantage of placing actual characteristics of the geoset into evidence, thus enhancing both the graphical presentation and the geostatistical analysis. For this reason it seems to be generally preferred over other ways of partitioning the geoset. We shall note only cursorily some elementary partition methods that can be employed and considered as geostatistical analogues to "quantiles" (quartiles, deciles, ...) used for partitioning one-dimensional distributions.

Given an ordered distribution of the values x_1 of a one-dimensional variable x, we conceive of the partition factor x_P as a value that divides the distribution into P and $1 - P$. Geostatistical analogues of x_P can be obtained in various ways, including:

1. Working separately on each of the two coordinates of the elements of the geoset, we can obtain the quantile desired for each of them. This method has the obvious shortcoming of leading to results that may change with the orientation of the system of the coordinates used. Sometimes it may be of interest if the system of coordinates that is used has a clear meaning in the context of our study. For example, dealing with the west–east and south–north axes of the coordinates is sometimes justified. In other cases, the use of the principal axes as axes of coordinates might be suggested.
2. We can draw a circle over the entire geoset that has as its center the center of the geoset and a radius selected so as to include the great majority of the elements. We can then calculate radii for concentric centers, including the sought after quantilic proportions.
3. An approach analogous to (2) can be followed by working with concentric ellipses, built within an ellipse having the same center and oblongity as the geoset.

Owing to the irregularity, complexity, and multimodality of many of the actual geosets, interpretation of the findings of these partition methods may be difficult.

11.11. Some Remarks on Parameters of Asymmetry

Symmetry is an important property in many fields of science and art and is the object of a wide and complex literature. It is doubtful whether examples of perfect symmetry can be found in territories of the types discussed in this book. It is even less likely to find symmetry in population geosets that can be influenced by the asymmetry of gross territory and by complexities of factors acting upon the population in various parts of the habitat. However, reference to symmetrical models as terms of comparison can sometimes be of considerable use.

The attention of researchers in centrography and related fields has thus been concentrated on lack of symmetry (skewness) and the parameters that measure it. One of the incentives for working on these topics has been desirability of finding a geostatistical analogue for measuring the skewness calculated on distributions of one-dimensional variables. Neft (1966, pp. 114–120) and Manson (1981) are prominent among the authors who have contributed important proposals to the measure of skewness.

The development of geostatistical parameters presented in this book seems to offer an additional route to the measure of asymmetry, which we describe below in a rather sketchy fashion, referring to only one type of asymmetry—central asymmetry.

Let us consider a territorial geoset ${}^gT$ and/or a population geoset ${}^gP$. We assume that we rotate these geosets through the center and thus obtain new rotated geosets to be denoted by ${}^gT'$ and ${}^gP'$, respectively. Central asymmetry can then be measured by comparing ${}^gT'$ to ${}^gT$ and/or ${}^gP'$ to ${}^gP$. The elements of ${}^gT'$ and ${}^gT$ that do not match (surpluses of ${}^gT'$ over ${}^gT$ or of ${}^gT$ over ${}^gT'$) are due to the asymmetry. The same applies to ${}^gP'$ and ${}^gP$. Parameters of asymmetry can be obtained: (a) by dividing the distance-variance of ${}^gT$ (or ${}^gP$) between the symmetrical and asymmetrical parts; and (b) by measuring the minimal distance between ${}^gT$ and ${}^gT'$ (or ${}^gP$ and ${}^gP'$).

As the understanding of (b) will be facilitated by the study of the properties of minimal distance, we postpone the presentation of the parameters of central asymmetry and will deal with them after Part IV in Appendix H. There we shall also hint at measures of asymmetry around the principal axes. The discussion of asymmetry is far less developed than we originally planned as it has not been possible, for technical reasons, to test the methods we propose on large-scale applications, as we have done for other geostatistical parameters. Thus there is no point in comparing the merits or defects of the approach we propose with those of other researchers.

Chapter 12

Parameters of Position of Geosets

Coincidence, Congruence, and Similarity

12.1. Introduction

The characterization of geosets is based mainly on the study of dimension, shape and position. Thus, the determination of parameters of the planor position of geosets is of great importance, both in the analysis of each individual geoset and in comparison between them. In the comparative analysis of pairs of geosets (such as ${}^gP$ and ${}^gQ$), we must often address issues such as:

1. Whether geosets ${}^gP$ and ${}^gQ$ differ (a) because they have different positions in the plane, or (b) because they have different dimensions, or (c) because their forms are different, or (d) because of some combination of these factors.
2. Discrepancies between geosets can be measured by the minimal distance between them (see Part IV). If the positions of ${}^gP$ and ${}^gQ$ are different, we might want to be able to measure the effects of the difference on the distance between ${}^gP$ and ${}^gQ$.
3. In certain cases, we might want to eliminate, as far as possible, the effect of the difference in position over distance and to measure separately the residual effects over the distance between ${}^gP$ and ${}^gQ$ of differences in dimensions and patterns.
4. If we have to compare multiple geosets, e.g., the distribution of commercial establishments over the territory of many cities (reduced to the same standard size), it may be useful to present all the maps of the cities, or of the establishments, in the same position.

Some important aspects of position were discussed earlier when we dealt with centers and with the orientation of geosets along the major axis. In this chapter we shall introduce a more complete battery of parameters of position and we start by discussing (a) the concepts of congruent and coinciding geosets, and (b) an additional parameter of position known as the quadrant setting.

12.2. Coincident and Congruent Geosets

In order to introduce the concepts of coincidence and congruence of two geosets in a simple fashion, we shall deal here with the example illustrated in Figure 12.1, which shows a small arbitrary geoset ${}^gP$ formed by $N = 31$ elements.

We assume that we know the exact location of each of its elements $P_i = (i = 1,2,\ldots, 31)$, as indicated by its coordinates (X_i^P, Y_i^P). The location of each element is shown by a point. The numerical list of the coordinates is given in Table 12.1. We generally deal with large populations distributed over a defined territory, divided into cells of equal size and shape, for which relative weights based on aggregated data are available. We recall here the comments made in Sections 3.1.3 and 3.4 with regard to point geosets versus geosets based on large populations, which may help to solve problems or clear up doubts in the application of the concepts of coincidence and congruence to point and other types of geosets.

The coordinates (X_i^P, Y_i^P) are measured in Figure 12.1 along the conventional axes (X, Y), taking an arbitrary point (0,0) as the origin, and an arbitrary segment $\bar{u}$ as a unit of length. The geoset is shown within an arbitrary square frame with vertices at (0,0), (0,9.5), (9.5,0), and (9.5,9.5).

Figure 12.2 shows a geoset ${}^gQ$ that has been obtained from ${}^gP$ through rigid motion. ${}^gP$ has been transformed by three controlled operations: a rotation through 120°; a reflection around a line located at level $Y = +2$; parallel translation by -5 units along the axis of the abscissas and by $+8$ along the axis of the ordinates. Knowing these operations, we can easily calculate the coordinates (X_i^Q, Y_i^Q) of each one of the points Q_i that correspond in ${}^gQ$ to the points P_i of Figure 12.1.

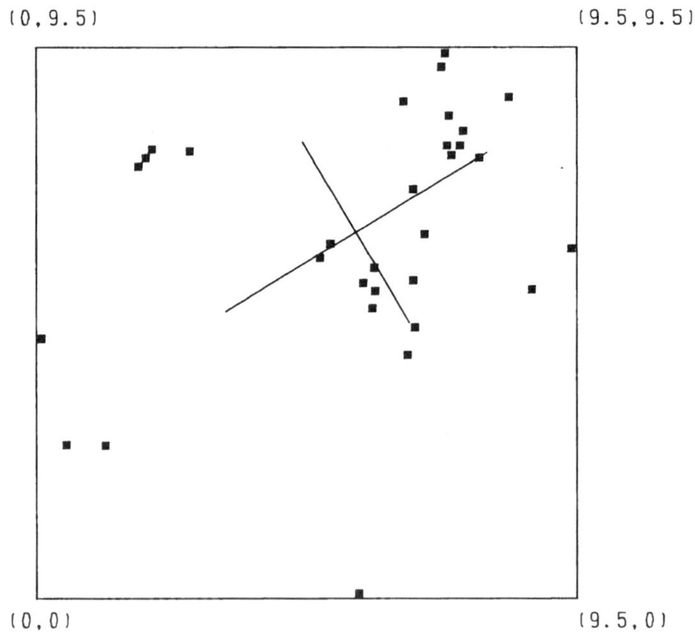

Figure 12.1. Original geoset ${}^gP$.

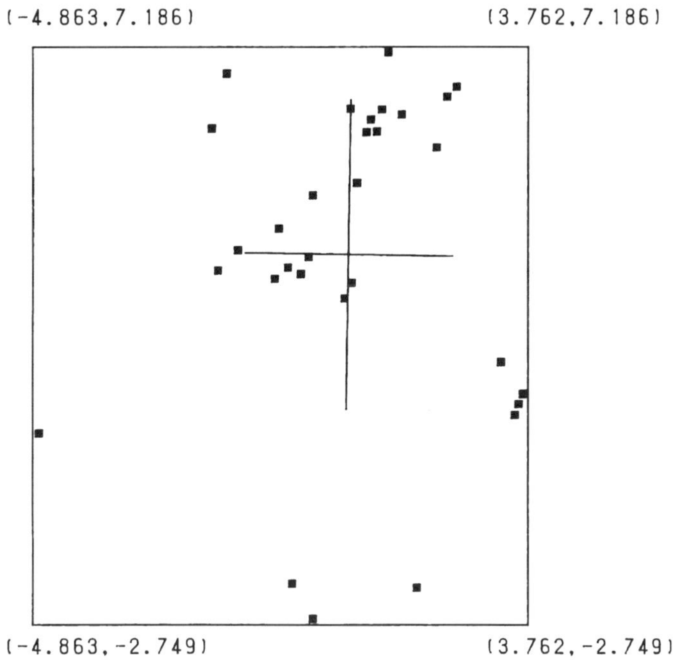

Figure 12.2. Transformed geoset ${}^gQ$.

Owing to the types of transformations used for moving from ${}^gP$ to ${}^gQ$, we obviously find that the coordinates of ${}^gP_i$ and ${}^gQ_i$ differ ($X_i^P \neq X_i^Q$; $Y_i^P \neq Y_i^Q$) and the location of each P_i over the plane differs from that of Q_i (see Table 12.2).

Let us now carry out operations inverse to those described above over geoset ${}^gQ$, namely parallel translation by $+5$ along the axis of the abscissa, and by -8 along the axis of the ordinates; rotation through $-120°$, and reflection around a line located at a level $Y = -2$. We thus obtain a new geoset (denoted by ${}^gR$) that is equal to ${}^gP$. Each of the points of ${}^gR$ has a location identical to that of the corresponding point in ${}^gP$. Two geosets that are duplicates of one another (such as ${}^gP$ and ${}^gR$) are called coincident.

We have already noted that in comparing ${}^gQ$ to ${}^gP$ we find no one-to-one coincidence between the locations of their elements. However, we know that ${}^gQ$ has been obtained from ${}^gP$ by changing only the position and that we can obtain ${}^gR = {}^gP$ by changing the position of ${}^gQ$. Pairs of geosets such as (${}^gP$, ${}^gQ$) and (${}^gQ$, ${}^gR$) that differ only in position and can be made coincident by a change of position are called congruent. However, suppose that we know neither the genesis of ${}^gQ$ as obtained by a change of position of ${}^gP$ nor that of ${}^gR$ as obtained by a change of position of ${}^gQ$. We may then ask whether we can find a way to discover the congruence of pairs of geosets such as (${}^gP$, ${}^gQ$) or (${}^gQ$, ${}^gR$); in other words, whether there exists at least one position in the plane to which the two geosets can be brought by rigid motion and where they will coincide.

Table 12.1. Coordinates of the Elements of Original Geoset ^{g}P

Point	X	Y
1	5.983	5.728
2	9.500	6.072
3	2.683	7.770
4	6.000	5.320
5	5.950	5.019
6	5.792	5.459
7	1.750	7.512
8	5.208	6.158
9	7.283	7.877
10	7.250	9.500
11	6.508	8.662
12	6.583	4.202
13	0.450	2.622
14	7.367	7.705
15	1.167	2.611
16	8.800	5.352
17	1.883	7.662
18	7.183	9.264
19	6.717	4.686
20	7.517	7.877
21	0.000	4.492
22	2.000	7.813
23	6.683	5.513
24	5.717	0.000
25	5.033	5.911
26	7.575	8.135
27	8.392	8.737
28	6.683	7.114
29	6.892	6.330
30	7.867	7.662
31	7.325	8.404

12.3. Preparatory Steps for Comparing the Locations of the Elements of Congruent Geosets

We can progress toward comparing the locations of the elements of congruent geosets by following a simple procedure:

1. We take as the origin of the coordinates (0,0) their mean center.
2. We rotate the geoset around its center so that the major axis M coincides with the conventional axis X. The orthogonal axes X and Y form four quadrants, which are denoted as $j = 1,2,3,4$ following the usual trigonomet-

Table 12.2. Coordinates of the Elements of Transformed Geoset ${}^{g}Q$

Point	X	Y
1	−0.36	3.596
2	−1.496	6.813
3	3.383	1.759
4	−0.398	3.406
5	−0.633	3.212
6	−0.172	3.295
7	3.626	0.822
8	0.724	3.139
9	1.176	5.796
10	2.598	6.579
11	2.243	5.517
12	−1.657	3.352
13	0.041	−2.749
14	0.985	5.782
15	−0.326	−2.134
16	−1.770	5.847
17	3.690	1.012
18	2.462	6.403
19	−1.305	3.710
20	1.059	5.998
21	1.886	−2.204
22	3.762	1.188
23	−0.572	4.094
24	−4.863	0.501
25	0.598	2.864
26	1.253	6.178
27	1.366	7.186
28	0.815	4.895
29	0.031	4.683
30	0.698	6.194
31	1.611	6.096

rical conventions. We denote the elements included in quadrant j by the symbol ji.

3. We calculate for each quadrant j the average of the squares of the distances of the elements ji from the mean center of the entire geoset ${}^{g}P$. We designate this mean as deviance of j and denote it by $(V_j^P)^2$. The total of the deviances is given by $(V^P)^2$, which is equal to the distance-variance d_P^2 of the geoset ${}^{g}P$.
4. To render $(V_j^P)^2$ a pure number we divide each $(V_j^P)^2$ by $(V^P)^2$. The ratio obtained is designated as the deviance ratio for quadrant j and denoted by v_j^P. Whenever there is no risk of misunderstanding, this symbol can be simplified to v_j. Obviously, $\Sigma v_j = 1$ and v_j can be given as a percentage with reference to the distance-variance of ${}^{g}P$.

Table 12.3. The Eight Quadrant Settings[a]

Ordinal number of quadrants settings	Principal axes along conventional axis of the abscissas		Sums of deviance ratios for quadrants along the diagonal	Deviance ratios for quadrants along the diagonal
	M	m		
a	abscissa	ordinate	$(v_1 + v_3) > (v_2 + v_4)$	$v_1 \geqslant v_3$
b	abscissa	ordinate	$(v_2 + v_4) > (v_1 + v_3)$	$v_2 \geqslant v_4$
c	abscissa	ordinate	$(v_1 + v_3) > (v_2 + v_4)$	$v_3 \geqslant v_1$
d	abscissa	ordinate	$(v_2 + v_4) > (v_1 + v_3)$	$v_4 \geqslant v_2$
e	ordinate	abscissa	$(v_1 + v_3) > (v_2 + v_4)$	$v_1 \geqslant v_3$
f	ordinate	abscissa	$(v_2 + v_4) > (v_1 + v_3)$	$v_2 \geqslant v_4$
g	ordinate	abscissa	$(v_1 + v_3) > (v_2 + v_4)$	$v_3 \geqslant v_1$
h	ordinate	abscissa	$(v_2 + v_4) > (v_1 + v_3)$	$v_4 \geqslant v_2$

[a] If $(v_1 + v_3) = (v_2 + v_4)$, we select the diagonal in which one of the v_j exceeds the other. Coordinates for points in settings $j = a, b, \ldots, h$ are denoted by ${}^jM_i$; ${}^jm_i$.

Under the conditions indicated above (origin of coordinates at the center of ${}^gP$; axis M coinciding with conventional axis X), we can generally distinguish four quadrant settings, which are conventionally designated as q.s.(a), q.s.(b), q.s.(c) and q.s.(d):

q.s.(a) if $(v_1 + v_3) > (v_2 + v_4)$ and $(v_1 \geqslant v_3)$

q.s.(b) if $(v_2 + v_4) > (v_1 + v_3)$ and $(v_2 \geqslant v_4)$

q.s.(c) if $(v_1 + v_3) > (v_2 + v_4)$ and $(v_3 \geqslant v_1)$

q.s.(d) if $(v_2 + v_4) > (v_1 + v_3)$ and $(v_4 \geqslant v_2)$

Four other settings (e, f, g, h) can be indicated whenever we consider the possibility that axis M coincides with axis Y (see Table 12.3).

12.4. Principal Position of a Geoset

Conventionally we define a principal position of any geoset ${}^gP$ as a position having the following properties:

1. The mean center is taken as the origin (0, 0) of the coordinates.
2. The major axis M coincides with the conventional axis of the abscissas and the minor axis m coincides with the conventional axis of the ordinates.
3. The geoset has quadrant setting (a), i.e., $(v_1 + v_3) > (v_2 + v_4)$ and $(v_1 \geqslant v_3)$.

If the actual geoset ${}^{g}P$ under survey is not in a principal position, we can place it there through the following operations:

1. Parallel translation for taking the mean center as the origin.
2. Rotation around the center to put M and m along the axes X and Y, respectively.
3. Reflection around m or M or both in order to move from q.s.(b), q.s.(c), or q.s.(d) to q.s.(a).

A geoset ${}^{g}P$ put in a principal position is denoted by the symbol ${}^{g}P^{*}$. The coordinates of the elements of ${}^{g}P^{*}$ along the principal axes M and m are denoted by M_i^{*P}, m_i^{*P}, or simply by M_i^{*} and m_i^{*}.

We have placed the geosets ${}^{g}P$ and ${}^{g}Q$ in their principal positions (see Section 12.2) and the results are shown in Figure 12.3 and Table 12.4. It is found that the principal geosets corresponding to ${}^{g}P$ and ${}^{g}Q$ are practically identical, i.e., that ${}^{g}P^{*} = {}^{g}Q^{*}$. We can generalize this result and state that:

1. Geosets corresponding to two congruent geosets are equal.
2. If two geosets have equal principal geosets, they are congruent.

In our example, the number of elements of ${}^{g}P$ and ${}^{g}Q$ is small and we can ascertain their equality by inspecting their coordinates, as shown in Table 12.4.

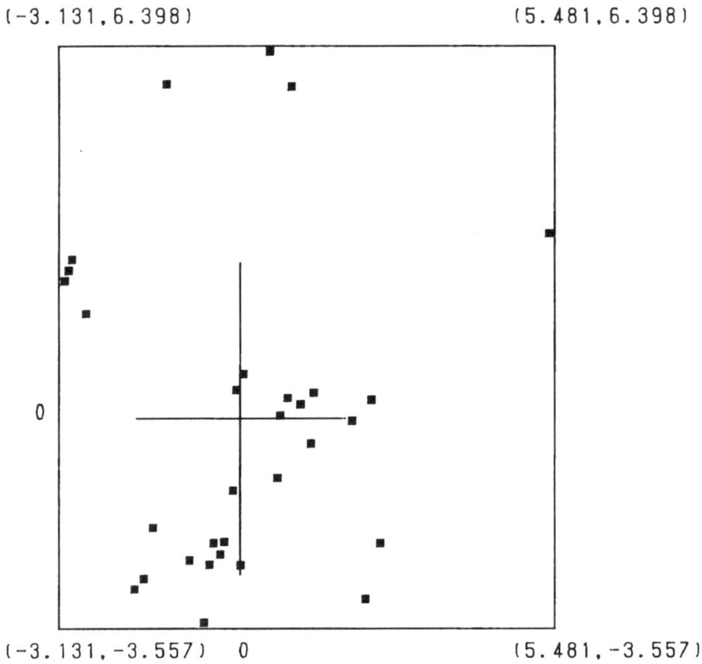

Figure 12.3. Geosets ${}^{g}P$ and ${}^{g}Q$ (now congruent) in their principal positions.

Table 12.4. Coordinates of the Original and Transformed Geoset in Principal Positions

	Original geoset ${}^gP$		Transformed geoset ${}^gQ$	
Point	X	Y	X	Y
1	0.704	0.055	0.704	0.055
2	2.216	−3.138	2.216	−3.138
3	−2.743	1.837	−2.743	1.837
4	1.063	0.251	1.063	0.251
5	1.295	0.448	1.295	0.448
6	0.836	0.358	0.836	0.358
7	−3.002	2.771	−3.002	2.771
8	−0.063	0.499	−0.063	0.499
9	−0.472	−2.164	−0.472	−2.164
10	−1.881	−2.969	−1.881	−2.969
11	−1.543	−1.902	−1.543	−1.902
12	2.321	0.325	2.321	0.325
13	0.526	6.398	0.526	6.398
14	−0.282	−2.147	−0.282	−2.147
15	0.903	5.789	0.903	5.789
16	2.474	−2.168	2.474	−2.168
17	−3.062	2.579	−3.062	2.579
18	−1.713	−2.791	−1.713	−2.791
19	1.975	−0.038	1.975	−0.038
20	−0.352	−2.364	−0.352	−2.364
21	−1.310	5.823	−1.310	5.823
22	−3.131	2.402	−3.131	2.402
23	1.248	−0.435	1.248	−0.435
24	5.481	3.227	5.481	3.227
25	0.059	0.776	0.059	0.776
26	−0.544	−2.547	−0.544	−2.547
27	−0.640	−3.557	−0.640	−3.557
28	−0.126	−1.258	−0.126	−1.258
29	0.654	−1.033	0.654	−1.033
30	0.012	−2.554	0.012	−2.554
31	−0.902	−2.471	−0.902	−2.471

However, for comparison between principal geosets with a large number of elements, we may prefer to use the minimal distance (see Section 13.5). For congruent geosets, such as ${}^gP$ and ${}^gQ$ in our example, we should find ${}^m\bar{D}(P^*, Q^*) = 0$.

12.5. Parameters Describing the Position of a Geoset Uniquely

Suppose that we wish to implant a given geoset of points in the plane in a given position or that we wish to compare the positions of any two geosets. It would be useful to have at our disposal a small battery of parameters describing the position of the geoset uniquely, and based on the discussion above, this battery is easily obtained. The position of a geoset is defined by:

1. The actual coordinates of the mean center (X, Y) over the relevant map.
2. The angle α formed by the principal axes and corresponding conventional axes.
3. The actual quadrant setting of the geoset.

12.6. Similar Geosets and Reduced Principal Coordinates

Two geosets, say, ${}^gP$ and ${}^gQ$, can be designated as similar if there is one position at least in which they can be obtained one from the other by dilation. To envisage this situation in a simple way, let us consider once more, without loss of generality, two geosets formed by a number N of elements i $(i = 1,2,\ldots, N)$ for which the exact location is known. This location can be shown on the map by a point. Let us denote by x_{P_i} and y_{P_i}) the deviations of the coordinates of the location of each point P_i from the center of ${}^gP$. An analogous definition is given to x_{Q_i}, and y_{Q_i}.

Let us denote any real and positive number by h. Suppose that in one of the infinite number of positions of ${}^gP$ in the plane, the coordinates (x_Q, y_Q) of the elements of ${}^gQ$ are linked to those of the corresponding elements of ${}^gP$ by the relation

$$x_{Q_i} = hx_{P_i} \qquad y_{Q_i} = hy_{P_i}$$

Then ${}^gP$ and ${}^gQ$ can be considered to be similar: their centers are equal; their standard deviations are the same (d_P and $d_Q = hd_P$); the angle between their principal and conventional axes is equal in ${}^gP$ and in ${}^gQ$:

$$\begin{aligned}\text{arc tan } 2\alpha_P &= \text{cov}(x_P, y_P)/[\sigma_{xP}^2 - \sigma_{yP}^2]\\ \text{arc tan } 2\alpha_Q &= \text{cov}(x_Q, y_Q)/[\sigma_{xQ}^2 - \sigma_{yQ}^2]\\ &= [h^2 \text{cov}(x_P, y_P)]/[h^2(\sigma_{xP}^2 - \sigma_{yP}^2)]\\ &= \text{arc tan } 2\alpha_P\end{aligned}$$

Suppose that the geosets ${}^gP$ and ${}^gQ$ are given in any arbitrary positions in the plane and we wish to determine whether they are similar or not. We calculate the principal coordinates $({}^{\text{q.s.}}M_P^*, {}^{\text{q.s.}}m_P^*)$ and $({}^{\text{q.s.}}M_Q^*, {}^{\text{q.s.}}m_Q^*)$, for four quadrant settings (a, b, c, d). If the geosets are indeed similar we should find a constant value for $({}^{\text{q.s.}}M_Q^*/{}^{\text{q.s.}}M_P^*)$ and $({}^{\text{q.s.}}m_Q^*/{}^{\text{q.s.}}m_P^*)$ for all points i. This would enable us to answer in the affirmative the question of whether ${}^gP$ and ${}^gQ$ are similar.

If the number N of the elements of ${}^gP$ and ${}^gQ$ is large, there may be some difficulty in identifying the corresponding points P_i and Q_i, which can be overcome in various ways:

1. We divide the principal coordinates of the elements of each geoset by the standard distance of the geoset. We thus obtain the reduced principal

coordinates:

$$\text{For geoset } {}^gP\text{: } {}^{\text{q.s.}}M^*_{P_i}/d_P,\ {}^{\text{q.s.}}m^*_{P_i}/d_P$$

and

$$\text{For geoset } {}^gQ\text{: } {}^{\text{q.s.}}M^*_{Q_i}/d_Q,\ {}^{\text{q.s.}}m^*_{Q_i}/d_Q$$

If ${}^gP$ and ${}^gQ$ are similar, we should find for each i:

$$ {}^{\text{q.s.}}M^*_{P_i}/d_P = {}^{\text{q.s.}}M^*_{Q_i}/d_Q $$

$$ {}^{\text{q.s.}}m^*_{P_i}/d_P = {}^{\text{q.s.}}m^*_{Q_i}/d_Q $$

2. The similarity of ${}^gP$ and ${}^gQ$ implies that when ${}^gP$ and ${}^gQ$ are placed in the same position, elements P_i and Q_i are linked in the calculation of minimal distance (see Section 16.4.1).

The average of the distances between linked elements becomes $(1 - h)^2 d_P^2$. If both geosets are transformed by calculating their principal reduced coordinates, then the minimal distance between the transformed geosets is 0.

Determining whether two geosets ${}^gP$ and ${}^gQ$ are similar can be of considerable theoretical interest, especially if the approach is extended to territorial geosets, as described above, as well as to models or to geometrical figures. While geostatistical analysis of actual geosets almost never involves examples of similar geosets, this does not mean that the calculation of reduced principal coordinates has no value. On the contrary, by eliminating the effects of both position and dimensions over the geosets being compared it allows us to focus on the shapes of territories or on patterns of distributions of populations. Examples of such calculations are given in Chapter 16.

Part IV

Summary Comparison of Geosets via Calculation of the Minimal Distance between Them

Chapter 13

Minimal Distance between Geosets

Objectives, Methods of Calculation, and Selected Examples

13.1. Comparing a Pair of Geosets by Means of a Summary Parameter

Until now we have discussed various approaches—some of them purely descriptive and some analytical—that enable us to compare pairs of geosets in detail, e.g., graphical presentation of two geosets by means of rational maps; presentation through maps showing surpluses of one geoset over another and matches between them; and geostatistical parameters of the characteristics of primary and derived geosets and the measures of the relationships among these parameters.

However, there is often a need for a summary parameter of the discrepancy between the geosets being compared. As we deal with territorial distributions, it seems natural to build our summary parameter on the consideration of how far apart the geosets are from one another, i.e., to judge discrepancy on the basis of the distance between them. Distances can be measured in many different ways: Here, we shall use the minimal (Euclidean) distance that must be covered to transform one geoset into another. This parameter may contribute to the solution of many of the research problems illustrated by the examples here and in Chapters 14–16, but is only an additional research tool and is not to be considered as an alternative to more detailed approaches.

Summary measurements of discrepancy between territorial distributions are often obtained by using different methods and we note some of them here, following a classification suggested by White (1986). Suppose we consider the distribution of two populations A and B over a territory T, or the distribution of two characteristics A and B of a population over T. We divide T into a network of territorial units i ($i = 1, 2, \ldots, N$ regions) and present the data in the form of proportions a_i and b_i, respectively, of the total amount of A and B found in i. White has classified the summary measurements of discrepancy between the A and the B distributions, or of segregation between A and B found in the literature as follows: (1) indexes of dissimilarity, Δ; (2) Gini's inequality measures; (3) entropy indexes; (4) interaction or exposure indexes; (5) Atkinson's indexes; (6) contingency table measures; (7) analysis of variance measures; (8) spatial measures.

Measures 1–7 refer to territorial units i as separate categories. As these categories are used only for classification purposes, the classification is actually based on a normal scale (see Section 4.8). Geographical characteristics such as shape, size, and reciprocal positions of the regions i over territory or distances between regions are completely disregarded. In consequence, all these measures are insensitive to geography, which leaves them open to criticism if they are used for geographical description. This has been known for a long time (see, e.g., Duncan and Duncan, 1955; Duncan *et al.*, 1961; Taeuber and Taeuber, 1965), and to understand this criticism thoroughly the reader is referred back to Section 3.9. Class 8 includes the only summary measures of discrepancy between A and B that are sensitive to geography. As this book is devoted to the study of geographical distributions as such, we shall restrict our attention here to this class, which according to White, includes some proposals by Jakubs (1981), Morgan (1982, 1983) Morgan and Norburg (1981), and White (1983, 1984). Actually we proposed distance-based parameters of discrepancy between geographical distributions independently much earlier (1962*a*, *b*, 1963; see also 1984, 1989*a*, *b*). The study of minimal distance as we proposed in 1962 and 1963 has recently been further developed with the help of a sophisticated computer program. This approach is discussed in detail in Chapters 13–16.

Presumably, the measure most widely used among classes 1–7 is Δ. We have already discussed this parameter as a measure of concentration of population over territory (see Chapter 8). It may also be employed in the study of segregation between populations, provided its meaning is clearly defined and its insensitivity to geography is acceptable for the study being pursued. An algebraic relationship between Δ and minimal distance will be shown and found to be very useful.

To introduce the minimal distance approach, we start with the very simple setting in which the elements of the two geosets being compared are aligned over a straight line. We then go on to consider two-dimensional geosets giving the individual locations of each of their elements. The approaches thus suggested will later be generalized to all types of paired geosets.

The minimal distance has important properties that help us to interpret it and, in particular, enable us to analyze relationships between triplets of geosets (see Chapter 14). The minimal distance between two geosets depends on the differences in their positions, dimensions, and forms. Taking an average of squared minimizing distances enables us to measure separately the squared distance between the centers and the squared difference between the dispersions of the geosets and the residual components (see Chapter 15). Further transformations, sometimes relevant for more in-depth analyses of minimizing distances, are discussed in Chapter 16.

In some of the applications of minimal distances we are interested in their absolute values, while in others we want to render them or their components dimensionless in order to facilitate comparisons between findings obtained for different populations, territories, and time periods. Methods for reaching these objectives are discussed in Chapters 13 and 16.

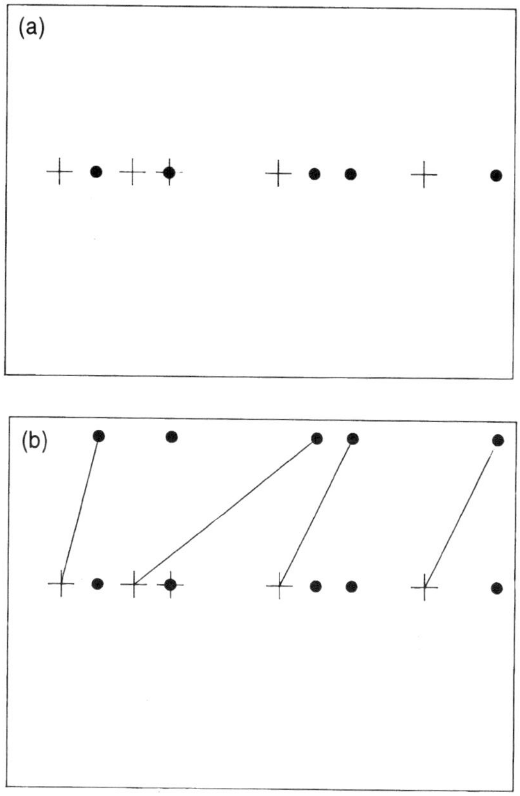

Figure 13.1. Geoset ${}^{g}P$ (bicycles) and geoset ${}^{g}Q$ (motorcycles): (a) the locations of ${}^{g}P$ and ${}^{g}Q$ scattered along a rectilinear country road; (b) comparison between ${}^{g}P$ and ${}^{g}Q$.

13.2. Minimal Distance between Two Aligned Geosets

13.2.1. Analogy to the Distance between Two Arrays of Cograduated Values

To start our discussion of the minimal distance between two geosets ${}^{g}P$ and ${}^{g}Q$, let us consider a very simple example (shown in Figure 13.1a). We assume that ${}^{g}P$ and ${}^{g}Q$ are, respectively, the geosets of the locations of N elements of P and of N elements of Q, which are scattered along the conventional axis of the abscissas. These locations are denoted by X_i^P $(i = 1, 2, \ldots, N)$ and X_j^Q $(j = 1, 2, \ldots, N)$, respectively; each location has a relative weight $1/N$. To figure out a setting of this type (albeit in an artificial way), we can assume that X_i^P and X_j^Q are, respectively, the locations of N bicycles (${}^{g}P$) and of N motorcycles (${}^{g}Q$), scattered along a rectilinear country road. The locations were identified at a given moment by an aerial photo. A symbol marks each location in the figure. Motorcycles are shown by small crosses and bicycles are identified by black circles. Suppose that we wish to compare ${}^{g}Q$ and ${}^{g}P$ by moving the motorcycles so as to render ${}^{g}Q$ equal to ${}^{g}P$. We can thus measure the mean distance covered by the motorcycles so that each of them reaches the same abscissa as a bicycle. Keeping the locations of the bicycles unchanged, we can associate a motorcycle with a bicycle in $N!$ different ways (permutations h). As we wish to perform our operation at minimal cost, we can select a permutation that minimizes the total of paired distances and thus their average.

Let us denote this average in an abbreviated form by ${}^m\bar{D}(P, Q)$, and the location of the motorcycle that is to be associated with the location $X^P_{h,i}$ of bicycle i in any permutation h by $X^Q_{h,j}$. The average distance to be covered by the motorcycle in permutation h to render ${}^gQ$ equal to ${}^gP$ is thus $(1/N)\Sigma|X^P_{h,i} - X^Q_{h,j}|$. The "best" solution is to select a permutation h such that ${}^m\bar{D}(P, Q) = \min[(1/N)\Sigma|X^P_{h,i} - X^Q_{h,j}|]$. (It may happen that more than one permutation leads to this result.)

However, if N is not small, the identification of ${}^m\bar{D}(P, Q)$ through the calculation of $N!$ averages of distances (one for each of the permutations) may be cumbersome, so we look for other methods for calculating ${}^m\bar{D}(P, Q)$. A very simple solution can be found by using a method suggested by Gini as early as 1914, when he was dealing with a formally analogous problem, namely the comparison of two sets of homogeneous values of a one-dimensional variable, such as income, age, or height.

Suppose, e.g., that P and Q are two populations formed by N individuals each, and that the respective ages of each element of P and of each element of Q are indicated by X^P_i $(i = 1, 2, \ldots, N)$ and X^Q_j $(j = 1, 2, \ldots, N)$. Gini suggested measuring the dissimilarity (in Italian: *dissomiglianza*) between the age distributions of P and Q by calculating $(1/N)\Sigma|X^P_i - X^Q_j|$, where pairing values of P and Q is done in the following way: We arrange the values of both X^P_i and X^Q_j in increasing order, so that

$$\ldots X^P_{i-1} \leqslant X^P_i \leqslant X^P_{i+1} \leqslant \ldots$$

$$\ldots X^Q_{j-1} \leqslant X^Q_j \leqslant X^Q_{j+1} \leqslant \ldots$$

Gini called values such as X^P_i and X^Q_j, having the same rank $(i = j)$ in the respective array, "cograduated," and showed that the average of absolute differences between them yields a minimizing solution.

Figure 13.1b shows a graphical method often employed in order to compare two linear sets P and Q containing the same number N of homogeneous values. In our example we have utilized the same numerical values given in Figure 13.1a. In Figure 13.1b the values X^P_i are given along one axis; the cograduated values of X^Q_j are given along another axis that is parallel to the first. Pairing each of the values X^P_i to the cograduated X^Q_j is done as indicated by the lines connecting them. As the values are cograduated, no connecting lines intersect.

The comparison between cograduated arrays can be extended to arrays of different sizes or formed by aggregates of values. Methods for carrying out these operations have been suggested by, among others, Leti (1961, 1982), Salvemini and Girone (1981). Samuel and Bachi (1964), and Leti showed that the minimizing solution in the comparison between two one-dimensional arrays of values can also be obtained by utilizing the algorithm of the transport problem (see Section 13.3.2). However, the comparison between cograduated values is simpler.

To return to our example of the comparison between two linear geosets: there, too, the pairing of each motorcycle with the cograduated bicycle seems to constitute the preferable solution. The idea that the first motorcycle should be moved to meet the first bicycle, the second motorcycle should meet the second bicycle, and so on seems to be acceptable intuitively as well.

To consider a numerical example, suppose that $N = 5$ and that X_i^P takes values 2, 4, 5, 8, 12 and X_j^Q takes values 3, 5, 9, 10, 14 (see Figure 13.1). Then

$$\sum |X_i^P - X_j^Q| = 10 \qquad \text{and} \qquad (1/N) \sum |X_i^P - X_j^Q| = 2 \tag{13.1}$$

A simplification in the calculation of the minimal distance between aligned geosets ${}^gP$ and ${}^gQ$ can be obtained as follows: We first identify the matches between them and eliminate matching elements from the calculation. In our numerical example ${}^gM(P, Q) = \{5\}$. We then restrict our concern to surpluses ${}^gSPQ = \{2, 4, 8, 12\}$ and ${}^gSQP = \{3, 9, 10, 14\}$, which are also cograduated: $\Sigma |X_i^{SPQ} - X_j^{SQP}| = 10$. The average distance covered by a nonmatched motorcycle is 2.5. Multiplying this by the proportion of nonmatched over total motorcycles ($\Delta = 4/5$), we again obtain $\Delta(1/N_{NM}) \Sigma |X_i^{SPQ} - X_j^{SQP}| = 2 = (1/N) \Sigma |X_i^P - X_j^Q|$, where N_{NM} is the number of nonmatching elements.

13.2.2. Arithmetic and Quadratic Average of Differences between Cograduated Values

The parameter ${}^m\bar{D}(X^P, X^Q) = (1/N) \Sigma |X_i^P - X_j^Q|$ enjoys the properties of a metric (see Leti, 1961, 1982, p. 101). Accordingly:

1. ${}^m\bar{D}(X^P, X^Q) \geqslant 0$, i.e., it is never negative
2. ${}^m\bar{D}(X^P, X^Q) = 0$ if, and only if, all $X_i^P = X_j^Q$
3. ${}^m\bar{D}(X^P, X^Q) = {}^m\bar{D}(X^Q, X^P)$ (symmetrical property)
4. ${}^m\bar{D}(X^P, X^R) \leqslant {}^m\bar{D}(X^P, X^R) + {}^mD(X^R, X^Q)$, where X^R is also an array of values homogeneous to those of X^P and X^Q defined over the X-axis.

Gini suggested as an alternative to the simple dissimilarity index the quadratic average of the differences between cograduated values of P and Q, namely $\sqrt{[(1/N) \Sigma (X_i^P - X_j^Q)^2]}$. This value is also minimal among all possible quadratic averages of differences between pairs of values.

We (Samuel and Bachi, 1964) demonstrated that this index also has the properties of a metric and discussed its use both for descriptive purposes and for testing hypotheses of goodness of fit, as well as in two sample problems. When working on the quadratic average of the distance between P and Q we cannot introduce a simplification similar to the split between Δ and ${}^m\bar{D}(SPQ, SQP)$ introduced in Section 13.2.1, as can be deduced from the numerical example given there.

13.3. Minimal Distance between Two-Dimensional Geosets

13.3.1. Introducing Two-Dimensional Minimal Distance in a Simple Setting

Let us now go on to the study of minimal distance between two-dimensional geosets. Here, too, we start by considering a very simple setting as shown in Figure 13.2, where ${}^gP$ and ${}^gQ$ are assumed to be two geosets of N locations of unweighted

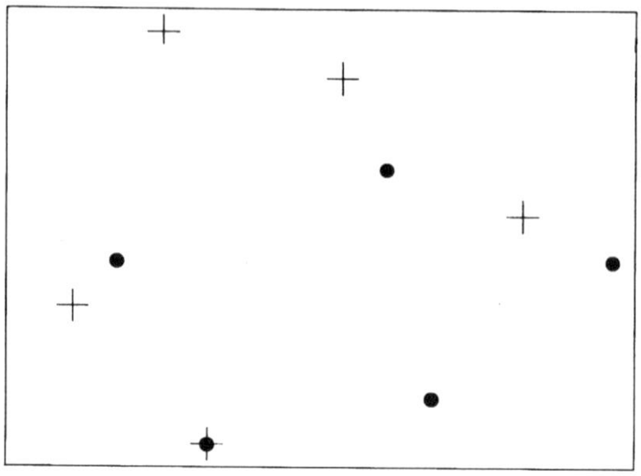

Figure 13.2. Geoset ${}^gP$ (Chinese) and geoset ${}^gQ$ (Japanese).

and dimensionless population elements over a plane T. These locations are represented in Figure 13.2 by different symbols. To fix the ideas, suppose that ${}^gP$ and ${}^gQ$ represent the distribution of two ethnic groups in an American city (take, e.g., P = Chinese and Q = Japanese). We wish to measure the distance between the corresponding geosets ${}^gP$ and ${}^gQ$. To simplify the example we have taken a very small value of N (=5). The coordinates of the elements of ${}^gP$ and ${}^gQ$ are shown by Table 13.1.

We have assumed that there is one element of ${}^gP$ and one element of ${}^gQ$ that intersect (say, found at the same address).

Let us indicate by p_i and q_j, the elements of ${}^gP$ and ${}^gQ$, respectively, and let us assume that we know the exact location of each of them. Elements of ${}^gP$ are indicated by the black circles and elements of ${}^gQ$ by the small crosses. The location common to elements of ${}^gP$ and ${}^gQ$ is indicated by a cross over a circle. The geoset ${}^gP$ is then defined as follows:

$$ {}^gP = \{X_i^P, Y_i^P; w_i^P = 1/N\} \tag{13.2} $$

where (X_i^P, Y_i^P) are the coordinates of the location of element p_i of P. Similarly,

$$ {}^gQ = \{X_j^Q, Y_j^Q; w_j^Q = 1/N\} \tag{13.3} $$

We adopt here the same basic concept used in Section 13.2 and assume that each element p_i $(i = 1, 2, \ldots, N)$ of ${}^gP$ is shifted to meet an element of ${}^gQ$. In other words, each Chinese pays a visit to one Japanese. The visits are organized in a way that minimizes the total distance covered; however, while in Section 13.2 we considered one-dimensional space, we are now considering two-dimensional space. We must minimize the total distance between paired elements p_i and q_j by taking

$$ {}^md_i = \min_j \sqrt{(X_i^P - X_j^Q)^2 + (Y_i^P - Y_j^Q)^2} \tag{13.4} $$

use of the symbol ${}^md_i$ emphasizes the fact that pairing is done in the way that minimizes the total distances. Here, also, we indicate the minimal distance by

Table 13.1. Coordinates of Fictitious Geosets ^gP and ^gQ

Geoset ^gP			Geoset ^gQ		
Ordinal number	X	Y	Ordinal number	X	Y
1	2	4	6	3	5
2	4	10	7	5	1
3	5	1	8	9	7
4	8	9	9	10	2
5	12	6	10	14	5

$^m\bar{D}(P, Q) = (1/N) \Sigma\, ^md_i$, which symbolizes the minimal average of distances between paired elements of two-dimensional ^gP and ^gQ. The basic problem to be solved is that of identifying a minimizing solution to the linking of each element of ^gP with an element of ^gQ.

In Section 13.2, we saw the simple method proposed by Gini for calculating minimal distance between one-dimensional sets by associating cograduated elements of the sets. However, this technique cannot be applied to two-dimensional distributions because cograduation, which is meaningful along a linear space (as the X-axis), is generally not meaningful over a two-dimensional space. (An exception to this is found in the special case in which all the elements of one of the geosets are distributed along a line and all the elements of the other geoset are distributed along a different line.)

As we cannot identify cograduated elements in two-dimensional space, we can, theoretically, identify the "best" solution to minimization, by (a) keeping ^gP fixed; (b) listing all the $N!$ permutations of the elements of ^gQ; (c) calculating the distance between pairs of elements of ^gP and ^gQ in all possible permutations; (d) selecting the set of linkages that minimizes total distances. Following this approach, we have found, for example, the minimizing linkages shown in Figure 13.3.

However, for any N that is not very small, this approach would require a great deal of work, which can be avoided by adopting the method described below.

13.3.2. Adapting the Transport Problem Algorithm

We can find a solution for larger N by adapting the method in Section 13.2, which is used to solve the transport problem. This problem was introduced in 1941 by Hitchcock in a paper that clearly indicates its aim: "Distribution of product from several sources to numerous localities." It was later formulated as a linear programming problem by Danzig (1951) in a paper entitled "Application of the Simplex Method to a Transportation Problem." It has since become a classical problem about which there is an enormous quantity of literature. To consider just one of its many applications, suppose we know (a) the quantities S_i of some kind of merchandise supplied in each factory or store i; (b) the quantities D_j of the same merchandise demanded in each consumer market or in each shopping center j; (c) the matrix of costs for transporting the merchandise from each origin i to each destination j.

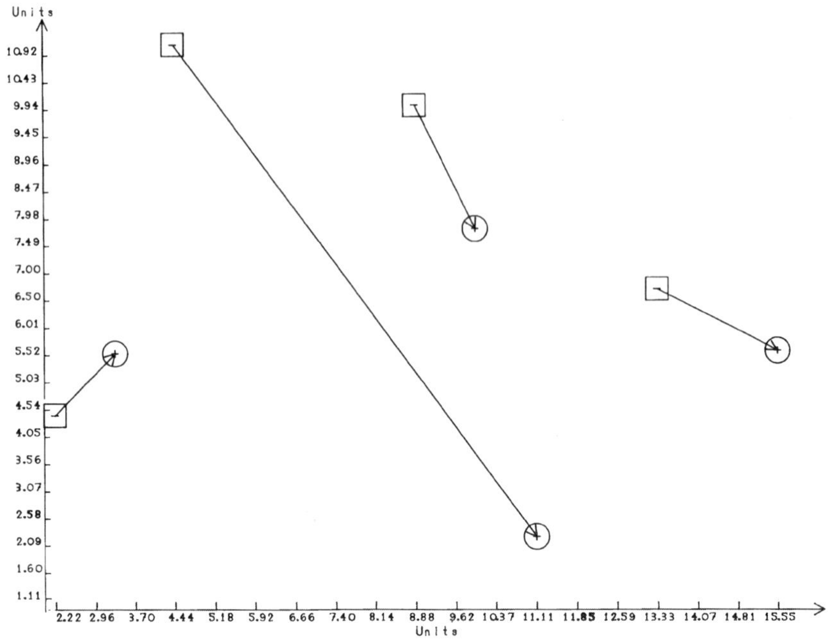

Figure 13.3. The "optimal transportation scheme" (minimizing linkages) between ${}^{g}P$ (Chinese) and ${}^{g}Q$ (Japanese). Key: squares = origins; circles = destinations. The arrows indicate the direction of the shifted units; the areas of the circles and squares are proportional to the number of shifted units.

Suppose also that $\Sigma S_i = \Sigma D_j$. Solutions to the transport problem enable us to calculate what amount of merchandise should be sent from each origin to each destination in a way that minimizes the total cost.

Similar problems are found in connection with, e.g., cash flow, airline scheduling, shipping, water resource optimization, personnel planning, machine loading, and telecommunications. We (Shimony, 1963) implemented some of the first applications of the measurement of minimal distance between geosets, and others were done by Yam (1968, 1969). Later applications were devised by Jakubs (1981), Morgan (1982, 1983), and White (1984), among others.

In the program utilized for the calculations in this volume, we have adopted an algorithm for the transport problem proposed by Syslow *et al.* (1983), which was adapted to the specific problem of distance between two geosets G and J by Ever-Hadani (1980).

In the simple setting described at the beginning of this section, we assume that at each point of origin and at each destination, we find $1/N$ of the total (here N is analogous to $\Sigma S_i = \Sigma D_j$). The matrix of costs is substituted by a matrix of distances, which can be, e.g., simple Euclidean distances between pairs of points, squared distances, or road distances.

In this volume we have utilized simple Euclidean distances almost exclusively. Let us again represent the coordinates of any element of P and Q by a point. The program suggests to which element of P each element of Q should be linked in order to obtain minimization of total distance (or optimal linkage). Let us indicate for the sake of simplicity the element of Q linked in this optimal way to element p_i of P as q_j. Let us indicate the distance between p_i and q_j as ${}^m d_i$ in order to stress that we are considering a linkage that leads to minimization. The program selects the minimizing linkages and then calculates

$$ {}^m d_i = \min \sqrt{(X_i^P - X_j^Q)^2 + (Y_i^P - Y_j^Q)^2} \tag{13.5} $$

For short, we shall indicate the distances ${}^m d_i$ as minimizing distances.

13.3.3. Minimizing Totals of Distances and Squares of Distances

The arithmetic average of minimizing distances $\Sigma\, {}^m d_i/N$ is designated, for short, as minimal distance (between geosets ${}^g P$ and ${}^g Q$) and expressed as ${}^m \bar{D}(P, Q) = (\Sigma {}^m d_i/N)$. This is the main summary measure of discrepancy between ${}^g P$ and ${}^g Q$ used in this volume. In the example of Figure 13.2, we find ${}^m \bar{D}(P, Q) = 4.413$, as we found in the calculation based on 120 permutations ($5! = 120$) and by the linkages shown in Figure 13.3.

There might be some interest in the alternative or concurrent calculation of the quadratic average of distances, as we saw in Section 13.2.2 for the case of one-dimensional sets. This measure could be based on linkages between each element of ${}^g P$ with an element of ${}^g Q$, arranged in the way that minimizes the sum of the squares of distances. Let us indicate a point in ${}^g P$ and a point in ${}^g Q$ linked in this way by P_k and Q_k, respectively; the distance between these points by ${}^{m2} d_k = \sqrt{[\Sigma (X_k^P - X_k^Q)^2 + (Y_k^P - Y_k^Q)^2]}$; and the quadratic average of the minimizing square distances as ${}^{m2} \ddot{D}(P, Q) = \sqrt{(\Sigma\, {}^{m2} d_k^2/N)}$. We studied this parameter in the early 1960s (Bachi, 1962 and Shimony, 1963).

We have already indicated that ${}^{m2} \ddot{D}(P, Q)$ can be calculated by means of the same algorithm used for calculating ${}^m \bar{D}(P, Q)$. For this purpose it is sufficient to introduce squared distances into the matrix of transport costs. The main reasons that ${}^m \bar{D}(P, Q)$ seems to be generally preferable to ${}^{m2} \ddot{D}(P, Q)$ are:

1. We saw in Section 13.2.2 that in the calculation of distances between one-dimensional sets, the linkages bringing about minimization of the total of squared distances are equal to those bringing about minimization of total distances. However, this may not be true when the calculation is carried out over geosets.
2. The parameter ${}^{m2} \ddot{D}(P, Q)$ does not share the important properties of ${}^m \bar{D}(P, Q)$, which are described in Sections 14.1–14.3.
3. From an intuitive standpoint it is easier to refer to minimization of total distances than to minimization of total of squares of distances.
4. Whenever we deal with practical problems, e.g., the average distance between populations served and services given or between production and its factors, reference to simple distances seems preferable: costs often appear to be, presumably, proportional to distances and not to square distances.

5. It is true that the utilization of squares enhances the possibilities of algebraic manipulations [such as separate consideration of abscissas and ordinates or splitting of squared distances into components (see Chapter 15)]. However, these manipulations can be carried out on the average of squares of simple minimizing distances ${}^{m}\ddot{D}^2(P, Q) = \Sigma\, {}^{m}d_i^2/N$ without having to refer to the different linkages that minimize the total of squared distances.

In the rest of this volume, if it is not otherwise indicated, when speaking of minimal distance between geosets, we refer to the arithmetic average of minimizing distances between a pair of two-dimensional geosets of any type. In speaking of the average of squared minimizing distances we refer to the same minimizing distances.

13.3.4. Generalization of the Calculation of the Minimal Distance among All Types of Paired Population Geosets

Earlier we introduced the calculation of minimal distance in the simple setting in which we compare two geosets ${}^{g}P$ and ${}^{g}Q$, indicating the locations of individual elements of the two populations. The approach was to pair each element of ${}^{g}P$ to an element of ${}^{g}Q$ assuming that both populations contained the same number of elements.

We can generalize the calculation of minimal distances to the comparison of all the types of pairs of geosets presented in Chapters 3 and 4. For example, geosets ${}^{g}P$ and ${}^{g}Q$ might consist of locations of individual elements to which a weight is attached (as in the case of hotel beds considered in Section 3.1.3) or of aggregates of locations (by cells or by regions); a geoset ${}^{g}P$ might be compared to the geoset ${}^{g}U$ of the uniform distribution; calculation of minimal distances might be carried out not only on primary geosets but also on population and territorial geosets derived from them; the primary geosets being compared might refer to territories; or we might compare component geosets to their compound.

The generalization of the calculation of minimal distance to all these settings is based on the following approach: Let us indicate any pair of geosets over which the minimal distance is to be calculated by ${}^{g}G$ and ${}^{g}J$. Geoset ${}^{g}G$ furnishes the relative weights w_i^G to be attached to the location of any element i. Let us multiply w_i^G by any suitable coefficient Φ and indicate the nearest integer by Φ_i^G. In a similar way, let us obtain Φ_i^J for the element i of geosets ${}^{g}J$ linked to element i of geosets ${}^{g}G$. We thus get the distribution of Φ elements in ${}^{g}G$ and Φ elements in ${}^{g}J$, and calculate the minimizing distances between them.

It should be noted that owing to the complexity of the computer program for the calculation of minimizing distances, the size of Φ is rather limited. A value of $\Phi = 200$ is often recommended.

13.3.5. Remarks Concerning the Notation for Minimal Distance

The following conventions are introduced to simplify the notation in minimal distance:

1. Minimal distance between any two geosets ${}^{g}G$ and ${}^{g}J$ is indicated by ${}^{m}\bar{D}(G, J)$. The use of the same letter as the prefix superscript g to indicate geoset in

general and G for a specific geoset is unfortunate, but is a result of the fact that there are not a sufficient number of letters available for simple symbols.
2. Minimal distance between the geosets of the surpluses ${}^{g}SGJ$ and ${}^{g}SJG$ is indicated by ${}^{m}\bar{D}(SGJ, SJG)$ or by ${}^{m}\bar{D}(SG, SJ)$.
3. The proportion of movers for transforming ${}^{g}G$ into ${}^{g}J$ is usually given by Δ_G^J. However, when there is no danger of misunderstanding, this is shortened to Δ.
4. In the textual presentation, minimal distances are often abridged into m-distances or even only distance.
5. Distances between an element of ${}^{g}G$ and the elements of ${}^{g}J$ linked to it are often termed minimizing distances for short.

13.3.6. Graphical Presentation of Lines of Minimizing Distances

The graphical method for representing the lines of minimizing distances as shown by Figure 13.3 can also be extended to larger geosets, as shown in Figures 13.4 and 13.5.

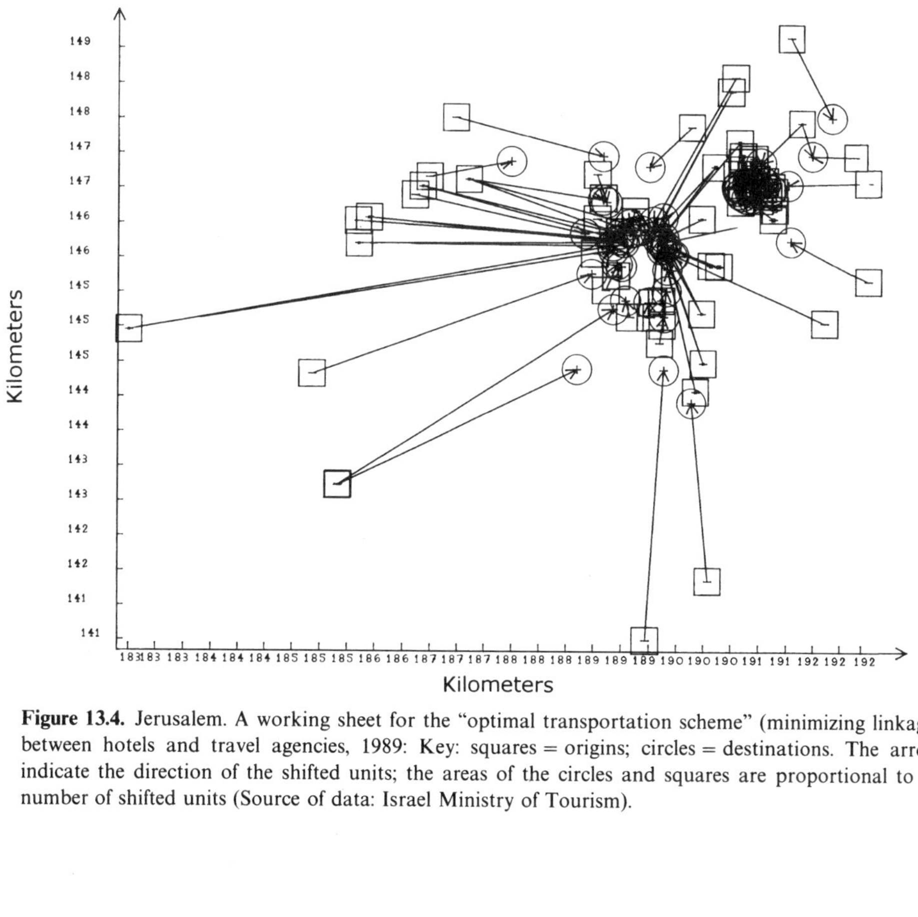

Figure 13.4. Jerusalem. A working sheet for the "optimal transportation scheme" (minimizing linkages) between hotels and travel agencies, 1989: Key: squares = origins; circles = destinations. The arrows indicate the direction of the shifted units; the areas of the circles and squares are proportional to the number of shifted units (Source of data: Israel Ministry of Tourism).

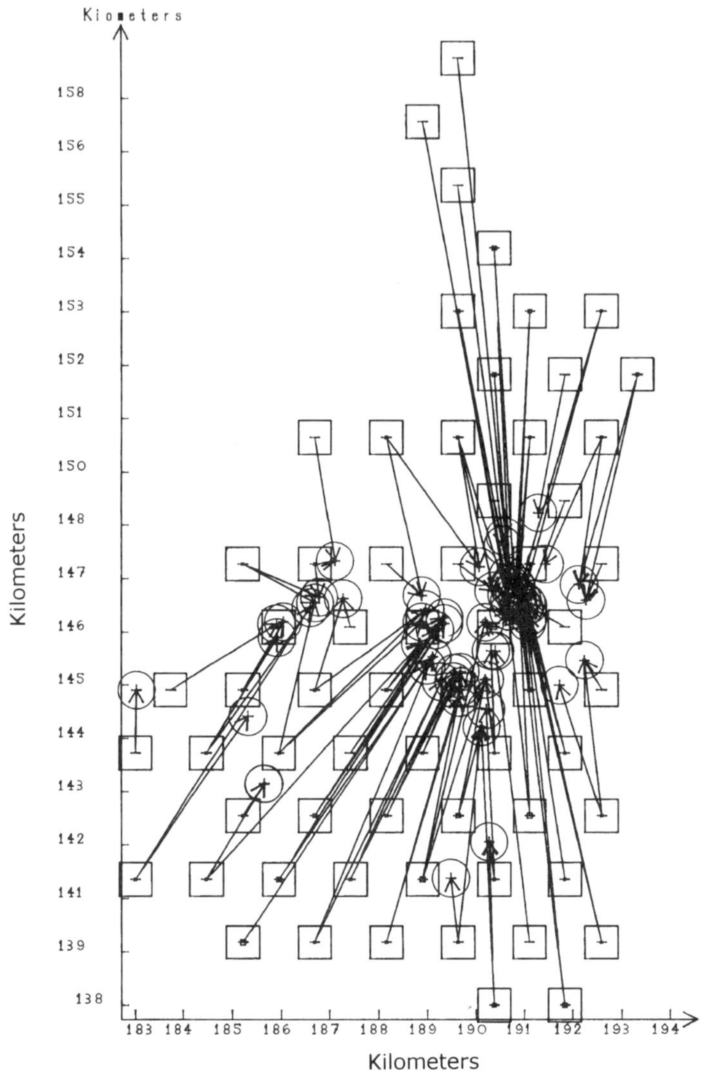

Figure 13.5. Jerusalem. A working sheet for the "optimal transportation scheme" (minimizing linkages) between hotels and uniform distribution, 1989: Key: squares = origins; circles = destinations. The arrows indicate the direction of the shifted units; the areas of the circles and squares are proportional to the number of shifted units (Source of data: Israel Ministry of Tourism).

Figure 13.4 relates to the calculation of the minimal distance between the hotels and the tourist agencies in Jerusalem (see Section 3.3). The distributions of the hotels, ${}^{g}P$, and of the tourist agencies, ${}^{g}Q$, are shown in Figures 3.7 (Section 3.2) and 3.13 (Section 3.3), respectively. The geosubsets derived from them and representing ${}^{g}M(P, Q)$, ${}^{g}SQP$, and ${}^{g}SPQ$, respectively, are shown in Figures 3.14–3.16 (Section 3.3).

Figure 13.4 shows the minimizing linkages between the origins of shifts (${}^{g}SPQ$) and their destinations (${}^{g}SQP$). The proportion Δ of surplus elements is 100%. The average of the distances covered by them is 1.154 km. Matching elements (Figure 3.14) do not contribute to the distance covered. Multiplying Δ by the average distance covered by shifting elements, we get a figure of 1.154 km as the minimal distance between hotels and agencies.

Similarly Figure 13.5 shows the minimizing distances between hotels and uniform distributions (see Figures 3.7 and 3.5 in Section 3.2). Here the average distance covered by shifting elements is 3.47 km; $\Delta = 100\%$; the minimal distance between hotels and uniform distribution is 3.47 km. As expected, this value is much greater than the minimal distance between the hotels and the agencies.

In the examples we have given, the lines connect relevant points of the geosets under consideration. By using colors and symbols (which unfortunately cannot be reproduced here) we would be able to show the weight to be attributed to connecting lines.

In Figures 13.3–13.5, the lines of minimizing distances do not cross, which is a general rule for minimizing distances for all pairs of geosets. This property is not necessarily shared by linkages that minimize the total of the squares of distances between pairs of geosets.

13.3.7. Expressing Minimal Distances as Pure Numbers

A very important application of minimal distances is the comparison of geographical discrepancies between pairs of geosets that refer to different territories or different time periods. For this purpose, we may wish to transform the absolute values of the minimal distances (expressed, e.g., in km or miles) into relative values, netted from the effects of the dimensions of the territories over which the minimal distances are calculated.

This problem, which is rather complex, will be discussed in Chapter 16, but anticipating this discussion, we shall adopt a simple approach below and calculate the ratio between each minimal distance and a measure of dispersion of the relevant territory T. Whenever available, we shall take the arithmetic average of the distances between all possible pairs of the elements of T as the measure of dispersion. When this measure is not available, we shall use the quadratic average of these distances, which is equal to $\sqrt{(2d_T)}$ (d_T being the standard distance of the elements of the territory).

13.4. Examples of Minimal Distance between Actual Geosets and the Uniform Distribution

13.4.1. Minimal Distance from the Uniform Distribution of Selected Geosets for Italy

The detailed comparison between actual geosets and uniform distribution constitutes a central theme of this book, and we have already discussed it at some length. The calculation of the minimal distance between the actual geoset and the

Table 13.2. Minimal Distance between Selected Geosets and Uniform Distribution

Geoset	Figure	Minimal distance between geoset and uniform distribution (km) (a)	Percentage of elements to be shifted (Δ) (b)	Minimal distance covered by shifts (km) (c)	Ratio of 100(a) to 451 km[a] (d)
Population 1987	1.3	54.21	39.28	138.0	12.0
All economic activities, 1981	1.6	119.8	45.2	264.8	25.0
Petrol industry, 1981	1.9	125.2	78.6	159.2	27.8
Sheep, 1982	9.4	198.4	46.6	425.9	44.0
Illiterates, 1981	9.9	243.0	42.6	570.6	53.9
Citrus, 1982	1.11	551.9	81.2	679.6	122.4

[a]451 km is the average of all possible distances between pairs of elements of the territory.

uniform distribution enables us to supplement the detailed study with a synthetic tool.

As an example of this approach, we list the minimal distance from uniform distribution for seven selected geosets illustrated in various figures in this book in Table 13.2. The values are taken from the *Statistical Atlas of Italy*. Readers who undertake to compare these distances with the corresponding figures will find that the results make sense and enable us to place the geosets in progressive order of their discrepancy from the uniform distribution over the territory. The distances found range from a minimum for the population to a maximum for citrus.

The minimal distances of Table 13.2 are given in km. To evaluate their meaning, we can compare them to the average of the distances between all the pairs of cells of the territory of Italy (451 km). The corresponding quadratic average is 527 km. The ratio in percent of each minimal distance to 451 km is given in the last column of the table.

13.4.2. Minimal Distance from the Uniform Distribution of 202 Geosets for Italy

The *Statistical Atlas of Italy* lists the minimal distances from the uniform distribution for 202 geosets taken from the censuses of population, housing, agriculture, industry, trade, and services. A detailed comparison of the items on this long list with each of the maps for the corresponding geosets fully confirms the value of the synthetic information provided by the minimal distances as a way to study discrepancy from the uniform distribution over the territory. A perusal of the minimal distances clearly confirms what we have said above, i.e., that the discrepancy of actual geosets from uniform distribution is a very general feature.

Considering 165 nonagricultural geosets separately, we find that no geoset has a distance less than 43 km from the uniform distribution; only one-tenth of them have a distance of less than 54 km; and only half have a distance less than 97 km. The average distance is 119.2 km, with a standard deviation of 65.0 km; one-quarter

Table 13.3. Italy (1981). Arithmetic Averages and Standard Deviations of Minimal Distance from Uniform Distribution

Groups of geosets considered		Minimal distance between geoset and uniform distribution (km)[a] (a)	Percentage of elements to be shifted (Δ) (b)	Minimal distance covered by shifts (km) (c)	Ratio of 100(a) to 451 km[b] (d)
165 nonagricultural geosets	AV	119.2	47.3	243.7	326.4
	SD	65.0	11.5	99.3	144.3
37 agriculture geosets	AV	134.3	36.4	337.2	29.8
	SD	106.4	14.3	150.1	23.6

[a]AV = Arithmetic averages; SD = standard deviations.
[b]451 km is the average of all possible distances between pairs of elements of the territory.

have a distance of over 156 km; one-tenth have a distance of over 221 km. A comparatively big distance from uniform distribution is found for farms classified by types of cultivation and for individuals employed in industry classified by sector. The relevant data are shown in Table 13.3. These findings suggest that specific types of cultivation and specific industrial sectors have very peculiar distributions.

13.4.3. Minimal Distance between Populations and the Uniform Distribution for Various Countries, Cities, and Time Periods

A very interesting application of the minimal distance is the measure of the discrepancy between the actual distribution of the human population of a country or a city and the uniform distribution over that territory. A few examples follow for which we give both the absolute distance in km and a standardized value used to make the distances for different countries or cities comparable. The standardization is achieved by dividing the minimal distance by a measure of the dispersion of the uniform distribution over the territory. For technical reasons, we have adopted the quadratic average of all possible distances between elements of the territory, $\sqrt{(2d_T)}$, where d_T is the standard distance of the territory. The values of $\sqrt{(2d_T)}$ that have been used are: 527 km for Italy; 428 km for France; and 1146 km for the United States. In France and in Italy the concentration, as measured by Δ, has grown markedly during the periods considered. Presumably the main factor has been increasing urbanization. In France, the increase of the minimal distance from uniform distribution is also due primarily to an increase in the shares of several regions that are rather far from the center of the territory. In Italy, the minimal distance of the population from the uniform distribution has oscillated owing to factors that have been identified but cannot be gone into here. The standardized minimal distance from the uniform distribution is greater in France than in Italy, and even larger in the United States (see Table 13.4).

Table 13.4. Italy, France, and the United States. Minimal Distance between the Population and the Uniform Distribution over the Territory

	Minimal distance between population and uniform (km)	Percentage of elements to be shifted (Δ)	Minimal distance covered by shifts (km)	Standardized minimal distance between population and uniform distribution (%)
Italy				
1871	53.2	29.4	180.9	10.1
1901	47.7	30.3	157.4	9.1
1931	51.2	31.8	161.0	9.7
1961	50.5	35.7	141.5	9.6
1981	54.2	39.3	137.9	10.3
France				
1801	48.5	14.1	344.6	11.3
1884	59.4	18.5	321.4	13.9
1954	90.6	29.8	304.4	21.2
Conterminous United States				
1980	809.7	48.86	1657.0	70.7
1989	770.0	49.07	1569.1	67.2
1994	753.2	49.06	1535.1	65.7

13.5. Examples of Minimal Distances between Pairs of Geosets: Comparisons to Populations or Other Paradigmatic Geosets

13.5.1. Comparisons between Pairs of Geosets

The comparison between pairs of geosets distributed over the same territory is also a central theme of geostatistical research, in which consideration of summary measures of minimal distance is a very good supplement to the detailed study of comparative maps and parameters of primary and derived geosets.

Experience in the field has shown that in the systematic and extensive study of geosets of demographic, economic, and sociological phenomena of countries and cities, considerable advantages accrue from the analysis of the minimal distances between these geosets and one or more paradigmatic ones, the latter being those that concern aspects of the geographical distribution that can be expected to have a marked influence on many other geosets, sometimes being "leaders." Characteristics of paradigmatic geosets can often be expected to be widely known.

The distribution of the population is, quite often, a very important paradigm, and systematic comparison between it and subpopulations delineated by ethnic, demographic, socioeconomic, or other criteria has proved very useful, especially when it is accompanied by study of the distances between pairs or groups of subpopulations, and it is particularly efficacious if the study is oriented toward substantive objects of research.

In some instances, the distance from the population distribution can also be studied in the context of other phenomena, particularly those derived from the population (such as births or deaths) and from services (e.g., health or social) offered to the population or to subpopulations.

Other useful paradigms, especially in the study of geosets regarding economic phenomena, are, e.g. geosets of income produced, of the value of consumption, or geosets for the total number of individuals engaged in industry and trade, or in services and all tertiary occupations. We shall give some examples of these approaches further on in this chapter.

From a methodological viewpoint, we consider here mainly minimal distances ${}^{m}\bar{D}(P, Q)$ between two geosets ${}^{g}P$ and ${}^{g}Q$, while in Chapter 14, we will deal primarily with studies in which we look at three or more geosets together, say, ${}^{g}P$, ${}^{g}Q$, and ${}^{g}R$ through the comparison of ${}^{m}\bar{D}(P, Q)$, ${}^{m}\bar{D}(P, R)$, and ${}^{m}\bar{D}(Q, R)$.

13.5.2. Minimal Distances between Geosets of Populations and Subpopulations Classified by Sex, Age, and Marital Status

Table 13.5 shows minimal distances that help us in the comparison of the geosets of the population of Italy and of its subpopulations, classified by sex, age, and marital status. Some of the main findings are given below.

As expected, the territorial distributions of men and women are almost identical, the minimal distance between them being extremely small. Also, within each marital status, the geosets of males and females are quite close, with the lowest minimal distance being that between married men and women.

Comparing the distribution of each age group with that of the entire population, we see that the values of Δ (proportion of shifting elements) are rather small, which suggests that the general distribution of the population is a basic factor in the distribution of various age groups. However, the distances covered by shifting elements are far from negligible. These distances are due to peculiarities in the distribution of the various age groups, which in turn are functions of complex factors. Among them, we shall note only the process of population-aging, which has affected earlier, and in a much stronger measure, the population of several northern and central regions, as compared to those of the southern regions and the islands, which are generally less developed economically and socially. As a consequence, the minimal distance from the total population is rather large for individuals aged 0–14 and 65+, while it is smaller for intermediate age groups (15–24, 25–54, and 55–64).

Comparing marital status with total population for individuals, we see that the minimal distance is much smaller for the single and the married than for the separated and the divorced. This is due to—among other factors—the higher ratios of divorce in cities of more developed regions, where the process of modernization started earlier and has been more pronounced, weakening the influence of the Catholic religion on family life.

As a criterion for comparison of the levels of the minimal distances discussed here, the last two lines of Table 13.5 show the average and standard deviation of the minimal distance between the population and the 164 nonagricultural geosets obtained from demographic and economic censuses of Italy (1981).

Table 13.5. Italy (1981). Minimal Distance between Population and Subpopulations by Sex, Age, and Marital Status

Geosets compared	Minimal distance between geosets (km) (a)	Percentage of elements to be shifted (Δ) (b)	Minimal distance covered by shifts (km) (c)	Ratio of 100(a) to 451 km[a] (d)
Men aged 15+ vs. Women aged 15+				
(1) All marital statuses	3.56	1.19	299.5	0.79
(2) Single	11.19	3.63	308.1	2.48
(3) Married	0.96	0.30	321.1	0.21
(4) Widowed	33.10	6.99	473.6	7.34
(5) Separated	15.65	3.97	394.3	3.47
(6) Divorced	21.37	7.86	271.8	4.74
Men aged 15+: each marital status vs. total				
(7) Single	9.83	3.53	278.6	2.18
(8) Married	2.83	1.36	207.7	0.63
(9) Widowed	17.14	8.23	208.2	3.80
(10) Separated	100.70	22.00	457.3	22.33
(11) Divorced	93.00	23.00	403.5	20.62
Women aged 15+: each marital status vs. total				
(12) Single	22.38	5.66	395.3	4.96
(13) Married	4.18	1.79	233.2	0.93
(14) Widowed	43.95	7.47	588.4	9.75
(15) Separated	83.10	21.80	380.8	18.43
(16) Divorced	73.30	25.60	286.2	16.25
Each age group vs. total population				
(17) Aged 0–14	57.5	40.6	141.7	12.7
(18) Aged 15–24	52.9	40.0	132.1	11.7
(19) Aged 25–54	61.1	40.4	151.0	13.5
(20) Aged 55–64	62.0	38.5	161.1	13.7
(21) Aged 65+	68.8	36.5	188.4	15.3
Population vs. each of 164 geosets taken from *Demographic and Economic Censuses of Italy (1981)*				
Arithmetic average	92.2			20.4
Standard deviation	69.6			15.4

[a] 451 km is the average of all possible distances between pairs of elements of the territory.

Table 13.6. Italy (1981). Minimal Distance between the Labor Force and Those Engaged in Commerce

	Minimal distance between geosets (km) (a)	Percentage of surplus elements to be shifted (Δ) (b)	Minimal distance covered by shifts (km) (c)	Ratio of 100(a) to 451 km[a] (d)
All ten sectors	40.0	7.0	572.0	8.9
Agriculture	140.2	33.1	423.1	31.0
Building	103.4	16.0	643.1	22.9
Other eight sectors				
Average	14.0	7.0	205.4	3.1
Standard deviation	7.7	3.0	72.2	1.7

[a]451 km is the average of all possible distances between pairs of elements of the territory.

13.5.3. Minimal Distance between the Labor Force and Individuals Engaged in Commerce in Italy

Data on individuals active in each of ten economic sectors in Italy were collected in 1981 through two channels: (a) the demographic census dealing with the labor force; and (b) the census of individuals employed in industry, trade, and services.

The question was asked whether the geographical distributions according to (a) and (b) give a consistent picture. It was expected (for reasons of definition and for other reasons which we cannot go into here) that the reply would be rather negative with regard to agriculture and to building and rather positive for other sectors. The minimal distances between (a) and (b) confirmed this hypothesis, as can be seen in Table 13.6.

13.5.4. Minimal Distances between the Population and Ethnic Subpopulations in the United States

The geographical distribution of several ethnic subpopulations in the United States is shown in Figures 8.5, 8.7, and 8.9 (Section 8.8). Table 13.7 answers in a summary way the question of how far the subpopulation considered is from the total population and how far the subpopulations are from each other. We find, as is to be expected, that the minimal distance from the total population is lowest for the White majority, followed by the Afro-Americans. The Hispanics and the Asians are more distant from the total population. However, they are comparatively close to each other. The Afro-Americans have the smallest distance from the Whites and the largest distance from the Hispanics and the Asians.

A look at the maps will explain these findings.

Table 13.7. Conterminous United States (1980). Minimal Distances between Population and Subpopulations

Population/subpopulation	Minimal distance between geosets (km) (a)	Percentage of surplus elements to be shifted (Δ) (b)	Minimal distance covered by shifts (km) (c)	Ratio of 100(a) to 1146 km[a] (d)
Total population vs.:				
Whites	22.4	5.2	428.0	2.0
Afro-Americans	250.5	32.4	773.7	21.9
Hispanics	566.8	49.4	1,148.0	49.5
Asians	608.9	41.7	1,461.4	53.2
Whites vs.:				
Afro-Americans	258.3	36.4	703.9	22.5
Hispanics	579.9	51.9	1,117.5	50.6
Asians	617.1	44.2	1,293.8	53.8
Afro-Americans vs.:				
Hispanics	744.8	52.8	1,409.8	65.0
Asians	823.1	46.0	1,788.6	71.8
Hispanics vs.:				
Asians	273.3	30.1	906.8	23.8

[a] 1146 km is the quadratic average of the distances between all possible pairs of territorial elements.

13.6. Minimal Distance between Territories

We illustrated in Chapter 4 how the study of territories might be required in geostatistical research. Sometimes it is of interest to calculate the minimal distance between territories $T_1, T_2, \ldots,$ which can be shown on the same map. Here we consider a few examples:

1. Assume that two territories T_1 and T_2 are disjoint. To measure ${}^m\bar{D}(T_1, T_2)$ we can distribute M dots uniformly over T_1 and T_2. Each element of T_1 is linked with an element of T_2. The linkage is carried out in the minimizing way. We have used this method, in a simplified and somewhat inaccurate manner, for the study of the distance between the three altimetric regions of Italy, shown in Plate 1. Let us denote the mountains by 1, the hills by 2, and the plain by 3. We obtain ${}^m\bar{D}(1, 2) = 207.0$ km; ${}^m\bar{D}(2, 3) = 221.4$ km; and ${}^m\bar{D}(1, 3) = 120.1$ km. To evaluate these distances, we can take into account that the average of the distances between all pairs formed 1000 cells of the territory of Italy shown in Figure 1.1 (Section 1.1) is of 451 km. Figure 13.6 shows the minimizing distances in the calculation of ${}^m\bar{D}(1, 3)$ between the mountains and the plain.

2. Let us compare two partly overlapping territories T_1 and T_2. Suppose that T_1 and T_2 are equal in area, or have been transformed so as to be equal. Let us denote the intersection between T_1 and T_2 by T_3; the part of T_1 that is not common to T_2 by T_4; and the part of T_2 that is not common to T_1 by T_5; and the corresponding areas by A_1, A_2, A_3, A_4, and A_5, where $A_1 = A_2$ and $A_4 = A_5$; the area of the union by $A = A_3 + A_4 + A_5$; the proportion of the area of surpluses

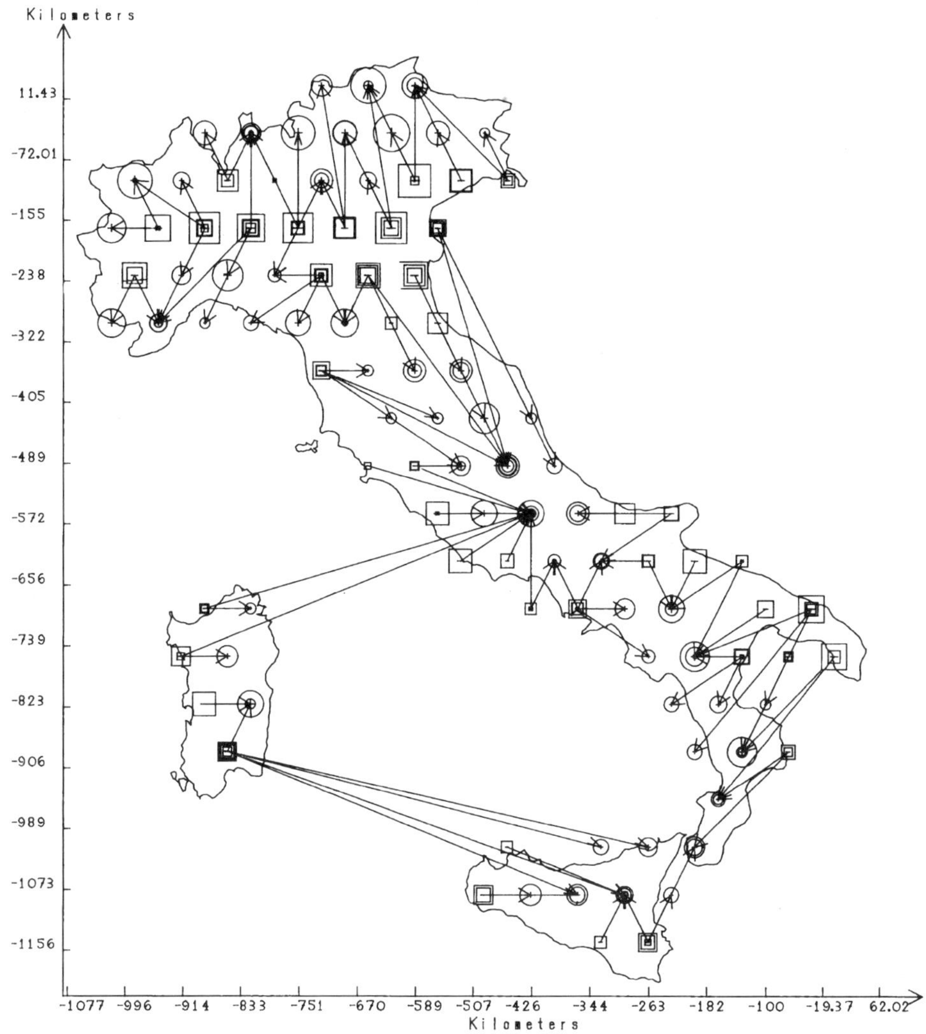

Figure 13.6. Italy. The "optimal transportation scheme" (minimizing linkages) between the mountains and the plain: Key: squares = origins; circles = destinations. The arrows indicate the direction of the shifted units; the areas of the circles and squares are proportional to the number of shifted units.

over that of the union by Δ, where $\Delta = A_4/A = A_5/A$. We assume that we distribute M dots uniformly over T_4 and M dots uniformly over T_5 to link them in the minimizing way, so as to obtain ${}^m\bar{D}(A_4, A_5)$. Multiplying this value by Δ we obtain ${}^m\bar{D}(1, 2) = \Delta {}^m\bar{D}(4, 5)$.

3. The approach described under (2) can be used—among other applications—in the measurement of asymmetry (see Appendix H).

4. The approaches described under (1) and (2) can be used for the summary comparison between any pair of two geometrical figures kept apart from each other or made concentric and rotated around their common center.

Table 13.8. Italy (1981). Minimal Distances by Aggregation Levels

Aggregation levels	Average of minimal distance between pairs of geosets (km)	Percentage of surplus elements to be shifted (Δ)	Minimal distance covered by shifts (km)
250 cells	93	29.7	315
95 provinces	93	26.3	358
20 regions	89	19.0	458

13.7. The Influence of Aggregation of Input Data over Minimal Distance

We discussed in Section 8.3 the well-known shortcoming of concentration parameters (as Δ) of changing their level with changing type and refinement of the network on which the input data are based. Exploratory work done over various geosets suggests that this defect is by far less significant in regard to minimal distance. Some examples follow:

1. Napolitano (1990) calculated the 325 minimal distances between pairs formed by 26 geosets for tertiary activities in Italy (1981) at three alternative levels of aggregation: 250 hexagonal cells, 95 provinces, and 20 regions. The findings are shown in Table 13.8. Passing from aggregations by cells to aggregations by provinces: (a) the average level of minimal distance between pairs of geosets remains practically unaffected, and the cogradation is practically perfect. On the other hand, the average level of Δ decreases. Passing to the very rough aggegation by 20 regions, Δ decreases very significantly, while the average level of minimal distance is by far less affected.
2. Studies done for some cities comparing results of aggregation by cells versus aggregation according to centers of regions also show considerable stability of levels of minimal distance.
3. Suppose we calculate the minimal distance between two geosets ${}^{g}P$ and ${}^{g}Q$ that have the same number of elements, as we did in the examples of Section 13.3. In this setting, each element of ${}^{g}P$ is linked to one element of ${}^{g}Q$. The problem of aggregation effects disappears completely and minimal distance is perfectly accurate.

Chapter 14

Properties of Minimal Distance

Triangle of Distances, Joint Analysis of Several Minimal Distances, and Measuring Inconsistency

14.1. Minimal Distance as a Metric

The minimal distance ${}^{m}\bar{D}(G, J)$ between two-dimensional geosets ${}^{g}G$ and ${}^{g}J$ has important properties, analogous to those of the minimal distance between a pair of one-dimensional sets. Napolitano (1988, 1990) demonstrated that ${}^{m}\bar{D}(G, J)$ is a metric, which implies that:

1. ${}^{m}\bar{D}(G, J)$ is never negative.
2. ${}^{m}\bar{D}(G, J) = 0$ if, and only if, the two geosets ${}^{g}G$ and ${}^{g}J$ are equal, so that ${}^{m}\bar{D}(G, J) = {}^{m}\bar{D}(G, G) = {}^{m}\bar{D}(J, J)$; conversely, if ${}^{m}\bar{D}(G, J) = 0$ then we can conclude that the two geosets are coincident.
3. ${}^{m}\bar{D}(G, J) = {}^{m}\bar{D}(J, G)$: when we refer to the simple scheme of shifting elements of one of two geosets in order to make the two geosets coincident, this equality implies that it does not matter whether elements of ${}^{g}G$ are shifted to make ${}^{g}G$ coincident with ${}^{g}J$, or elements of ${}^{g}J$ are shifted to make ${}^{g}J$ coincident with ${}^{g}G$.
4. If we consider three geosets ${}^{g}G$, ${}^{g}J$, and ${}^{g}K$ and the minimal distance between each pair of them, we find that

$${}^{m}\bar{D}(G, K) + {}^{m}\bar{D}(K, J) \geqslant {}^{m}\bar{D}(G, J) \tag{14.1a}$$

$${}^{m}\bar{D}(G, J) + {}^{m}\bar{D}(J, K) \geqslant {}^{m}\bar{D}(G, K) \tag{14.1b}$$

$${}^{m}\bar{D}(K, G) + {}^{m}\bar{D}(G, J) \geqslant {}^{m}\bar{D}(K, J) \tag{14.1c}$$

14.2. Minimal Distances between Three Geosets

14.2.1. Triangularity or Collinearity and Measures of Inconsistency

Let us consider any of the equations given above, e.g., equation (14.1a). We can distinguish between two basic configurations of the three minimal distances

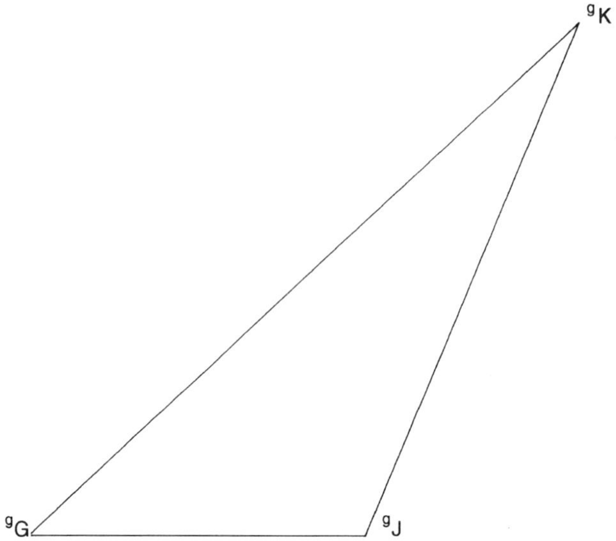

Figure 14.1. Fictitious example of triangle of minimal distance.

considered:

$$^m\bar{D}(G, K) + {}^m\bar{D}(K, J) > {}^m\bar{D}(G, J) \tag{14.2a}$$

$$^m\bar{D}(G, K) + {}^m\bar{D}(K, J) = {}^mD(G, J) \tag{14.2b}$$

In configuration (14.2a) we have the triangular inequality, which can be represented graphically as shown in Figure 14.1.

The basic idea underlying the presentation of configuration (14.2a) over a triangle is as follows: We consider the three minimal distances $^m\bar{D}(G, J)$, $^m\bar{D}(G, K)$, and $^m\bar{D}(K, J)$. If a triangle is constructed with sides proportional to these distances its shape is uniquely determined. The geosets ^gG, ^gJ, and ^gK are represented as points that are the vertices of the triangle. The scale and position of the triangle are determined as a matter of convenience depending on the nature of the particular geosets and the objectives of the research.

If $^m\bar{D}(G, K) + {}^m\bar{D}(K, J) = {}^m\bar{D}(G, J)$, we obtain configuration (14.2b), in which the three points representing ^gG, ^gJ, and ^gK lie on the same line. The triangle of minimal distances thus degenerates to a line of minimal distances. Conversely, if ^gG, ^gJ, and ^gK lie on the same line, then $^m\bar{D}(G, K) + {}^m\bar{D}(K, J) = {}^m\bar{D}(G, J)$. Such geosets can be termed as collinear. Perfect or approximate collinearity of geosets has been found empirically in many examples.

14.2.2. Triangle of Minimal Distances

By way of illustration we consider the minimal distances between the three geosets shown in Figures 14.2 and 14.3. In these figures we deal with the distributions

Figure 14.2. Italy. Minimal distance between births, deaths, and population, 1982 (Source of data: ISTAT).

of births (denoted by ${}^gB$, in place of ${}^gG$) and deaths (denoted by ${}^gD$, in place of ${}^gJ$) in Italy in 1982. In Figure 14.2 ${}^gB$ and ${}^gD$ are compared to the geoset of population (denoted by ${}^gP$, in place of ${}^gK$). In Figure 14.3, ${}^gB$ and ${}^gD$ are compared to the uniform distribution over the territory of Italy (denoted by ${}^gU$, in place of ${}^gK$).

The minimal distances (in km) represented in Figure 14.2 are ${}^m\bar{D}(B, P) = 59.5$, ${}^m\bar{D}(P, D) = 21.1$, and ${}^m\bar{D}(B, D) = 80.5$. Since ${}^m\bar{D}(B, P) + {}^m\bar{D}(P, D) = 80.6$, the configuration represented by Figure 14.2 shows perfect collinearity of ${}^gB$, ${}^gP$, and ${}^gD$. This collinearity is explained by the facts that (a) the distribution of births and deaths are both rather closely connected to that of the population; (b) birth rates (say, per 1000 of population) are much higher in several less developed and more traditional-religious regions of the south and the islands; (c) death rates are higher in several regions of the north and the center, where the proportions of elderly people are higher; (d) because (b) and (c) produce opposite settings of births and deaths with respect to population, there is perfect consistency between ${}^m\bar{D}(B, D)$ and ${}^m\bar{D}(B, P) + {}^m\bar{D}(P, D)$.

In Figure 14.3, instead of comparing births and deaths with population, we have compared them with the uniform distribution over the territory and have obtained: ${}^m\bar{D}(B, U) = 54.9$, ${}^m\bar{D}(D, U) = 48.5$, ${}^m\bar{D}(B, U) + {}^m\bar{D}(U, D) = 103.4$, as compared with ${}^m\bar{D}(B, D) = 80.5$.

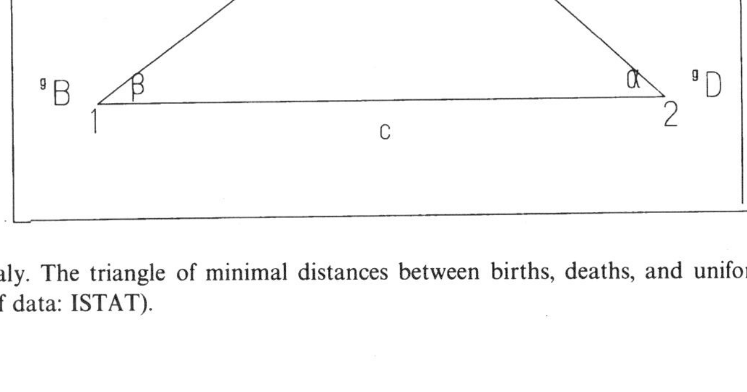

Figure 14.3. Italy. The triangle of minimal distances between births, deaths, and uniform distribution, 1982 (Source of data: ISTAT).

Let us abbreviate the notation by replacing ${}^{m}\bar{D}(B, U)$, ${}^{m}\bar{D}(D, U)$, and ${}^{m}\bar{D}(B, D)$ with a, b, and c, respectively. Figure 14.3 shows a, b, and c by means of a triangle whose vertices are conventionally denoted by 1, 2, and 3. Vertices 1, 2, and 3 represent ${}^{g}B$, ${}^{g}D$, and ${}^{g}U$, respectively. The sides $\overline{13}$, $\overline{23}$, and $\overline{12}$ of the triangle have lengths proportional to a, b, and c, respectively. The symbols a' and b' denote the ratios a/c and b/c.

Let us denote the angles $1\hat{2}3$, $2\hat{1}3$, and $1\hat{3}2$ by α, β, and γ, respectively. The size of these angles can be obtained as follows: $\cos\alpha = (b^2 + c^2 - a^2)/2bc$, $\cos\beta = (a^2 + c^2 - b^2)/2ac$, and $\cos\gamma = (a^2 + b^2 - c^2)/2ab$. Accordingly, in our comparison of ${}^{g}B$ and ${}^{g}D$ to ${}^{g}U$ we obtain $\alpha = 41.80°$, $\beta = 36.2°$, and $\gamma = 102.0°$.

We can measure inconsistency: (a) by means of the ratio to $(a' + b' - 1) = 28.5\%$; (b) by means of the inconsistency angle $180° - \gamma = \alpha + \beta = 78°$ (in our example). Consistency is measured by $\gamma = 102°$. In the comparison of ${}^{g}B$ and ${}^{g}D$ with ${}^{g}P$ shown in Figure 14.2, both the inconsistency ratio and the inconsistency angle are 0.

Given the minimal distances a, b, and c, the shape of their triangle is uniquely determined. However, position and scale can be selected at will, depending on expediency and the objectives of the research.

In the example of Figure 14.3, side $\overline{12}$ of the triangle is described along the conventional axis of the abscissas; the coordinates of the vertices are: $X_1 = 0$; $Y_1 = 0$; $X_2 = c$; $Y_2 = 0$; $X_3 = b\cos\alpha$; $Y_3 = b\sin\alpha$.

The scale selected in Figure 14.3 shows ratios a', b', and 1 of (a, b, and c) to c. This presentation enables us to compare the consistency or inconsistency of the distances numerically. We might at some point be interested in using a scale that enables us to read the absolute values of a, b, and c. In other settings, we might prefer to present ratios of a, b, and c to the total $a + b + c$, thus obtaining a "unitary triangle." In still other instances, we might want to compare triangles for various geosets, netting them from the influence of the size of the territories T over which each one of the triplets of geosets is distributed. When this is the case, we can calculate the ratios of a, b, and c to a general measure of territorial dispersion as $\sqrt{(2d_T)}$, where d_T is the standard distance of the territory.

14.2.3. Consistencies and Inconsistencies in Minimal Distances of Ethnic Subpopulations in the United States

Let us return to the material in Section 13.5.4. We denote the geosets of any two subpopulations of the United States by ${}^{g}G$ and ${}^{g}J$ and the geoset of the total population by ${}^{g}K$.

Table 13.7 (Section 13.5) shows the values of all the ${}^{m}\bar{D}(G, K)$, ${}^{m}\bar{D}(J, K)$, and ${}^{m}\bar{D}(G, J)$. We have compared $[{}^{m}\bar{D}(G, K) + {}^{m}\bar{D}(J, K)]$ to each ${}^{m}\bar{D}(G, J)$, and Table 14.1 shows inconsistency ratios and inconsistency angles obtained from these comparisons.

For four comparisons the degree of consistency is quite remarkable and suggests that distances between the total population and each subpopulation are the basic factors of the distance between pairs of subpopulations. However, this is not true for

Table 14.1. United States (1980). Inconsistency of Distances between Subpopulations and Total Population

Geosets compared	Inconsistency ratio (%)	Consistency angle (degrees)
White, Total, Afro-Americans	5.6	71.8
White, Total, Hispanic	3.3	33.7
White, Total, Asian	2.3.	69.3
Afro-Americans, Total, Hispanic	8.9	50.8
Afro-Americans, Total, Asian	4.4	36.9
Asian, Total, Hispanic	328.00	153.4

the distance between Hispanics and Asians as compared to the sum of the distances between (a) Hispanics and total and (b) Asians and total, and the reasons are immediately apparent when we look at Figures 8.7 and 8.9 (Section 8.8).

14.2.4. Minimal Distances of a Geoset from Two Paradigmatic Distributions

The study of a geoset can often be enhanced by analyzing the minimal distance between it and meaningful and important geosets taken as paradigms in our research. We referred to this approach in Sections 13.4 and 13.5, and there are examples of it in Appendix IV, of the second volume of the *Statistical Atlas of Italy* (1992) where hundreds of minimal distances are listed for demographic, economic, and social geosets compared to the geosets of: (a) the uniform distribution over the territory of Italy (${}^{g}U$); (b) the distribution of the population of Italy according to the census of 1981 (${}^{g}P$); and (c) the distribution of persons employed in the economic activities enumerated in the census of industry, trade, and services of 1981 (${}^{g}E$).

The geoset ${}^{g}U$ depends on the shape and dimensions of the territory. Examples of minimal distances from ${}^{g}U$ were discussed in Section 13.4.2. Here we relate to distances from ${}^{g}P$ and ${}^{g}E$. The meaning of ${}^{g}P$ is obvious; ${}^{g}E$ can be considered as a global indicator of the territorial distribution of economic activities. Table 14.2 lists examples of distances from ${}^{g}P$ and ${}^{g}E$ for 10 geosets, generally denoted as ${}^{g}K$. ${}^{m}\bar{D}(K, P)$ and ${}^{m}\bar{D}(K, E)$ have been standardized by dividing them by ${}^{m}\bar{D}(P, E)$, i.e., by the distance between the population and economic activities.

The geosets are ordered by increasing inconsistency ratio

$$[{}^{m}\bar{D}(K, P) + {}^{m}\bar{D}(K, E) - {}^{m}\bar{D}(P, E)]/{}^{m}\bar{D}(P, E)$$

The first three geosets have no or almost no inconsistency and are perfectly or almost perfectly included between the geosets of population and economic activities. Retail trade and individuals in the labor force are nearer to the population than to economic activities. This is shown graphically for the first geoset in Figure 14.4, where the dots representing the three geosets compared are collinear, and ${}^{g}K$ is quite close to ${}^{g}P$. Geosets 4, 5, 6, and 7 also appear to be influenced by both ${}^{g}P$ and ${}^{g}E$. However, their inconsistency ratios are much higher than those for geosets 1, 2, and 3. Geosets 4, 5, and 6 are close to the economic activities: for obvious reasons

Table 14.2. Italy (1981). Minimal Distances of Selected Geosets from the Population Geoset and from the Geoset of Economic Activities[a]

Geoset	Distance from population $^m\bar{D}(K, P)/^m\bar{D}(P, E)$	Distance from economic activities $^m\bar{D}(K, E)/^m\bar{D}(P, E)$	Inconsistency ratio (%)	Inconsistency angle (degrees)
1. Total labor force	0.32	0.68	0.0	0
2. Retail trade	0.22	0.79	1.0	19.6
3. Tertiary activities	0.43	0.59	2.0	23.0
4. Divorced women	0.99	0.36	35.0	98.9
5. Financial institutions	1.19	0.35	54.0	130.3
6. Wholesale trade	1.40	0.45	85.0	157.3
7. Public administration	0.58	1.54	112.0	163.0
8. Petrol industry	1.32	1.98	230.0	153.1
9. Metal and refined mech.	2.44	1.45	289.0	175.7
10. Illiterates	2.94	3.94	588.0	180

[a]Ratios to $^m\bar{D}(P, E) = 87.8$ km.

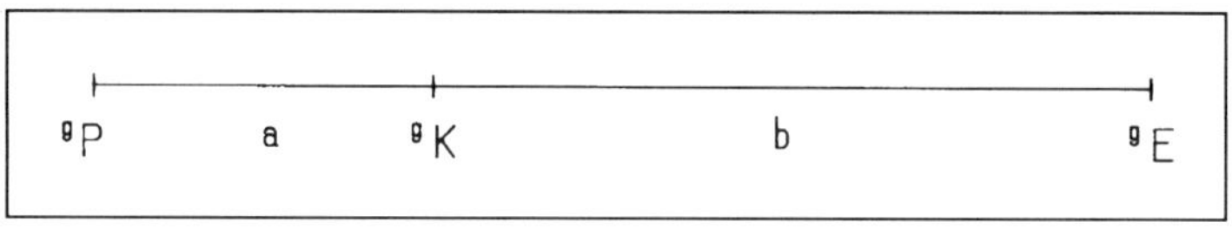

Figure 14.4. Italy. Minimal distance between population (P), labor force (K), and economic activities (E), 1982 (Source of data: ISTAT).

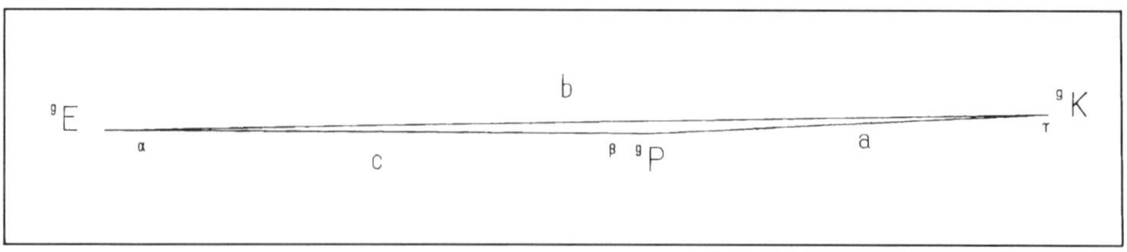

Figure 14.5. Italy. The triangle of minimal distances between population (P), illiterates (K), and economic activities (E), 1982 (Source of data: ISTAT).

relating to financial institutions and wholesale trade. The closeness to ${}^gE$ of divorced women is less obvious but can be explained by the fact that there are relatively more divorces in less religious, very urban, and more economically developed regions. Geosets 8, 9, and 10, which have very peculiar distributions, show little, if any, links with either ${}^gP$ or ${}^gE$.

The big inconsistency between the distances from illiterates is shown by the triangle of Figure 14.5, and the reasons for it are evident from the maps of ${}^gK$, ${}^gP$, and ${}^gE$, given in Figures 9.9, 1.16, and 1.6, respectively.

The analysis of minimal distances for paradigmatic geosets thus appears to be a useful instrument for research, as it may suggest or prove hypotheses with regard to factors influencing the geosets being surveyed.

14.2.5. Minimal Distances between Three Ordered Geosets

Let us now consider the very important special configuration of minimal distances between ordered geosets, i.e., geosets that refer to sets forming a natural sequence, such as, e.g., a population considered at successive times, classes of a population by income, classes of families by the number of children, or classes of farms by size.

Table 14.3. Italy. Minimal Distance between the Distributions of the Population on Two Successive Censuses[a]

Successive censuses	a	a_i	Nonsuccessive censuses	b_i	Inconsistency ratio[b]
1951 and 1961	6.7				
1961 and 1971	13.6				
		(a_1) 20.3	1951 and 1971	(b_1) 19.1	6.3
1961 and 1971	13.6				
1971 and 1981	4.8				
		(a_2) 18.4	1961 and 1981	(b_2) 10.0	84.0
1971 and 1981	4.8				
1981 and 1987	6.8				
		(a_3) 11.6	1971 and 1987	(b_3) 11.5	0.9
1951 and 1971	19.1				
1971 and 1987	11.5				
		(a_4) 30.7	1951 and 1987	(b_4) 13.0	136.2

[a] a = minimal distance (km) and a_i = sum of distances (km). b_i = minimal distance between two nonsuccessive censuses.
[b] The inconsistency ratio $= 100[(a_i - b_i)]/b_i]$.

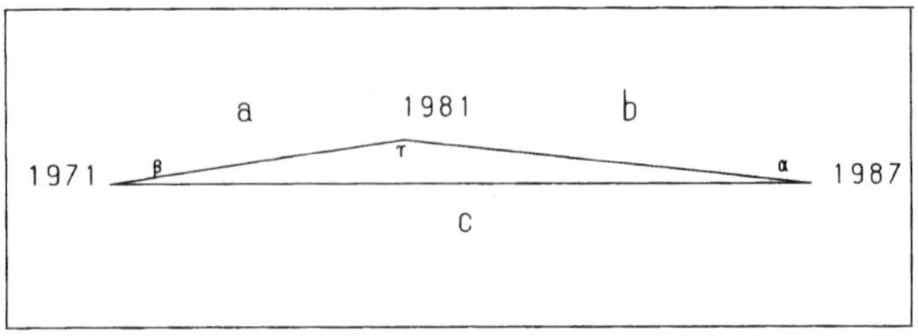

Figure 14.6. Italy. The triangle of minimal distances between populations in 1971, 1981, and 1987 (Source of data: ISTAT).

We shall consider here, as an example, triplets of geosets showing the distribution of a population at three successive times (1, 2, 3). This study might have several different aims. For example we might (i) be interested in the analysis of the actual degree of change that has occurred in the distribution during the period in question; (ii) want to evaluate the relative extent of changes of population distribution by comparing them to proper measures of the dispersion of the actual population or to the uniform distribution over territory (an approach that enables us to compare the degree of variability of territorial distributions in different countries, cities, periods); (iii) wish to determine the degree of agreement or disagreement of changes in the

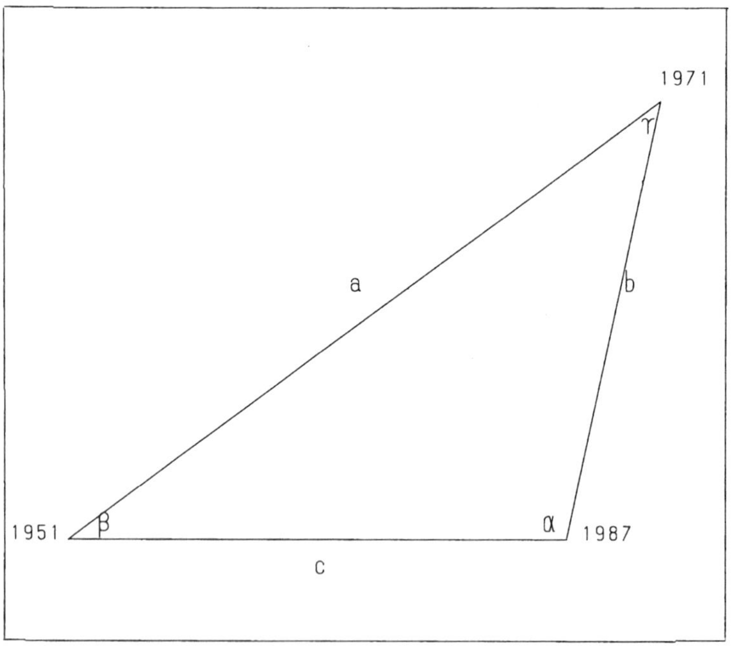

Figure 14.7. Italy. The triangle of minimal distances between population of 1951, 1971, and 1987 (Source of data: ISTAT).

distribution that have occurred in the period (1, 2) and in the period (2, 3); (iv) wish to analyze study (iii) for a long chain of periods.

Here we shall concentrate on approach (iii). Table 14.3 shows such triplets of minimal distances for the population of Italy at various periods. These parameters have been singled out from a study covering the evolution of the population of Italy from 1871 to 1987 (see the *Statistical Atlas of Italy* (1992), Vol. 2, pp. 109–119).

From the specific examples in Table 14.3 it appears that the minimal distance between 1951 and 1971 (19.1 km) is just a little less than the sum (20.3) of the minimal distance between 1951 and 1961 plus the minimal distance between 1961 and 1971.

We can build a measure of the degree of inconsistency between the changes that occurred in 1951–1961 and 1961–1971 by subtracting 1 from the ratio (1.063) between 20.3 and 19.1 km. This ratio is close to 0 (which would indicate a perfect consistency between the evolution of the distribution in the two periods considered). For 1971–1981 and 1981–1987, the consistency is almost perfect, which is seen clearly in Figure 14.6. By contrast, between 1951–1971 and 1971–87 there is a strong disagreement measured by a ratio of 136% (see Figure 14.7).

The change in trends that occurred between the periods 1951–1971 and 1971–1987 corresponds very well with the expectations based on the detailed analysis of the determinants of change given in the *Atlas*. This analysis shows that the determinants of distribution changes of the population of Italy from 1951 to 1987 were extremely complex. For example, the modern trends of nuptiality and fertility have led to a significant decline in the birthrate, which has been much more pronounced in the more developed regions of the northwest regions than in those of more traditionalist (Catholic) and economically less progressive regions in the south and the islands; mortality has declined everywhere, but the deathrate is considerably lower in the south owing to a smaller proportion of elderly people; the tendency to emigrate, which was once very marked in the south, weakened greatly during the

Table 14.4. Rome (1981). Minimal Distance between Geosets of Dwellings According to the Period of Building

	(1) Up to 1919	(2) 1919–1945	(3) 1946–1960	(4) 1961–1970	(5) 1971–1981
A. Minimal distance between geosets (meters)					
(2)	2563				
(3)	3975	2122			
(4)	4060	2406	1130		
(5)	5201	3871	2111	1874	
B. Rough estimate of average distance between periods of building compared					
(2)	32				
(3)	53.5	21.5			
(4)	66	34.0	12.5		
(5)	76	44.0	22.5	10	

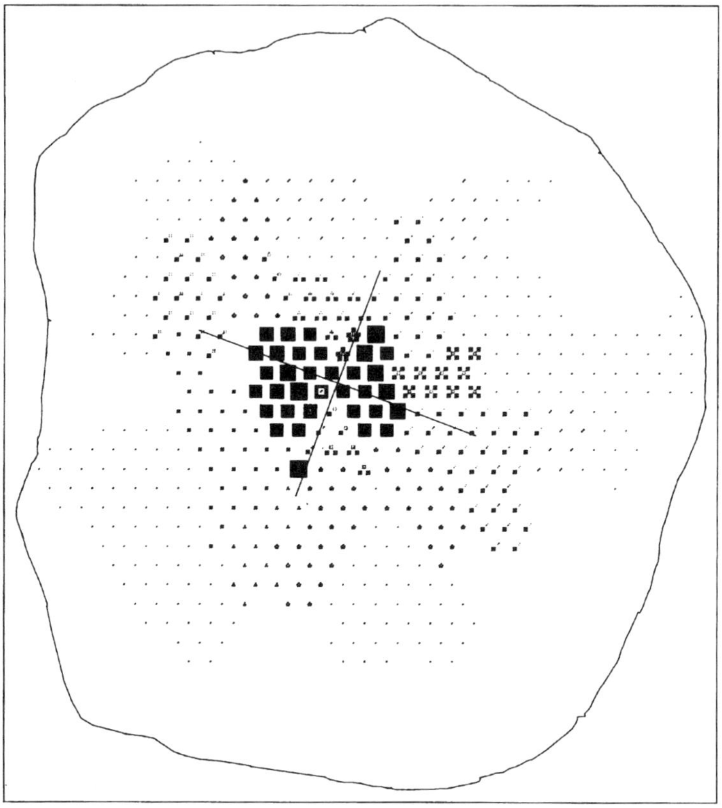

Figure 14.8. Rome. Dwellings built up to 1919: . 5 ⠂56 ⠴118 ⧫292 ■500+ (Source of data: Comune di Roma, Segretariato Generale il Direzione, Ufficio di Statistica e Censimento).

period considered. However, the strong contrast between 1951–1971 and 1971–1987 is largely explained by the reversal in two dominant phenomena: the enormous internal migrations during 1951–1971 period from the poorer south to the quickly developing northwestern metropolitan areas were drastically reduced after 1971, and at the same time the strong movement from rural to urban zones, which held sway for over a century and was very marked in 1951–1971, subsided or was even reversed after 1971.

14.2.6. Minimal Distances between Many Ordered Geosets

For any pair (K_i, K_j) of N geosets K_i $(i = 1, 2, \ldots, N)$ that refer to different levels V_i of a variable (such as income or age), we can measure the minimal distance ${}^m\bar{D}(K_i, K_j)$ and the difference between the levels (V_i, V_j) of the variable. Joint analyses of spatial distances and variations of levels might be of considerable interest.

As an example of this approach, Table 14.4 refers to the geosets of five classes of dwellings in Rome (1981), classified according to when they were built. The

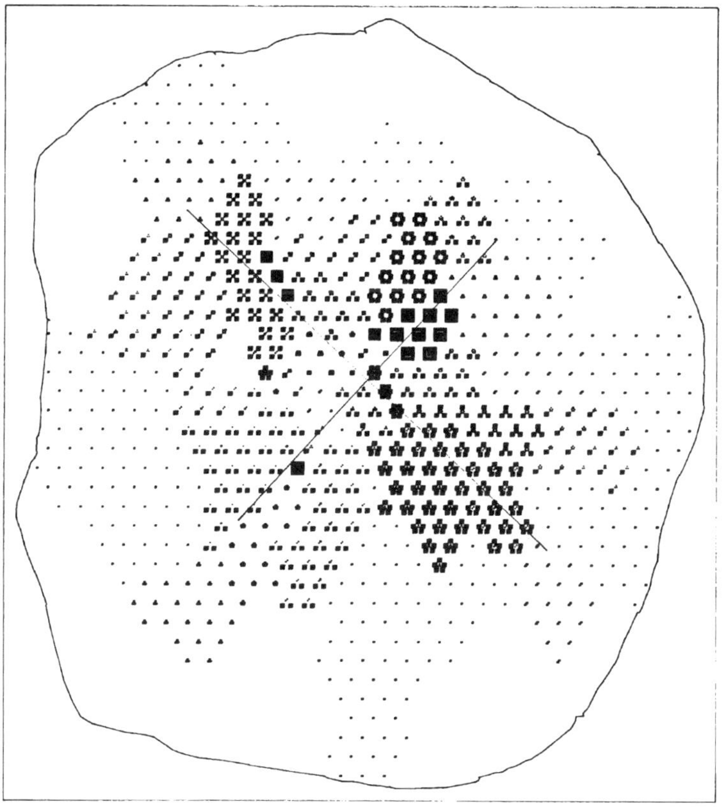

Figure 14.9. Rome. Dwellings built between 1919 and 1945: . 12 ▪ 113 ∴ 265 ✿ 656 ■ 1200+ (Source of data: ISTAT).

relevant maps are given by Figures 14.8–14.12, respectively. Part A of the table shows the minimal distance between each of the 10 pairs of geosets. Part B gives a rough evaluation of the average difference in the time of construction of the dwellings compared for each pair.

The correlation between territorial and time differences is very high ($r = 0.962$), which suggests that the matrix of distances can be interpreted in a very simple way: changes in the territorial distribution of dwellings tend to occur over the course of time somewhat consistently.

14.3. Collinearity of Three Geosets, ^gK, ^gG, and ^gJ, Related Through $^gK = v_G^{Kg\,g}G + v_J^{Kg\,g}J$

In Sections 3.6. and 3.8, we gave examples of triplets of geosets of territories or populations (say, ^gG, ^gK, and ^gJ) that are related through $^gK = v_G^{Kg\,g}G + v_J^{Kg\,g}J$ (for $v_G^K + v_J^K = 1$). Minimal distances between (G, K), (K, J), and (G, J) are related in a

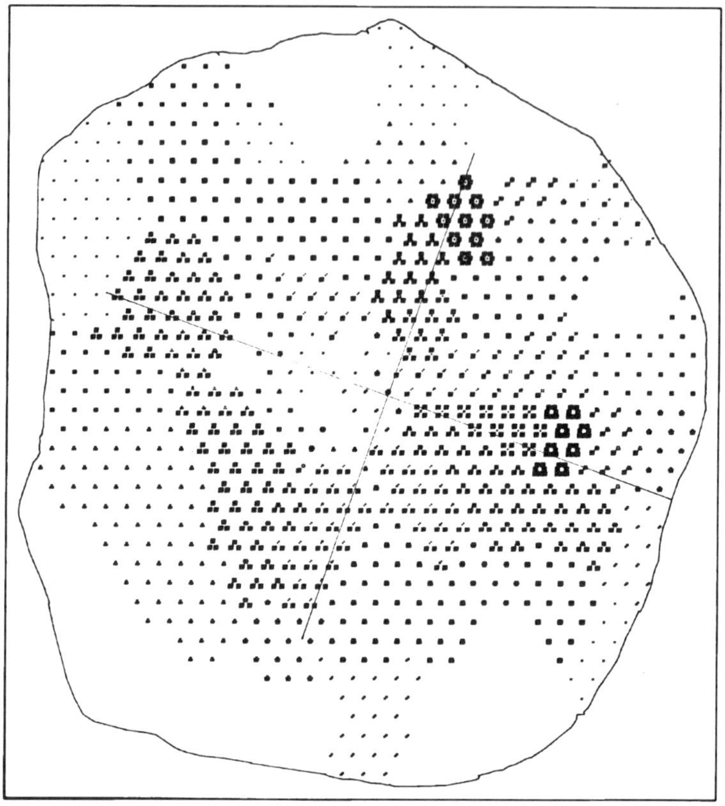

Figure 14.10. Rome. Dwellings built between 1946 and 1960: ·26 ▪260 ⁘486 ⁂1562 ▣1793 (Source of data: Comune di Roma, Segretariato Generale il Direzione, Ufficio di Statistica e Censimento).

very simple way, as shown by the hypothetical example shown in Figure 14.13. In this figure ${}^{g}G$, ${}^{g}K$, and ${}^{g}J$ are represented by three points, which are collinear.

If we assume that ${}^{m}\bar{D}(G, J) = 10$ arbitrary linear units, and $v_G^K = 0.3$ and $v_J^K = 0.7$, we find that:

$$ {}^{m}\bar{D}(G, K) = v_J^{K\,m}\bar{D}(G, J) = 7 \tag{14.3a}$$

$$ {}^{m}\bar{D}(K, J) = v_G^{K\,m}\bar{D}(G, J) = 3 \tag{14.3b}$$

$$ {}^{m}\bar{D}(G, K) + {}^{m}\bar{D}(K, J) = {}^{m}\bar{D}(G, J) \tag{14.3c}$$

These three geosets can thus be considered as collinear (see Section 14.2).

In equations (14.1a) and (14.1b) v_J^K and v_G^K are inverted with respect to ${}^{g}K = v_G^{K\,g}G + v_J^{K\,g}J$. The ratios v_G^K and v_J^K can have different meanings depending on the nature of the geosets considered. One very important setting is that in which these values denote the proportion of the sizes of ${}^{g}G$ and ${}^{g}J$ as compared with that

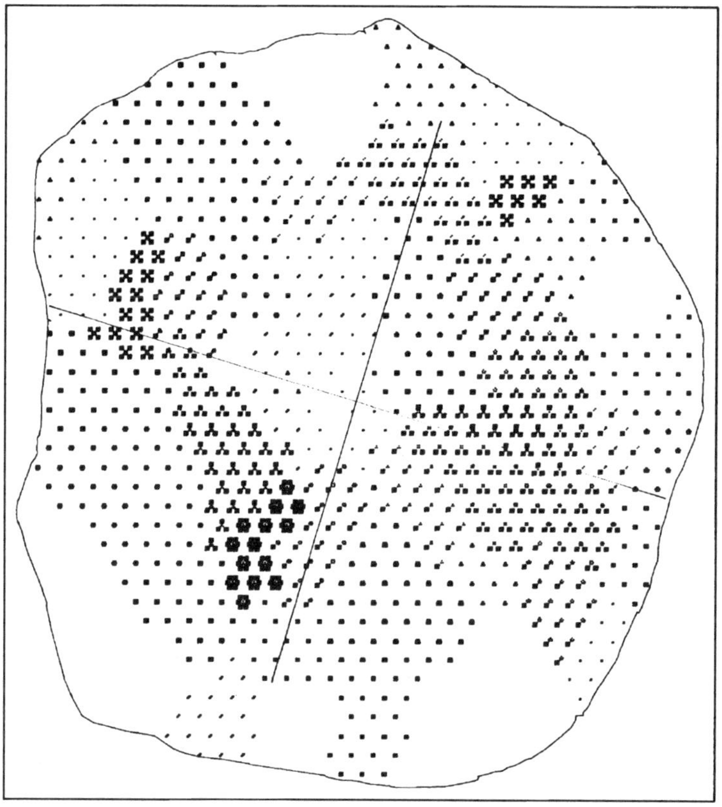

Figure 14.11. Rome. Dwellings built between 1961 and 1970: . 24 ▪ 245 466 834 1546 (Source of data: Comune di Roma, Segretariato Generale il Direzione, Ufficio di Statistica e Censimento).

of ${}^{g}K$ ($v_G^K = F^G/F^K$ and $v_J^K = F^J/F^K$ for $F^G + F^J = F^K$). Graphically seen, the segment representing ${}^{m}\bar{D}(G, J)$ is to be divided into two parts whose lengths are proportional to v_J^K and v_G^K, respectively.

Figure 14.14 shows some examples of collinear triplets of geosets related as explained above. All these examples were taken from the analysis of geosets of dwellings enumerated in Rome (1981) and divided into two classes (built up to 1945 and between 1946 and 1981). These geosets are denoted by ${}^{g}G$ and ${}^{g}J$, respectively, while the geoset of all dwellings is denoted by ${}^{g}K$. Characteristics of these geosets were commented on in Section 3.6. The relevant maps are shown in Figures 3.17–3.19 and 3.23–3.25 (Section 3.7).

The minimal distances represented were obtained directly from the computer program and the algebraic check of the findings using our equations gave very satisfactory results, apart from some very small irregularities, which are possibly due to the relatively small number of input points (79 districts). It should be noted that the use of a finer resolution (200 points) would have led to almost identical results.

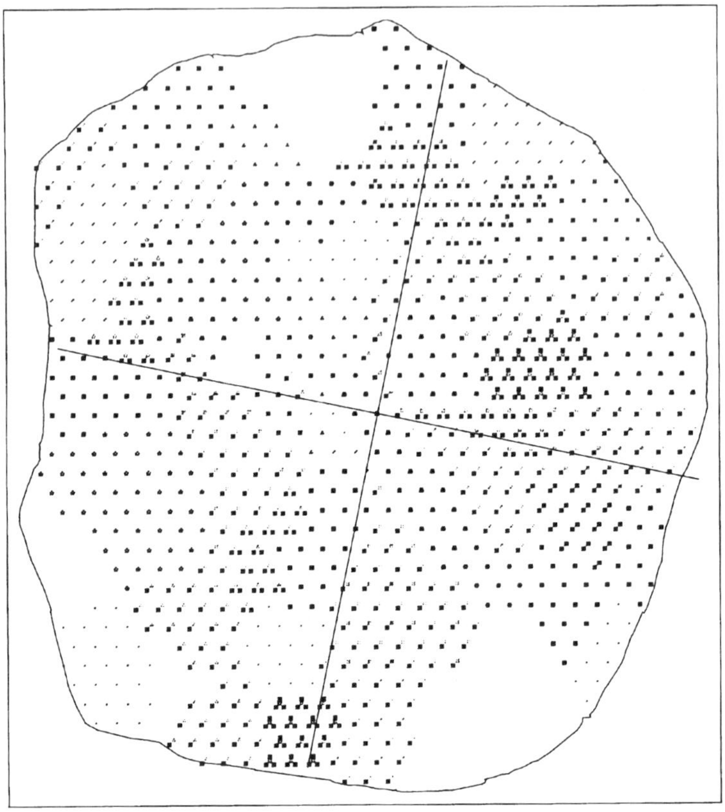

Figure 14.12. Rome. Dwellings built between 1971 and 1981: • 9 ▪ 93 ▪ 187 ▪ 290 ▪ 356 (Source of data: Comune di Roma, Segretariato Generale il Direzione, Ufficio di Statistica e Censimento).

Figure 14.14 shows the following lines of geosets:

(A) ${}^{g}G$, ${}^{g}K$, ${}^{g}J$. Here ${}^{m}\bar{D}(G, K) = 1929$ m; ${}^{m}\bar{D}(K, J) = 603$ m; ${}^{m}\bar{D}(G, J) = 2532$ m; $v_G^K = F^G/F^K = 23.81\%$; $v_J^K = F^J/F^K = 76.19\%$.

(B) ${}^{g}SGJ$, ${}^{g}G$, ${}^{g}M(G, J)$. Here ${}^{m}\bar{D}[(SGJ, M(G, J)] = \overline{13} = 3181$ m; ${}^{m}\bar{D}(SGJ, G) = \overline{14} = 1396$ m; ${}^{m}\bar{D}[G, M(G, J)] = \overline{43} = 1785$ m; Δ = proportion of surplus elements assumed to be shifted, $\Delta = 56.12\%$; $(1 - \Delta)$ = proportion of matching elements = 43.87%.

(C) ${}^{g}SJG$, ${}^{g}J$, ${}^{g}M(J, G)$. Here ${}^{m}\bar{D}[SJG, M(J, G)] = \overline{23} = 2296$ m; ${}^{m}\bar{D}(SJG, J) = \overline{25} = 1007$ m; ${}^{m}\bar{D}[J, M(J, G)] = \overline{53} = 1289$ m; and $(1 - \Delta)$ as indicated under (B).

(D) and (E) To make comparisons possible, we have added segments proportional to ${}^{m}\bar{D}(G, J) = \overline{45} = 2532$ m and ${}^{m}\bar{D}(SGJ, SJG) = \overline{12} = 4511$ m, respectively.

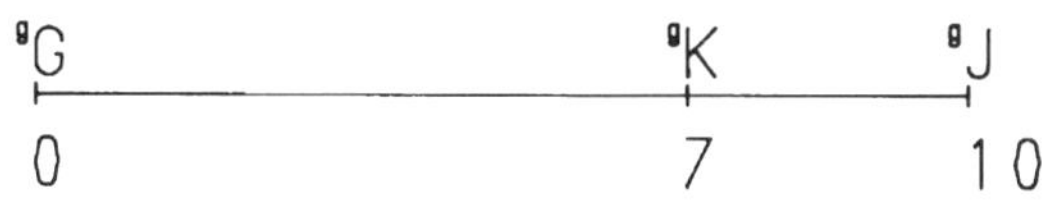

Figure 14.13. Fictitious example of ${}^m\bar{D}(G, K)$ and ${}^m\bar{D}(K, J)$ for ${}^m\bar{D}(G, J) = 10$, $V_G^K = 0.3$, and $V_J^K = 0.7$.

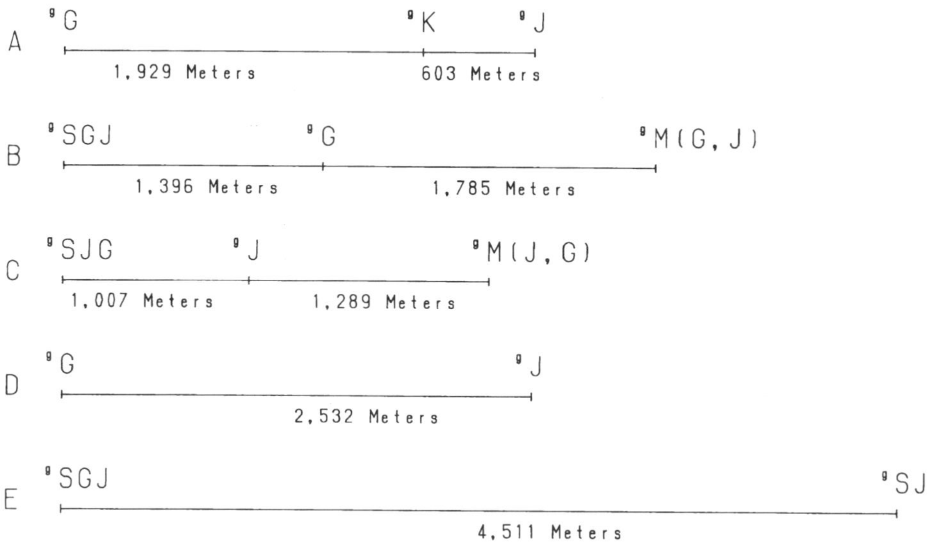

Figure 14.14. Rome. Example of collinear triplets of related geosets: Dwellings built up to 1945 (G), dwellings built between 1946 and 1981 (J), and total dwellings (K) (Source of data: Comune di Roma, Segretariato Generale il Direzione, Ufficio di Statistica e Censimento).

14.4. Triangle of Minimal Distances between Two Primary Geosets and Three Geosets Derived from Them

Our basic approach for a summary comparison between any two primary geosets ${}^gG$ and ${}^gJ$ is to measure the minimal distances between them and between the geosets of surpluses and matches derived from them. The graphical presentation suggested in Section 14.3 may be of some help here, but we might prefer a more comprehensive figure in which ${}^gG$, ${}^gJ$, ${}^gSGJ$, ${}^gSJG$, and ${}^gM(G, J)$ are all shown together in an integrated way. The distances between them can then all be made clear at the same time.

Let us return to Figure 14.14. Lines B and C bind the points representing

$$[{}^gSGJ, {}^gG, {}^gM(G, J)] \qquad \text{and} \qquad [{}^gSJG, {}^gJ, {}^gM(J, G)]$$

respectively. Because ${}^gM(G, J) = {}^gM(J, G)$, the two lines have the point that represents the geoset of the matches in common. This point can be taken as a vertex (labeled 3) of a triangle of distances, as shown in Figure 14.15. The two other vertices (labeled 1 and 2) represent ^gSGJ and ^gSJG. The distance between them is equal to that indicated by E in Figure 14.14. A line is drawn that is parallel to $\overline{12}$, has a length proportional to $^m\bar{D}(G, J)$, and is equal to D in Figure 14.14. The end points of this line are denoted by 4 and 5, and represent geosets ^gG and ^gJ, respectively.

Triangles 123 and 453 are similar. Thus, $\overline{43}/\overline{13}=\overline{53}/\overline{23}=\overline{45}/\overline{12}=\Delta$ and $\overline{14}/\overline{13}=\overline{25}/\overline{23} = (1 - \Delta)$, where Δ is the proportion of elements to be shifted. We can interpret our triangle, as follows: the direct object of our study is the distance $^m\bar{D}(G, J)$ as represented by $\overline{45}$. Part of these geosets ($^gG, {}^gJ$) is formed by the geoset $^gM(G, J) = 3$ of matching elements. These elements do not influence the distance $^m\bar{D}(G, J)$. The lines labeled $\overline{13}$ and $\overline{32}$ show synthetically the behavior of nonmatching elements as compared to matching elements.

The nonmatching elements are supposed to be shifted in order to make ^gG and ^gJ coincident. It is assumed that geoset ^gG represents the origins of the shifts, while geoset ^gJ represents their destinations. Comparing $(\overline{13} + \overline{32})$ to $\overline{12}$, we obtain a summary measure of the inconsistencies between the origins and the destinations of the shifts as compared to the matches.

In our example, the degree of inconsistency measured by $[(\overline{13}+\overline{32}-\overline{12})/\overline{12}] = 21.46\%$ suggests that the inconsistency is rather moderate. This fits well with the configurations of ^gG, ^gJ, ^gSGJ, ^gSJG and $^gM(G, J)$ as shown by Figures 3.18, 3.19 (Section 3.6) and 3.23–3.25 (Section 3.7) and with the history of construction in Rome during the last century: as a result of the somewhat concentric development, the dwellings built from 1946 to 1981 form a sort of periphery around those built up to 1945. To evaluate these findings, we take into consideration the nonlinearity. If consistency were perfect, origins (1) and destinations (2), would be linked along line $\overline{12}$ without the detour through vertex 3. In this case the inconsistency ratio would

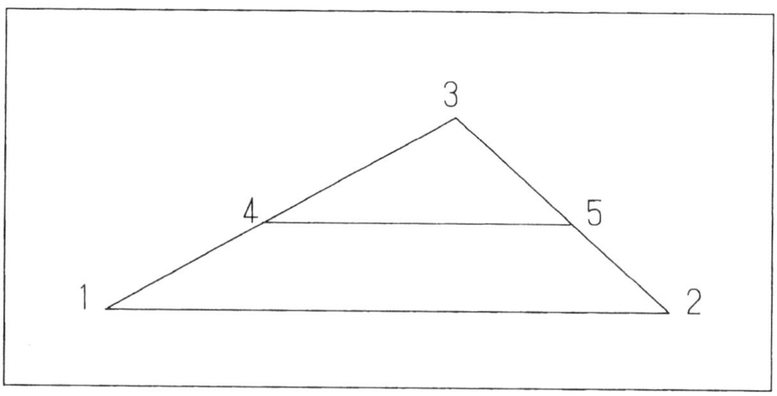

Figure 14.15. Rome. The triangle of minimal distances: Dwellings built up to 1945 (G), dwellings built between 1946 and 1981 (J) and total dwellings (K) (Source of data: Comune di Roma, Segretariato Generale il Direzione, Ufficio di Statistica e Censimento).

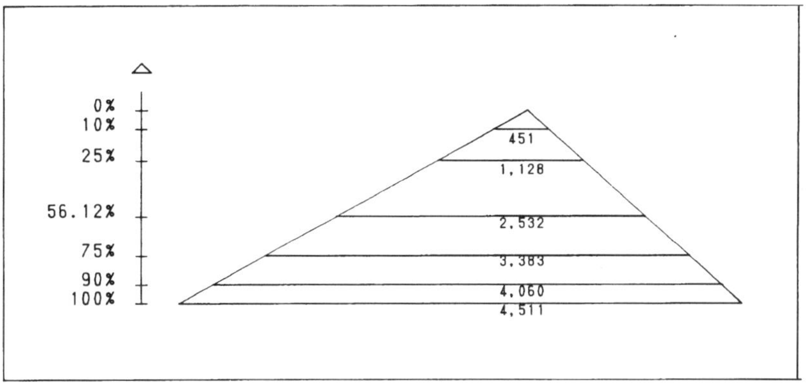

Figure 14.16. The triangle of minimal distances with a scale of percentage values of Δ (Source of data: Comune di Roma, Segretariato Generale il Direzione, Ufficio di Statistica e Censimento).

be 0 and the consistency angle would be 180°. If the dwellings built before and after 1945 had been distributed equally, we would find ${}^{g}G = {}^{g}J = {}^{g}M(G, J)$ and ${}^{m}\bar{D}(SGJ, SJG) \cong {}^{m}\bar{D}(G, J) \cong 0$.

The comparison of $\overline{45}$ to $\overline{12}$ enables us to envisage the value of $\overline{45}/\overline{12} = \Delta$, which shows the percentage of the elements to be shifted in order to make ${}^{g}G$ and ${}^{g}J$ coincident. In this example the percentage is considerable ($\Delta = 56.12\%$), as shown in Figure 14.16.

In Figure 14.16 we have kept the triangle (1, 2, 3) constant, but have added a scale of percentage values of Δ. In the correspondence of selected increasing values of Δ, we show on the triangle increasing values of the distance $\overline{45}$ between the origins ${}^{g}G$ and the destinations ${}^{g}J$, of the assumed shifts. In our example ${}^{m}\bar{D}(SGJ, SJG) = 4511$ m is constant and we have different values of Δ: 0, 10, 25, 56.12, 75, 90, and 100%. Corresponding values of ${}^{m}\bar{D}(G, J)$ are, respectively, 0, 451, 1128, 2532, 3383, 4060, and 4511. From this example and from the much wider experience of many analogous triangles used in various studies, we have seen that this graphical approach can be of considerable help in our analyses.

In Figures 14.15 and 14.16, the distances are expressed in absolute values (meters). For the purpose of comparing different types of geosets, different territories, or various times, we can use alternative values. For example:

1. If we wish to focus on measures of inconsistencies, independently of the absolute values of distances, we can give all distances as ratios to ${}^{m}\bar{D}(SGJ, SJG)$.
2. In some cases we might wish to show ratios to the length $(\overline{13} + \overline{32} + \overline{21})$ of the perimeter of the triangle (unitary triangle).
3. In comparisons among, e.g., different cities or different countries, we might prefer to standardize the relevant distances by a measure of dispersion of the corresponding territory.

14.5. Minimal Distance between Two Geosets as a Product of the Proportion of Surpluses and the Distance between Them

14.5.1. The General Rule $^m\bar{D}(G, J) = \Delta^m\bar{D}(SGJ, SJG)$

The equality $\overline{45}/\overline{12} = \Delta$, given in Section 14.4, can be written in the important form:

$$^m\bar{D}(G, J) = \Delta^m\bar{D}(SGJ, SJG) \tag{14.4}$$

which has already been illustrated by many examples.

We have explained this equation repeatedly by noting that if we transform ^gG into ^gJ, a proportion $(1 - \Delta)$ of the elements do not shift (or shift at distance 0), while the minimal distance between the origins and the destinations of the $(1 - \Delta)$ shifting elements is $^m\bar{D}(SGJ, SJG)$.

The minimal distance between ^gG and ^gJ can thus be obtained as a weighted average:

$$^m\bar{D}(G, J) = (1 - \Delta)0 + \Delta^m\bar{D}(SGJ, SJG) = \Delta^m\bar{D}(SGJ, SJG) \tag{14.5}$$

We saw in Figure 14.16 how changes in Δ can affect $^m\bar{D}(G, J)$ for constant $^m\bar{D}(SGJ, SJG)$. Also, for distances between cograduated values of one-dimensional sets, we found an analogous property in Section 13.2.

Splitting $^m\bar{D}(G, J)$ into Δ and $^m\bar{D}(SGJ, SJG)$ was found to be a very important tool for the analysis of minimal distances and also proved to be of considerable help in reducing the complex computer calculations.

Actually, the computer feature used in the Italian *Atlas* and that used for many examples prepared by means of the newer computer program are based on calculations at various stages. It starts by identifying elements that have identical locations in ^gG and ^gJ, namely the geoset of matches $^gM(G, J)$ and its relative size $(1 - \Delta)$. In a second stage, the calculation of minimizing distances is restricted to elements of ^gG and ^gJ that do not match. This leads to the calculation of $^m\bar{D}(SGJ, SJG)$, which is then multiplied by Δ. For high values of $(1 - \Delta)$ this way of calculating may lead to a considerable reduction of work.

14.5.2. Interpreting $^m\bar{D}(SGJ, SJG)$ and Δ in the Study of $^m\bar{D}(G, J)$

The two factors Δ and $^m\bar{D}(SGJ, SJG)$ of the minimal distance $^m\bar{D}(G, J)$ between the primary geosets ^gG and ^gJ are not functionally related. In order to use them as analytical tools, it is of interest to compare their mutual behavior on the basis of empirical data. This study can be done at different levels as, for example:

1. Following a very detailed and sometimes even microgeographic approach, we have used the hundreds of examples of minimal distances furnished by the *Statistical Atlas of Italy*. For the pairs of geosets compared we had at our disposal the values of $^m\bar{D}(G, J)$ of $^m\bar{D}(SGJ, SJG)$ and of Δ, as well as the relevant maps and the geostatistical parameters of the geosets, together with comments by experts on the substantive meaning of the available material.

It would be impossible to relate here to examples of this detailed analysis, but the interested reader can refer to Tables 11.14 (Section 11.7) and 13.7 (Section 13.5) and to the relevant maps in this book. The general conclusions of our extensive analysis are that: (a) in all cases the various elements available for joint analysis were found to fit well together and to make sense; (b) both Δ and ${}^{m}\bar{D}(SGJ, SJG)$ were found to contribute independently to the understanding and to the analysis of ${}^{m}\bar{D}(G, J)$.

2. At a less detailed level of analysis, it is possible to form groups of geosets and investigate empirically for each whether Δ and ${}^{m}\bar{D}(SGJ, SJG)$ are related in some way and their respective impacts.

14.6. Minimal Distances between Two Components and the Compound Geoset

14.6.1. Triangle of Minimal Distances between Two Components and Their Compound Geoset

We now go on to study the minimal distances between component and compound geosets, as defined in Sections 3.6 and 3.8. We start by considering the very important case of a compound geoset formed by two subsets. Dealing specifically with population sets we consider a population K divided into two subpopulations P and Q, for which we know the corresponding geosets ${}^{g}K$, ${}^{g}P$, and ${}^{g}Q$. Again, we take the example given in Sections 3.6, 14.3, and 14.4: $P =$ dwellings in Rome built up to 1945; $Q =$ dwellings in Rome built from 1946 to 1981; $K =$ all dwellings. In order to stress that we are dealing with populations, we have used the symbols P and Q, which are less general than G and J as employed in Sections 14.3 and 14.4. We have already seen that geosets ${}^{g}P$ and ${}^{g}Q$, as well as geosets derived from them, can be represented graphically by means of the triangles of Figures 14.15 and 14.16. In these figures, each point represents a geoset and the lines binding any pair of dots are proportional to the minimal distance between the geosets represented by these dots.

Figure 14.17 reproduces the triangle with vertices 1, 2, and 3 and with the dots 4 and 5 of Figure 14.15. Here, dots 1, 2, and 3 represent geosets ${}^{g}SPQ$, ${}^{g}SQP$, and ${}^{g}M(P, Q)$, respectively. Points 4 and 5 denote geosets ${}^{g}P$ and ${}^{g}Q$, respectively. The triangle was constructed on the basis of the minimal distances

$$ {}^{m}\bar{D}(P, Q), \qquad {}^{m}\bar{D}[(SPQ, M(P, Q)], \qquad {}^{m}\bar{D}[SQP, M(Q, P)] $$

and ${}^{m}\bar{D}(P, Q) = \Delta{}^{m}\bar{D}(SPQ, SQP)$ as calculated by the computer program. Numerical values of our example were given, checked, and commented on in Section 14.4.

The triangle of Figure 14.15 shows the relationships between the component geosets ${}^{g}P$ and ${}^{g}Q$ and the geosets derived from them.

To complete the graphical presentation, we also introduce in our triangle the dots representing the compound geoset ${}^{g}K$ and the distances from it. This additional information is given by the dots labeled 6, 7, 8, and 9, respectively, over the triangle of Figure 14.17.

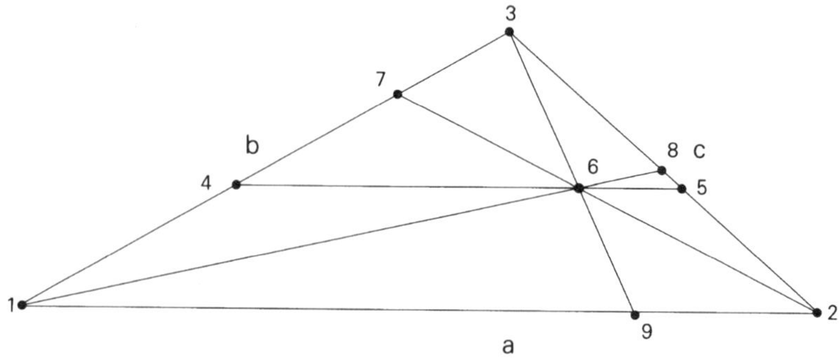

Figure 14.17. Rome. The triangle of minimal distances between two components, their compound geoset, and the geosets derived from them: dwellings built up to 1945 (P), dwellings built between 1946 and 1981 (Q), total dwellings (K) (Source of data: Comune di Roma, Segretariato Generale il Direzione, Ufficio di Statistica e Censimento).

Dot 6 represents the geoset ${}^gK$, and is found over the line $\overline{45}$ between ${}^gP$ and ${}^gQ$ and cuts the length ${}^m\bar{D}(P, Q)$ of this line into two parts proportional, respectively, to ${}^m\bar{D}(P, K) = v_Q^{K\,m}\bar{D}(P, Q)$ and to ${}^m\bar{D}(K, Q) = v_P^{K\,m}\bar{D}(P, Q)$, where v_P^K and v_Q^K are the proportions of the sizes of ${}^gP$ and ${}^gQ$ to the total size of ${}^gK$. It can be shown that ${}^gSPQ = {}^gSPK = {}^gSKQ$. Accordingly, vertex 1 represents all of these three geosets and similarly vertex 2 represents ${}^gSQP = {}^gSKP = {}^gSQK$. Consequently, $\overline{12} = {}^m\bar{D}(SPQ, SQP) = {}^m\bar{D}(SPK, SKP) = {}^m\bar{D}(SKQ, SQK)$. From dot 6, we draw the lines $\overline{61}$, $\overline{62}$, and $\overline{63}$ binding ${}^gK$ to vertices 1, 2, and 3, respectively and extend these lines to dots 8, 7, and 9, which are crossing points with sides $\overline{23}$, $\overline{13}$, and $\overline{12}$, respectively.

Dots 7, 8, and 9 can be interpreted as follows:

1. As line $\overline{12}$ is parallel to line $\overline{45}$, dot 9 cuts line $\overline{12}$ into two parts $\overline{19}$ and $\overline{92}$, the lengths of which are proportional, respectively, to the lengths of $\overline{46}$ and $\overline{65}$. We thus obtain $\overline{19} = v_Q^{K\,m}\bar{D}(SPQ, SQP) = v_Q^{K\,m}\bar{D}(SPK, SKP) = v_Q^{K\,m}\bar{D}(SKQ, SQK)$. Accordingly, considering triangle $1\hat{9}3$, we obtain

$$ {}^m\bar{D}(K, P) = v_Q^K \Delta_P^{Q\,m}\bar{D}(SPQ, SQP) $$

   In a similar way, considering triangle $2\hat{9}3$, we get:

$$ {}^m\bar{D}(K, Q) = v_P^K \Delta_P^{Q\,m}\bar{D}(SPQ, SQP) $$

2. Dot 7 represents the geoset of the match as ${}^gM(K, P)$ between ${}^gK$ and ${}^gP$. The triangle having vertices 1, 2, and 7 can be considered as a triangle of distances between geosets $1 = {}^gSKP$, $2 = {}^gSPK$, and $7 = {}^gM(K, P)$; the length of line $\overline{46}$ parallel to line $\overline{12}$ measures the distance between ${}^gK$ and ${}^gP$.
3. Dot 8 represents the geoset of the matches as ${}^gM(K, Q)$ between ${}^gK$ and ${}^gQ$. The triangle having vertices $1 = {}^gSKQ$, $2 = {}^gSQK$, and $8 = {}^gM(K, Q)$ can be interpreted as a triangle of minimal distances between these geosets; line $\overline{65}$ measures the distance between ${}^gK$ and ${}^gQ$.

14.6.2. An Excursus: Equations for the Triangle of the Minimal Distances between Components and Compound Geosets

We saw above that knowing ${}^{m}\bar{D}(SPQ, SQP)$, ${}^{m}\bar{D}[SPQ, M(P, Q)]$, ${}^{m}\bar{D}[SQP, M(Q, P)]$, Δ, and v_P^K we can construct the detailed triangle of distances shown in Figure 14.17. Tables 14.5 and 14.6 give the equations that enable us to obtain the values of the coordinates of the dots representing each of the nine geosets we are concerned with and the minimal distance between each pair of relevant geosets.

To simplify the presentation we introduce here some conventions regarding the symbols:

1. The three basic minimal distances are denoted by the following letters

$${}^{m}\bar{D}(SPQ, SQP) = a, \quad {}^{m}\bar{D}[SPQ, M(P, Q)] = b, \quad \text{and} \quad {}^{m}\bar{D}[SQP, M(Q, P)] = c$$

2. The coordinates of ${}^{g}SPQ$ are taken as $(0, 0)$; the coordinates of ${}^{g}SQP$ are taken as $[{}^{m}\bar{D}(SQP, SPQ); 0]$.
3. Some of the relevant angles are denoted by Greek letters, as follows: $\alpha = 2\hat{1}3$; $\beta = 1\hat{2}3$; $\gamma = 1\hat{3}2$; $\delta = 5\hat{6}8$; $\varepsilon = 6\hat{8}5$; $\eta = 4\hat{6}7$; $\zeta = 4\hat{7}6$.
4. The ratio Δ_P^Q is denoted by Δ.
5. v_P^K is denoted by v.

The values that are obtained by using the expressions in Tables 14.5 and 14.6 have been checked by direct computer calculation of the minimal distances between all the pairs of relevant geosets. This check shows a perfect concordance, apart from very minor irregularities that are explained by the fact that some of the calculations are based on a comparatively small number of cells.

Table 14.5. Coordinates of Dots Representing Geosets in the Triangle of Distances

		Expression for	
Dot	Sets	Abscissa X	Ordinate Y
1	${}^{g}SPQ$, ${}^{g}SPK$, ${}^{g}SKQ$	0	0
2	${}^{g}SQP$, ${}^{g}SKP$, ${}^{g}SQK$	a	0
3	${}^{g}M(P, Q)$	$b\cos\alpha$	$b\sin\alpha$
4	${}^{g}P$	$(1-\Delta)b\cos\alpha$	$(1-\Delta)b\sin\alpha$
5	${}^{g}Q$	$(1-\Delta)b\cos\alpha+\Delta a$	$(1-\Delta)b\sin\alpha$
6	${}^{g}K$	$(1-\Delta)b\cos\alpha+(1-v)\Delta\alpha$	$(1-\Delta)b\sin\alpha$
7	${}^{g}M(K, P)$	$\left[(1-\Delta)b+(1-v)a\dfrac{\sin\eta}{\sin\zeta}\right]\cos\alpha$	$\left[(1-\Delta)b+(1-v)\Delta a\dfrac{\sin\eta}{\sin\zeta}\right]\sin\alpha$
8	${}^{g}M(K, Q)$	$\left[\sqrt{X_6^2+Y_6^2}+v\Delta a\dfrac{\sin\beta}{\sin\varepsilon}\right]\cos\delta$	$\left[\sqrt{X_6^2+Y_6^2}+v\Delta a\dfrac{\sin\beta}{\sin\varepsilon}\right]\sin\delta$
9	$v{}^{g}SPQ+(1-v){}^{g}SQP$	$(1-v)a$	0

Table 14.6. Expressions for Minimal Distances

Ordinal number of minimal distance	Length of line in triangle of distances	Corresponding minimal distance	Expression
1st	$\overline{19}$	${}^{m}\bar{D}(SPQ, P)$	$(1-\Delta)b$
2nd	$\overline{47}$	${}^{m}\bar{D}[P, M(K, P)]$	$(1-v)\Delta a \sin\eta/\sin\varepsilon$
3rd	$\overline{73}$	${}^{m}\bar{D}[M(K, P), M(P, Q)]$	$\Delta[b-(1-v)a \sin\eta/\sin\zeta$
4th	$\overline{43}=\overline{47}+\overline{73}$	${}^{m}\bar{D}[P, M(P, Q)]$	Δb
5th	$\overline{13}=\overline{14}+\overline{43}$	${}^{m}\bar{D}[SPQ, M(P, Q)]$	b
6th	$\overline{25}$	${}^{m}\bar{D}(SQP, Q)$	$(1-\Delta)c$
7th	$\overline{58}$	${}^{m}\bar{D}[Q, M(K, Q)]$	$v\Delta a \sin\delta/\sin\varepsilon$
8th	$\overline{83}$	${}^{m}\bar{D}[M(K, Q), M(P, Q)]$	$\Delta(c - va \sin\delta/\sin\varepsilon)$
9th	$\overline{53}=\overline{58}+\overline{83}$	${}^{m}\bar{D}[Q, M(Q, P)]$	Δa
10th	$\overline{23}=\overline{25}+\overline{53}$	${}^{m}\bar{D}[SQP, M(P, Q)]$	c
11th	$\overline{46}$	${}^{m}\bar{D}(P, K)$	$(1-v)\Delta a$
12th	$\overline{65}$	${}^{m}\bar{D}(K, Q)$	$v\Delta a$
13th	$\overline{45}=\overline{46}+\overline{65}$	${}^{m}\bar{D}(P, Q)$	Δa
14th	$\overline{16}$	${}^{m}\bar{D}(SKQ, K)$	$(1-v\Delta)a \sin\beta/\sin(\beta+\delta)$
15th	$\overline{68}$	${}^{m}\bar{D}[K, M(K, Q)]$	$v\Delta a \sin\beta/\sin(\beta+\delta)$
16th	$\overline{18}$	${}^{m}\bar{D}[SKQ, M(K, Q)]$	$a \sin\beta/\sin(\beta+\delta)$
17th	$\overline{26}$	${}^{m}\bar{D}(SKP, K)$	$[1-(1-v)\Delta]a \sin\alpha/\sin(\alpha+\eta)$
18th	$\overline{67}$	${}^{m}\bar{D}[K, M(K, P)]$	$(1-v)\Delta a \sin\alpha/\sin(\alpha+\eta)$
19th	$\overline{27}$	${}^{m}\bar{D}[SKP, M(K, P)]$	$a \sin\alpha/\sin(\alpha+\eta)$
20th	$\overline{19}$	$(1-v)^{m}\bar{D}(SPQ, SQP)$	$(1-v)a$
21st	$\overline{92}$	$v^{m}\bar{D}(SPQ, SQP)$	va
22nd	$\overline{12}=\overline{19}+\overline{92}$	${}^{m}\bar{D}(SPQ, SQP)$	a

14.6.3. Practical Rules Concerning Minimal Distances between Components and Compound Geosets

The equations in Section 14.6.2 are of help in the detailed study of compound geosets formed by two components. We will now give some practical rules resulting from the equations that have been found to be particularly meaningful. To simplify the presentation we indicate minimal distance between movers or between shifts as distance between surpluses.

1. The distance between surpluses is equal in the comparison between the two components and to the comparison between one component and the compound:

$$a = {}^m\bar{D}(SPQ, SQP) = {}^m\bar{D}(SPK, SKP) = {}^m\bar{D}(SQP, SPQ) \qquad (14.6)$$

2. Knowing the proportion of surpluses between the two components and the proportion v_P^K, v_Q^K of the elements of each component in the compound, we can obtain the properties of surpluses between each component and the compound:

$$\Delta_P^K = \Delta_Q^K : v_P^K \qquad (14.7)$$

3. Inversely, knowing the proportion of surpluses between one component and the proportion of its elements in the compound, we can obtain the proportion of surpluses between the two components:

$$\Delta_P^Q = \Delta_P^K : v_P^K \qquad (14.8)$$

4. Adding the proportions of surpluses between each component and the compound, we obtain the proportions of surpluses between the components:

$$\Delta_P^Q = \Delta_P^K + \Delta_Q^K \qquad (14.9)$$

5. The distance between each component and the compound can be obtained as the product between the proportion of surpluses and the distance between them:

$${}^m\bar{D}(P, K) = \Delta_Q^K a = v_Q^K \Delta_P^Q a \qquad (14.10a)$$

$${}^m\bar{D}(Q, K) = \Delta_P^K a = v_P^K \Delta_P^Q a \qquad (14.10b)$$

6. Adding the distances between each component and the compound, we obtain the distance between the two components:

$${}^m\bar{D}(P, Q) = v_Q^K \Delta_P^Q a + v_P^K \Delta_Q^P - \Delta_P^Q a = {}^m\bar{D}(Q, P) \qquad (14.11)$$

7. Knowing the distance between the two components and the proportion of elements of each component in the compound, we can obtain the distance between each component and the compound:

$${}^m\bar{D}(P, K) = {}^m\bar{D}(P, Q) : v_Q^K \qquad (14.12a)$$

$${}^m\bar{D}(Q, K) = {}^m\bar{D}(P, Q) : v_P^K \qquad (14.12b)$$

8. Inversely, knowing the distance between one component and the compound and the proportion of the elements of the component over the compound, we can obtain the distance between the two components:

$${}^m\bar{D}(P, K) = v_Q^{K\,m}\bar{D}(P, K) \qquad (14.13a)$$

$${}^m\bar{D}(P, Q) = v_P^{K\,m}\bar{D}(Q, K) \qquad (14.13b)$$

14.6.4. A Special Configuration: Nonintersecting Components

Consider the special, but not unusual, situation, in which a compound geoset ${}^gK$ is formed by two nonintersecting geosubsets ${}^gP$ and ${}^gQ$. In this case, the triangle of minimal distances reduces to a line of length ${}^m\bar{D}(P, Q) = {}^m\bar{D}(SPQ, SQP) = a$. Here, $\Delta_P^Q = 1$ and $(1 - \Delta_P^Q) = 0$.

- Rules 1, 7, and 8 of Section 14.6.3 still hold.
- Rules 2 and 3 are reduced to $\Delta_P^K = v_Q^K$ and $\Delta_Q^K = v_P^K$.
- Rule 4 is reduced to $\Delta_P^K + \Delta_Q^K = 1$.
- Rule 5 is reduced to ${}^m\bar{D}(P, K) = {}^m\bar{D}(SPK, SKP)$.
- Rule 6 is reduced to ${}^m\bar{D}(P, Q) = a$.

14.6.5. Extensions to Other Types of Geosets

The findings and rules given in Sections 14.6.1–14.6.4 can be extended immediately, or with some minor adjustments, to the geosets discussed in Section 3.8.4 (simple average of two geosets), Section 3.8.5 (weighted average of two geosets), and Section 3.8.8 (intermediate geosets).

A compound geoset can be formed by many component geosubsets (as described in Section 3.8.2), but in this case, we cannot determine simple general relationships of minimal distances between components and compound.

Let us indicate by ${}^gP, {}^gQ, \ldots$ two out of many components of a geoset ${}^gK$. In the study of segregation between the population geoset ${}^gK$ and the geosets of many subpopulations, ${}^gP$, ${}^gQ$, ${}^gR, \ldots$, it might be convenient to compare systematically the geosets of each subpopulation to the geoset of its complement, e.g., to compare ${}^gP$ to ^{g}(Non-P), ${}^gQ$ to ^{g}(Non-Q), or ${}^gR$ to ^{g}(Non-R). This approach enables us to evaluate the segregation of each subpopulation directly, independently of its weight over the population.

From a computational point of view, it may be expedient to obtain the distances between the geosets of each subpopulation and its complement in the following way. We calculate the distances between the geoset of the population and those of the subpopulations and then divide them by the weight of the complements of the subpopulations over the population:

$$ {}^m\bar{D}(P, \text{Non-}P) = {}^m\bar{D}(P, K) : (1 - v_P^K) \qquad (14.14a) $$

$$ {}^m\bar{D}(Q, \text{Non-}Q) = {}^m\bar{D}(Q, K) : (1 - v_Q^K) \qquad (14.14b) $$

$$ {}^m\bar{D}(R, \text{Non-}R) = {}^m\bar{D}(R, K) : (1 - v_R^K) \qquad (14.14c) $$

14.7. An Excursus: Indirect Calculation of Minimal Distance between Any Two Geosets Based on Their Distances from Paradigmatic Distributions

14.7.1. Possible Approaches to the Problem

Suppose we want to study a large number N of geosets and have to explore possible similarities or differences between them. Theoretically, the best approach would be to calculate a matrix of the $N(N - 1)/2$ minimal distances between all the possible pairs of these geosets. However, if N runs into the hundreds or thousands, the computation may become extremely laborious and costly. Indirect approaches may then become preferable, even at the cost of loss of accuracy in the results.

Let us indicate various paradigmatic distributions such as those considered in Section 14.2.2 by $K^i = K^1, K^2, \ldots$ For example K^i may be: (a) the uniform distribution over the territory considered, or (b) the geosets of the total population or of a subpopulation having important demographic, social, economic or ethnic characteristics such as (i) main labor force classes; (ii) economic activities (measured by the gross value of total production, by the value added, or by the number of individuals employed; (iii) farms or their lands; or (iv) individuals working on farms, their incomes, and expenditures. It is assumed that the geosets of the K^i selected can have a positive or a negative impact on the geosets ${}^gG, {}^gJ, \ldots$, considered.

We assume that we have at our disposal distances ${}^m\bar{D}(G, K^i)$, ${}^m\bar{D}(J, K^i), \ldots$ between the geosets ${}^gG, {}^gJ, \ldots$ to be studied and some paradigmatic distribution $K^i = K^1, K^2, \ldots$. In Sections 14.7.2 and 14.7.3, two approaches are suggested for obtaining either the values of ${}^m\bar{D}(G, J)$ for all desired pairs of geosets studied, or the coordinates $X({}^gG)$, $Y({}^gG)$; $X({}^gJ)$, $Y({}^gJ)$; … enabling us to represent the geosets graphically.

The methods proposed can be utilized to construct the entire matrix of all possible distances or selected pairs of distances or coordinates of selected geosets. One important application might be a case in which available sources put at our disposal only the minimal distances between the geosets studied and a few selected paradigmatic distributions. As there would be little detailed information on the geosets themselves, a direct calculation of the minimal distances between them would not be possible, and an indirect method would have to be used.

In order to introduce the methods of indirect calculation of minimal distances, let us consider a practical example, as shown in Table 14.7. Here, some of the minimal distances for geosets in Italy (1981) from paradigmatic distributions that were given in Table 14.2 are repeated. We consider three paradigmatic distributions: ${}^gK^1$ = Uniform distributions over the territory of Italy (${}^gU$); ${}^gK^2$ = Distribution of the total population, (${}^gP$); ${}^gK^3$ = Distribution of economic activities (${}^gE$), measured by the total number of individuals employed in industry, trade, and services (1981). The geosets considered concern individuals employed in public administration, retail trade, wholesale trade, and market research (see Table 14.7).

Table 14.7. Italy (1981). Minimal Distance from ${}^gU$, ${}^gP$, and ${}^gE$ to Four Geosets Concerning Individuals Employed

	Minimal distance (km)		
	From ${}^gU$	From ${}^gP$	From ${}^gE$
Public administration	59.6	50.7	135.5
Retail trade	59.1	19.6	69.9
Wholesale trade	155.7	123.1	39.5
Market research	219.9	191.0	107.9

14.7.2. Calculating the Limits to the Minimal Distance

Our object is to calculate limits, as narrow as possible, within which ${}^m\bar{D}(G, J)$ for any pair of geosets is contained. This is done on the basis of the knowledge of ${}^m\bar{D}(K^i, G)$ and ${}^m\bar{D}(K^i, J)$ for the three standard distributions $K^i(K^1, K^2, K^3 \ldots)$ considered.

Suppose that we wish to calculate limits to the minimal distance between the geosets ${}^gG$ of retail trade and ${}^gJ$ of wholesale trade. Consider any paradigmatic distribution K^i, such as ${}^gK^2 = {}^gP$. This setting is shown in Figure 14.18, where ${}^gK^2 = {}^gP$ is represented by dot. We know that a dot representing ${}^gG$ is found at a distance ${}^m\bar{D}(G, P) = 19.6$ km from the dot representing ${}^gP$. Therefore, this dot is found over a circle centered at ${}^gP$ and having as radius $r_P = 19.6$ km. A dot representing ${}^gJ$ is found over a circle centered at ${}^gP$ and having a radius $R_P = 123.1$ km. The subscript P indicates that the radii are referenced to ${}^gP$. The upper limit of ${}^m\bar{D}(G, J)$ is thus $M_P = |R_P + r_P| = 142.7$ km, while the lower limit to ${}^m\bar{D}(G, J)$ is $m_P = |R_P - r_P| = 103.5$ km.

In a similar way, referring respectively to ${}^gK^1 = {}^gU$ and ${}^gK^3 = {}^gE$ we obtain: as lower limits $m_U = 96.6$ and $m_E = 30.4$; and as upper limits, $M_U = 214.8$ and $M_E = 109.4$. Thus we get:

$$96.6 < {}^m\bar{D}(G, J) < 214.8$$

$$103.5 < {}^m\bar{D}(G, J) < 142.7$$

$$30.4 < {}^m\bar{D}(G, J) < 109.4$$

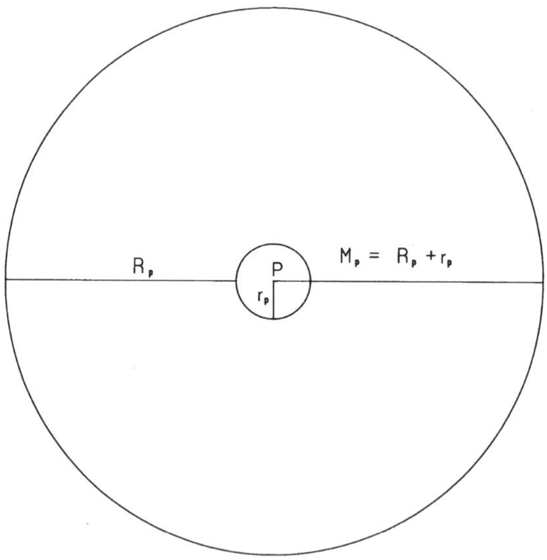

Figure 14.18. Italy. Limits of minimal distance between the geoset of retail trade (G) and the geoset of wholesale trade (J) based on distance from the geoset of population (P), 1981 (Source of data: ISTAT).

Table 14.8. Italy (1981). Evaluating Minimal Distance and the Relative Range of the Estimates

	Retail trade	Wholesale trade	Market research
	(a) Estimates of minimal distances (km)		
Public administration	68.0	136.0	201.0
Retail trade	—	106.5	174.6
Wholesale trade	—	—	108.4
	(b) $\|(M - m)/2\|$		
Public administration	2.35	38.85	40.7
Retail trade	—	2.95	3.2
Wholesale trade	—	—	39.5
	% Ratios of (b) over (a)		
Public administration	3.5	28.6	20.2
Retail trade	—	2.8	1.8
Wholesale trade	—	—	35.4

The highest lower limit, indicated by m, is thus $m = m_P = 103.5$. The lowest upper limit, indicated as M, is thus $M = M_E = 109.4$. We can conclude that ${}^m\bar{D}(G, J)$ is included between 103.5 and 109.4, i.e., $103.5 < {}^m\bar{D}(G, J) < 109.4$.

Suppose we take as an estimate of ${}^m\bar{D}(G, J)$ the average $(m + M)/2 = 106.5$, to be indicated as Est ${}^m\bar{D}(G, J)$. The maximum possible error involved in adopting such an estimate would be

$$|M - \text{Est}[{}^m\bar{D}(G, J)]| = |m - \text{Est}[{}^m\bar{D}(G, J)]| = |(M - m)/2| = 2.95$$

Various criteria can be suggested in order to judge whether a maximum error of this order is acceptable. A very simple method would be to compare the maximum error $|(M - m)/2|$ to the estimated average $|(M + m)/2|$. According to this criterion, the error in our example $100(m - M)/(m + M) = 2.8\%$ would be considered very small.

Table 14.8. shows the estimates of the minimal distances $(M + m)/2$, the ranges $(M - m)/2$, and the ratios $(M - m)/(M + m)$ for all the pairs of geosets considered. The estimates appear to be very good for distances between public administration and retail trade and retail trade and market research, but the other estimates have relative ranges that might be judged as not very reliable.

14.7.3. Calculating the Coordinates of Dots Representing the Geosets Studied on the Basis of Minimal Distances from Paradigmatic Distributions

Let us denote any of the geosets to be studied by ${}^gG$ and three paradigmatic distributions by ${}^gU$, ${}^gP$, and ${}^gE$ (as we did in Sections 14.7.1 and 14.7.2). We can construct a triangle (labeled 1) the vertices of which represent the paradigmatic distributions and the sides of which (a, b, c) are proportional to ${}^m\bar{D}(U, P)$, ${}^m\bar{D}(U, E)$, and ${}^m\bar{D}(P, E)$, respectively, as shown in Figure 14.19, where G is retail trade. As a matter of convenience we take 0,0 as the coordinates of ${}^gU$ and $[{}^m\bar{D}(U, P); 0]$ as the coordinates of ${}^gP$. Taking $a = {}^m\bar{D}(U, P)$ to be constant, we construct a triangle (labeled 2) having sides a, d, and e, where d and e are proportional to ${}^m\bar{D}(G, U)$ and ${}^m\bar{D}(G, P)$, respectively. We construct a triangle (labeled 3) of sides b, f, and g where f and g are proportional to ${}^m\bar{D}(G, U)$ and ${}^m\bar{D}(G, E)$, respectively, and a triangle (labeled 4) of sides c, h, and l, where h and l are proportional to ${}^m\bar{D}(G, P)$ and ${}^m\bar{D}(G, E)$, respectively. In Figure 14.19, sides d, e, f, g, h, and l are shown with thinner lines. We see from this figure that we obtain three different estimates of the point representing ${}^gG$. The numerical values used are: $a = 54.21$; $b = 119.76$; $c = 87.81$; $d = 155.72$; $e = 123.10$; $f = 155.72$; $g = 39.63$; $h = 123.10$; and $l = 39.63$. The resulting

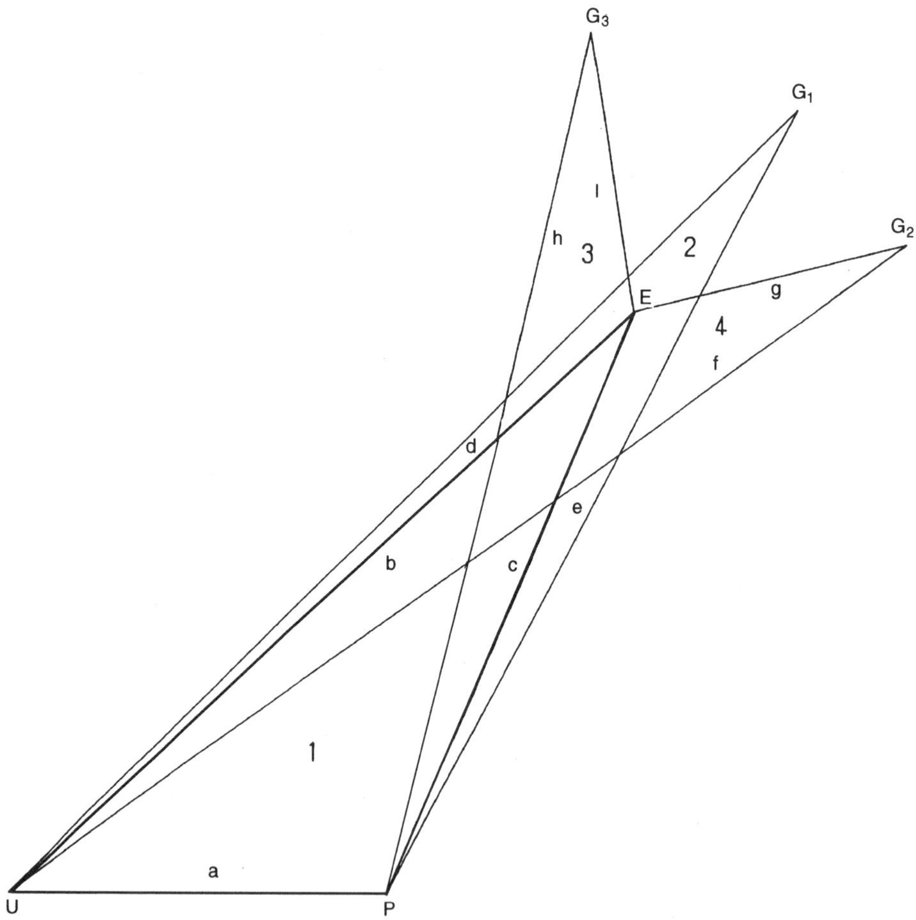

Figure 14.19. Italy. Estimate of retail trade (G) on the basis of minimal distances of three paradigmatic distributions: uniform distribution (U), population (P), and economic activities (E) (Source of data: ISTAT).

coordinates of the three estimates of the dots representing ${}^{g}G$ are:

$$X_1 = 111.00 \qquad Y_1 = 109.21$$
$$X_2 = 126.68 \qquad Y_2 = 90.56$$
$$X_3 = 81.68 \qquad Y_3 = 120.00$$

If these estimates are not too dispersed, we can attribute the following average coordinates to ${}^{g}G$:

$$X = (X_1 + X_2 + X_3):3 = 106.45$$
$$Y = (Y_1 + Y_2 + Y_3):3 = 106.59$$

Various criteria can be proposed for evaluating the degree of confidence we can have in our estimate of X_G, Y_G. Here we compare (a) the average of the distances (36.15) between G_1 and G_2, G_1 and G_3, G_2 and G_3 to (b) the average of the distances

$$(a + b + c + d + e + f + g + h + l)/9 = 99.85$$

In this example, the ratio 100(36.15:99.85) = 36.2% would suggest that the approximations of X_G, Y_G are to be considered rather unsatisfactory. On the other hand, for other geosets we might find very satisfactory ratios; e.g., for the retail trade, we have 4.70:62.09 = 7.57%.

Obviously, having obtained the estimates of the coordinates (X_G, Y_G), $(X_J, Y_J), \ldots$ of the points representing geosets ${}^{g}G, {}^{g}J, \ldots$ we can calculate estimates of the minimal distances ${}^{m}\bar{D}(G, J), \ldots$.

14.7.4. An Experiment for Evaluating the Minimal Distances Calculated Indirectly

We can utilize the estimates of minimal distances by the methods explained in Sections 14.7.2 and 14.7.3 together and the average of the two should yield better results. An experiment designed to ascertain what degree of confidence we can have in distances obtained indirectly was conducted as follows. A matrix of $(26 \times 25):2$ minimal distances calculated directly was available from a paper by Napolitano (1990) on 26 geosets of tertiary activities in Italy. We compared these values with the minimal distances for the same pairs of geosets calculated indirectly (as averages of estimates obtained through both the methods explained above $(m + M)/2$, distances between coordinates of dots representing geosets). It was found that the absolute differences (in km) between the estimates and the actual values were distributed as follows:

Distance (km)	<5	5–9.99	10–19.99	20–29.99	30–49.99	50+	Total
% of distribution	39.6	27.3	16.4	8.4	6.9	1.4	100

If we take into account that the average of the 325 minimal distances considered is 93.56 km with a standard deviation of 53 km, the results given above appear to be satisfactory.

14.8. Minimal Distance between Net Geosets

14.8.1. Geosets Netted from the Influence of a Common Determinant

We assume that the minimal distance ${}^{m}\bar{D}(G, J)$ between two geosets ${}^{g}G$ and ${}^{g}J$ is found to be small (according to any of the criteria we have suggested here for rendering minimal distances dimensionless). There may be good reason to suspect that this propinquity between ${}^{g}G$ and ${}^{g}J$ is due to the fact that both ${}^{g}G$ and ${}^{g}J$ are strongly influenced by a third geoset ${}^{g}L$ (possibly to be considered as "leading"). We may then wish to eliminate the influence of ${}^{g}L$ from both ${}^{g}G$ and ${}^{g}J$. Following is a simple method to achieve this.

Let us calculate the geosets ${}^{g}SGL$ and ${}^{g}SJL$ of the surpluses of ${}^{g}G$ and ${}^{g}J$, respectively, over ${}^{g}L$. We consider ${}^{g}SGL$ and ${}^{g}SJL$ as net geosets. The minimal distance ${}^{m}\bar{D}(SGL, SJL)$ between them can be taken as a measure of the territorial discrepancy between ${}^{g}G$ and ${}^{g}J$ after the influence of ${}^{g}L$ has been removed. Taking a simple example, we consider the distribution of children and of elderly people in Rome, as illustrated in Section 7.1. We indicate children under fifteen by G, individuals aged sixty-five and over by J, and people of all ages by L.

We find that ${}^{m}\bar{D}(G, J) = 1.409$ km, which can be divided into $\Delta = 22.76\%$ and ${}^{m}\bar{D}(SGJ, SJG) = 6.19$ km. Δ is thus rather low and accordingly $(1 - \Delta)$ is rather high, while ${}^{m}\bar{D}(SGJ, SJG)$ is high. The high proportion of elements of ${}^{g}G$ and ${}^{g}J$ that match may suggest that the two distributions are subject to some strong influence of common determinants. On the other hand, the high value of ${}^{m}\bar{D}(SGJ, SJG)$ may suggest that there are strong systematic differences between ${}^{g}SGJ$ and ${}^{g}SJG$. The wider study of the distributions of all age groups both in Rome and in other cities suggests a possible general hypothesis. We may posit that the territorial distribution of all age groups is strongly influenced, in a similar way, by common determinants, which find their expression in the characteristics of geoset ${}^{g}L$. However, the territorial distribution of each age group has its own peculiarities, which are expressed mainly by geosets such as ${}^{g}SGL$ and ${}^{g}SJL$. Actually we find low values for both ${}^{m}\bar{D}(G, L) = 0.512$ km and ${}^{m}\bar{D}(J, L) = 0.96$ km.

This confirms propinquity of both ${}^{g}G$ and ${}^{g}J$ to ${}^{g}L$, which is due mainly to the high proportion of matchings between ${}^{g}G$ and ${}^{g}L$ and between ${}^{g}J$ and ${}^{g}L$ ($1 - \Delta_G^L = 92.36\%$; $1 - \Delta_J^L = 84.83\%$). The fact that there are considerable peculiarities in each age group with respect to the general population distribution is confirmed by the high values of ${}^{m}\bar{D}(SGL, SLG) = 6.7$ km and ${}^{m}\bar{D}(SJL, SLJ) = 6.33$ km.

14.8.2. Netting Each Geoset from the Influence of a Different Determinant

The approach suggested above can be extended to settings in which we net ${}^{g}G$ from the influence of a determinant ${}^{g}A$ and ${}^{g}J$ from the influence of another determinant ${}^{g}B$. As an example let us consider the findings of a study designed to measure territorial concordance or dominance between the distribution of fertility and the distribution of people with strong religious orientation conducted over the Jewish population of Jerusalem in 1981, based on many small electoral districts.

We denote the geoset of people who voted for a very religiously oriented party (Agudath Israel) by ^{g}G; the geoset of all voters by ^{g}A; the geoset of all children ever born to the married women enumerated in the population census by ^{g}J; and the geoset of these women by ^{g}B. All four geosets (^{g}G; ^{g}A; ^{g}J; ^{g}B) were reduced to the same conventional size of $\Phi = 10{,}000$.

We can consider: ^{g}SGA as the geoset of surpluses of religiously oriented voters; ^{g}SAG as the geoset of deficits of religiously oriented voters; ^{g}SJB as the geoset of surpluses of fertility; and ^{g}SBJ as the geoset of deficits of fertility.

The minimal distances found were as follows (in meters):

	Minimal distance between		
	Surpluses of religiously oriented voters	and	Deficits of religiously oriented voters
Surpluses of fertility	$295 = a$		$3395 = b$
Deficits of fertility	$3404 = c$		$707 = d$

To simplify the presentation the distances are denoted above by a, b, c, and d, respectively. Short distances between geosets of surpluses and between geosets of deficits suggest concordance between geosets of religious voters and geosets of fertility.

We can suggest calculating as follows:

Dimensionless index of concordance for the case in which $(b+c) > (a+d)$; index of concordance is $[(b+c) - (a + d)]/(b+c)$, which in our example is 0.85.

This calculation has been extended to eight major political parties, classified as follows: 1 — Agudath Israel. 2, 3, 4 — Other religious parties, ordered by decreasing degree of religious orientation. 5, 6, 7, 8 — Nonreligious parties ordered by increasing tendency to secularity. For cases in which $(a+d) > (b+c)$ the index becomes negative and measures discordance from fertility. The results are: 1: $+0.85$; 2: $+0.75$; 3: $+0.48$; 4: $+0.34$; 5: -0.19; 6: -0.23; 7: -0.52; and 8: -0.50.

Chapter 15

Decomposition of the Average of the Squares of Minimizing Distances

15.1. Introduction

The minimal distance between two geosets G and J depends on the difference in position, dimensions, and shape of ${}^gG$ and ${}^gJ$. The interpretation of ${}^m\bar{D}(G, J)$ might be facilitated if we could measure and analyze the impact of each of these determinants separately. This is not easy, but we can make considerable progress toward it by following the two approaches discussed in this chapter and the one that follows.

We first utilize the method of squaring the minimizing distances ${}^md(G_i, J_i)$ between ${}^gG$ and ${}^gJ$ and averaging these squares. The average of squared minimizing distances (ASMD) is expressed as ${}^m\ddot{D}^2(G, J) = \Sigma\, {}^md^2(G_i, J_i)/N$. Owing to its algebraic properties, it can be broken down into three components:

1. The (squared) distance between the centers of ${}^gG$ and ${}^gJ$—this important aspect of position will be indicated for short (as in earlier chapters) as eccentricity and symbolized by ${}^cD^2(G, J)$.
2. The difference between the dispersions of ${}^gG$ and ${}^gJ$—this aspect of the difference in dispersions between ${}^gG$ and ${}^gJ$ will be measured by $(d_G - d_J)^2$, i.e., the squared difference of standard distances of ${}^gG$ and ${}^gJ$, and can look at d_G and d_J as measures of an important dimension of the geosets considered.
3. The residual components, including difference in shape between ${}^gG$ and ${}^gJ$, together (eventually) with other determinants.

This approach has the advantages of (a) giving separate parameters for each of three components, (b) measuring the relative impact of each of them on ${}^m\ddot{D}^2(G, J)$; (c) linking the analysis of minimizing distances to the study of centers and standard distances of ${}^gG$, ${}^gJ$, ${}^gSGJ$, ${}^gSJG$, and ${}^gM(G, J)$, introduced earlier; (d) clarifying the relationships that exist between compound and component geosets; (e) enabling us to present some of the parameters in the form of ratios varying between definite limits; (f) furnishing dimensionless measures that enable us to carry out comparisons of minimal distances between pairs of geosets that refer to different territories, different time periods, or different types of populations.

The methods presented in this chapter are based on a hypothesis that we have already discussed: we assume that ${}^gG$ is transformed into ${}^gJ$ by linking each element of ${}^gG$ to an element of ${}^gJ$: linkage is accomplished by minimization and these distances are squared and averaged. As part of this process, we assume that ${}^gG$ can be transformed by parallel translation in such a way that the centers of ${}^gG$ and ${}^gJ$ become coincident. The concentric geosets obtained are denoted as ${}^gG^\circ$ and ${}^gJ^\circ$, and they may be easier to analyze than ${}^gG$ and ${}^gJ$.

In Chapter 16 we shall introduce further transformations, which are sometimes (but not always), relevant for deeper analysis of minimizing distances: We shall assume that (a) ${}^gG$ and ${}^gJ$ are rotated and reflected so that they have the same direction and orientation, which may eliminate or at least reduce the impact of position over minimizing distances between them; (b) we transform ${}^gG$ or ${}^gJ$ or both so that their dimensions become equal. These operations will enable us to study the impact of shape over ${}^m\ddot{D}^2(G, J)$ independently of the other two components. For these transformed geosets the decomposition of the squares of minimizing distances will also be found to be useful.

In the presentation of ${}^m\ddot{D}^2(G, J)$ we can, without any loss of generality, refer to the simple setting in which ${}^gG$ and ${}^gJ$ are geosets formed by N points (G_i, J_i). Applications of ${}^m\ddot{D}^2(G, J)$ to other settings (in which the sizes of the primary geosets differ or the points are, e.g., weighted or are centers of regions or of cells or of other elements of territory) are easily implemented. To help the reader follow the steps of the calculation and decomposition of ${}^m\ddot{D}^2(G, J)$ a fully worked out numerical example for two small geosets (with $N = 3$), is given in Appendix I.

In Section 13.3.3, we indicated the possibility of calculating the average of the squares of distances between ${}^gG$ and ${}^gJ$ by utilizing linkages that minimize the total of the squares of distances. However this approach does not seem to lead to encouraging results, so the use of ${}^md^2(G_i, J_i)$ calculated for linkages that minimize the total of distances is preferred. In utilizing ${}^m\ddot{D}^2(G, J)$ one should bear in mind that this value is not necessarily always lower than the average of the squares of distances between the points of ${}^gG$ and ${}^gJ$ linked in a way that is different from that which renders ${}^m\bar{D}(G, J)$ minimal.

Before entering into a detailed discussion of ASMD, it is appropriate to note that we do not in this case have hundreds of examples of applications as we did for the methods described in earlier chapters. However, the applications made to many pairs of geosets seem to be encouraging, as they help in the analysis and interpretation of these geosets. For the example of dwellings built in Rome up to 1945 versus those built between 1946 and 1981, a detailed presentation of ASMD is given in Section 15.3. Other examples are given in Section 15.4 in an abridged format.

15.2. First Approach to Decomposition of ASMD

15.2.1. Developing ASMD

Let us now develop

$$
\begin{aligned}
{}^m\ddot{D}^2(G, J) &= \frac{\Sigma\, {}^m d^2(G_i, J_i)}{N} = \frac{\Sigma\,[(X_{G_i} - X_{J_i})^2 + (Y_{G_i} - Y_{J_i})^2]}{N} \\
&= \frac{\Sigma\,(X_{G_i}^2 + X_{J_i}^2 + Y_{G_i}^2 + Y_{J_i}^2)}{N} - \frac{2\Sigma\,(X_{G_i}X_{J_i} + Y_{G_i}Y_{J_i})}{N} \qquad (15.1)
\end{aligned}
$$

We subtract from and add to ${}^m\ddot{D}^2(G, J)$ the square of the distances between the centers of ${}^gG$ and ${}^gJ$,

$$
\begin{aligned}
{}^cD^2(G, J) &= (\bar{X}_G - \bar{X}_J)^2 + (\bar{Y}_G - \bar{Y}_J)^2 \\
&= (\bar{X}_G^2 + \bar{X}_J^2 + \bar{Y}_G^2 + \bar{Y}_J^2) - 2(\bar{X}_G\bar{X}_J + \bar{Y}_G\bar{Y}_J) \qquad (15.2)
\end{aligned}
$$

Let us denote

$$
\begin{aligned}
x_{G_i} &= X_{G_i} - \bar{X}_G & x_{J_i} &= X_{J_i} - \bar{X}_J \\
y_{G_i} &= Y_{G_i} - \bar{Y}_G & y_{J_i} &= Y_{J_i} - \bar{Y}_J \\
d_{G_i} &= \sqrt{x_{G_i}^2 + y_{G_i}^2} & d_{J_i} &= \sqrt{x_{J_i}^2 + y_{J_i}^2} \\
\Sigma\, d_{G_i}^2/N &= d_G^2 & \Sigma\, d_{J_i}^2/N &= d_J^2
\end{aligned}
$$

where d_G^2 and d_J^2 are the distance-variances of the geosets ${}^gG$ and ${}^gJ$, respectively.

We thus obtain the important decomposition equation

$$
\begin{aligned}
{}^m\ddot{D}^2(G, J) &= \Sigma\, {}^m d^2(G_i, J_i)/N \\
&= {}^cD^2(G, J) + (d_G^2 + d_J^2) - 2[{}^m\text{cov}(X_G, X_J) + {}^m\text{cov}(Y_G, Y_J)] \qquad (15.3)
\end{aligned}
$$

where

$$
{}^m\text{cov}(X_G, X_J) + {}^m\text{cov}(Y_G, Y_J) = \Sigma\,(x_{G_i}x_{J_i} + y_{G_i}y_{J_i})/N \qquad (15.4)
$$

To simplify the notation, when there is no danger of misunderstanding, we can write

$$
{}^m\text{cov}(G, J) = (\Sigma\, x_{G_i}x_{J_i} + \Sigma\, y_{G_i}y_{J_i})/N \qquad (15.5)
$$

and term it *m*-covariance between G and J, which indicates that elements of G and J are linked in such a way that the sum of the distances between them is a minimum. Then we can present the decomposition of ${}^m\ddot{D}^2(G, J)$ in the following abridged form:

$$
{}^m\ddot{D}^2(G, J) = {}^cD^2(G, J) + d_G^2 + d_J^2 - 2\,{}^m\text{cov}(G, J) \qquad (15.6)
$$

We thus see that ${}^m\ddot{D}^2(G, J)$, which is always nonnegative, increases with increasing eccentricity and dispersions of G and J and with decreasing covariance.

Let us suppose that we shift ${}^gG$ or ${}^gJ$ or both by parallel translation, so that their centers coincide, and that we denote the concentric geosets thus obtained by ${}^gG^\circ$ and ${}^gJ^\circ$. As ${}^cD^2(G^\circ, J^\circ) = 0$, from equation (15.6) we get

$$ {}^m\ddot{D}^2(G^\circ, J^\circ) = d_G^2 + d_J^2 - 2\,{}^m\text{cov}(G, J) \tag{15.7} $$

Anticipating the classification of pairs of geosets given below in Section 15.3, we can say that if ${}^m\ddot{D}^2(G, J) = 0$, then G and J are coincident and vice versa; and if ${}^m\ddot{D}^2(G, J) \neq 0$, then ${}^m\ddot{D}^2(G, J)$ measure their discrepancy.

15.2.2. Interpreting $d_G^2 + d_J^2$

We now generalize the possible linkages between each of the N elements G_i of G and the elements of J without seeking to minimize the sum of the distances. If there are no ties, there are $N!$ possible ways for doing this. Let us designate any of the $N!$ permutations of the ordinal numbers $1, 2, \ldots, N$ by h. We can link each element G_i to an element $J_{h,i}$, where h, i is the ith ordinal number in permutation h. Let us denote the average of the squares of distances between elements of G and J thus linked by ${}^h\ddot{D}^2(G, J)$. We can develop ${}^h\ddot{D}^2(G, J)$ in a manner analogous to that followed in developing ${}^m\ddot{D}^2(G, J)$ and obtain

$$ {}^h\ddot{D}^2(G, J) = {}^cD^2(G, J) + d_G^2 + d_J^2 - 2\,{}^h\text{cov}(G, J) \tag{15.8} $$

where ${}^h\text{cov}(G, J)$ indicates the covariance in linkage h, i.e.,

$$ {}^h\text{cov}(G, J) = (\Sigma\, x_i x_{h,i} + \Sigma\, y_i y_{h,i})/N $$

Equations (15.6) and (15.8) have the terms ${}^cD^2(G, J)$ and $d_G^2 + d_J^2$ in common while their third term (covariance) may differ, and very often actually does. It can be shown that the expected value of the covariances ${}^h\text{cov}(G, J)$ is 0 and that, accordingly, the expected value of ${}^h\ddot{D}^2(G, J)$ reduces to ${}^cD^2(G, J) + d_G^2 + d_J^2$:

$$ E[{}^h\text{cov}(G, J)] = 0 \tag{15.9} $$

$$ E[{}^h\ddot{D}^2(G, J)] = {}^cD^2(G, J) + (d_G^2 + d_J^2) \tag{15.10} $$

Accordingly,

$$ E[{}^h\ddot{D}^2(G^\circ, J^\circ)] = d_G^2 + d_J^2 \tag{15.11} $$

We can thus interpret $(d_G^2 + d_J^2)$ as the expected value of the average of the squares of the distances between elements of G° and J°, linked in all possible ways. This is a very meaningful measure of joint dispersion of the geosets G° and J°, rendered concentric.

We might find it useful to present this joint dispersion in a somewhat different way: If G° and J° are reduced to having the same size and the same center, then $(d_G^2 + d_J^2)/2$ can be interpreted as the distance-variance of the compound geoset formed by taking the elements of ${}^gG^\circ$ and the elements of ${}^gJ^\circ$ together. Let us indicate this concentric compound geoset by ${}^gK$ and its distance-variance as $d_K^2 = (d_G^2 + d_J^2)/2$. The average of the squares of distances between all possible pairs of elements of ${}^gK$ is $2d_K^2 = d_G^2 + d_J^2$.

15.3. Decomposition of ASMD into Nonnegative Components

15.3.1. Absolute and Percentage Components of ASMD

The analysis of ${}^{m}\ddot{D}^2(G, J)$ can often be improved if we break down $d_G^2 + d_J^2$ as follows:

$$d_G^2 + d_J^2 = (d_G - d_J)^2 + 2d_G d_J \tag{15.12}$$

Accordingly, ${}^{m}\ddot{D}^2(G, J)$ can be broken down into the following three nonegative components: (i), (ii), and (iii):

$$\,{}^{m}\ddot{D}^2(G, J) = \underset{\text{(i)}}{{}^{c}D^2(G, J)} + \underset{\text{(ii)}}{(d_G - d_J)^2} + \underset{\text{(iii)}}{2[d_G d_J - {}^{m}\text{cov}(G, J)]} \tag{15.13}$$

which measure the impact over ${}^{m}\ddot{D}^2(G, J)$ of (i) an important aspect of difference in position of G and J, namely the (squared) distance between the centers of ${}^{g}G$ and ${}^{g}J$; (ii) an aspect of difference in dimensions of ${}^{g}G$ and ${}^{g}J$, namely the (squared) difference between the standard distances of ${}^{g}G$ and ${}^{g}J$; and (iii) the residual components, already mentioned, including aspects of difference in shape and eventually of other determinants, which will be analyzed further in Chapter 16.

The term $2d_G d_J$ can be interpreted as a measure of joint dispersion of ${}^{g}G$ and ${}^{g}J$ netted from the effect of the difference in dispersion between ${}^{g}G$ and ${}^{g}J$. Since ${}^{m}\text{cov}(G, J) \leqslant d_G d_J$, component (iii) is nonegative.

Table 15.1 gives an example of the decomposition of ${}^{m}\ddot{D}^2(G, J)$. The geosets considered refer, respectively, to dwellings in Rome (1981) built up to 1945 (G) and those built from 1946 to 1981 (J). The maps of these geosets were given earlier in Figures 3.18 and 3.19 and show as we have noted before that dwellings built before World War II were far more concentrated in the central parts of the city, while those built in the second period are much more peripheral.

In Table 15.1, the components of ${}^{m}\ddot{D}^2(G, J)$ are shown in three ways:

1. As absolute values—in column (c).
2. As percentages of ${}^{m}\ddot{D}^2(G, J)$—in column (d).
3. As ratios of each component over joint dispersion $(d_G^2 + d_J^2)$—in column (e).

In Table 15.1, as in others of similar nature, the values are given as they result from direct computer calculations and there may be small discrepancies from values obtained by applying functional relations.

The calculation of absolute values of the components is obviously a necessary step for our analysis. However, the interpretation of these absolute values is generally not easy and in any event should be limited to types of research in which the pairs of geosets ${}^{g}G$ and ${}^{g}J$ are calculated with reference to a predetermined territory, the characteristics of which are supposed to be generally known.

The analysis of percentages of each component over ${}^{m}\ddot{D}^2(G, J)$ is generally supposed to be carried out after the order of magnitude of ${}^{m}\ddot{D}^2(G, J)$ has been evaluated. A study of the percentages owing to each component may contribute

Table 15.1. Rome (1981). Components of ASMD (G, J): G = Dwellings Built up to 1945; J = Dwellings Built from 1946 to 1981[a]

Components (squared) of ASMD (G, J)				
Character measured (a)	Expression for the absolute component (b)	Absolute component (km) (c)	Percentage of component over ASMD $[{}^m\ddot{D}^2(G, J)]$ (d)	Ratio of component to $d_G^2 + d_J^2$ (e)
(1) Squared eccentricity	${}^cD^2(G, J)$	0.74	5.6	0.01
(2) Dispension difference	$(d_G - d_J)^2$	4.34	33.2	0.08
(3) Residual	$2[d_G d_J - {}^m\text{cov}(G, J)]$	8.02	61.2	0.14
(4) ASMD = discrepancy (1) + (2) + (3)	${}^m\ddot{D}^2(G, J)$	13.10	100.0	0.23
Components of ASMD between G° and J° (geosets rendered concentric)				
(5) Sum of distance-variances	$d_G^2 + d_J^2$	57.39	RC[b]	1.00
(6) Minus twice covariance	$-2{}^m\text{cov}(G, J)$	−44.90	0.85	−0.78
(7) ASMD between $(G^\circ, J^\circ) = (2) + (3) = (5) + (6)$	$d_G^2 + d_J^2 - 2{}^m\text{cov}(G, J) = {}^m\ddot{D}^2(G^\circ, J^\circ)$	12.49		0.22

[a] Calculations based on the division of Rome (within the Raccordo Anulare) into 200 hexagonal cells. The results are almost identical to those of calculations based on 79 centers of regions.
[b] RC = relative covariance = ${}^m\text{cov}(G, J)/d_G d_J$.

important additional information on the relative impact of each of the components over minimizing distances, as shown by the examples given below in Section 15.4.

To simplify the notation, percentages of each component over ${}^m\ddot{D}^2(G, J)$ are indicated by adding a prefix %. For example,

$$ {}^{\%c}D^2(G, J) = 100\,{}^cD^2(G, J)/{}^m\ddot{D}^2(G, J) \tag{15.14} $$

In the specific case of the example of Table 15.1, the percentage contribution of eccentricity is very low, namely only 5.6% of ${}^m\ddot{D}^2(G, J)$, which is explained by the concentric nature of development in Rome. As dispersion differs significantly between the rather compact dwellings built prior to 1945 and the far more dispersed ones built after World War II the contribution of dispersion difference (33.2%) is very strong. Even stronger (61.2%) is the contribution of residual components, which is explained by the very different patterns of distribution of the dwellings built in these two periods. A setting of this type seems to be rather typical of various concentric urban distributions, but is very different from that found for other types of geosets (see Section 15.4).

15.3.2. Ratios of ASMD and Its Components to the Joint Dispersion of ${}^gG$ and ${}^gJ$

The ratios of ASMD and its components to the joint dispersion of ${}^gG$ and ${}^gJ$ might be of use for comparing parameters for, e.g., different territories, different types of population, or different time periods. We have already seen that in order to obtain dimensionless geostatistical parameters comparable across, e.g., space, time, and type of population we can divide absolute geostatistical parameters by a measure of dispersion, and we will use a similar approach here. Dealing with the sum of the squares of minimizing distances between ${}^{g}G$ and ${}^{g}J$, we can take $d_G^2 + d_J^2$ as the standard measure of joint dispersion of ${}^{g}G$ and ${}^{g}J$ for the reasons explained above in Section 15.2.2. Examples of ratios to $d_G^2 + d_J^2$ are given in Column (e) of Table 15.1. Some of the ratios to $d_G^2 + d_J^2$ can also be calculated for each of the components of ${}^{m}\ddot{D}^2(G, J)$. Some of these dimensionless components have clearly definable minimum and maximum values, as explained below.

In order to simplify the notation, we shall use the prefix L to denote the dimensionless parameters obtained through division of absolute parameters by $d_G^2 + d_J^2$. For example:

$$ {}^{L.c}D^2(G, J) = {}^{c}D^2(G, J)/(d_G^2 + d_J^2) \tag{15.15} $$

Dimensionless Squared Eccentricity: ${}^{L.c}D^2(G, J)$

In a study of the distance between the centers of ${}^{g}G$ and ${}^{g}J$, we might wish to construct a relative measure included between 0 and 1 by calculating ${}^{c}D^2(G, J)/\max{}^{c}D^2(G, J)$. However, apart from special cases, it is generally difficult to evalute the maximum possible value of the eccentricity in a meaningful fashion, so we suggested above measuring the dimensionless squared eccentricity by dividing eccentricity by $d_G^2 + d_J^2$, taken as a yardstick of the joint dispersion of G and J. In the example given in Table 15.1, ${}^{c}D^2(G, J)$ is extremely small as compared to values found in other examples (see Section 15.4), which fits well with the setting studied (a city developing in concentric manner).

Dimensionless Difference between Dispersions

The dimensionless difference between dispersions ${}^{L}(d_G - d_J)^2$ can be transformed as follows:

$$ {}^{L}(d_G - d_J)^2 = 1 - \frac{2}{d_G/d_J + d_J/d_G} \tag{15.16} $$

It has a minimum value of 0 for $d_G = d_J$, but if there is a large difference between d_G and d_J, it tends to 1.

In the example given in Table 15.1, ${}^{L}(d_G - d_J)^2 = 0.08$, and this value depends on the more peripheral character of the dwellings built from 1946 to 1981 (${}^{g}J$) as compared to those built earlier (${}^{g}G$).

Dimensionless Residual

The dimensionless residuals $2^{L}[d_G d_J - {}^{m}\text{cov}(G, J)]$, has minimum value of 0, which is reached when the relative covariance ${}^{m}\text{cov}(G, J)/d_G d_J$ is equal to 1. If the relative covariance is 0, then the relative residual is $2d_G d_J$. In the example in Table 15.1, the relative covariance is 0.85. The level of the relative covariance can also be evaluated, by examining a triangle whose sides are proportional to d_G, d_J, and ${}^m\ddot{D}(G, J)$ with the angle opposite ${}^m\ddot{D}(G, J)$ proportional to the relative covariance (see Section 15.3.5).

Comparing relative covariance for the dwellings of Rome built up to 1945 and from 1946 to 1981, we see that the level of 0.85 can be considered high in comparison to the levels found in many other pairs of geosets (see Section 15.4).

Dimensionless ASMD between Concentric Geosets

The dimensionless ASMD between concentric geosets (${}^gG^\circ, {}^gJ^\circ$), can be presented as follows:

$$ {}^{L,m}\ddot{D}^2(G^\circ, J^\circ) = [1 - 2\,{}^m\mathrm{cov}(G, J)]/(d_G^2 + d_J^2) \qquad (15.17)$$

If the relative covariance ${}^m\mathrm{cov}(G, J)/d_G d_J = 1$, this parameter reduces to $(d_G - d_J)^2$. If, in addition, $d_G = d_J$, it reduces to 0. If relative covariance is 0, it becomes 1. In the rare case in which the covariance is negative, it can exceed 1.

Dimensionless ASMD

The dimensionless ASMD

$$ {}^{L,m}\ddot{D}^2(G, J) = {}^{L,c}D^2(G, J) + {}^L(d_G - d_J)^2 + 2\,{}^L[d_G d_J - {}^m\mathrm{cov}(G, J)]/(d_G^2 + d_J^2) \qquad (15.18)$$

reaches its lowest limit of 0 if all its components are 0, which happens for ${}^gG = {}^gJ$. As ${}^{L,c}D^2(G, J)$ has no upper limit, ${}^{L,m}\ddot{D}^2(G, J)$ too has no upper limit. However, its value can be well evaluated by breaking it down into its dimensionless components.

15.3.3. Decomposition of the ASMD between Surpluses *SGJ* and *SJG*

The equations given above for ${}^m\ddot{D}^2(G, J)$ between pairs of geosets ${}^gG$ and ${}^gJ$ can be used immediately for the study of minimal distances between the geosets of surpluses ${}^gSGJ$ and ${}^gSJG$. It is sufficient for this purpose to substitute ${}^m\ddot{D}^2(SGJ, SJG)$ for ${}^m\ddot{D}^2(G, J)$ in equation (15.18).

This particular application is important for several reasons, including:

1. The analysis of minimal distances can be performed in two steps. We first examine the proportion Δ of elements of G and J included in the surpluses, indicating the extent of the shifts necessary in order to transform ${}^gG$ into ${}^gJ$. In the second stage, the analysis focuses on the determinants of ${}^m\ddot{D}^2(SGJ, SJG)$ for the surpluses.
2. This two-stage approach has been found to be very useful (see Section 14.5). It has accordingly been extended to the program for the calculation of the components of the average of the squares of minimizing distances.

As an example of this approach we show in Table 15.2 the components of ${}^m\ddot{D}^2(SGJ, SJG)$ for the geosets of the dwellings in Rome introduced in Table 15.1.

Table 15.2. Rome (1981). Components of ASMD (SGJ, SJG): G = Dwellings Built up to 1945; J = Dwellings Built from 1946 to 1981[a]

Components of ASMD (SGJ, SJG): Character measured (a)	Expression for the absolute component (b)	Absolute component (km) (c)	Percentage of component over ${}^{m}\ddot{D}^2(SGJ, SJG)$ (d)	Ratio of component to $d_G^2 + d_J^2$ (e)
(1) Squared eccentricity	${}^{c}D^2(SGJ, SJG)$	2.51	10.4	0.04
(2) Dispension difference	$(d_{SGJ} - d_{SJG})^2$	15.75	65.2	0.26
(3) Residual	$2[d_G d_J - {}^{m}\text{cov}(G, J)]$	5.91	24.4	0.10
(4) ASMD = discrepancy (1) + (2) + (3)	${}^{m}\ddot{D}^2(SGJ, SJG)$	24.18	100.0	0.40
Components of ASMD between geosets rendered concentric				
(5) Sum of distance-variances	$d_{SGJ}^2 + d_{SJG}^2$	60.89	RC[b]	1.00
(6) Minus twice covariance	$-2\,{}^{m}\text{cov}(SGJ, SJG)$	−38.77	0.86	−0.64
(7) ASMD between (SGJ°, SJG°) = (2) + (3) = (5) + (6)	${}^{m}\ddot{D}^2(SGJ^\circ, SJG^\circ)$	22.12		0.36

[a]Calculations based on the division of Rome (within the Raccordo Anulare) into 200 hexagonal cells. The results are almost identical to those of calculations based on 79 centers of regions.
[b]RC = relative covariance = ${}^{m}\text{cov}(G, J)/d_G d_J$.

From the decomposition of ${}^{m}\ddot{D}^2(SGJ, SJG)$, as given in Table 15.2, it appears that here the difference between the dispersions is the dominant component (65.2%): this percentage is much higher than that found in the decomposition of ${}^{m}\ddot{D}^2(G, J)$, given in Table 15.1. The reason for this is quite clear when we compare the standard distances by measuring the dispersion of the geosubsets of surpluses ${}^{g}SGJ$ and ${}^{g}SJG$ to the standard distances measuring the dispersion of geosets ${}^{g}G$ and ${}^{g}J$:

	$d_G = 4.21$	$d_{SGJ} = 3.16$
	$d_J = 6.30$	$d_{SJG} = 7.13$
Absolute difference:	2.09	3.97
Squared difference:	4.34	15.75

As we have noted several times the dwellings built from 1946 to 1981 are much more dispersed than those built earlier. Dealing with the surpluses, we eliminate from the calculation the dwellings built during the different periods but having the same territorial distribution. Therefore, for the period after 1945 the surpluses present the same characteristic of wider dispersion as for the preceding period, as we saw it for the primary geosets. However, this characteristic is more marked for the surpluses and, consequently, $(d_{SGJ} - d_{SJG})^2$ is much higher than $(d_G - d_J)^2$.

For the surpluses, too, the contribution of eccentricity is very limited (10.4%), although somewhat greater than that for the primary geosets (5.6%).

Table 15.3. Expressing Components of ${}^m\ddot{D}^2(G, J)$ in Terms of Components of ${}^m\ddot{D}^2(SGJ, SJG)$

Absolute components	Equation
(1) Squared eccentricity	${}^cD^2(G, J) = \Delta^{2c}D^2(SGJ, SJG)$
(2) AMSD for concentric geosets	${}^m\ddot{D}^2(G^\circ, J^\circ) = \Delta[d^2_{SGJ} + d^2_{SJG} - 2{}^m\text{cov}(SGJ, SJG)] + (\Delta - \Delta^2){}^cD^2(SGJ, SJG)$
(3) Discrepancy (1) + (2)	${}^m\ddot{D}^2(G, J) = \Delta{}^m\ddot{D}^2(SGJ, SJG)$
Relative contribution of each component	
(1) Squared eccentricity	${}^cD^2(G, J)/{}^m\ddot{D}(G, J) = \Delta{}^cD^2(SGJ, SJG)/{}^m\ddot{D}^2(SGJ, SJG)$
(2) ASMD for concentric geosets	${}^m\ddot{D}^2(G^\circ, J^\circ)/{}^m\ddot{D}^2(G, J) = {}^m\ddot{D}^2(SGJ^\circ, SJG^\circ)/{}^m\ddot{D}^2(SGJ, SJG) + (1 - \Delta){}^cD^2(SGJ, SJG)/{}^m\ddot{D}^2(SGJ, SJG)$

15.3.4. An Excursus: Algebraic Relationships between Components of ASMD for ${}^gG$ and ${}^gJ$ and ${}^gSGJ$ and ${}^gSJG$

We shall now go on to give the equations that bind the components of ${}^m\ddot{D}^2(G, J)$ to the corresponding components of ${}^m\ddot{D}^2(SGJ, SJG)$. As already indicated in Section 15.3.3, some of these equations are used in the computer program that enables us to obtain, first, some parameters for the surpluses, and then, to derive from them the parameters for the primary geosets ${}^gG$ and ${}^gJ$.

1. Average of squared distances:

$$ {}^m\ddot{D}^2(G, J) = \Delta{}^m\ddot{D}^2(SGJ, SJG) \tag{15.19}$$

2. Squared eccentricity:

$$ {}^cD^2(G, J) = \Delta^{2c}D^2(SGJ, SJG) \tag{15.20}$$

3. Sum of distance of variances of ${}^gG$ and ${}^gJ$:

$$ d_G^2 + d_J^2 = \Delta(d^2_{SGJ} + d^2_{SJG}) + 2(1-\Delta)d^2_M + (\Delta - \Delta^2)[{}^cD^2(M, SGJ) + {}^cD^2(M, SJG)] \tag{15.21}$$

where M indicates geoset of matches between ${}^gG$ and ${}^gJ$.

4. Sum of covariances of ${}^gG$ and ${}^gJ$:

$$ {}^m\text{cov}(G, J) = \Delta[{}^m\text{cov}(SGJ, SJG) + (1-\Delta)d^2_M + \Delta(\bar{X}_{SGJ}\bar{X}_{SJG} + \bar{Y}_{SGJ}\bar{Y}_{SJG} + (1-\Delta)(X^2_M + Y^2_M) - (\bar{X}_G\bar{X}_J + \bar{Y}_G\bar{Y}_J)] \tag{15.22}$$

5. (Sum of distance-variances) − 2 × (Sum of covariances) for ${}^gG$ and ${}^gJ$

$$ (d_G^2 + d_J^2) - 2{}^m\text{cov}(G,J) = \Delta[d^2_{SGJ} + d^2_{SJG} - 2{}^m\text{cov}(SGJ,SJG)] + (\Delta - \Delta^2){}^cD^2(SGJ, SJG) \tag{15.23}$$

The program for the calculation of ${}^m\bar{D}(G, J)$ furnishes directly centers and distance-variances of ${}^gSGJ$, ${}^gSJG$, ${}^gM$, ${}^gG$, and ${}^gJ$. With regard to the average of squares of minimizing distances, it calculates ${}^m\ddot{D}^2(SGJ, SJG)$ directly and obtains ${}^m\ddot{D}^2(G, J)$ from the product $\Delta{}^m\ddot{D}^2(SGJ, SJG)$.

With regard to covariances, the program starts by calculating those between surpluses $[{}^m\text{cov}(X_{SGJ}, X_{SJG}) + {}^m\text{cov}(Y_{SGJ}, Y_{SJG})]$. It then obtains the covariances between ${}^gG$ and ${}^gJ$ by using equation (15.22). This equation is based on the following considerations: A proportion Δ of the elements belong to the surpluses SGJ and SJG, while a proportion $1 - \Delta$ of elements intersect. The sum of the products between the coordinates of linked elements of SGJ and SJG is

$$\Sigma X_{SG_i} X_{SJ_i} + \Sigma Y_{SG_i} Y_{SJ_i} = N_S[{}^m\text{cov}(X_{SGJ}, X_{SJG}) + {}^m\text{cov}(Y_{SGJ}, Y_{SJG}) + \bar{X}_G \bar{X}_J + \bar{Y}_G \bar{Y}_J] \tag{15.24}$$

where N_S is the number of elements in surplus. Matching elements of ${}^gG$ and ${}^gJ$ have the same coordinates. The product of their abscissas and of their ordinates is therefore

$$\Sigma X_{Mk}^2 + \Sigma Y_{Mk}^2 = N_M(d_M^2 + \bar{X}_M^2 + \bar{Y}_M^2) \tag{15.25}$$

where N_M is the number of elements in matching.

The sum of covariances between the coordinates of linked elements of ${}^gG$ and ${}^gJ$ is obtained from equation (15.22), which corresponds to

$$\begin{aligned} &{}^m\text{cov}(X_G, X_J) + {}^m\text{cov}(Y_G, Y_J) \\ &= \Delta\left[\sum_{i=1}^{N_S} X_{SG_i} X_{SJ_i} + \sum_{i=1}^{N_S} Y_{SG_i} Y_{SJ_i} + (1-\Delta)\sum_{i=1}^{N_S} (X_{Mk}^2 + Y_{Mk}^2) - (\bar{X}_G \bar{X}_J + \bar{Y}_G \bar{Y}_J)\right] \end{aligned} \tag{15.26}$$

The program gives the absolute values of the components of ${}^m\ddot{D}^2(G, J)$ and ${}^m\ddot{D}^2(SGJ, SJG)$ as well as the proportion of each component over the sum of distance-variances and over the average of squared distances.

Some of the absolute and relative components of ${}^m\ddot{D}^2(G, J)$ can be expressed in a simple way in terms of the corresponding components of ${}^m\ddot{D}^2(SGJ, SJG)$, as shown in Table 15.3.

15.3.5. Graphic Presentation of ${}^m\ddot{D}^2(G, J)$ and Its Components

The components of ${}^m\ddot{D}^2(G, J)$ can be presented graphically in a simple way, as shown in Figure 15.1. To introduce this presentation, let us start by considering again the decomposition of ${}^h\ddot{D}^2(G^\circ, J^\circ)$ for two geosets (G and J rendered concentric). We assume that the elements of G° and J° are associated through any arbitrary permutation h (see Section 15.2.2):

$${}^h\ddot{D}^2(G^\circ, J^\circ) = d_G^2 + d_J^2 - 2\,{}^h\text{cov}(G, J) \tag{15.27}$$

Let us suppose that we construct a triangle whose sides are proportional to d_G, d_J, and ${}^h\ddot{D}(G^\circ, J^\circ)$, respectively, and that we use γ to denote the angle opposite the side of length ${}^h\ddot{D}(G^\circ, J^\circ)$. In this triangle,

$${}^h\ddot{D}^2(G^\circ, J^\circ) = d_G^2 + d_J^2 - 2d_G d_J \cos\gamma \tag{15.28}$$

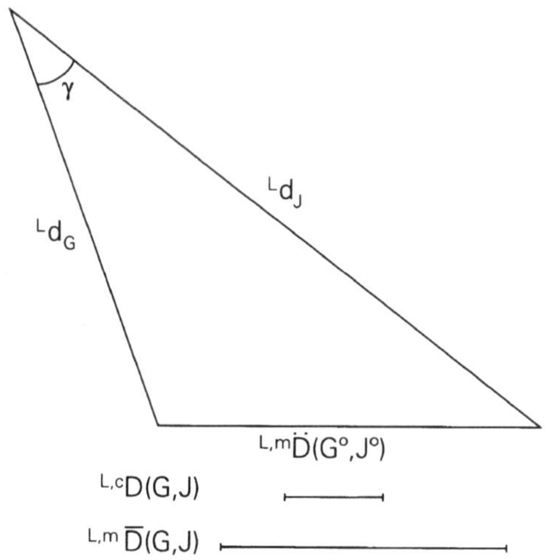

Figure 15.1. Graphic presentation of ${}^m\ddot{D}^2(G, J)$ and its components.

Comparing equation (15.27) and (15.28), we see that

$$ {}^h\text{cov}(G, J) = d_G d_J \cos\gamma \tag{15.29} $$

Let the relative h-covariance $[{}^{r,h}\text{cov}(G, J)]$ be expressed as

$$ {}^{r,h}\text{cov}(G, J) = {}^h\text{cov}(G, J)/d_G d_J = \cos\gamma \tag{15.30} $$

As $-1 \leqslant \cos\gamma \leqslant +1$, we get

$$ -d_G d_J \leqslant {}^h\text{cov}(G, J) \leqslant +d_G d_J \tag{15.31} $$

We now use these findings in the specific case in which $h = m$, and take as an example G denoting dwellings of Rome built up to 1945, and J those built from 1946 to 1981. The same triangular presentation can be used with a simple change of scale for the study of absolute components of ${}^m\ddot{D}^2(G, J)$, for percentage components, and for dimensionless components (see Sections 15.3.1 and 15.3.2).

Dimensionless parameters are shown in Figure 15.1. The side of the triangle proportional to ${}^{L,m}\ddot{D}(G^\circ, J^\circ)$ is given along the x-axis. The other two sides are proportional to ${}^Ld_G$ and ${}^Ld_J$, respectively. As ${}^Ld_G^2 + {}^Ld_J^2 = 1$, we can easily compare the impact of the difference in dispersion over ${}^{L,m}\ddot{D}(G^\circ, J^\circ)$ in the various triangles to be studied. As the pairing between elements of G and J is done by minimization, very often we have ${}^{r,m}\text{cov}(G, J) \geqslant 0$. Examples of ${}^{r,m}\text{cov}(G, J) = 0$ are given by geosets formed by collinear points along orthogonal lines.

If ${}^{r,m}\text{cov}(G^\circ, J^\circ) = 0$, then

$$ {}^{r,m}\ddot{D}^2(G^\circ, J^\circ) = {}^{L,m}d_G^2 + {}^{L,m}d_J^2 = 1 \tag{15.32} $$

In the rare cases in which the covariance is negative, ${}^{r,m}\ddot{D}^2(G^\circ, J^\circ)$ exceeds 1, but it cannot exceed 2.

If ${}^{r,m}\mathrm{cov}(G^\circ, J^\circ) = 1$, then

$$ {}^{r,m}\ddot{D}^2(G^\circ, J^\circ) = ({}^L d_G - {}^L d_J)^2 \tag{15.33} $$

If, in addition, ${}^L d_G = {}^L d_J$, then ${}^{r,m}\ddot{D}^2(G^\circ, J^\circ) = 0$.

To complete the graphic presentation of components of ${}^m\ddot{D}^2(G, J)$, one line proportionl to ${}^{L,c}D(G, J)$ is drawn under the triangle, and a second line is added representing—for comparative purposes—${}^{L,m}\bar{D}(G, J)$, which is often the main object of our research:

$$ {}^{L,m}\bar{D}(G, J) = {}^m\bar{D}(G, J)/\sqrt{d_G^2 + d_J^2} \tag{15.34} $$

Looking at angle γ we are immediately able to judge the degree of relative covariance between associated elements of G and J. If angle γ is 90°, then ${}^{r,m}\mathrm{cov}(G, J)$ is 0 (and vice versa). Smaller sizes of γ suggest positive covariance between elements of G and J. In the example in Figure 15.1, we have $\gamma = 32.11°$, which corresponds to ${}^{r,m}\mathrm{cov}(G, J) = 0.847$, and ${}^L d_G = 0.546$, ${}^L d_J = 0.837$, ${}^{L,m}\ddot{D}(G^\circ, J^\circ) = 0.469$, ${}^{L,c}D(G, J) = 0.114$, and ${}^{L,m}\bar{D}(G, J) = 0.3397$. This can be compared to ${}^{L,m}\ddot{D}(G, J) = 0.48$. If $\gamma = 0$ the triangle degenerates into a line that is perpendicular to the X-axis. In this case, the relative covariance reaches $+1$.

The graphic presentation introduced here can also be applied to ${}^m\ddot{D}^2(SGJ, SJG)$, i.e., to the distances between surpluses ${}^gSGJ$ and ${}^gSJG$. In Section 15.4, we shall see many examples of this method that will help in interpreting and comparing the determinants of ${}^m\ddot{D}^2(G, J)$ for different populations, time periods, and territories.

15.3.6. The Quadratic Average of Minimizing Distances

When comparing the graphic presentations of parameters of minimizing distances between pairs of geosets ${}^gG$ and ${}^gJ$ and between their surpluses ${}^gSGJ$ and ${}^gSJG$, we should note particularly that the graphs show linear parameters, such as (a) standard distances, (b) distance between centers, (c) arithmetic average of minimizing distances, and (d) quadratic average of minimizing distances. Research interest is largely focused on (a), (b), and (c), but as algebraic relationships hold for their squares, they permit the neat decomposition of ${}^m\ddot{D}^2(G, J) = \Delta^m\ddot{D}^2(SGJ, SJG)$ discussed in this chapter.

The side of the triangle that is given on the X-axis is parameter (d), which is the quadratic average $\sqrt{[\mathrm{ASMD}(G, J)]}$ in the comparison between ${}^gG$ and ${}^gJ$, and $\sqrt{[\mathrm{ASMD}(SGJ, SJG)]}$ in the comparison between surpluses. We indicate the graphical method introduced in this Section as the triangle of the quadratic average of minimizing distances, in order to distinguish it from the triangle of the minimal distances discussed in Chapter 13.

Let us consider now various pairs of geosets that have a constant distribution of minimizing distances between surpluses but a varying Δ. For these pairs both ${}^m\bar{D}(SGJ, SJG)$ and $\sqrt{[\mathrm{ASMD}(SGJ, SJG)]}$ are constant. However, ${}^m\bar{D}(G, J) = \Delta^m\bar{D}(SGJ, SJG)$ varies in proportion to changing values of Δ, while $\sqrt{[\mathrm{ASMD}(G, J)]} = \sqrt{\Delta}\sqrt{[\mathrm{ASMD}(SGJ, SJG)]}$ varies with changing values of $\sqrt{\Delta}$. As our research interest is mainly focused on ${}^m\bar{D}(G, J)$, uncritical consideraton of

$\sqrt{[\text{ASMD}(G, J)]}$ might be misleading. Under these circumstances, it might be advisable to take into account the product of $\sqrt{[\text{ASMD}(SGJ, SJG)]}$ by Δ. Obviously for constant ${}^{m}\bar{D}(SGJ, SJG)$ and constant $\sqrt{[\text{ASMD}(SGJ, SJG)]}$, both ${}^{m}\bar{D}(G, J)$ and $\Delta\sqrt{[\text{ASMD}(SGJ, SJG)]}$ change in proportion to changes in Δ.

15.3.7. Decomposition of ASMD for Component and Compound Geosets

Let us denote two component geosets by ${}^{g}G$ and ${}^{g}J$ and their compound geoset by ${}^{g}K$. If v_G^K and v_J^K are the proportions of elements respectively of ${}^{g}G$ and ${}^{g}J$ over the elements of ${}^{g}K$, we have ${}^{g}K = v_G^K\,{}^{g}G + v_J^K\,{}^{g}J$.

Let us examine the values of ${}^{m}\ddot{D}^2(G, J)$, ${}^{m}\ddot{D}^2(G, K)$, and ${}^{m}\ddot{D}^2(J, K)$. We shall again take as an example dwellings in Rome: G = dwellings built up to 1945; J = dwellings built from 1946 to 1981; K = all dwellings.

If we use the following abbreviated notation, $(1) = {}^{g}SGJ$; $(2) = {}^{g}SGK$; $(3) = {}^{g}SKJ$; $(4) = {}^{g}SJG$; $(5) = {}^{g}SKG$; $(6) = {}^{g}SJK$, we see, as before, that

$$(1) = (2) = (3) \qquad \text{and} \qquad (4) = (5) = (6)$$

Consequently, the geosets of minimizing distances between (1) and (4), (2) and (5), and (3) and (6) are equivalent and so are the geosets of the squares of these distances. Consequently, the corresponding parameters are equal. We have already found that ${}^{m}\bar{D}(1, 4) = {}^{m}\bar{D}(2, 5) = {}^{m}\bar{D}(3, 6)$. Similarly, ${}^{m}\ddot{D}^2(1, 4) = {}^{m}\ddot{D}^2(2, 5) = {}^{m}\ddot{D}^2(3, 6)$.

The components of ${}^{m}\ddot{D}^2[(1), (4)]$, ${}^{m}\ddot{D}^2[(2), (5)]$, and ${}^{m}\ddot{D}^2[(3), (6)]$ are also equal:

(a) For squared eccentricities:

$${}^{c}D^2[(1), (4)] = {}^{c}D^2[(2), (5)] = {}^{c}D^2[(3), (6)].$$

(b) For standard distances we find: $d[(1) - (4)] = d[(2) - (5)] = d[(3) - (6)]$ and $d[(4) - (1)] = d[(5) - (2)] = d[(6) - (3)]$; consequently, for squared differences of dispersions:

$$[d(1-4) - d(4-1)]^2 = [d(5-2) - d(2-5)]^2 = [d(3-6) - d(6-3)]^2$$

(c) For minimizing covariances:

$$\begin{aligned}{}^{m}\text{cov}\{[(1)-(4)], [(4)-(1)]\} &= {}^{m}\text{cov}\{[(2)-(5)], [(5)-(2)]\}\\ &= {}^{m}\text{cov}\{[(3)-(6)], [(6)-(3)]\}\end{aligned}$$

Taking this into account as well as the equality of standard distances indicated under (b), we find that residual components are equal in the comparisons between $[(1)-(4)]$ and $[(4)-(1)]$; $[(2)-(5)]$ and $[(5)-(2)]$; and $[(3)-(6)]$ and $[(6)-(3)]$.

The equalities indicated above hold for absolute, relative, and dimensionless components. Values found empirically in many applications correspond well to these general laws, apart from some discrepancies owing to the smallness of the number of points or cells on which the calculations are based. Table 15.4 gives, as an example, the results of calculations for comparison between (2) and (5) and (3) and (6). They can be compared to the findings of the comparison between (1) and (4) given in Table 15.2.

Table 15.4. Rome (1981). Components of ${}^{m}\ddot{D}^2(SGK, SKG)$ and ${}^{m}\ddot{D}^2(SJK, SKJ)$: G = Dwellings Built up to 1945; J = Dwellings Built from 1946 to 1981; K = Total Dwellings

Characteristics measured	Components of ${}^{m}\ddot{D}^2(SGK, SKG)$			Components of ${}^{m}\ddot{D}^2(SGK, SKJ)$		
	Absolute value	Percentage	Ratio to sum of distance-variance	Absolute value	Percentage	Ratio to sum of distance-variance
(1) Eccentricity	2.51	10.4	0.04	2.51	10.3	0.04
(2) Dispersion difference	15.75	65.4	0.26	15.75	64.2	0.26
(3) Residual	5.82	24.2	0.10	6.26	25.5	0.10
(4) Discrepancy = (1) + (2) + (3)	24.09	100.0	0.40	24.53	100.0	0.40
	Components of discrepancy between G and K rendered concentric			Components of discrepancy between J and K rendered concentric		
(5) Sum of distance-variance	60.89	RC[a]	1.00	60.89	RC[a]	1.00
(6) Minus twice covariance	−38.86	0.86	−0.64	38.42	0.85	−0.63
(7) Unlikeness = (2) + (3) = (5) + (6)	22.03		0.36	22.47		0.37

[a] RC = relative covariance.

From the components of ${}^{m}\ddot{D}^2(SGK, SKG)$, we obtain the components of ${}^{m}\ddot{D}^2(G, K)$ by following the same procedure that was suggested in Section 15.3.4 for passing from components of ${}^{m}\ddot{D}^2(SGJ, SJG)$ to the components of ${}^{m}\ddot{D}^2(G, J)$. The same procedure can be followed for obtaining components of ${}^{m}\ddot{D}^2(J, K)$.

From the components of ${}^{m}\ddot{D}^2(G, J)$, as given in Table 15.3 we can easily obtain the components of ${}^{m}\ddot{D}^2(G, K)$ and ${}^{m}\ddot{D}^2(J, K)$. Algebraic equations and actual findings in our numerical example for the dwellings of Rome are given below.

Let us note that: (a) the proportion Δ_G^K of surpluses in the comparison between ${}^gG$ and ${}^gK$ is $\Delta_G^K = v_J^K \Delta_G^J$, where v_J^K is the proportion of elements of J over the total of elements of K; (b) similarly, $\Delta_J^K = v_G^K \Delta_G^J$; (c) $\Delta_G^K + \Delta_J^K = \Delta_G^J$; (d) $v_G^K + v_J^K = 1$; (e) in our numerical example: $v_G^K = 0.2381$; $v_J^K = 0.7619$; $\Delta_G^K = 0.4129$; $\Delta_J^K = 0.1290$; and $\Delta_G^J = 0.5419$.

We can utilize the following equations:

1. For (G, K): (a) Squared eccentricity: ${}^cD^2(G, K) = (v_J^K)^2\, {}^cD^2(G, J)$; (b) discrepancy: ${}^{m}\ddot{D}^2(G, K) = v_J^K\, {}^{m}\ddot{D}^2(G, J)$; and (c) discrepancy between geosets rendered concentric: ${}^{m}\ddot{D}^2(G^\circ, K^\circ) = \text{(b)} - \text{(a)}$. The numerical values thus obtained are (a) 0.428; (b) 9.98; and (c) 9.55.
2. The analogous calculations for (J, K) give the following results: (a) ${}^cD^2(J, K) = (v_G^K)^2\, {}^cD^2(G, J) = 0.0418$; (b) ${}^{m}\ddot{D}^2(J, K) = v_G^K\, {}^{m}\ddot{D}^2(G, J) = 3.12$; and (c) discrepancy between geosets rendered concentric = 3.078.

15.4. Examples of Decomposition of ASMD

15.4.1. Types of Examples Presented

No extensive body of the empirical applications of the many equations introduced in the previous sections is available. Under these circumstances, it would have been tempting to discuss such applications to the pairs of minimal distances discussed in Chapters 13 and 14. However, this would have demanded a discussion of too many geographical details and of the complex relationships among the many parameters proposed. Accordingly, we shall discuss below only applications to a few selected examples and for these we shall only show: (a) the percentages of the three main (squared) components of ASMD between pairs of geosets (eccentricity, difference in dispersion, residual); (b) the percentages of the components of ASMD between the corresponding surpluses; (c) the relative covariance for geosets rendered concentric ${}^{m}\text{cov}(G^\circ, J^\circ)/d_G d_J$; and (d) the relative covariance between corresponding surpluses.

15.4.2. Components of ASMD between Subpopulations in the United States

Table 15.5 shows the components of ASMD between subpopulations in the United States. Several of the characteristics of the geosets of these subpopulations were discussed in Sections 8.8, 9.5, and 11.9 and maps of these geosets were reproduced in Figures 8.1, 8.5, 8.7, and 8.9 (Section 8.8).

Table 15.5. United States (1980). Components of ASMD between Population and Subpopulations

ASMD between	Percentage of squared components of ASMD			
	Eccentricity	Dispersion difference	Residual	Relative covariance
Total population and Whites	2.5	0.3	97.2	0.99
Same: ASMD between surpluses	46.9	7.0	46.1	0.81
Total population and Afro-Americans	14.6	7.1	78.4	0.79
Same: ASMD between surpluses	53.2	24.5	22.4	0.62
Total population and Hispanics	34.8	2.2	63.0	0.67
Same: ASMD between surpluses	70.5	4.4	25.2	0.48
Total population and Asians	32.4	4.5	63.1	0.63
Same: ASMD between surpluses	77.6	2.4	20.0	0.49
Whites and Afro-Americans	18.3	7.0	74.7	0.80
Same: ASMD between surpluses	50.2	23.5	26.3	0.58
Whites and Hispanics	36.1	2.4	61.5	0.67
Same: ASMD between surpluses	69.6	4.4	26.0	0.50
Whites and Asians	33.1	4.8	62.0	0.63
Same: ASMD between surpluses	74.9	3.4	21.7	0.52
Afro-Americans and Hispanics	40.8	6.5	52.7	0.51
Same: ASMD between surpluses	77.2	4.6	18.2	0.19
Afro-Americans and Asians	41.6	8.9	49.5	0.47
Same: ASMD between surpluses	90.3	0.9	8.8	0.22
Hispanics and Asians	13.7	2.5	83.8	0.87
Same: ASMD between surpluses	45.5	15.8	38.7	0.47

Table 15.6. Italy (1981). Components of ASMD According to Sex and Marital Status: Women of Each Marital Status versus Aged 15+

	Percentage of squared components of ASMD		
	Eccentricity	Dispersion difference	Residual
Single	65.1	8.1	26.8
Same: ASMD between surpluses	66.3	6.2	27.5
Married	24.0	2.4	73.6
Same: ASMD between surpluses	47.3	2.7	50.0
Widowed	86.0	6.1	7.9
Same: ASMD between surpluses	86.5	2.1	11.4
Separated	56.2	6.4	37.4
Same: ASMD between surpluses	55.9	5.9	38.2
Divorced	45.8	4.7	49.5
Same: ASMD between surpluses	45.7	7.4	46.9

Table 15.7. Italy (1981). Components of ASMD According to Sex and Marital Status: Men of Each Marital Status versus Aged 15+

	Percentage of squared components of ASMD		
	Eccentricity	Dispersion difference	Residual
Single	24.5	10.2	65.3
Same: ASMD between surpluses	33.4	22.0	44.6
Married	10.7	17.2	72.1
Same: ASMD between surpluses	9.0	15.6	75.4
Widowed	32.3	17.2	50.5
Same: ASMD between surpluses	33.0	8.2	58.8
Separated	63.9	7.8	28.3
Same: ASMD between surpluses	63.8	5.0	31.1
Divorced	58.7	4.9	36.4
Same: ASMD between surpluses	58.3	6.5	35.2

As might be expected, we see from Table 15.5 that the distance between centers plays a minor role in the ASMD of the total population and the Whites, of Hispanics and Asians, and of Whites and Afro-Americans. It plays a major role in the ASMD of Whites and Hispanics, Afro-Americans and Hispanics, and Afro-Americans and Asians. In the ASMD between surpluses, the proportion owing to distance between centers is much greater than that found in the corresponding ASMD between primary geosets.

Table 15.8. Italy (1981). Components of ASMD within Each Marital Status: Men and Women Aged 15+

	Percentage of squared components of ASMD		
	Eccentricity	Dispersion difference	Residual
Single	50.8	2.2	47.0
Same: ASMD between surpluses	51.7	0.9	47.3
Married	70.2	0.7	29.1
Same: ASMD between surpluses	40.2	39.2	20.6
Widowed	68.2	2.3	29.5
Same: ASMD between surpluses	76.4	1.5	22.0
Separated	64.4	12.9	22.7
Same: ASMD between surpluses	66.7	1.4	31.9
Divorced	54.1	1.9	44.0
Same: ASMD between surpluses	53.6	0.2	46.2
All marital statuses (M 15+ versus F 15+)	55.5	2.3	42.2
Same: ASMD between surpluses	51.1	2.4	46.5

Table 15.9. Rome (1981). Components of ASMD between Dwellings According to the Period of Building

	Percentage of squared components of ASMD			
Periods compared	Eccentricity	Dispersion	Residual	Relative covariance
Up to 1945/1946–1981	5.7	55.2	61.2	0.85
Same: ASMD between surpluses	10.4	65.2	24.4	0.86
Up to 1945/Total	4.3	27.9	67.8	0.86
Same: ASMD between surpluses	10.4	65.4	24.2	0.86
1946–1981/Total	1.3	5.5	93.2	0.96
Same: ASMD between surpluses	10.3	64.2	25.5	0.84

15.4.3. Components of ASMD According to Sex and Marital Status

Minimal distances between subpopulations distinguished by sex and marital status were presented and commented upon in Section 13.5.2. Tables 15.6–15.8 are given in order to complete the picture, through the analysis of components of ASMD, broken down by sex and marital status for Italy (1981).

15.4.4. ASMD for a City Growing Concentrically (Rome)

Table 15.9 relates to the comparison between the geosets of dwellings of Rome built from early times to 1945 and from 1946 to 1981. We have already seen how the dwellings that were built earlier are more clustered in central zones, while those built later are more peripheral.

Accordingly, we find that the distance between the centers of the geosets considered plays a minor role in the determination of ASMD, but that a very important impact on ASMD is due to the difference between the dispersions. The ASMD between surpluses show analogous characteristics. As might be expected, the impact of eccentricity is somewhat weaker, and the impact of dispersion differences

Table 15.10. Rome (1981). Components of ASMD According to Age Groups (%)

	Percentage of squared components of ASMD			
ASMD between	Eccentricity	Dispersion	Residual	Relative covariance
Aged 0–14 and total population	0.3	2.3	97.4	0.96
Same: ASMD between surpluses	3.5	35.0	61.5	0.56
Aged 65+ and total population	0.6	3.3	96.1	0.85
Same: ASMD between surpluses	3.9	23.2	72.9	0.12
Aged 0–14 and aged 65+	0.8	5.1	94.1	0.80
Same: ASMD between surpluses	3.5	24.6	71.9	0.20

is even stronger than that of the ASMD between the primary geosets. Table 15.10 presents the analysis of ASMD between extreme age groups (0–14; 65+). Here the impact of eccentricity is almost nil and dispersion difference plays a certain role, but the main component is the residual. The analysis for pairs of other periods of buildings and of other ages shows similar results.

It can be surmised, on the basis of data for other cities, that this type of structure of ASMD is common to populations of cities that grow concentrically and for territorial distributions of other phenomena that are affected by general concentric development.

Tables 15.5 and 15.6 show that the relative covariances between the geosets compared is very high. This finding, too, corresponds to what might be expected.

Chapter 16

Minimal Distance between One Geoset and Another Obtained by Transforming the Coordinates of the First

Eliminating Influences of Position and Dimensions from Minimal Distance

16.1. Introduction

In this chapter we deal briefly with several problems, related in part to those discussed in Chapter 12: (a) Given a geoset ${}^{g}G$, we transform the coordinates of its elements by parallel translation, dilatation of deviations from the center, rotation, reflection, or other changes of position and dimensions and thus obtain a new geoset ${}^{g}J$; we then investigate the minimal distance between ${}^{g}G$ and ${}^{g}J$. (b) Considering a pair of geosets ${}^{g}G$ and ${}^{g}J$ we study how ${}^{m}\bar{D}(G, J)$ and ${}^{m}\ddot{D}^{2}(G, J)$ change when we transform the coordinates of both geosets in the same way.

The results of these studies help us to find solutions to practical problems such as: (a) How to measure the distance between geosets when we wish to compare them independently of their position, their dimensions, or both; this might then pave the way for comparative research dealing with pairs of geosets of different kinds and referred to different time periods or distributed over different territories. (b) How to use ${}^{m}\bar{D}(G, J)$ and ${}^{m}\ddot{D}^{2}(G, J)$ to classify pairs of geosets, ${}^{g}G$ and ${}^{g}J$, as being coincident, congruent, similar, or possessed of opposite properties. To introduce the approach to these problems we use below simple examples of geosets ${}^{g}G$ and ${}^{g}J$ formed by a small number of elements. The extension to other types of geosets is easy.

16.2. Minimal Distance between One Geoset and Another Obtained by Parallel Translation of the First

16.2.1. Parallel Translation

Let us consider a small arbitrary geoset ${}^{g}G$ formed by 10 elements i, with coordinates (X_{G_i}, Y_{G_i}) as shown by squares in Figure 16.1. We assume that ${}^{g}G$ is

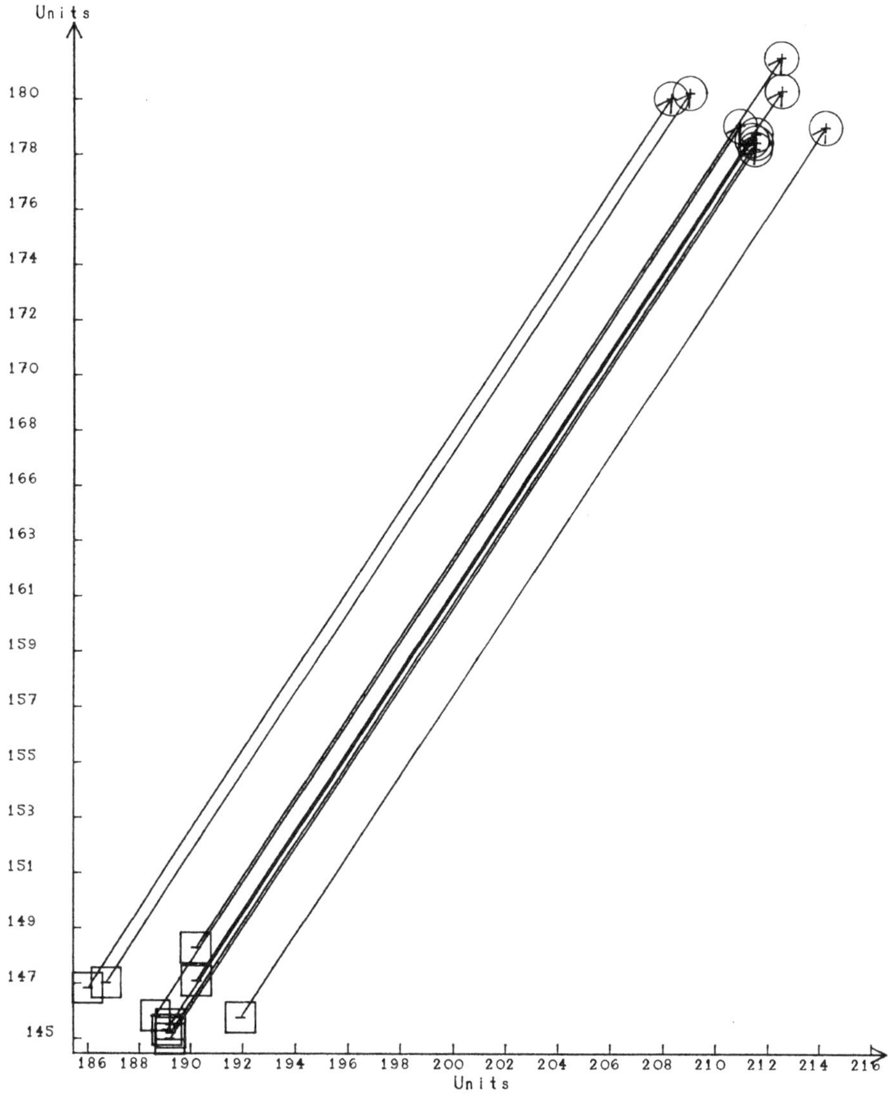

Figure 16.1. Optimal transportation scheme (minimizing linkages) between geosets ${}^gG$ and ${}^gJ$. Key: squares = origins; circles = destinations. The arrows indicate the direction of the shifted units and the areas of the circles and squares are proportional to the number of shifted units.

translated by a lengths of an arbitrary element along the axis of the abscissas and by b units along the axis of the ordinates. Let ${}^gJ$ denote the translated set having as coordinates of any point J_i: $X_{J_i} = X_{G_i} + a$, $Y_{J_i} = Y_{G_i} + b$. Geoset ${}^gJ$ is shown in the figure by circles.

Figure 16.1 shows minimizing linkages between geoset ${}^gG$ and geoset ${}^gJ$. We see that each point J_i is linked to its corresponding point G_i, so that we obtain the following minimizing distances:

$$d(G_i, J_i) = \sqrt{(X_{J_i} - X_{G_i})^2 + (Y_{J_i} - Y_{G_i})^2} = \sqrt{a^2 + b^2} \tag{16.1}$$

Obviously the simple average and the average of the squares of these distances are, respectively,

$$\Sigma\, d(G_i, J_i)/N = \sqrt{a^2 + b^2} \tag{16.2}$$

and

$$\Sigma\, d^2(G_i, J_i)/N = a^2 + b^2 = {}^m\ddot{D}^2(G, J) \tag{16.3}$$

Under the assumption that $X_{J_i} = X_{G_i} + a$ and $Y_{J_i} = Y_{G_i} + b$ we see that the minimal distance between ${}^gG$ and ${}^gJ$ is

$${}^m\bar{D}(G, J) = \sqrt{a^2 + b^2} \tag{16.4}$$

and that the square root of the quadratic ASMD between ${}^gG$ and ${}^gJ$ is

$$\sqrt{{}^m\ddot{D}^2(G, J)} = {}^m\ddot{D}(G, J) = \sqrt{a^2 + b^2} \tag{16.5}$$

16.2.2. Translation of a Geoset to Its Center

Let us indicate by $(\bar{X}_G, \bar{Y}_G)$ the coordinates of the center of ${}^gG$. A special case of parallel translation is the translation of the geoset ${}^gG$ so that it is centered at the origin, i.e., a shift by $a = -\bar{X}_G$ and $b = -\bar{Y}_G$. Let us denote the translated geoset by ${}^gG^\circ$. The coordinates of its elements are

$$X_{G_i^\circ} = X_{G_i} - \bar{X}_G = x_{G_i} \tag{16.6a}$$

$$Y_{G_i^\circ} = Y_{G_i} - \bar{Y}_G = y_{G_i} \tag{16.6b}$$

where x_{G_i} and y_{G_i} are the deviations of the coordinates of the elements G_i from the respective coordinates of the center.

Since the center of ${}^gG^\circ$ is (0, 0) and the center of ${}^gG$ is $(\bar{X}_G, \bar{Y}_G)$,

$${}^cD(G, G^\circ) = \sqrt{\bar{X}_G^2 + \bar{Y}_G^2} \tag{16.7}$$

The minimal distance between G and G° is

$$\begin{aligned} {}^m\bar{D}(G, G^\circ) &= \sum\sqrt{[(X_{G_i} - X_{G_i^\circ})^2 + (Y_{G_i} - Y_{G_i^\circ})^2]}/N \\ &= \sqrt{\bar{X}_G^2 + \bar{Y}_G} \end{aligned} \tag{16.8}$$

so that in this special case

$${}^cD(G, G^\circ) = {}^m\bar{D}(G, G^\circ) = {}^m\ddot{D}(G, G^\circ) \tag{16.9}$$

16.2.3. Comparing Two Geosets Translated to Their Centers

When we compare any two geosets ${}^gG$ and ${}^gJ$, we find that the simple relationships shown in equation (16.9) hold between ${}^gG$ and ${}^gG^\circ$ and between ${}^gJ$

and ${}^{g}J^{\circ}$:

$$^{c}D(G, G^{\circ}) = {}^{m}\bar{D}(G, G^{\circ}) = {}^{m}\ddot{D}(G, G^{\circ}) \tag{16.10a}$$

$$^{c}D(J, J^{\circ}) = {}^{m}\bar{D}(J, J^{\circ}) = {}^{m}\ddot{D}(J, J^{\circ}) \tag{16.10b}$$

where G° and J° are referred to the same center, with coordinates (0, 0). However, knowing ${}^{m}\bar{D}(G^{\circ}, J^{\circ})$ does not help in the study of the minimal distance ${}^{m}\bar{D}(G, J)$ for there is no simple algebraic relationship between these geosets. On the other hand, working on the squares we find the simple relation

$$^{m}\ddot{D}^{2}(G^{\circ}, J^{\circ}) + {}^{c}D^{2}(G, J) = {}^{m}\ddot{D}^{2}(G, J) \tag{16.11}$$

which is useful for both theoretical and practical reasons. We saw in Chapter 15 that it may be convenient to translate two geosets ${}^{g}G$ and ${}^{g}J$ to the same center (0, 0). We can thus measure ${}^{m}\ddot{D}^{2}(G^{\circ}, J^{\circ})$ between these concentric sets and eliminate the effect of the squared distance between the centers. Calculating ${}^{m}\ddot{D}^{2}(G, J) - {}^{c}D^{2}(G, J)$ we isolate the effects of other components of ${}^{m}\ddot{D}^{2}(G, J)$.

16.3. Minimal Distance between One Geoset and Others Obtained from It Through Rotation and Reflection

In order to simplify the analysis, let us look at a geoset ${}^{g}G$ transformed into ${}^{g}G^{\circ}$ by considering the center of ${}^{g}G$ as the origin of the coordinates of its elements. We assume that ${}^{g}G^{\circ}$ is rotated around the center by an angle γ resulting in a new geoset ${}^{\gamma}G^{\circ}$. We then calculate the minimal distance ${}^{m}\bar{D}(G^{\circ}, {}^{\gamma}G^{\circ})$ between the original G° and ${}^{\gamma}G^{\circ}$. Very often ${}^{m}\bar{D}(G^{\circ}, {}^{\gamma}G^{\circ}) \neq 0$, thus showing that G° and ${}^{\gamma}G^{\circ}$ do not coincide.

It does not seem possible to find general laws for minimal distances ${}^{m}\bar{D}(G^{\circ}, {}^{\gamma}G^{\circ}) \neq 0$, as their values differ depending on the characteristics of G° and the angle γ. However, in certain cases it may be of some interest to determine how ${}^{m}\bar{D}(G^{\circ}, {}^{\gamma}G^{\circ})$ changes with γ. This study can be based on empirical results, considering, e.g., variations of ${}^{m}\bar{D}(G^{\circ}, {}^{\gamma}G^{\circ})$ for any angle $\gamma = 0^{\circ}, 1^{\circ}, 2^{\circ}, \ldots, 180^{\circ}$, or to save work, say, $\gamma = 0^{\circ}, 2^{\circ}, 4^{\circ}, \ldots, 180^{\circ}$, or $\gamma = 0^{\circ}, 5^{\circ}, 10^{\circ}, \ldots, 180^{\circ}$, etc.

It may be noted that in such a setting ${}^{c}D^{2}(G^{\circ}, {}^{\gamma}G^{\circ}) = 0$ and $d_G = d_{G^{\circ}} = d_{{}^{\gamma}G^{\circ}}$. Using ${}^{m}\ddot{D}^{2}(G^{\circ}, {}^{\gamma}G^{\circ})$ for analytical purposes we find, according to the findings of Section 15.2 that

$$^{m}\ddot{D}^{2}(G^{\circ}, {}^{\gamma}G^{\circ})/2 = d_G^2 - {}^{m}\text{cov}(G, {}^{\gamma}G^{\circ}) \tag{16.12}$$

where d_G^2 is the distance variance of G.

It might be convenient to utilize relative minimal distance by dividing equation (16.12) by d_G^2 (as $d_{G^{\circ}}^2 = d_{{}^{\gamma}G^{\circ}}^2 = d_G^2$), by which we obtain ${}^{r,m}\ddot{D}^{2}(G^{\circ}, {}^{\gamma}G^{\circ}) = 1 - {}^{m}\text{cov}(G, G^{\circ})$. Analysis by means of this coefficient is much simplified by the fact that it can only vary between 0 and 1.

It has been found from examples (which are not referred to here) that the variations of relative covariance with changes in γ are rather regular in the special cases considered, which suggests that this approach might be of use for comprehensive research on asymmetry. The analysis of the effects of rotation through 180° can be used for the study of central asymmetry, while for axial asymmetry referred to principal axes, the study of effects of reflection around these axes may be of value. The analysis of changes of minimal distance between a geoset G° kept in the same position and another geoset J° that is supposed to rotate around their common center (0, 0) not discussed here. It is interresting to note, however, the special case in which both G° and J° are rotated in such a way that their respective major (minor) axes coincide.

16.4. Comparing Geosets Found over Different Territories

16.4.1. Minimal Distance between Geosets in the Same Position

Suppose we have a collection of geosets indicating distributions of comparable populations over different territories, e.g., maps of habitats of human populations in different cities or distributions of industrial areas over different urban territories. These distributions differ from each other owing to: (a) their different positions in the respective planes, (b) their different dimensions, (c) the shape of their containers or (d) the patterns of distribution of their elements.

We have shown in Chapter 12 that the position of any geoset ${}^gG$ in the plane is completely described by means of three parameters: its center $(\bar{X}_G, \bar{Y}_G)$, the angle α between its principal axes and the conventional axes, and the quadrant setting. A first simple step for calculating the minimal distance between two geosets ${}^gG$ and ${}^gJ$ is to render them concentric, an operation that we have already discussed. A second step is to give the geosets the same orientation: we rotate them so that their major axes coincide. It may be convenient to place these axes along the conventional axis of the abscissas (or ordinates).

16.4.2. Checking Coincidence and Congruence of Two Geosets

We refer here to the concepts of coincident and congruent geosets, as introduced in Chapter 12. Let us calculate ${}^m\bar{D}(G, J)$ between any two geosets. If they are coincident, ${}^m\bar{D}(G, J) = 0$. Conversely, if ${}^m\bar{D}(G, J) = 0$, they are coincident. If ${}^m\bar{D}(G, J) \neq 0$, then the minimal distance is a global measure of noncoincidence.

Let us term two geosets ${}^gG$ and ${}^gJ$ that are coincident in at least one position, congruent. The methods alluded to in Section 16.4.1 enable us to explore whether such a position exists.

We assume that we transform the coordinates of ${}^gG$ and ${}^gJ$ into principal positions (see Section 12.4) and thus obtain geosets to be designated as ${}^gG^*$ and ${}^gJ^*$. If the geosets ${}^gG$ and ${}^gJ$ are congruent, the corresponding geosets ${}^gG^*$ and ${}^gJ^*$ must be coincident; then ${}^m\bar{D}(G^*, J^*) = 0$ and we can affirm that ${}^gG$ and ${}^gJ$ are congruent. If ${}^m\bar{D}(G^*, J^*) \neq 0$, we can generally consider this parameter as a measure of incongruence between ${}^gG$ and ${}^gJ$. However, for low values of ${}^m\bar{D}(G^*, J^*)$ a more thorough investigation may be required in order to exclude the hypothesis of congruence.

16.5. Minimal Distance between Similar, Reduced, and Standardized Geosets

16.5.1. Comparing a Geoset to a Similar Set Obtained from It by Dilatation

To simplify the presentation of this problem without any loss of generality, we assume a geoset ${}^{g}G^*$ formed by N points and placed in principal position. A geoset ${}^{g}J^*$ is formed by taking $X_{J_i^*} = kX_{G_i^*} = kx_{G_i}$ and $Y_{J_i^*} = kY_{G_i^*} = ky_{G_i}$, where k is any real and positive number and x_{G_i} and y_{G_i} are deviations of points G_i from the center along conventional axes.

Geoset ${}^{g}J^*$ is thus obtained from geoset ${}^{g}G^*$ by dilatation through a coefficient k. These geosets differ only in their dispersion and can be called similar.

Let us indicate the arithmetic average of the distances of the elements ${}^{g}G$ (or of ${}^{g}J$) from their center as $\bar{d}_G$ and $\bar{d}_J$:

$$\bar{d}_G = \bar{d}_{G*} = \sum \sqrt{(X_{G_i} - \bar{X}_G)^2 + (Y_{G_i} - \bar{Y}_G)^2}/N \tag{16.13a}$$

$$\bar{d}_J = \bar{d}_{J*} = k\bar{d}_G \tag{16.13b}$$

the standard distance of ${}^{g}G$ and ${}^{g}J$ by d_G and d_J:

$$d_G = d_{G*} = \sqrt{\sum [(X_{G_i} - \bar{X}_G)^2 + (Y_{G_i} - \bar{Y}_G)^2]/N} \tag{16.14a}$$

$$d_J = d_{J*} = kd_G \tag{16.14b}$$

so,

$$k = \bar{d}_{J*}/\bar{d}_{G*} = d_{J*}/d_{G*} \tag{16.15}$$

It can be shown that the minimal distance ${}^{m}\bar{D}(G^*, J^*)$ is given by

$$|1 - k|d_G = |\bar{d}_G - \bar{d}_J| \tag{16.16}$$

and that the square root of the quadratic average of minimizing distances

$${}^{m}\ddot{D}(G^*, J^*) = |1 - k|d_G \tag{16.17}$$

16.5.2. Minimal Distance between Geosets Transformed in the Same Way

Consider a pair of geosets ${}^{g}G$ and ${}^{g}J$ with a minimal distance ${}^{m}\bar{D}(G, J)$ and a quadratic average of minimizing distances ${}^{m}\ddot{D}^2(G^*, J^*)$. Suppose that we have transformed both geosets in the same way, thus obtaining geosets ${}^{g}G'$ and ${}^{g}J'$, respectively. It may be of interest to study possible relationships between ${}^{m}\bar{D}(G', J')$ and ${}^{m}\bar{D}(G, J)$ and between ${}^{m}\ddot{D}^2(G', J')$ and ${}^{m}\ddot{D}^2(G, J)$. For example, we may find that in the case of equal parallel translation, rotation, or reflection, ${}^{m}\bar{D}(G', J') = {}^{m}\bar{D}(G, J)$ and ${}^{m}\ddot{D}^2(G', J') = {}^{m}\ddot{D}^2(G, J)$.

If G' and J' are obtained by taking $x_{G'} = kx_G$, $y_{G'} = ky_G$, $x_{J'} = kx_J$, and $y_{J'} = ky_J$ we find that

$${}^{m}\bar{D}(G', J') = k\,{}^{m}\bar{D}(G, J) \tag{16.18a}$$

and

$$ {}^{m}\ddot{D}^2(G', J') = k^2\,{}^{m}\ddot{D}^2(G, J) \tag{16.18b} $$

16.5.3. Checking Whether Two Reduced Geosets Are Similar

We have seen that the analysis of geosets can be facilitated by placing them in the same position and dividing the coordinates of their elements by the standard distance. Geosets transformed in this manner are termed reduced geosets, and have center 0 and standard distance 1. For example, given a geoset ${}^{g}G$, with coordinates of elements X_{G_i}, Y_{G_i}, we obtain a reduced set ${}^{r}G^*$ by taking $X({}^{r}G^*) = X_G/d_G$ and $Y({}^{r}G^*) = Y_G/d_G$ as the coordinates of its elements.

If the two original sets are similar, then the quadratic average of the distance between the reduced sets is

$$ {}^{m}\ddot{D}^2({}^{r}G^*, {}^{r}J^*) = 0 \tag{16.19} $$

Inversely, if ${}^{m}\ddot{D}^2({}^{r}G^*, {}^{r}J^*) = 0$, then the sets are similar. If ${}^{m}\ddot{D}^2({}^{r}G^*, {}^{r}J^*) \neq 0$, we can consider this parameter a measure of dissimilarity.

We saw in Section 15.2 that the ASMD between two concentric geosets G° and J° is at most $d_{G^\circ}^2 + d_{J^\circ}^2$. In the special case of two reduced geosets, this becomes 2, because $d^2({}^{r}G^*) = d^2({}^{r}J^*) = 1$. Taking ${}^{m}\ddot{D}^2({}^{r}G^*, {}^{r}J^*)$ as a measure of dissimilarity, we find that this parameter is included between 0 and 2.

16.5.4. Comparing Standardized Geosets and Checking Affine Equivalence

It is sometimes convenient to transform a geoset ${}^{g}G$ into a standardized geoset ${}^{g}G^*$ by putting it in principal position and dividing the deviations from the center, along the principal axis, by the respective standard deviations. The coordinates of the elements of this standardized geoset are $({}^{M}G_i^*/\sigma_M, {}^{m}G_i^*/\sigma_m)$. In this geoset, the center is 0, the standard deviations are 1, and the standard distance is $\sqrt{2}$.

Let us consider two standardized geosets ${}^{\sigma}G^*$, ${}^{\sigma}J^*$. The quadratic average of the distance between these sets reduces to

$$ {}^{m}\ddot{D}^2({}^{\sigma}G^*, {}^{\sigma}J^*) = 2(1 - ({}^{m}r^{M} + {}^{m}r^{m})/2) \tag{16.20} $$

where ${}^{m}r^{M}$ and ${}^{m}r^{m}$ denote correlation between the coordinates of the elements of the ${}^{\sigma}G^*$ and ${}^{\sigma}J^*$ associated by the minimizing linkages $\boldsymbol{m}$. These correlations are measured, respectively, along the major axes (M_{G^*}, M_{J^*}) and the minor axes (m_{G^*}, m_{J^*}). The minimum value of ${}^{m}\ddot{D}^2({}^{\sigma}G^*, {}^{\sigma}J^*) = 0$ is obtained when the correlation is ${}^{m}r^{M} = {}^{m}r^{m} = 1$. In this case, the two geosets are coincident (${}^{\sigma}G^* = {}^{\sigma}J^*$), which means that ${}^{g}G$ and ${}^{g}J$ are affinely equivalent: it is possible to find a position (in our case the principal position) in which, by expansion or compression, they can be rendered coincident.

Suppose that ${}^{m}\bar{D}({}^{\sigma}G^*, {}^{\sigma}J^*) = 0$. The maximum value of ${}^{m}\ddot{D}^2({}^{\sigma}G^*, {}^{\sigma}J^*) = 2$ is obtained for ${}^{m}r^{M} = {}^{m}r^{M} = 0$. A value of 2 indicates the least possible affine equivalence. A dimensionless value of ${}^{m}\ddot{D}^2({}^{\sigma}G^*, {}^{\sigma}J^*)$ can be obtained by dividing this parameter by the maximum value 2. We thus obtain a coefficient $1 - ({}^{m}r^{M} + {}^{m}r^{m})/2$, which varies between 0, for the case of affine equivalence, to 1 for total lack of affine equivalence.

16.6. Converting the Minimal Distance to a Pure Number

16.6.1. Studies in Which Dimensionless Minimal Distance May Be Required

Minimal distances can be given in absolute values, measured in units such as kilometers, meters, or miles. We have seen several examples of such a presentation in preceding chapters, which is fully justified in studies for which we actually require a measure indicating, say, the distance between, e.g., distributions of health, educational, social, or other services and the population served, or between distributions of factors of production and production centers. It may also be useful when we deal with comparisons among many distributions over well-defined and meaningful territories. Many examples of this approach are given in the second volume of the *Statistical Atlas of Italy* (1992).

However, in many other cases the use of the absolute minimal distance is unsatisfactory and the need for dimensionless measures of distance between geosets is acutely felt. For example:

1. We may calculate minimal distances between many subpopulations, as measures of degree of mutual segregation. Then we need a yardstick to measure among what subpopulations the segregation is "large" or "small."
2. We may use minimal distance as a measure of segregation between subpopulations defined according to the same characteristics j (e.g., j = income, age or race) in many cities i at different times t. We need methods that make it possible to compare the degrees of segregation, e.g., for a characteristic j in different cities i; average segregation for all characteristics j in each of the cities i; average segregation for each characteristic j in all cities considered together; segregation at each time separately, or average segregation during certain periods of time.
3. We can employ minimal distance to check hypotheses on similarity among certain distributions $(\ldots {}^{g}G, \ldots, {}^{g}J, \ldots, {}^{g}R \ldots)$ in one country or city or in many countries or cities. We may need to determine whether, say, ${}^{g}G$ and ${}^{g}J$ can be considered as belonging to the same "family" while neither ${}^{g}G$ and ${}^{g}R$ nor ${}^{g}J$ and ${}^{g}R$ do.

16.6.2. An Ideal Solution

The ideal solution in all the cases described above as well as in many others is to transform the minimal distance ${}^{m}\bar{D}(G, J)$ between any pair of geosets ${}^{g}G$ and ${}^{g}J$ or the corresponding quadratic average ${}^{m}\ddot{D}^{2}(G, J)$ into a coefficient that varies between 0 and 1.

We saw in Chapters 14 and 15 that this approach can be followed in the comparison between geosets that have been made concentric, using ${}^{m}\ddot{D}^2(G^\circ, J^\circ)$. Coefficients rendering this parameter and its components dimensionless were suggested in Section 15.3. Dimensionless coefficients for the study of reduced geosets were introduced in Section 16.5.3. The possibility of obtaining these simple coefficients, varying between 0 and 1, derives from the facts that: (1) for concentric sets the average of squares of minimizing distances has a well-defined maximal value: $d_G^2 + d_J^2$; (2) if effects of discrepancy between dispersions measured by $(d_G - d_J)^2$ are eliminated, then the residual component ${}^{m}R^2(G^\circ, J^\circ)$ also has a defined maximal value $2d_G d_J$.

16.6.3. Problems Arising from the Distance between Centers and a Solution

When we consider the problem of making any ${}^{m}\bar{D}(G, J)$ or ${}^{m}\ddot{D}^2(G, J)$ (not rendered concentric) dimensionless we are faced with the difficulty that no such simple maximum values are available. In fact, attempts (Bachi, 1962*b*, 1984); Jakubs, 1981) to find such values did not show promising results. This failure can be attributed to the facts that: (a) ${}^{m}\bar{D}(G, J)$ and ${}^{m}\ddot{D}^2(G, J)$ are often very significantly affected by the distance ${}^{c}D(G, J)$ between the centers of the geosets; and (b) comparison of ${}^{g}G$ and ${}^{g}J$ does not directly suggest an easily definable maximum value for ${}^{c}D(G, J)$.

A general way to overcome this difficulty seems to be to consider ${}^{g}G$ and ${}^{g}J$ as being derived from, or included in, a basic geoset ${}^{g}B$, to be defined in a way that is adapted to the actual setting of the particular study, as indicated below. We calculate the average of the distances between all the possible pairs of elements of ${}^{g}B$. We then obtain a parameter measuring the dispersion of the elements of ${}^{g}B$ in a way that is adapted to comparison with ${}^{m}\bar{D}(G, J)$. While ${}^{m}\bar{D}(G, J)$ averages minimizing distances, the average for ${}^{g}B$ takes into account all possible distances. As both averages are expressed in the same units of measure (e.g., kilometers or miles) their ratio is a pure number.

We give below a list of examples of various settings, with suggestions for the selection of a proper basic geoset ${}^{g}B$ in each case:

1. Geosets ${}^{g}P, {}^{g}Q, \ldots$ (and eventually many others) are distributed over a predetermined territory T, and it is reasonable to assume that such geosets can be found in any part of T. Under these conditions we can take ${}^{g}B$ as being the uniform distribution over T. Consider, e.g., the large-scale study of the distribution of the many demographic, social, and economic phenomena in Italy described in Chapter 1. The *Statistical Atlas of Italy* gives examples of some of these phenomena in almost every location in T. The decision to designate the uniform distribution over T as ${}^{g}B$ is especially justified with regard to phenomena such as agricultural distributions, which are directly connected with the territory.
2. For some countries, it might be more reasonable to exclude from ${}^{g}B$ parts of the territory over which it is even potentially impossible to find geosets ${}^{g}P, {}^{g}Q, \ldots$, e.g., extensive desert or mountainous areas. Consideration of the

joint habitat of all the geosets studied might be of use. Alternatively, for agricultural distributions, we may take ${}^gB$ as being the distribution of arable land. The need for a large-scale comparison among many geosets in various countries might warrant the investment in research necessary for obtaining such reduced territories.

3. In many cases, it might be acceptable to take ${}^gB$ as the distribution of the general population, or the active population, or other population segments.
4. In urban research we may select as ${}^gB$ the territory T to which reference is made in the study of geosets ${}^gG$ and ${}^gJ$ or only the habitat of the distribution of the general population.
5. In some cases, we deal with a compound set ${}^gK$ formed by several subsets ${}^gP, {}^gQ, \ldots$. For example, K = all individuals in the labor force, $P, Q, \ldots$ = age, occupational, ethnic, or income groups in the labor force. In a case of this type, selection of ${}^gK$ as ${}^gB$ seems to be a good solution.
6. Suppose that in the comparison between geosets ${}^gG$ and ${}^gJ$, G is derived from J (as G = deaths, births, marriages; J = population or adult population). In such a setting, we might take ${}^gJ$ as ${}^gB$.
7. If there is no other solution for ${}^gB$, we might take the union between ${}^gU$ and ${}^gJ$ or its convex hull as ${}^gB$.

16.6.4. Rendering ${}^m\bar{D}(G, J)$ and ${}^m\ddot{D}^2(G, J)$ Dimensionless Through Comparison with the Average Distance between Pairs of Elements of the Basic Set

We noted above that ${}^m\bar{D}(G, J)$ and ${}^m\ddot{D}^2(G, J)$ are based on the respective calculation of the quadratic and the arithmetic average of minimizing distances. To render these values dimensionless we have proposed comparing them to the average of all possible distances, calculated between all the elements of the basic set.

With respect to the quadratic average of all possible distances, the calculation is very simple (see Risser, 1937, Bachi, 1962*b*, and Chapter 10).

Let us denote elements of the basic set (where $i = 1, 2, \ldots, k, \ldots, j, \ldots, N$) by B_i; the distance between elements B_k and B_j by $d_{kj} = \sqrt{[(X_{B_k} - X_{B_j})^2 + (Y_{B_k} - Y_{B_j})^2]}$; and the relative weight attached to elements B_k and B_j by w_k and w_j.

It is known that in general

$$\sum_{k=1}^{N} \sum_{j=1}^{N} w_k w_j d_{kj}^2 = 2d_B^2 \tag{16.21}$$

where d_B is the standard distance of the basic geoset ${}^gB$. The average of the squares of the distances, if taken "with repetition" (i.e., including in the calculation distances $d_{ii} = 0$), is equal to twice the distance-variance. Dividing ${}^m\ddot{D}^2(G, J)$ by $2d^2$ we obtain a dimensionless ASMD. It may be preferable to use the square root. This parameter is the dimensionless quadratic average of minimizing distances, and is termed the relative quadratic minimal distance (RQMD):

$$\text{RQMD} = {}^m\ddot{D}^2(G, J)/\sqrt{2}\, d_B \tag{16.22}$$

With respect to the arithmetic average of minimizing distances we can proceed in a similar way. We divide ${}^{m}\bar{D}(G, J)$ by the arithmetic average of the distances between all possible pairs of elements of the basic set,

$$\sum_{k=1}^{N} \sum_{j=1}^{N} w_k w_j d_{kj}$$

and obtain a relative minimal distance

$${}^{r,m}\bar{D}(G, J) = {}^{m}\bar{D}(G, J) \Big/ \sum_{k=1}^{N} \sum_{j=1}^{N} w_k w_j d_{kj} \tag{16.23}$$

There are no algebraic simplifications of the denominator on the right-hand side of equation (16.23). However, Neft (1966) found empirically that this parameter is close to the product of $\sqrt{2}$ and the arithmetic average $\bar{d}$ of the distances of the elements of a geoset and its center (see Chapter 10). Thus we may take

$${}^{r,m}\bar{D}(G, J) \cong {}^{m}\bar{D}(G, J)/\sqrt{2}\bar{d}_B \tag{16.24}$$

as a reasonable approximation.

Part V

Epilogue

Chapter 17

An Overview and Some Concluding Remarks

17.1. Introduction

Now that we have concluded our discussion of selected geostatistical methods, we move on to some general considerations. To preclude any misunderstanding, we repeat here that the term "geostatistical" refers to geographical–statistical methods and not to geological–statistical methods, as developed by Mathéron and his followers. The methods described in this volume have been selected in the hope that they will be of help in various fields of basic and applied research that deal with spatial distributions of populations, subpopulations, territories, and subterritories. They may contribute in exploration, first orientation, description, and analysis of geographical–statistical data and in the formulation and checking of hypotheses. They can also facilitate the condensation and presentation—micro- and macrogeographically—of the principal information obtainable from bulky data, specified by many small and irregular territorial units.

The methods presented here have different characteristics and will enable display of the findings graphically, algebraically, and/or numerically. They may achieve different degrees of detail, of abstraction, of generality, and of comparability, but they are closely and systematically integrated among themselves. Some of them are new, and some are known. However, the integrative approach on which this book is based is rather new. This consideration and the limitation of space may justify the lack of systematic comparison among all the methods discussed here and those scattered in the literature on territorial statistics and related fields, which have developed prodigiously in recent decades.

Our conviction that the methods herein are of real value is based on the wide and critical applications that are documented in the two volumes of the *Statistical Atlas of Italy* (1988, 1992), as well as on their successful implementation in comparative urban research and in other fields. The examples described in this volume are generally based on these applications. For many of them, the substantive meaning of the findings has been checked by analytical research conducted by experts in the relevant fields.

However we make no claim that the techniques proposed can be considered as constituting a universal panacea, nor that they are fully developed, and we stress the need for further and critical research. An important lesson gained by experience and partly (but not sufficiently) applied in this volume is the need to base geostatistical research on interdisciplinary approaches: cooperation in research is important among geostatisticians, general statisticians, quantitative geographers, and experts conversant with specific fields such as demography, sociology, and economics, to which the geostatistical data we worked with actually refer.

17.2. Principal Guidelines for the Geostatistical Methods Described in this Volume

Most of the methods and examples discussed in this volume are based on the following approaches:

1. The geostatistical data have been conceptualized and formalized as population geostatistical sets (abridged geosets).
2. By population P we mean any collection or set of, e.g., human beings, other living organisms, establishments, institutions, economic activities, or events.
3. By geoset (${}^gP$) we have denoted the list of locations of elements of P, to which we attach the relative weight to be given to each location, in order to prepare maps of geostatistical parameters. Locations are generally expressed here as Cartesian coordinates. If detailed information on the location of each element P_i of ${}^gP$ is available, the geoset is defined as a set of triplets: ${}^gP = \{(X_i^P, Y_i^P); w_i^P\}$. Here (X_i^P, Y_i^P) are the coordinates of the location of P_i and w_i^P is the relative weight to be attached to P_i.
4. In general, we have referred to large populations for which the primary information available is detailed by many irregular and small territorial units such as municipalities, counties, or census tracts. These data have been transformed into information detailed over a network of small regular hexagonal cells of equal area.
5. In general, we posit that each geoset is distributed over a defined gross territorial container T (see Section 17.3). The network of hexagonal cells is assumed to cover T approximately.
6. Accordingly, ${}^gP$ can be compared to the geoset ${}^gU$ of the theoretical uniform distribution over T. The comparison between ${}^gP$ and ${}^gU$ is made: (a) graphically, showing macrogeographical characteristics of ${}^gP$ and, if required, also showing microgeographical characteristics; (b) by comparing geostatistical parameters of ${}^gP$ and ${}^gU$; and (c) by calculating the minimal distance between ${}^gP$ and ${}^gU$ and analyzing its meaning.
7. The comparison between primary geosets ${}^gP$ and ${}^gU$ is based on the assumption that we transform ${}^gP$ into ${}^gU$. This is done by deriving the geoset of surpluses ${}^gSPU$, the geoset of surpluses ${}^gSUP$, and the geoset of matches ${}^gM(P, U)$ stepwise from ${}^gP$ and ${}^gU$. Derived geosets are treated and analyzed by methods similar to those used for the treatment and analysis of ${}^gP$ and ${}^gU$. It is assumed that elements of ${}^gM(P, U)$ are not shifted; ${}^gSPU$ and ${}^gSUP$ show the distributions of the origins and destinations of the elements being shifted.

8. Another central theme of geostatistical studies is the comparison between primary geosets (${}^gP, {}^gQ, \ldots$) distributed over the same territory T. This is done in ways similar to those used for the comparison between ${}^gP$ and ${}^gU$. Here we also derive the geosets of surpluses of ${}^gP$ over ${}^gQ$, of surpluses of ${}^gQ$ over ${}^gP$, and of matches between ${}^gP$ and ${}^gQ$ from the primary geosets ${}^gP$ and ${}^gQ$.
9. Primary and derived geosets are functionally related, which makes it much easier to interpret corresponding maps, their geostatistical parameters, and minimal distances.
10. Analogous functional relations have been found and utilized in many other settings (e.g., in the comparison between component geosubsets and their compound geoset and in the decomposition of various types of geosets). Typical examples of these relationships are collinearity of centers and of minimal distances utilized in the analysis of triangles of centers and triangles of minimal distances.
11. The operations on geosets have both similarities to and differences from operations on common sets. The comparative analysis of these operations has been found to be of use for geostatistical research.

17.3. Types of Territorial Containers

17.3.1. Gross Territorial Containers

As a rule, we assume that any geoset being studied is distributed over a territorial container T, the boundaries of which are known, have been digitized, and can be drawn on a map. In many cases the territorial container is predetermined and predefined according to political, administrative, social, economic, ethnic, or other criteria (such as the territory of a country, a federal state, the municipal area of a city, or a conurbation). If the distribution of the population is known but unbounded, we can determine the territorial container or territorial range of the geoset by drawing its convex hull, or a convex hull netted from the influence of outliers, or a convex hull including the major clusters of the population within it.

The determination of the gross territorial container T has been found to be very useful both conceptually and practically: it allows us to compare any geoset ${}^gP$ to the uniform distribution ${}^gU$ over T systematically and to understand that each population element covers a finite areal element, even if that element is very small. This, in turn, helps us to avoid the conceptual difficulties that stem from the usual convention, which considers the elements of population as dimensionless dots.

17.3.2. The Habitat as a Net Territorial Container

Almost all the geosets studied in this book differ very considerably from the corresponding uniform distribution, with their elements covering only part, and often a very limited part, of T. Considering the simple setting in which a population of small size F^P is distributed over a territory T divided into $N^c = F^P$ cells, we can easily distinguish (a) the cells in which elements of the population are found from (b) the cells in which elements of the population are absent. The set of cells (a) has been denoted as habitat H, the set of cells (b) has been denoted as nonhabitat NH. This dichotomy can be created for all types of geosets, and H and NH can be separated graphically over the map of T.

The geographical distribution of H and NH can also be studied through comparison of the geostatistical parameters of ${}^gH$, ${}^gNH$, ${}^gT$, and ${}^gP$. We can consider H as the net territorial container of the population, which answers the question, "where is the population P actually found." The geostatistical parameters of ${}^gH$ are equal to those of ${}^gM(P, U)$; those of ${}^gNH$ are equal to the geostatistical parameters of ${}^gSUP$. In a comparison between ${}^gP$ and ${}^gU$, we assume that surpluses ${}^gSPU$ (origins) are redistributed over the cells of ${}^gNH$ (destinations). Experience has shown that the analysis of ${}^gH$ and ${}^gNH$ is often very useful.

17.3.3. Other Types of Territorial Containers

In the comparison between ${}^gP$ and ${}^gU$, it is often useful to distinguish the portions of territory containing the surpluses ${}^gSPU$ from the residual part of the habitat. In the comparison between two geosets ${}^gP$ and ${}^gQ$ distributed over the same territory T, it is sometimes useful to divide T into four subterritories corresponding, respectively, to surpluses of ${}^gP$ over ${}^gQ$, to surpluses of ${}^gQ$ over ${}^gP$, to matches between ${}^gP$ and ${}^gQ$, and to the residual void portions of territory, where there are no elements of ${}^gP$ or ${}^gQ$.

17.3.4. Enumerative or Meaningful Units into which the Gross Territorial Container Is Divided

Very often the data available as input for the geostatistical research refer to a large number of small territorial units having a purely enumerative function as, e.g., ZIP code areas, census enumeration districts of a country, or census tracts of a city. This is the typical setting for which the transformation of T into a network of cells of equal size and area has been employed in this book. In general, there is no interest in the separate study of many thousands or even more of such enumeration units. However, experience has shown that detailed inputs helps in forming a cell network that can illustrate both microgeographical and macrogeographical characteristics of the geoset.

This approach is not in conflict with the separate and comparative analysis of meaningful regions having a clear political or administrative definition or determined for researh purposes. Often each of these regions can be subdivided, in turn, into regular cells. The input for this operation and for the study of within-region variations can be furnished by more detailed enumerative units.

17.4. Geostatistical Maps

17.4.1. General Considerations

In this book we have placed great emphasis on maps of geosets and graphical presentation of geostatistical parameters. We have done this out of our deep conviction that graphical presentation is generally an essential component of statistical research and that it must be based on well-defined scientific methods adapted to the nature of the statistics represented and the object of the representation. Accordingly, we started our discussion by highlighting the evolution and the present state of the art of statistical graphics in general and of statistical mapping in particular.

This leads us to discuss the present unbalanced situation of statistical graphics, which has been created, on the one hand, by the still prevailing insufficient interest on the part of professional statisticians on generally agreed principles of statistical graphics and, on the other, by the tremendous technical development resulting from automation. The proliferation of software for graphics unaccompanied by adequate training for users has created a situation in which a large proportion of the statistical maps currently being produced are of poor quality and can engender misinterpretations. This is especially unfortunate at the present stage of development as popularized statistical maps are going to play an increasing role as instruments for research, for decision-making, and for utilization of the enormous masses of detailed geographical data that are now easily accessible.

17.4.2. Presenting Geosets by Means of Graphical Rational Patterns

The central graphical problem discussed in this book has been the representation of primary and derived geosets on maps. The still largely prevailing method of presenting geostatistical data by means of choropleth maps has generally been found to be very unsatisfactory, and the best graphical method known, the Cartesian system, can be applied directly to statistical maps only to a very limited extent. Accordingly, an attempt has been made to: (a) formulate the properties to which the Cartesian system owes its excellence; (b) to seek a substitute applicable to maps and other graphical presentations which accords, to a considerable extent, with those properties.

Graphical rational patterns (GRP) enable us to obtain results answering to (b) and so have been used extensively in this book for representing primary and derived geosets. However, the implementation of the GRP used here is not the only one possible and is far from being perfect, and we particularly wish to stress the need for improvements and further research.

Colored GRP enable us, among other applications, to: (a) find solutions to the problems widely discussed in the literature of presenting two geosets, ${}^{g}P$ and ${}^{g}Q$, on the same map; (b) show surpluses ${}^{g}SPQ$ and ${}^{g}SQP$ together; (c) represent rates referred to the population of each meaningful territorial unit studied and the relative weight of the population of the unit together; and (d) construct maps of the products between standard scores used for calculating ecological correlations.

17.4.3. Other Graphical Presentations Used in Geostatistical Research

In addition to GRPs, we have used other types of maps in this book, especially the following:

1. Maps showing the boundaries of the gross territory over which each geoset studied is distributed and the corresponding habitat. The residual void territory shows the geographical distribution of the nonhabitat.
2. Maps of a territory over which some of the relevant geostatistical parameters are shown. Experience has suggested that among these graphical presentations the following are particularly useful: (a) the cross centered at the mean center with arms proportional to $2\sigma_M$ and $2\sigma_m$, which calls attention to important aspects of position, principal direction, dispersion, and oblongity of the geosets being studied; (b) The corresponding ellipse of equivalent scatter, which is sometimes a good substitute for the gross or net territorial container; and (c) the triangle of centers, which in the comparison between ${}^gP$ and ${}^gU$ shows the centers of ${}^gSPU$, ${}^gP$, and ${}^gM(P, U)$ equal to ${}^gH$, ${}^gT$, and, ${}^gNH$ and in the comparison between any pair of geosets ${}^gP$ and ${}^gQ$ shows the centers of ${}^gSPQ$, ${}^gP$, ${}^gM(P, Q)$, ${}^gQ$, and ${}^gSQP$; (d) Maps showing the mean center and/or the median center of a population at various times or of many populations or subpopulations being compared.

Among the geographical presentations of geostatistical parameters not shown on maps, we can note the triangles of minimal distances, which compare (a) primary geosets to geosets derived from them and (b) component geosets and compound geosets. Such triangles can help the reader to see the relationships between the geosets being compared at a glance.

17.5. Geostatistical Parameters

The third part of this volume was devoted to review, discussion, and selected examples of parameters indicating or measuring important characteristics of primary and derived geosets. The main characteristics that we looked at were: the size of the geoset, territorial coverage and density, position, dispersion, and shape. Obviously, there is some arbitrariness in this selection but experience has shown that these particular parameters are of considerable help in geostatistical analysis. For example, in the second volume of the *Statistical Atlas of Italy* (1992) these parameters have been calculated for almost 400 "populations"; for each population they have been applied to: (a) the comparison between each geoset and the uniform distribution over the territory of Italy [primary geosets ${}^gP$ and ${}^gU$, derived geosets ${}^gSPU$, ${}^gSUP = {}^gNH$, ${}^gM(P, U) = {}^gH$]; (b) the comparison between pairs of geosets (primary geosets ${}^gP$ and ${}^gQ$ and derived geosets ${}^gSPQ$, ${}^gSQP$, and ${}^gM(P, Q)$; (c) geosets of territories and subterritories of various types $T(SPQ)$, $T(SQP)$, $TM(P, Q)$, and $TV(P, Q)$. As a result tables, diskettes, maps, and diagrams have been prepared containing many tens of thousands of geostatistical parameters. The experts who have substantively analyzed the geostatistical characteristics of the populations studied have in each case selected the parameters relating to the specific questions of interest to them. Further research done after publication of the *Atlas* has also been based upon the selection and comparison of the appropriate geostatistical parameters from among those published in the *Atlas*.

In this research, particular attention has been paid to: (a) the joint study of the same parameter for various "populations" related substantively; (b) the analysis of a battery of parameters, which when combined describe or measure one or more geographical characteristics of a distribution of a population; (c) to the behavior of parameters of component and compound geosets or of geosets otherwise related.

Another important application of the large number of parameters calculated has been the creation of tables showing distribution by size and summary values such as quantiles, arithmetic average, and variability for each of the parameters. This has helped in exploring the general behavior of each of the various parameters studied.

17.5.1. Parameters of Territorial Coverage and Density

Having defined the net territory or habitat (H) over which a population is found and the nonhabitat (NH), we can easily obtain their respective areas A_H and A_{NH}. Obviously $A_H + A_{NH} = A_T$ (area of gross territory T). The ratios $A_H/A_T = 1 - \Delta_P^U$ and $A_{NH}/A_T = \Delta_P^U$ have been found to be very useful for comparative and conceptual purposes. The ratio $1 - \Delta_P^U$ measures the net territorial coverage of the geoset studied, which is a very important aspect of the territorial distribution: its values can be comparatively analyzed for, e.g., different populations, different countries or cities, or different time periods. Suppose we compare a geoset ${}^gP$ with the uniform distribution ${}^gU$ over territory T assuming that ${}^gP$ is transformed into ${}^gU$. We then find that: (a) a proportion $1 - \Delta_P^U$ of the elements of ${}^gP$ match elements of ${}^gU$ and are not to be shifted; (b) a proportion Δ_P^U of elements ${}^gP$ are in surplus versus the elements of the uniform distribution and are to be shifted; and (c) under certain conditions Δ_P^U coincides with Duncan's concentration indexes (usually denoted by Δ).

Duncan's Δ as well as other measures of territorial concentration have been criticized in the literature because they are insensitive to the geographical distribution of the elements to be shifted. The approach suggested in this book enables us to overcome or to mitigate this disadvantage because we can identify and analyze geographically the geoset of the origins of the elements to be shifted, ${}^gSPU$; the geoset of the destinations of these elements, ${}^gNH$; and the geoset of the elements not to be shifted, ${}^gM(P, U)$.

It is well known that gross population density (measured by the ratio of the size of the population to the area of the gross territory contained) is often an unsatisfactory parameter. Considerable refinements can be realized if we supplement it with the net density (the ratio of the size of the population to the area of the habitat) and the surplus density (the ratio of the total of population surpluses to the area covered by these elements).

In the comparison between two geosets ${}^gP$ and ${}^gQ$ distributed over the same territory we can calculate in a similar fashion the proportion $\Delta_Q^P = \Delta_P^Q$ of the elements of ${}^gP$ that are in surplus of those of ${}^gQ$. This ratio corresponds to Duncan's index of segregation between ${}^gP$ and ${}^gQ$. Here, too, the study of ${}^gSPQ$, ${}^gSQP$, and ${}^gM(P, Q)$ enables us to obtain geographical information that cannot be gotten from the ratio Δ_Q^P alone. In the comparison between ${}^gP$ and ${}^gQ$, it is also sometimes of interest to study the proportions of the given territory that are covered only by the surpluses of ${}^gP$ or of ${}^gQ$, or only by the matches or the void areas.

17.5.2. Parameters of Position

17.5.2.1. The Battery of Parameters of Position

The position of a geoset in a plane can be changed by rigid motion without changing its shape and dimensions. Accordingly, the determination of the parameters of position can constitute an important ingredient in the characterization of the geoset. These parameters are of use in, e.g., describing and analyzing the geoset, discovering whether two geosets are congruent or similar, and the control and eventual elimination of the effects of position on the minimal distance between two geosets.

The following three parameters constitute a convenient way to describe the position of a geoset: (a) the mean center; (b) the angle between each principal axis and the corresponding conventional axis; and (c) one of the eight possible quadrant settings, where the quadrants are formed by the principal axes. The proportion of the total deviance of the elements included in each quadrant from the center of the geoset is calculated. The comparison between these proportions generally enables a unique determination of the quadrant setting of the relevant geoset. One of the eight settings is conventionally considered as principal, and any geoset can be transformed into its principal setting by reflection around the principal axes. The original coordinates (X_i^P, Y_i^P) of the elements i of the original geoset can be thus transformed into principal coordinates (X_i^{P*}, Y_i^{P*}).

Two geosets ${}^gP$ and ${}^gQ$ that are exact replicas of one another are termed coincident. Two geosets that are coincident at least in one position are called congruent. Two geosets can be proved to be congruent by putting both of them in the principal position. The transformed geosets thus obtained are found to be coincident.

Two geosets can be designated as similar if there is at least one position in which one can be obtained from the other by dilatation. Similarity can be proved by putting both geosets in principal positions and comparing their principal coordinates after division by the respective standard distances.

17.5.2.2. The Mean Center

Obviously there is an infinite number of centers expressing an average of the locations of the elements of a geoset. However, in this book we have emphasized applications of the mean center, owing to its important algebraic and other properties. Briefly, these applications include:

1. Systematic comparison and interpretation of the relationships between the centers of primary geosets and the centers of geosets derived from them.

2. Application of the approach noted under (1) to the study of the relationships between the center of component geosets and the center of their compound geoset.
3. Measure of the relative eccentricity of a population or subpopulation. This value is obtained by calculating the distance between this center and the center of the territorial container T and then dividing this distance by the standard distance of T. This enables comparison of eccentricity for populations of, e.g., different countries or cities or different time periods.
4. Calculation of relative eccentricity along the major axis of the territory in special settings in which this axis is particularly meaningful.
5. Relative distance between the centers of the major clusters of a geoset.

17.5.3. Parameters of Dispersion and Oblongity

Parameters of territorial dispersion have been applied by us and by many others to various populations and subpopulations of different countries and cities. These applications—going back to early studies in social physics and centrography—have demonstrated the scientific and practical utility of this field of research. Many applications in the present work have been focused on parameters of: distance-variance, standard distance, oblongity, scatter, and ratios between parameters of dispersion of a population and parameters of the corresponding territory.

- Distance-Variance (d^2): is a comprehensive measure of areal dispersion obtained as the arithmetic average of the squares of the distances between the location of each element and the location of the center of the geoset. Distance-variance enjoys many important algebraic properties, which render it very flexible and meaningful. We have utilized mainly the following properties:

  1. d^2 is equal to the sum of the distance-variance within and the distance-variance between. This has been applied to the clusters of surplus elements of geosets; to dispersions of subpopulations versus dispersion of total populations; to the study of the component versus the compound geoset; to the analysis of distance-deviance of the elements included in each quadrant formed by the principal axes; to the joint analysis of distance-deviance of primary geosets and derived geosubsets.
  2. d^2 is equal to the sum of the variances of the elements of any geoset measured along any two orthogonal axes crossing at the mean center. The primary emphasis has been on the principal axes, but some attention has also been paid to other pairs of axes, including the equating axes.
  3. d^2 can be broken down into a parameter of oblongity and a parameter of dispersion netted from the effects of oblongity.

- Standard Distance (d): In practical applications of parameters of dispersion, it is often inadvisable to use the original value of d^2, which can run to very high levels difficult to interpret. To circumvent this problem, we often resort to the use of the standard distance d (the square root of d^2), which is a simple linear measure of territorial dispersion, being the quadratic average of the distances of the elements of the geoset from its mean center.

- Oblongity: In practical geostatistical analysis, we can almost always identify the principal axes through the center and the standard deviations σ_M and σ_m along them. We have used the term oblongity for a property that described a geoset in which $\sigma_M \neq \sigma_m$, and it has been found to be a very important characteristic: it is clearly discernible from the cross superimposed on the map; it can be measured by dimensionless and widely comparable parameters; and it may be of help in the study of relative dispersion and of looseness (noncompactness). Among the various possible parameters measuring oblongity and the impact of oblongity in dispersion we have employed mainly the following:

1. A linear parameter of oblongity
$$^{L}b = 1 - (\sigma_m/\sigma_M) \tag{17.1}$$
which can vary approximately between 0 and 1;
2. Working on squared values, we can measure the absolute oblongity by taking $(\sigma_M - \sigma_m)^2$, which can be rendered dimensionless through division by d^2. We then obtain
$$(\sigma_M - \sigma_m)^2/d^2 = 1 - (s^2/d^2) \tag{17.2}$$
where $s^2 = 2\sigma_M\sigma_m$.
3. Relative Oblongity $^{r}b^2 = d^2/s^2$. Obviously, these parameters are functionally related; $^{r}b^2$ is less sensitive than $1 - s^2/d^2$, but it has been found useful for analytical purposes.

- Scatter: This is a term we have chosen to denote the parameter $s^2 = 2\sigma_M\sigma_m$, which has been found to be a very useful measure of territorial dispersion netted from the impact of oblongity.
- Ratio of Parameters of Dispersion for a Population Geoset and the Corresponding Parameter for the Territory: These ratios help us in making comparisons between geosets referring to different populations, different territorial containers, and different time periods. They can be presented in various ways. A particularly meaningful one is the ratio between the scatters s_P^2/s_T^2, which can be very informative, as we know that it can vary between a minimum of near to 0 and a maximum, which under certain conditions, can be about 2.

17.5.4. Parameters of Shape

As a first step in the discussion of parameters of shape, we dealt with parameters measuring various aspects of the shapes of territories, considered primarily, but not solely, as containers of population geosets. A very important aspect of shape is lack of compactness, called looseness, which is measured by the relative distance-variance, $^{r}d_T^2 = 2\pi d_T^2/A_T$.

This approach has been suggested independently by many authors. In this volume, we have split $^{r}d_T^2$ into two factors: (a) the relative scatter $^{r}s_T^2 = 4\pi\sigma_M\sigma_m/A_T$

and (b) the relative oblongity ${}^{r}b^2$. This split facilitates the complex analysis of the looseness, as shown by many applications to actual territories of countries and cities. These applications have been facilitated by the previous study on the behavior of the relative scatter in simplified models, such as regular geometrical figures, other simply connected figures, territories with internal and/or external concavities, and fragmented territories.

Concepts and equations for the measures of looseness of territories have been extended with appropriate modifications to the study of looseness of primary and derived population geosets. Other aspects of shape for which parameters have been studied are clustering and asymmetry.

17.5.4.1. Parameters of Looseness of Territories

The relative distance-variance as a parameter of looseness of territories has been obtained by comparing the distance-variance per unit of area d_T^2/A_T of any territory T to the minimal value of this parameter, $\frac{1}{2}\pi$, which is found for the most compact geometrical figure, the circle. Similar territories have equal relative distance-variance and affinely equivalent territories have equal relative scatter. For example, all ellipses have ${}^{r}s^2 = 1$; for any territory T that is not an ellipse, ${}^{r}s_T^2 > 1$; the relative scatter of regular polygons is

$$ {}^{r}s^2 = \frac{\pi[\tan(180°/n) + 3\cot(180°/n)]}{3n} \tag{17.3} $$

where n is the number of sides of the polygon; the relative scatter of an isosceles trapezoid is

$$ {}^{r}s^2 = \frac{2\pi\sqrt{(1 + k^2)(1 + 4k + k^2)}}{3\sqrt{3}(1 + k)^2} \tag{17.4} $$

where k is the ratio between the lengths of the shorter and the longer base.

Actual territories do not, generally, have regular shapes. However, values of ${}^{r}s_T^2$ and ${}^{r}b_T^2$ actually found for a territory T may sometimes suggest an analogy to regular figures, especially some important characteristics of T such as the following: (a) convex territories have a value of ${}^{r}s_T^2$ that varies between 1 and 1.209; (b) a territory with ${}^{r}s_T^2 \leqslant 1.209$ is not necessarily convex but its departure from convexity cannot be great, and (c) territories with large internal gaps, large external concavities, and strong fragmentation are expected to have high values of ${}^{r}s_T^2$.

For the relative distance-variance of territories with internal gaps an equation has been found that connects it with the relative distance-variance of the external territorial container; the proportion of area of the gaps, and the distance between the center of the gaps and that of the container. This equation has helped in interpreting hundreds of examples of relative distance-variance of territories divided into habitat and nonhabitat. Equations for the study of the relative distance-variance of simply connected territories with external concavities and fragmented territories have been developed and used successfully in field studies.

17.5.4.2. Parameters of Looseness of Population Geosets

Concepts and equations of parameters of looseness of territories have been extended, through appropriately conceptual adaptation, to measure the looseness of primary and derived population geosets. It has been found that the relative distance-variance of the population geoset ^{g}P increases with increasing oblongity of ^{g}P, with increasing scatter of ^{g}P, and with decreasing net territorial coverage A_H/A_T. The analysis of relative distance-variance for hundreds of population geosets has proved the interest and utility of this parameter and its factors.

17.5.4.3. Parameters of Clustering and Asymmetry

Clustering is discussed in the statistical literature from many different perspectives. In this book we have dealt with this concept in a specific way that has been found to facilitate the description and analysis of important aspects of the shape of the distributions of populations and territories and may improve our understanding of the idea of territorial concentration. Clusters are defined as contiguous cells containing surpluses of actual population over uniform distribution. Major clusters can be identified, and for each of them, appropriate geostatistical parameters that describe their principal characteristics are calculated, as are overall parameters of the extent of clustering.

Another aspect of shape that can be measured by appropriate parameters, both for territories and for population geosets, is central asymmetry, and various approaches are possible. We have emphasized the minimal distance between a geoset ^{g}P and a geoset ^{g}Q obtained by rotating ^{s}P through 180° around the center.

17.6. Measuring the Territorial Discrepancy between Geosets by Means of the Minimal Distance between Them

17.6.1. Basic Concepts of Minimal Distance

In the first three parts of this book we discussed methods that enable us to compare primary geosets in detail through derived geosets, maps, and geostatistical parameters. Part IV was devoted to a discussion of another approach to comparison between geosets: a summary parameter of the territorial discrepancy between the geosets compared.

Consider, for instance, two primary geosets ^{g}P and ^{g}Q distributed over the same territorial container T (having an area A_T). To compare ^{g}P to ^{g}Q, we have reduced ^{g}P and ^{g}Q to the same number of elements. Then each element of ^{g}P has been linked to an element of ^{g}Q, and the distance between the linked elements has been measured. The linkage has been done in a way that minimizes the total of the distances and is carried out by adapting the algorithm of the transport problem to the setting.

Taking an average of all the distances between linked elements, we obtain the minimal distance $^{m}\bar{D}(P, Q)$ between the primary geosets ^{g}P and ^{g}Q, which can be interpreted as the distance to be covered to transform ${}^{g}P$ into ${}^{g}Q$. The calculation is simplified if we derive the geosubsets of the matches ${}^{g}M(P, Q)$ and the surpluses ${}^{g}SPQ$ and ${}^{g}SQP$ from the primary geosets. Matching elements of ${}^{g}P$ and ${}^{g}Q$ are at distance 0. If we denote the proportion of surpluses by Δ_{Q}^{P}, and the average of minimizing distances between surpluses, by ${}^{m}\bar{D}(SPQ, SQP)$, then we obtain the basic equation

$$ {}^{m}\bar{D}(P, Q) = \Delta_{Q}^{P}\,{}^{m}\bar{D}(SPQ, SQP) \tag{17.5} $$

which enable us to reduce the work of calculation and has been found to be of great help in the interpretation of ${}^{m}\bar{D}(P, Q)$.

Minimizing distances between ${}^{g}SPQ$ and ${}^{g}SQP$ can be shown graphically by lines connecting each surplus element of ${}^{g}P$ to the corresponding surplus element of ${}^{g}Q$. These lines do not cross each other.

17.6.2. Properties of Minimal Distance

The minimal distance has many important properties that facilitate its analysis, which has been proved by many hundreds of examples.

The minimal distance ${}^{m}\bar{D}(P, Q)$ is a metric. Accordingly,

$$ {}^{m}\bar{D}(P, P) = 0 \tag{17.6a} $$

$$ {}^{m}\bar{D}(P, Q) = {}^{m}\bar{D}(Q, P) \tag{17.6b} $$

$$ {}^{m}\bar{D}(P, Q) \leqslant [{}^{m}\bar{D}(P, K) + {}^{m}\bar{D}(K, Q)] \tag{17.6c} $$

Considering equation (17.6c), if we represent each of the geosets ${}^{g}P$, ${}^{g}Q$, and ${}^{g}K$ graphically by a point, we obtain two alternative settings, which may be very meaningful in substantive analysis of the geosets: ${}^{g}P$, ${}^{g}K$, and ${}^{g}Q$ may be collinear; or the lines proportional to the distances ${}^{m}\bar{D}(P, K)$, ${}^{m}\bar{D}(K, Q)$, and ${}^{m}\bar{D}(P, Q)$ may form a triangle of distances. Examples of collinear geosets include:

1. ${}^{g}K$ is a compound geoset formed by two component geosets ${}^{g}P$ and ${}^{g}Q$. Let v be the actual proportion of the elements of ${}^{g}P$ over the elements of ${}^{g}K$. Let w_{C}^{P}, w_{C}^{K}, and w_{C}^{Q} be the relative weights attached to cell c in geosets ${}^{g}P$, ${}^{g}K$, and ${}^{g}Q$, respectively. Then, $w_{C}^{K} = v w_{C}^{P} + (1 - v) w_{C}^{Q}$.
2. Let P, K, and Q be sets having a "natural" order such as representing a population on successive dates. Suppose that in all cells w_{C}^{K} is intermediate between w_{C}^{P} and w_{C}^{Q}. In this setting there is consistency between ${}^{g}P$ and ${}^{g}Q$, and we have found actual examples of such a situation in empirical studies. However, in many other cases, it has been found that

$$ [{}^{m}\bar{D}(P, K) + {}^{m}\bar{D}(K, Q)] > {}^{m}\bar{D}(P, Q) \tag{17.7} $$

which is interpreted as showing inconsistency between ${}^{g}P$ and ${}^{g}Q$. Inconsistency can be measured by the ratio

$$ [{}^{m}\bar{D}(P, K) + {}^{m}\bar{D}(K, Q) - {}^{m}\bar{D}(P, Q)]/{}^{m}\bar{D}(P, Q) $$

Study of inconsistency and its factors is one of the most useful applications of minimal distance.

3. If K is a compound set, formed by two components P and Q, then

$$^{m}\bar{D}(SPK, SKP) = {}^{m}\bar{D}(SQK, SKQ) = {}^{m}\bar{D}(SPQ, SQP) = {}^{m}\bar{D}S \tag{17.8}$$

and

$$\Delta_P^K = (1 - v)\Delta_P^Q \qquad \Delta_Q^K = v\Delta_P^Q \tag{17.9}$$

$$^{m}\bar{D}(P, K) = (1 - v)\Delta_P^{Q\,m}\bar{D}S \tag{17.10a}$$

$$^{m}\bar{D}(K, Q) = v\Delta_P^{Q\,m}\bar{D}S \tag{17.10b}$$

and thus,

$$^{m}\bar{D}(P, K) + {}^{m}\bar{D}(K, Q) = \Delta_P^{Q\,m}\bar{D}S \tag{17.10c}$$

These complex relationships can be shown graphically by the triangle of distances, on which the related geosets are represented by dots. Among the various types of such triangles, one of them can represent nine related geosets.

4. Consider the average of the squares of minimizing distances (ASMD), between elements of ^{g}P and ^{g}Q, $^{m}\ddot{D}^2(P, Q)$, can be broken down into three components: (a) square of distance between the centers, $(\bar{X}_P - \bar{X}_Q)^2 + (\bar{Y}_P - \bar{Y}_Q)^2$; (b) squared difference between the parameters of dispersion, $(d_P - d_Q)^2$; and (c) residual owing to the difference in shape and to other determinants.. This decomposition and others enable us to create a bridge between the analysis of minimizing distances and the analysis of other geostatistical parameters.

17.6.3. Some Applications of Minimal Distances

We have over 1500 examples available of minimal distances: (a) between a geoset ^{g}P and the uniform distribution ^{g}U over the territory; (b) between pairs of geosets ^{g}P and ^{g}Q; and (c) between surpluses ^{g}SPQ and ^{g}SQP or ^{g}SPU and ^{g}SUP. Most of them refer to Italy, others to the United States, France, and various cities. Some deal with specific selected fields such as: a comparison between a population and ethnic subpopulations for measuring segregation or a comparison by subgroups delineated according to special criteria (e.g., sex, age, and marital status). Others show the use of minimal distance for checking hypotheses (such as similarity between the distribution of a labor force and the distribution of individuals employed in economic establishments, classified by sectors).

Considerable attention has been devoted to the study of consistencies and inconsistencies for changes of a distribution over the course of time, and for distances between pairs of ethnic subpopulations and between the total population and each subpopulation. Distances of specific geosets from paradigmatic distributions have been utilized for two main purposes: (a) facilitating the identification of characteristics of the specific geosets analyzed; and (b) estimating the distance between two geosets ^{g}P and ^{g}Q on the basis of the distances of each of them from two paradigmatic distributions.

We have described examples of applications of methods for calculating minimal distance between two geosets netted from the influence of a common determinant or from the influence of different determinants and have outlined rules concerning distances between component geosets and their compound.

17.7. Differences in Detail, Generalization, Abstraction, and Comparability of Maps and Geostatistical Parameters

The geostatistical methods discussed here enable us to obtain the information we seek in very different degrees of detail: the GRP maps can help us to work at a microgeographic level or to suggest macrogeographical characteristics for any geoset. The battery of geostatistical parameters allows us to measure general aspects of dimensions, dispersion, position, and shape.

Measuring these aspects in parallel for various geosets distributed over the same territory enables important comparisons among them. Various methods have been suggested to render the geostatistical parameters dimensionless, which allows us to compare geosets that are distributed over different territories and relate to different objects and different time periods. The minimal distance gives us more synthetic comparative measures: overviews thus obtained can be very comprehensive, dealing with many geosets at the same time. Minimal distances can also be rendered dimensionless, thus extending the possibilities for comparisons over time and space.

Obviously, the increasing generalization is accompanied by increasing abstraction and by loss of detailed information. However, as the various methods are anchored in the same basic territorial information, they are interconnected and we can consistently obtain the degree of detail or generalization that we are looking for in each particular geostatistical study. In geostatistical methods (as opposed to other techniques) the fact that the primary geosets are based on information on locations of their elements goes far to reduce the dangers of bias owing to aggregation of basic data, an advantage that was noted by O.D. Duncan as early as 1961.

17.8. Mainframe Computer Programs Used in the *Statistical Atlas of Italy*

In this book we have relied largely on maps and geostatistical parameters that first appeared in the two volumes of the *Statistical Atlas of Italy* (1988, 1992). It is thus pertinent to indicate here that this material was prepared on the basis of programs for mainframe computers developed by Dr. Shai Ever-Hadani in close consultation with the author and the adaptation of those programs was carried out by the staff of the National Institute of Statistics of Italy (ISTAT). Drs. Pierpaolo Napolitano and Orietta Gargano were especially involved in this phase of the work.

17.9. Software for Processing Geosets, Constructing Maps, and Calculating Parameters

17.9.1. The Software Package

Ever-Hadani, in close consultation with the author, developed a software package, which enabled us to process the geostatistical data, construct the necessary maps, and calculate the geostatistical parameters for this volume. The input data for the package includes the digitized map of the gross territorial container and various population geosets distributed over the same gross container. For each geoset the input may indicate the location of each individual element (weighted or unweighted) or aggregate locations corresponding to inhabited places or territorial units.

The output data allow us to: (a) compare population geosets with a uniform distribution over the territorial container, (b) compare various population geosets, (c) compare primary and derived geosets by means of the minimal distances, and (d) achieve various other objectives.

17.9.2. Maps and Parameters Comparing Geosets with a Uniform Distribution

To compare geosets with a uniform distribution we start with the primary geosets ${}^{g}P$ and ${}^{g}U$ to derive the geosets of the surpluses ${}^{g}SPU$ and ${}^{g}SUP$ and of the matches ${}^{g}M(P, U)$ as well as those of habitat ${}^{g}H$ and nonhabitat ${}^{g}NH$. Primary and derived geosets ${}^{g}P$, ${}^{g}U$, ${}^{g}SPU$, ${}^{g}SUP$ and ${}^{g}M(P, U)$ are represented by GRP maps. The map representing habitat and nonhabitat shows inhabited tracts of territory by means of proportional hexagons and can be either in black and white or in color. The map of surpluses can be given in two colors showing the geosets of surpluses ${}^{g}SPU$ and ${}^{g}SUP$ separately and together. Any part of these maps can be magnified as needed. Maps identifying major clusters can also be generated, as well as maps showing centers of ${}^{g}P$, ${}^{g}U$, ${}^{g}SPU$, ${}^{g}SUP = {}^{g}NH$, and ${}^{g}H = {}^{g}M(P, U)$.

For each of the geosets noted above the program calculates 20 parameters: 4 of size, 5 of position (coordinates of mean center and/or median center, angle between principal axes and conventional axes); 8 of dispersion (standard deviation along principal and conventional axes, covariance between abscissas and ordinates of the locations of the elements along the conventional axes, mean distance from the median center, standard distance, scatter); 3 of shape (relative distance variance, relative scatter, relative oblongity). These parameters can also be calculated for each of the major clusters. Some general measures of clustering can also be given.

17.9.3. Maps and Parameters Comparing Two Geosets Distributed over the Same Gross Territorial Container

To compare two geosets distributed over the same gross territorial container we start from pairs of primary geosets (as ${}^{g}P$ and ${}^{g}Q$) and derive from them the geosubsets of the surpluses ${}^{g}SPQ$ and ${}^{g}SQP$ and the geosubset of the matches ${}^{g}M(P, Q)$.

Surpluses and matches are derived from ${}^{g}P$ and ${}^{g}Q$ after both these primary geosets have been reduced to the same conventional size (such as 10,000). The number of elements of surpluses is thus $10{,}000\Delta$, while the number of matching elements is $10{,}000(1 - \Delta)$. A special feature enables the inflation of ${}^{g}SPQ$, ${}^{g}SQP$, and ${}^{g}M(P, Q)$ to the same size (as 10,000). Another feature enables the comparison of the geosets of many populations, inflating them to a total size that corresponds to the conventional size of the largest geoset.

The software that we have developed enables us to obtain the subterritories corresponding to surpluses $T(SPQ)$ and $T(SQP)$, to matches $TM(P, Q)$, and to residual void tracts of territory. ${}^{g}P$, ${}^{g}Q$, ${}^{g}SPQ$, ${}^{g}SQP$, and ${}^{g}M(P, Q)$ are represented by GRP maps. ${}^{g}SPQ$ and ${}^{g}SQP$ can be shown on the same map in two different colors. A map can also be constructed for the centers of ${}^{g}P$, ${}^{g}Q$, ${}^{g}SPQ$, ${}^{g}SQP$, and ${}^{g}M(P, Q)$.

17.9.4. Minimal Distance between Two Geosets: Parameters and Maps

We can calculate the minimal distance between any pair of primary or derived geosets and generate the corresponding maps and detailed parameters. The geosets are reduced to the same number of elements. We denote the compared geosets generically by ${}^{g}P$ and ${}^{g}Q$. The matrix of the minimizing linkages between elements of ${}^{g}P$ and elements of ${}^{g}Q$ and the distance between each pair of linked elements are calculated. Minimizing linkages can be represented graphically by lines binding each "origin" in ${}^{g}P$ to each "destination" in ${}^{g}Q$.

The basic result of calculations over minimizing distances is given as

$$ {}^{m}\bar{D}(P, Q) = \Delta\, {}^{m}\bar{D}(SPQ, SQP) \tag{17.11} $$

In this equation Δ is the proportion of surplus elements and ${}^{m}\bar{D}(P, Q)$ and ${}^{m}\bar{D}(SPQ, SQP)$ are the minimal distances between ${}^{g}P$ and ${}^{g}Q$ and between the surpluses, respectively. Many other detailed parameters are given in parallel for minimizing distances between (a) ${}^{g}P$ and ${}^{g}Q$ and (b) distances between ${}^{g}SPQ$ and ${}^{g}SQP$. For both (a) and (b) the ASMD breaks down into three components: the squared distance between the centers; the square of the difference between standard distances; and the residual. The distribution of minimizing distances by size and their average and standard deviations can also be obtained.

17.9.5. Additional Features

The software for geostatistical programs has special features that provide for a range of objectives, including, among others:

1. *Processing of information for territories:* The input is given by coordinates of dots evenly distributed along the borders of a territory. The territory is simply connected or formed by separate fragments. In addition, two overlapping territories, T_1 and T_2, are considered over which the following subterritories are distinguished: subterritory belonging only to T_1, subterritory belonging only to T_2, and subterritory common to T_1 and T_2. This feature

provides parameters for a theoretical population uniformly distributed over the area of a simply connected territory. On the basis of this information, maps are drawn and the 20 geostatistical parameters described earlier are calculated for each territory and subterritory.

2. *Ellipse of equivalent scatter:* This feature enables us to draw an ellipse over a map of the territorial container that has the same center, same principal axes, and the same scatter as the population geoset being considered. The proportion of elements of the geoset included within the ellipse is calculated.
3. *Convex hull:* This feature shows the convex hull including the elements of the population geoset being considered.
4. Various types of *triangles of minimal distance.*
5. *Mapping rates and ratios:* Special features enable representation of rates or ratios according to optional scales and together with weights to be attached to each rate or ratio.
6. Maps corresponding to *ecological correlation* are graphical representations of the standard scores of two variables (say, V_j^A and V_j^B), over each of the cells *j*, for which the ecological correlation is calculated; use of colors enables us to distinguish between positive and negative scores for both V_j^A and V_j^B.
7. *Residual spatial analysis:* This feature enables us to analyze spatial residuals derived from any function operated on by the program. The two-color map shows the degree of the discrepancy by means of GRP, where these GRP are proportional to the absolute values of the residuals. This feature is illustrated in Plate 14. In this example we consider the income per capita (the dependent geoset *Y*) as a function of four population groups (the independent geosets): Whites (X_1); Afro-Americans (X_2); Hispanics (X_3); and Asians (X_4), in the United States in 1980. The data were classified by the 3109 counties of the United States. The function considered was a linear multiple regression. (This feature was developed by Dr. Shai Ever-Hadani.)

For more information about the software package see Appendix J.

Appendixes

Appendix A: Distance-Variance per Unit Area of a Circle

Consider a perfect circle of radius R with a center at $(0, 0)$. Let us assume that the area of the circle is divided into an infinite number of rectangular cells. The area of each cell is $dxdy$. The expression for the distance-variance for the circle is given by the double integration:

$$\int_y \int_x (x^2 + y^2)dxdy$$

Let r denote the distance from the origin $(0, 0)$ to a specific rectangle whose center is (x, y). From the Jacobian calculus it is known that the transformation from the polar to the Cartesian coordinate system is given by $dr = rdxdy$. Since $x^2 + y^2 = r^2$, the above double integral reduces to

$$\int_r rdr = R^2/2$$

Hence, the distance-variance of a circle of radius R is $R^2/2$. If we divide this by the area of the circle, which is πR^2, we obtain the distance-variance per unit area of the circle, $\frac{1}{2}\pi$, which is independent of the radius R. As the circle is the figure with the most compact shape, the value of $\frac{1}{2}\pi$ can serve as an indicator of the distance-variance per unit area of an actual distribution.

Appendix B: Selected Demographic Findings: Washington, D.C., Atlanta, and Austin*

B.1. Showing Geostatistical Parameters of Each Population Distribution by Means of an Ellipse

The method of drawing an ellipse to show the principal geostatistical parameters of a population distribution had already been used and found to be efficacious in earlier geostatistical studies at the U.S. Bureau of the Census.†

*These findings have been taken from unpublished research work that was carried out at the U.S. Bureau of the Census during June–October 1978 (see unpublished Geostatistical Paper, No. 14).

†See memo by J. Passel on ellipses for the population and subpopulation in the SMSA's of Akron and Columbus (1976) and Interim Report by Bachi on General Characteristics of the Geographical Distribution of the Industries in the United States (1977).

Table B.1. Percentage of Each Population Included in the Corresponding Ellipse

Percentages of population in the ellipse	United States	Washington, D.C.	Atlanta	Austin	Total	Per 100 populations
Less than 75%	—	—	—	—	—	—
75–79%	—	—	—	5	5	2.1
80–84%	2	4	31	36	73	30.3
85%–89%	4	56	28	19	107	44.4
90–94%	2	17	20	15	54	22.4
95–100%	—	—	—	2	2	0.8
Total	8	77	79	77	241	100.0
Median of percentages found	86	88	86	89	87	

The ellipse that is drawn has its center at the mean center of population. Its major and minor axes are the principal axes of the distribution. The lengths of the semiaxes of the ellipse are equal to twice the standard deviations along the principal axes, $2\sigma_{x'}$ and $2\sigma_{y'}$. There are good theoretical and practical reasons to use these ellipses: they enable us to: (a) see at a glance the important characteristics of each distribution, such as average location, main direction, dispersion, and oblongity; (b) compare different distributions over the same territory quickly (different subpopulations, same populations at different times, different levels of, e.g., income or education of a population).

Table B.1 shows the results obtained in the project. It is seen that the ellipses generally cover a very high proportion of the population. The median coverage is 87%; in only 2% of the cases studied the coverage is under 80%, and it has never been found to be less than 75%.

Suppose we compare the distribution of two populations 1 and 2. As the areas of their ellipses are given, respectively, by $4\pi\sigma_{x_1}\sigma_{y_1}$ and $4\pi\sigma_{x_2}\sigma_{y_2}$, we can simply calculate:

$$\frac{4\pi\sigma_{x'_1}\sigma_{y'_1}}{4\pi\sigma_{x'_2}\sigma_{y'_2}} = \frac{\sigma_{x'_1}\sigma_{y'_1}}{\sigma_{x'_2}\sigma_{y'_2}}$$

This ratio is an accurate comparison of the bivariate measures of dispersion, $2\sigma_{x'_1}\sigma_{y'_1}$ and $2\sigma_{x'_2}\sigma_{y'_2}$, for the two distributions. However, according to the findings of Table B.1, it can also be considered as a ratio between the areas over which the great majority of populations 1 and 2 are distributed.

B.2. Distribution of the Total Population in Three SMSAs (Standard Metropolital Statistical Areas): Density of Population

The following represents a few selected findings out of many that appeared from the initial survey of the available parameters.

Table B.2. Population Densities in Three SMSAs (individuals per sq mile)

	Washington, D.C.	Atlanta	Austin
Gross density	1180	787	285
Density over ellipse, not corrected	3965	1938	1138
Density over ellipse, corrected	3497	1626	964

We indicate the standard deviations of the populations of each SMSA along its principal axes by $\sigma_{x'_p}$ and $\sigma_{y'_p}$ and the standard deviation of the territory of the SMSA along its principal axes by $\sigma_{x'_t}$ and $\sigma_{y'_t}$. The latter parameters can be interpreted as being a measure of the distribution that the population would have if it were uniformly distributed over the territory. The ratio

$$\frac{\sigma_{x'_p}\sigma_{y'_p}}{\sigma_{x'_t}\sigma_{y'_t}} = \frac{4\pi\sigma_{x'_p}\sigma_{y'_p}}{4\pi\sigma_{x'_t}\sigma_{y'_t}}$$

indicates (a) to what extent the two-dimensional distribution of the actual population differs from that of a uniform population; (b) in the light of the findings of Section B.1 it also compares the area of the ellipse over which the bulk of actual population is found to the area of a comparable ellipse for a population uniformly distributed over the territory. A considerable advantage of these ratios is that they enable us to evaluate the type of dispersion of the population in different SMSAs, eliminating the effects of different sizes and shapes of their respective territories: $100 \times (\sigma_{x'_p}\sigma_{y'_p}/\sigma_{x'_t}\sigma_{y'_t})$ for Washington, D.C. = 28.17; for Atlanta = 37.56; and for Austin = 21.57. It is seen immediately that in the three SMSAs the bulk of the population is distributed over an area that is much smaller than what would be expected in the case of a uniform distribution.

Clearly there is a central area in the three SMSAs with considerable population density and wide peripheral areas with comparatively very low densities.

Under these circumstances, the average density ratio P/A (where P is the population and A is the area of the SMSA) provides very little information. However, we can note here that reference to the ellipse $4\pi\sigma_{y'_p}\sigma_{y'_p}$ improves the calculation of density very considerably.

Table B.2 shows some results of the following calculations for the three SMSAs included in the Project: (a) gross density, P/A; (b) ratio of total population to area of ellipse $P/4\pi\sigma_{x'_p}\sigma_{y'_p}$; (c) corrected ratio, $P'/4\pi\sigma_{x'_p}\sigma_{y'_p}$, where P' is the population actually found within the ellipse.

B.3. Differences between the Dispersions of Population Groups in SMSAs

Table B.3 shows ratios

$$\frac{\sigma_{x'_i}\sigma_{y'_i}}{\sigma_{x'_p}\sigma_{y'_p}} = \frac{4\pi\sigma_{x'_i}\sigma_{y'_i}}{4\pi\sigma_{x'_p}\sigma_{y'_p}}$$

Table B.3. Ratio of Distribution in Each Subpopulation *i* as Compared to Distribution in Corresponding Total Population ($\sigma_{x_i'}\sigma_{y_i'}/\sigma_{x_p'}\sigma_{y_p'}$)

	Washington, D.C.			Atlanta			Austin		
	White	Afro-American	Total	White	Afro-American	Total	White	Afro-American	Total
Population	1.16	0.39		1.16	0.36		1.06	0.31	
Spanish			0.77			0.62			0.56
Individuals									
under 18			1.15			1.08			1.15
18–24			0.86			0.87			0.55
25–64			0.95			1.00			1.12
65+			0.72			0.84			0.80
Number owner-occupied units			1.14			1.16			1.24
Number renter-occupied units			0.63			0.65			0.53
Number vacant year around			1.07			0.85			1.35
Number owner-occupied units value under $20,000			0.98			1.12			0.82
Number owner-occupied units value $20,000–$34,999			1.19			1.13			1.37
Number owner-occupied units value $35,000–$49,999			0.98			0.78			1.34
Number owner-occupied units value $50,000+			0.67			0.48			(1.03)
Dwellings lacking plumbing facilities			2.05			1.76			0.99
Number workers living in area, 1975						1.19			
Number employed in area, 1975						0.68			
Outside SMSA, 1965	1.02	1.34	1.13	0.86	0.80	0.94	1.05	0.93	1.08
Different house in SMSA, 1965	1.06	0.91	1.02	1.08	0.86	1.02	1.15	0.79	1.11
Age 5–15 enrolled in school	1.18	1.03	1.16	1.10	1.08	1.08	1.24	1.15	1.21
25+ under 8 years education			1.06			1.11			0.86
25+ with 8–11 years education			0.89			1.08			0.98
25+ with 12 years education			0.99			0.99			1.17

25+ with 13–18 years education			0.90			0.81			1.09
25+ with 16+ years education	0.72	0.79	0.81	0.50	0.44	0.68	0.99	(0.89)	1.02
Unemployed, 16+	0.85	0.66	0.74	0.93	0.81	0.84	0.67	0.61	0.66
Married women in labor force	0.96	0.96	0.96	1.02	0.98	1.04	1.02	0.91	1.02
Married women, husband present, child 18, labor force	1.10	1.06	1.07	1.13	1.04	1.13	1.14	0.91	1.13
Not married, "husband present, labor force"	0.61	0.65	0.58	0.69	0.86	0.66	0.62	0.55	0.60
Not married, "husband present, children under 18, labor force"	0.84	0.67	0.67	0.86	0.78	0.68	0.86	0.67	0.76
Employed free professions, managers		0.78	0.87		0.56	0.85		0.68	1.03
Employed sales, clerical		0.54	0.77		0.58	0.90		0.60	0.93
Employed crafts, operators		0.86	1.06		1.10	1.22		0.97	1.00
Employed laborers, farmers		1.51	1.36		1.15	1.02		1.15	0.97
Employed sources, private households		1.01	0.78		1.07	0.74		0.80	0.67
Families with incomes <$5,000	1.16	1.01	0.92	1.12	0.92	0.88	0.76	(1.08)	0.74
Families with incomes $5,000–$15,000	1.10	0.88	1.05	1.07	0.97	1.09	1.04	0.77	1.04
Families with incomes $15,000–$25,000	0.94	0.75	1.02	0.89	0.71	1.00	1.30	0.38	1.35
Families with incomes $25,000+	0.66	0.70	0.74	0.62	0.58	0.72	1.15	(—)	1.15
Families social security retired income	0.87	1.02	0.87	0.93	0.96	0.93	0.96	0.97	0.95
Public assistance welfare	1.02	0.63		1.14	0.99		0.74	0.96	
Families in poverty	1.26	1.03	0.91	1.18	0.97	0.85	0.76	1.26	0.74
Male heads, poverty	1.36	1.49	1.15	1.24	1.15	1.02	0.78	1.86	0.84

where $\sigma_{x_i'}$ and $\sigma_{y_i'}$ denote the distributions along the principal axes of a population group i, while $\sigma_{x_p'}$ and $\sigma_{y_p'}$ denote the dispersion of the total population.

These ratios enable us to judge to what extent the distribution of each group differs from that of the population to which it belongs. Again, they can also be interpreted as ratios between the ellipse over which the bulk of the population group is distributed as compared to the ellipse for the corresponding total population. The ratios enable comparison of the dispersion of parallel groups in three SMSAs independently of the sizes and shapes of their territories and of the general patterns of their population distribution.

1. *Ethnic Group.* The Afro-American population has a two-dimensional distribution of 31%, 36%, and 39% of that of total population in Austin, Atlanta, and Washington D.C., respectively. The Hispanic population is much less spread out than the total population but more spread out than the Afro-American population. The White population is more spread out than the total population, clearly having a considerable tendency toward peripheral areas.
2. *Age.* Distribution seems to decrease with increasing ages in the three SMSAs included in the project, but people of ages 18–24 have comparatively low dispersion. Possibly, many factors impact at the same time, such as lesser tendency of elderly persons to change their location; the tendency of unattached youngsters to live nearer to the center of town; and the tendency of families with children to move to the suburbs. However, these are mentioned only as possibilities. These phenomena have also been found in some cities outside of the United States. It can also be hypothesized that the somewhat low spread of families in receipt of social security and railroad pensions is connected with higher ages of recipients of this type of income and low distribution of the elderly.
3. *Education.* In Washington and Atlanta, the distribution appears to decrease with increasing educational level.
4. *Income.* In Washington and Atlanta, the distribution decreases for both Whites and Afro-Americans with increasing income. However, the unemployed appear to have low distribution.
5. *Occupation.* Occupation has some influence on distribution. For instance, with both the Afro-American and the total populations, sales people and clerical workers have low distributions, while laborers and farmers have high distributions, probably owing to the more peripheral distribution of farmers.
6. *Housing conditions and value of the house.* Owner-occupied homes have a higher distribution than those that are renter-occupied in the three SMSAs studied. Domiciles lacking some plumbing facilities have a higher distribution, probably being more peripheral in Washington and Atlanta. Owned dwellings with a value of over $50,000 have a rather low distribution.
7. *In-migrants and movers within SMSAs.* People who moved within the SMSA from 1965 to 1970 seem to be less dispersed than the corresponding total population among the Afro-Americans and more dispersed among the

Whites. However, these findings may be coincidental. No clear pattern appears for people who moved into the SMSA during this time period.

8. *Workers and residents.* For Atlanta, data are available that show that the resident population is more dispersed than the working population.

Appendix C: Selected Findings of Comparative Studies Regarding the Characteristics of the Territorial Distribution of the Jewish Minority in Various Cities and Time Periods

C.1. Research on Urban Ecology of the Jews in the Diaspora

Tables C.1 and C.2 give selected examples of geostatistical parameters calculated for the Jewish minority as compared to the total population in many cities of Europe. These data were included in a very detailed report prepared for the U.S.–Israel Bi-National Science Foundation in 1987. In this report (Bachi, DellaPergola, and Klaff, 1988), GRP maps were presented, very detailed parameters were posited and analyzed, and sociological and ecological interpretations were given for selected cities and for the evolution that has occurred over the course of time. Moreover, an attempt was made to reach some general conclusions regarding the urban ecology of Jewish minorities considered together. Some of these general conclusions will be described below.

1. *Proportion of Jews to total population (column 11):* In most of the cities and towns surveyed, the Jews constituted a small minority. Warsaw and Lodz were the only cities in which their number exceeded the 30% of the total population.*
2. *Territorial coverage:* The proportion of the habitat to the total territory of the city is smaller for the Jews than for the total population. Accordingly, the proportion of the habitat of the Jews to the total population (column 1) is generally low or very low. However, it approaches the ratio of Jews to total population under the conditions of extreme segregation created in Warsaw in the late 1930s.

*Consideration of (a) the proportion of the population of the groups studied to the total population is of importance not only in itself but also for judging the meaning of (b) the ratios of parameters of coverage and dispersion to those of the total population. Under conditions that we cannot go into here, it is conceivable that these parameters can be expected to vary between the two theoretical limits: (i) in a setting of complete segregation of a minority it is conceivable that ratios (b) will be more or less similar to ratio (a): (ii) under conditions of complete ecological mixture of the minority studied with the total population (i.e., constancy of proportion of the minority to the total population) coverage and dispersion of the minority are equal to those of the total population (i.e., ratios b tend to 1).

Table C.1. Selected Measures of Jewish and Total Population Distribution in European Cities

City, year and population	Percent of area inhabited: Equivalent population (1)	Percent of area inhabited: Surplus population (2)	Surplus population per 100 of equivalent population (3)	Standard distance[a]: Equivalent population (4)	Standard distance[a]: Surplus population (5)	Standard deviation principal axes[a]: σ_M (6)	Standard deviation principal axes[a]: σ_m (7)	Ratio of scatter of population to scatter of habitat (8)	Relative oblongity (9)	Angle between conventional and principal axes[b] (10)
ITALY										
Rome										
1526/27										
Jews	11.8	6.7	88.2	473	386	408	239	0.299	1.145	+40.4
Total	43.9	20.0	56.1	950	676	717	623	0.649	1.010	−28.9
1928										
Jews	19.0	7.6	81.0	2,260	1,470	1,527	1,514	0.335	1.000	+4.4
Total	—	—	—	3,130	—	2,610	2,149	—	1.092	—
1965										
Jews	37.6	14.5	62.4	3,510	2,717	2,690	2,255	0.609	1.016	−28.5
Total	63.9	28.0	36.1	4,975	4,353	3,863	3,229	0.881	1.022	−20.0
1985										
Jews	37.0	22.2	63.0	4,406	3,422	3,633	2,494	0.593	1.072	−28.1
Total	58.5	35.8	41.5	5,762	5,003	4,275	3,863	0.854	1.005	−18.8
Trieste										
1875										
Jews	12.6	5.7	87.4	956	787	721	628	0.383	1.010	+35.7
Total	34.5	12.9	65.5	2,514	1,029	1,791	1,195	0.462	1.083	+37.7
1965										
Jews	18.7	8.6	81.4	1,153	910	840	789	0.401	1.002	+19.5
Total	42.2	19.6	57.8	2,394	1,353	2,003	1,311	0.596	1.091	+31.0

Milan										
1965										
Jews	31.9	19.8	68.1	2,972	2,573	2,530	1,561	0.617	1.119	+1.8
Total	58.5	28.4	41.5	4,180	3,128	3,185	2,707	0.749	1.013	+3.2
Turin										
1965										
Jews	27.7	13.8	72.3	2,575	2,086	1,971	1,657	0.522	1.015	−21.4
Total	66.1	26.9	33.9	4,293	3,105	3,590	2,355	0.818	1.090	−28.0
FRANCE										
Paris Region										
1975										
Jews	24.2	12.1	75.8	15,832	7,921	11,561	10,816	0.394	1.002	+4.5
Total	43.8	12.1	56.2	21,693	9,200	15,631	15,042	0.609	1.001	+12.8
SWITZERLAND										
Zurich										
1970										
Jews	50.0	19.3	50.0	2,776	1,847	1,969	1,958	0.675	1.000	−42.8
Total	75.5	19.3	24.5	3,429	2,309	2,503	2,344	0.839	1.002	+1.2
BELGIUM										
Brussels										
1961										
Jews	56.2	21.4	43.8	1,951	1,411	1,493	1,256	0.714	1.015	−17.3
Total	68.6	21.4	31.4	2,179	1,482	1,614	1,464	0.796	1.005	−43.9
UNITED KINGDOM										
Greater London										
1970										
Jews	47.9	22.6	52.1	12,795	9,884	9,981	8,806	0.709	1.024	−5.2
Total	75.5	18.7	24.5	14,417	7,729	11,014	9,304	0.810	1.014	+0.8
AUSTRIA										
Vienna										
1910										
Jews	27.2	11.3	72.8	3,934	2,584	3,161	2,342	0.457	1.045	−2.4
Total	46.1	18.7	53.9	5,390	3,559	4,262	3,300	0.597	1.033	−13.0

(Cont.)

Table C.1. (*Cont.*)

City, year and population	Percent of area inhabited: Equivalent population (1)	Percent of area inhabited: Surplus population (2)	Surplus population per 100 of equivalent population (3)	Standard distance[a]: Equivalent population (4)	Standard distance[a]: Surplus population (5)	Standard deviation principal axes[a]: σ_M (6)	Standard deviation principal axes[a]: σ_m (7)	Ratio of scatter of population to scatter of habitat (8)	Relative oblongity (9)	Angle between conventional and principal axes[b] (10)
AUSTRIA										
1923										
Jews	27.9	11.0	72.1	4,057	2,598	3,252	2,426	0.466	1.043	+0.8
Total	49.7	19.4	50.3	5,587	3,648	4,367	3,484	0.610	1.026	−11.6
1971										
Jews	35.5	12.0	64.5	4,377	2,446	3,242	2,941	0.461	1.005	−2.3
Total	59.7	18.5	40.3	6,335	3,709	4,689	4,261	0.678	1.005	12.4
GERMANY										
Berlin										
1933										
Jews	28.8	12.9	71.2	7,968	5,921	6,489	4,625	0.458	1.058	+22.1
Total	52.5	17.1	47.5	10,850	6,311	8,463	6,790	0.621	1.024	+17.9
POLAND										
Warsaw										
1839										
Jews	57.8	31.5	42.2	1,623	1,160	1,310	957	0.787	1.050	+19.3
Total[c]	54.5	33.4	45.5	1,478	1,000	1,326	1,042	0.761	1.033	+40.2
1882										
Jews	53.9	38.8	46.2	1,600	1,096	1,169	1,090	0.790	1.002	−13.9
Total	65.1	52.1	34.9	1,705	1,306	1,333	1,063	0.874	1.026	−31.3
1897										
Jews	39.6	19.4	60.4	1,611	1,034	1,267	996	0.673	1.029	+11.4
Total	59.5	41.1	40.5	1,870	1,260	1,423	1,213	0.816	1.013	+3.0

1913										
Jews	40.7	26.0	59.3	1,812	1,193	1,499	1,017	0.698	1.076	+15.5
Total	57.5	38.1	42.3	2,083	1,515	1,661	1,258	0.837	1.039	+12.4
1921										
Jews	25.2	14.3	74.8	1,912	1,288	1,408	1,292	0.481	1.004	+13.7
Total	37.3	19.3	62.7	2,532	1,798	1,835	1,744	0.565	1.001	+22.9
1931										
Jews	18.4	11.2	81.6	1,648	1,193	1,189	1,141	0.351	1.001	−6.4
Total[c]	42.9	27.2	57.1	2,852	2,059	2,054	1,979	0.617	1.001	+44.2
1938										
Jews	18.5	10.3	81.5	1,674	1,168	1,220	1,147	0.338	1.002	−5.2
Total[c]	50.1	28.1	49.9	3,171	2,245	2,273	2,219	0.676	1.000	+22.3
Lodz										
1931										
Jews	34.2	20.7	65.8	2,050	1,673	1,773	1,027	0.684	1.153	+7.9
Total	64.5	24.1	35.5	2,745	2,083	2,384	1,361	0.831	1.161	+17.6
Median — Jews	30.3	14.1	69.7	2,155	1,572	1,650	1,453	0.502	0.016	13.8
Median — Total	57.7	21.4	42.3	3,150	2,245	2,384	2,219	0.749	1.014	18.8

[a] Meters.
[b] Regardless of sign.
[c] Non-Jews only.

Table C.2. Percentage Ratios between Selected Measures of Jewish and Total Population Distribution in European Cities

City and year	Percent of area inhabited: Equivalent population (1)	Percent of area inhabited: Surplus population (2)	Surplus population per 100 of equivalent population (3)	Standard distance: Equivalent population (4)	Standard distance: Surplus population (5)	Standard deviation principal axes[a]: σ_M (6)	Standard deviation principal axes[a]: σ_m (7)	Ratio of population scatter to scatter of habitat (8)	Relative oblongity (9)	Relative eccentricity (10)	Jews % total population (11)
ITALY											
Rome											
1526/27	26.9	33.5	157.2	49.8	57.1	56.9	38.4	46.1	113.4	60.1	3.2
1928	—	—	—	72.2	—	58.5	70.5	—	91.6	11.9	1.3
1965	58.8	51.8	172.9	70.5	62.4	76.6	69.8	69.1	99.4	39.2	0.5
1985	63.2	62.0	151.8	76.5	68.4	85.0	64.6	69.4	106.7	32.7	0.5
Trieste											
1875	36.5	44.2	133.4	44.4	76.5	40.3	52.6	82.9	93.3	18.3	3.6
1965	44.3	43.9	140.8	48.2	67.3	41.9	60.2	67.3	91.8	20.6	0.4
Milan											
1965	54.5	69.7	164.1	71.1	82.3	79.4	57.7	82.4	110.5	12.3	0.5
Turin											
1965	38.0	51.3	213.3	60.0	67.2	54.9	70.4	63.8	93.1	23.7	0.2
FRANCE											
Paris Region											
1975	53.3	100.0	135.7	73.0	86.1	74.0	71.9	64.7	100.1	14.5	2.9
SWITZERLAND											
Zurich											
1970	66.0	100.0	206.6	81.0	80.0	78.7	83.5	80.5	99.8	36.4	1.3
BELGIUM											
Brussels											
1961	81.9	100.0	142.0	89.5	95.2	92.5	85.8	89.7	101.0	12.8	1.2

UNITED KINGDOM											
Greater London											
1970	63.4	120.9	212.7	88.7	127.9	90.6	86.0	87.5	101.0	45.6	3.5
AUSTRIA											
Vienna											
1910	59.0	60.4	135.1	73.0	72.6	74.2	71.0	76.5	101.2	32.4	8.6
1923	56.1	56.7	143.6	72.6	71.2	74.5	69.6	76.4	101.7	28.0	10.8
1971	59.5	64.9	160.0	69.1	65.9	69.1	69.0	68.0	100.0	12.1	0.5
GERMANY											
Berlin											
1933	54.9	75.4	149.9	73.4	93.8	76.7	68.1	73.8	103.3	25.1	3.8
POLAND											
Warsaw											
1839[b]	106.1	94.3	92.7	109.8	116.0	98.8	91.8	103.4	101.6	21.1	23.4
1882	82.8	74.5	132.4	93.8	83.9	87.7	102.5	90.4	97.7	30.9	33.4
1897	66.6	47.2	149.1	86.1	82.1	89.0	82.1	82.5	101.6	37.9	35.0
1913	70.5	68.2	140.2	87.0	78.7	90.3	80.8	83.4	103.6	32.8	37.6
1921	67.6	74.1	119.3	75.5	71.6	76.7	74.1	85.1	100.3	32.0	33.1
1931[b]	42.9	41.1	142.9	57.8	57.9	57.9	57.7	56.9	100.0	33.0	30.1
1938[b]	36.9	36.7	163.3	52.8	52.0	53.7	51.7	50.0	100.2	34.0	29.1
Lodz											
1931	53.0	85.9	185.4	74.7	80.3	74.4	75.5	82.3	99.3	34.2	33.5
Median	58.8	64.9	149.1	73.0	76.5	76.6	70.4	76.5	100.3		

[a]Meters.
[b]Jewish – Non-Jewish ratios.

3. *Concentration in "surplus areas":* The Jews have in general very high proportions of their population in surplus areas. On the other hand, the proportion of territory covered by these areas is generally low. The resulting picture is that their concentration is generally much stronger than that of the total population.
4. *Relative eccentricity of the Jewish population versus the total population:* This has been calculated by taking the ratio $\sqrt{[(\bar{X}_P-\bar{X}_Q)^2+(\bar{Y}_P-\bar{Y}_Q)^2]}/d_Q$ between (a) the distance between the center of the Jewish popultion P and that of the total population Q and (b) the standard distance of the total population. We see that everywhere the eccentricity of the Jews is considerable: in two-thirds of the observations it ranges between 20 and 40%.
5. *Dispersion:* All the parameters of dispersion point to a much smaller territorial dispersion of the Jews compared to that of the total population. This is true both for all the population elements and for elements in surplus versus uniform distribution.
6. *Shape:* The Jews appear to have a comparatively low dispersion and actually inhabit a comparatively restricted territory. The resulting parameters of shape (barring exceptional cases of compensation) are expected to point to differences between the Jews and the total population. The differences could be in the direction of either stronger or weaker compactness. We have given here only (a) the ratio of the population scatter to the habitat scatter and (b) the relative oblongity (columns 8 and 9). Ratios (a) are generally smaller for the Jews than for the total population. For (b) there is no general rule.

From this overview and from parallel data for Jewish populations in cities in other continents, it appears that these populations have some basic ecological similarities across space and over time. However, there are also significant differences and changes over the course of time (which we cannot go into here).

C.2. Minimal Distances between Various Religious–Ethnic Groups in Toronto

Tables C.3–C.6 (taken from the report referred to in Section C.1) show the minimal distances between various subpopulations of Toronto, Canada. The geosets that are compared are denoted generically by P and Q. The minimal distances have been calculated as a subsidiary tool in the study of segregations between subpopulations and the total population.

We see from these tables that:

1. Segregation between Protestants, Anglicans, and the United Church is very low. To render the distribution of any pair (P, Q) of these groups coincident, only 8.6% of the elements of these groups should be shifted (on the average). Also, the distance ${}^m\bar{D}(SPQ, SQP = 5.1$ km to be covered by shifting elements is moderate. Therefore, the distance ${}^m\bar{D}(P, Q)$ between pairs of these distributions is, on the average, very small (0.43 km). Furthermore, the distance

Table C.3. Parameters of Segregation between Pairs of Population Distributions P and Q (Δ = proportion of "surplus" elements to be shifted in order to render P and Q coincident)

	Distribution of P				
Distribution of Q	Protestants	Anglicans	United Church	Catholics	Jews
Anglicans	0.064				
United Church	0.082	0.111			
Catholics	0.275	0.301	0.314		
Jews	0.713	0.707	0.718	0.753	
Total population	0.146	0.171	0.194	0.171	0.683

Table C.4. Mean of Minimal Distances [$^m\bar{D}(SPQ, SQP)$] between Surplus Elements SPQ and SQP [to be shifted in order to render distributions P and Q coincident (km)]

	Distribution of SPQ				
Distribution of SQP	Protestants	Anglicans	United Church	Catholics	Jews
Anglicans	4.35				
United Church	6.92	4.06			
Catholics	12.94	11.52	12.70		
Jews	14.76	14.60	15.24	13.27	
Total population	11.35	10.27	11.51	10.54	12.73

Table C.5. Minimal Distance between P and Q [$^m\bar{D}(P, Q) = \Delta{}^m\bar{D}(SPQ, SQP)$ (km)] (Δ = proportion of surplus elements to be shifted in order to render P and Q coincident)

	Distribution of P				
Distribution of Q	Protestants	Anglicans	United Church	Catholics	Jews
Anglicans	0.28				
United Church	0.57	0.45			
Catholics	3.56	3.47	3.98		
Jews	10.52	10.32	10.95	9.99	
Total population	1.66	1.76	2.23	1.80	8.70

between each of these distributions and that of the total population is small (on the average, 1.86 km).

2. The Catholics are considerably segregated from the Protestants, Anglicans, and the United Church: Δ is almost 30%, and the average distance to be covered by shifting elements is 12.4 km, i.e., more than double the figure cited above (5.1 km). In consequence, the distance between Catholics and Reformed groups is 3.6 km.

Table C.6. Minimal Distances between Various Subpopulations

Average segregation	Dissimilarity	Distance between *SPQ* and *SQP*	Distance between *P* and *Q*
Between			
Protestants, Anglicans and United Church	0.086	5.11	0.43
Catholics vs. other Christian groups	0.296	12.39	3.67
Christian groups vs. total population	0.171	10.92	1.86
Jews vs. other groups	0.723	14.47	10.45
Jews vs. total population	0.683	12.73	8.70

3. The Jews are strongly segregated from all the other groups. To render their distribution coincident with that of any other group, 72% of the respective elements should be shifted, on an average; shifting elements should cover a distance of some 14.5 km. Under the impact of both these factors, the average distance of the Jews from other groups is comparatively very large.

C.3. Evolution of Minimal Distance between the Jews and Other Population Groups in New York

Tables C.7 and C.8 based on data collected by Ritterband shows the evolution in the minimal distance between the Jews and other population groups in New York. Table C.8 shows the minimal distances between the distribution of the Jews at various times.

Table C.7. Minimal Distances for New York Jews

Year	Population group	Proportion of surpluses, Δ	Minimal distance between surplus elements ${}^{m}\bar{D}(SPQ, SQP)$ (in miles)	Minimal distance between the sets *P* and *Q* (in miles)
1930	Non-Jewish Whites	0.500	2.28	1.14
1957	Non-Jewish Whites	0.402	2.14	0.86
1980	Non-Jewish Whites	0.266	2.26	0.60
1940	Afro-Americans	0.750	5.11	3.83
1957	Afro-Americans	0.629	3.46	2.18
1980	Afro-Americans	0.692	1.38	1.99
1930	Other Non-Whites	0.672	6.20	4.17
1957	Other Non-Whites	0.550	7.10	3.91
1930	Total population	0.377	2.41	0.91
1957	Total population	0.302	2.24	0.68

Table C.8. Minimal Distances between the Distribution of the Jews on Two Different Dates

Year	Year	Proportion of surpluses (Δ)	Distance between surpluses	Distance between *P* and *Q*
1930	1940	0.124	1.61	0.20
1940	1950	0.104	3.38	0.35
1950	1957	0.123	3.62	0.44
1957	1980	0.402	2.51	1.01

It appears from the parameters in these tables that the distribution of the Jews differs very strongly from that of Afro-Americans and even more from that of other non-Whites, while it differs less strongly from that of non-Jewish Whites. All these discrepancies appear to weaken over the course of time. Processes of redistribution of all the groups from 1930 to 1980 have been considerable, particularly from 1957 to 1980. For the Jews this is shown by Table C.8.

Detailed maps and geostatistical parameters enable us to study in detail the very complex types of distribution involved (generally multiclustered) and the shifts and processes of dispersion that occurred in the periods studied.

Appendix D: Definitions and Symbols Used in this Book

1. Population sets

A set of population elements for which no reference is made to their location is denoted by capital letters as A, B, P, Q, ...

2. Geosets

A geoset is the distribution of relative frequencies (or other weights) of the locations of elements of a population or of a territory. A geoset is denoted by one or more capital letters preceded by a superscript lower-case g. For example: a primary population geoset is denoted by ${}^gG, {}^gJ, \ldots$. ${}^gK$ is generally used for a compound geoset. A uniform distribution over territory T is denoted by ${}^gU$. Two geosets whose centers become coincident are concentric and are denoted by ${}^gG^\circ$ and ${}^gJ^\circ$.

Geosubsets obtained by comparison of two primary geosets distributed over the same territory T are denoted by groups of letters:

- Surplus of P over Q is ${}^gSPQ$, matched between P and Q is ${}^gM(P, Q)$, and similarly, ${}^gSPU$, ${}^gSUP$, ${}^gM(P, U)$ etc.
- Territorial geosets are ${}^gT$, eventually followed by specifications such as ${}^gT(P)$, ${}^gT(Q)$, ${}^gT(SPQ)$, ${}^gTM(P, Q)$, and ${}^gTV(P, Q)$.
- Habitat geoset is ${}^gH[={}^gT(P)]$; Nonhabitat geoset is ${}^gNH[={}^gT(SUP)]$.
- ${}^gZ$ is used for any geoset (primary or derived, referred to population or to territory).

3. *Elements of sets and geosets*

Individual elements of sets or geosets are denoted in two alternative ways: (a) by a lower-case letter such as *h* for hotel, *p* for population elements, etc.; (b) by *Z* for any set and by ${}^{g}Z$ for any geoset with elements $Z_i, Z_j, Z_c, \ldots$, where $i = 1, 2, \ldots$, $j = 1, 2, \ldots, c = 1, 2, \ldots$, and *c* is a cell in a network of hexagons.

4. *Relative weight to be attributed to locations of elements of a geoset*

We indicate the relative weight by w_i^z, where *z* stands for any geoset and *i* may be substituted by *j*, *c*, etc.

5. *Coordinates of the locations of the elements of a geoset*

The coordinates of the locations are given for any elements Z_i (or Z_c, etc.) along two orthogonal axes. The main axes considered are:

- Conventional axes: *X*, *Y*.
- Conventional axes with origin at the center: *x*, *y*.
- Principal axes (with specification of relationship with conventional): X', Y'.
- Principal axes with reference to major or minor: M, m.

In general the coordinates are given; e.g., as X_i^z, Y_i^z; $X_i'^z$, $Y_i'^z$; and M_i^z, m_i^z.

6. *Main parameters*

- The structure of the symbol of a parameter

  The symbol of a parameter is formed by two parts: one denoting the meaning of the parameter (as abscissa of the mean center, *X*, or standard distance, *d*), the other the geoset(s) to which the parameter refers (as *Z*).

- Size

  N number of cells
  F frequency or size of the population considered
  A area of the territory considered
  M or Φ size of the equivalent geoset
  v_P^K relative weight of a component geoset (say, ${}^{g}P$) over a compound geoset (say, ${}^{g}K$)
  Δ_P^U the proportion of surpluses in the comparison of two geosets (say, *P* and *U*)
  ${}^{1}R, {}^{2}R, {}^{3}R$ main types of rates per population
  V_j the values of variables over each of the cell *j*
  S_j or S_c the approximate value (e.g., area, density, and intensity) of the graphical symbols used to represent V_j

  The values (inked-in areas) of the GRP are proportional both to the units they represent and to the actual drawn size.

- Centers

  Mean center is denoted by C and its coordinates by $\bar{X}, \bar{Y}$
  Median center is denoted by $\tilde{C}$ and its coordinates by $\tilde{X}, \tilde{Y}$

- Dispersion

  Scatter: s^2 Relative scatter: ${}^r s^2$
  Distance-variance: d^2 Relative distance-variance: ${}^r d^2$
  Standard distance: d
  Standard deviation: σ
  Dispersion difference: $(d_A - d_B)^2$
  Residual: $2[d_A d_B - {}^m\text{cov}(A, B)]$

- Oblongity

  Absolute oblongity: Ω
  Relative linear oblongity: ${}^L b$
  Relative squared oblongity: ${}^r b^2$

- Distances between two geosets $Z = 1$, $Z = 2$

  Eccentricity (the distance between the centers of two geosets): ${}^c D(1, 2)$
  Squared eccentricity: ${}^c D^2(1, 2)$
  Minimal distance: ${}^m \bar{D}(1, 2)$
  Discrepancy = Average of squared minimizing distances (ASMD):
  ${}^m \ddot{D}^2(1, 2)$

Appendix E: Relationships between the Mean Centers of Primary and Derived Geosets

In the following we prove the relationships in Section 9.4.2 between the mean centers of the primary geosets: ${}^g P$, ${}^g U$, and their derived geosets: ${}^g H$, ${}^g NH$, ${}^g M(P, U)$, ${}^g M(U, P)$, ${}^g SPU$, and ${}^g SUP$.

We consider a standard representation of all these geosets by means of the same set of coordinates (X_i, Y_i); $i = 1, \ldots, N$, where $N = N^P = F^P$ is both the number of points in the territory and the size of the population, and the points (X_i, Y_i) represent the uniform distribution over the territory. For the calculation of their mean centers all of the above geosets can be considered as different weights assigned to this same set of points.

Let F_i^P be the population at point (X_i, Y_i)—possibly 0—and let $\delta(F_i^P) = 1$ for $F_i^P \geqslant 1$ and 0 otherwise (e.g., if $F_i^P = 0$). Then the weights for the above geosets are, respectively:

$$w_i^P = F_i^P/N \qquad w_i^U = 1/N \qquad w_i^H = w_i^M = \delta(F_i^P)/N^H$$

$$w_i^{NH} = w_i^{SUP} = [1 - \delta(F_i^P)]/N^{NH} \qquad w_i^{SPU} = [F_i^P - \delta(F_i^P)]/N^{NH} \tag{E.1}$$

where w_i^M are the weights for both ${}^gM(P, U)$ and ${}^gM(U, P)$, and the sizes of the populations are given as:

$$N = N^P = F^P = \Sigma_i F_i^P \qquad N^H = N^M = \Sigma_i \delta(F_i^P) \tag{E.2}$$

$$N^{NH} = N^{SUP} = N^{SPU} = \Sigma_i [1 - \delta(F_i^P)] = \Sigma_i [F_i^P - \delta(F_i^P)] \tag{E.3}$$

so that $N = N^H + N^{NH}$.

From equation (E.3) it is easy to see that

$$w_i^P = \Delta w_i^{SPU} + (1 - \Delta) w_i^H \tag{E.4}$$

and

$$w_i^U = \Delta w_i^{SUP} + (1 - \Delta) w_i^M \tag{E.5}$$

where $\Delta = N^{NH}/N$ and $1 - \Delta = N^H/N$.

Equation (E.4) implies that the centers of ${}^gM(P, U)$, ${}^gP$, and ${}^gSPU$ are collinear [(1) of 9.4.2]. Similarly, equation (E.5) implies that the centers of ${}^gM(U, P)$, ${}^gU$, and ${}^gSUP$ are collinear [(2) of 9.4.2]. The equality of the centers of ${}^gH$ and of ${}^gM(P, U)$ [(3) of 9.4.2] follows from the equality of the relevant weights, as does the equality of the centers of ${}^gNH$ and of ${}^gSUP$ [(4) of 9.4.2].

From equation (E.4) it also follows that the center of ${}^gP$ is on the line between ${}^gH$ and ${}^gSPU$ and cuts the line into segments that are proportional, respectively, to Δ and $1 - \Delta$ [(5) of 9.4.2]. Similarly, it follows from equation (E.5) that the center of ${}^gU$ is on the line between ${}^gH[{}^gM(P, U)]$ and ${}^gNH[{}^gSUP]$ and also cuts the line into segments proportional, respectively, to Δ and $1 - \Delta$ [(6) of 9.4.2].

We note further that:

$$w_i^U - w_i^P = (1 - F_i^P)/N \qquad \text{and} \qquad w_i^{NH} - w_i^{SPU} = (1 - F_i^P)/N^{NH} \tag{E.6}$$

so that

$$(w_i^U - w_i^P)/(w_i^{NH} - w_i^{SPU}) = N^{NH}/N = \Delta \tag{E.7}$$

which implies that the line between the centers of ${}^gP$ and ${}^gU$ is parallel to the line between ${}^gNH$ and ${}^gSPU$ [(7) of 9.4.2].

From equation (E.7) it can easily be seen that

$$\frac{\bar{X}_U - \bar{X}_P}{\bar{X}_{NH} - \bar{X}_{SPU}} = \frac{\bar{Y}_U - \bar{Y}_P}{\bar{Y}_{NH} - \bar{Y}_{SPU}} = \Delta \tag{E.8}$$

where $(\bar{X}_U, \bar{Y}_U)$, $(\bar{X}_P, \bar{Y}_P)$, $(\bar{X}_{NH}, \bar{Y}_{NH})$, and $(\bar{X}_{SPU}, \bar{Y}_{SPU})$ are the mean centers of ${}^gP$, ${}^gU$, ${}^gNH$, and ${}^gSPU$, respectively. Thus, the minimum distance between ${}^gP$ and ${}^gU$, ${}^m\bar{D}(P, U)$ equals the product of Δ and the distance between ${}^gNH$ and ${}^gSPU$, ${}^m\bar{D}(SPU, NH)$ since

$$\begin{aligned} {}^m\bar{D}(P, U) &= [(\bar{X}_U - \bar{X}_P)^2 + (\bar{Y}_U - \bar{Y}_P)^2]^{1/2} \\ &= [(\bar{X}_{NH} - \bar{X}_{SPU})^2 + (\bar{Y}_{NH} - \bar{Y}_{SPU})^2]^{1/2} \\ &= \Delta\, {}^m\bar{D}(SPU, NH) = \Delta\, {}^m\bar{D}(SPU, SUP) \end{aligned} \tag{E.9}$$

Appendix F: Iterative Process for Calculating the Two-Dimensional Median Point*

Let there be given N points $[x(i), y(i)]$ with weights $w(i) \geqslant 0$, $i = 1, \ldots, N$, in a two-dimensional space D. Let

$$F(X, Y) = \sum_{i=1}^{N} \{w(i)\sqrt{[x(i) - X]^2 + [y(i) - Y]^2}\} \tag{F.1}$$

The problem is to find a unique point (X_m, Y_m) that minimizes $F(X, Y)$; i.e.,

$$\min F(X, Y) = \sum_{i=1}^{N} \{w(i)\sqrt{[x(i) - X_m]^2 + [y(i) - Y_m]^2}\} \tag{F.2}$$

THE SOLUTION:

Determine the first approximation (X_g, Y_g). A sensible choice is to choose the center of the set, i.e.,

$$X_g = \frac{\sum_{i=1}^{N} |w(i)x(i)|}{\sum_{i=1}^{N} w(i)} \qquad Y_g = \frac{\sum_{i=1}^{N} |w(i)y(i)|}{\sum_{i=1}^{N} w(i)} \tag{F.3}$$

Let

$$w(i, 1) = \frac{w(i)}{\sqrt{[x(i) - X_g]^2 + [y(i) - Y_g]^2}} \tag{F.4}$$

be the new weights of the set in the first iteration. We assume that

$$[x(i) - X_g]^2 + [y(i) - Y_g]^2 > 0, \qquad \text{otherwise } w(i, 1) = 0$$

Let (X_1, Y_1) denote the center of the set with the weights $w(i, 1)$, i.e.,

$$X_1 = \frac{\sum_{i=1}^{N} |w(i, 1)x(i)|}{\sum_{i=1}^{N} w(i, 1)} \qquad Y_i = \frac{\sum_{i=1}^{N} |w(i, 1)y(i)|}{\sum_{i=1}^{N} w(i, 1)} \tag{F.5}$$

Let e denote a very small number such as 0.001. If $|X_1 - X_g| < e$ and $|Y_1 - Y_g| < e$ then the point (X_1, Y_1) is a solution. Otherwise, continue with

$$w(i, 2) = \frac{w(i, 1)}{\sqrt{[x(i) - X_1]^2 + [y(i) - Y_1]^2}} \tag{F.6}$$

as the new weights of the set in the second iteration. We assume that

$$[x(1) - X_1]^2 + [y(1) - Y_1]^2 > 0, \qquad \text{otherwise } w(i, 2) = 0$$

*By Dr. Shai Ever-Hadani. Source: *Statistical Atlas of Italy*, 1992, p. 848.

Let (Y_2, Y_2) denote the center of the set with the weights $w(i, 2)$. If $|X_1 - X_2| < e$ and $|Y_1 - Y_2| < e$ then the point (X_2, Y_2) is a solution.

The iterative process is limited to 50 iterations, i.e., if the condition

$$\text{If } !X(t+1) - X(t)| < e \text{ and } |Y(t+1) - Y(t)| < e$$

is fulfilled, then the point $[X(t+1), Y(t+1)]$ is a solution, and we then proceed until $t + 1 = 50$.

The process converges nicely. It seems that after 10 iterations a solution is found.

Appendix G: Minimal Standard-Distance Set with Given Oblongity*

THEOREM:

Let Ω be the set of simply connected plane sets A that satisfy the conditions:

$$\iint_A x\,dx\,dy = \iint_A y\,dx\,dy = \iint_A xy\,dx\,dy = 0 \tag{G.1}$$

$$\iint_A x^2 dx\,dy = c^2 \iint_A y^2 dx\,dy \tag{G.2}$$

$$\iint_A dx\,dy = S \tag{G.3}$$

where c and S are constants. Then:

$$\min_{A \in \Omega} \iint_A (x^2 + y^2) dx\,dy = \iint_{A^*} (x^2 + y^2) dx\,dy \tag{G.4}$$

where A^* is the ellipse defined by $A^* = \{(x, y): x^2 + c^2 y^2 \leqslant cS/\pi\}$.

PROOF:

Using the transformation $z = cy$, let Ω' be the set of transformed sets A'. Since equations (G.1)–(G.3) are equivalent to

$$\iint_{A'} x\,dx\,dz = \iint_{A'} z\,dx\,dz = \iint_{A'} xz\,dx\,dz = 0 \tag{G.5}$$

$$\iint_{A'} x^2 dx\,dz = c \iint_A x^2 dx\,dy = c^3 \iint_A y^2 dx\,dy = \iint_{A'} z^2 dx\,dz \tag{G.6}$$

$$\iint_{A'} dx\,dz = c \iint_A dx\,dy = cS \tag{G.7}$$

*By Prof. Gad Nathan (Hebrew University of Jerusalem).

Ω' is the set of simply connected plane sets A' that satisfy the conditions of equations (G.5)–(G.7). Then, obviously,

$$\min_{A' \in \Omega'} \iint_{A'} (x^2 + z^2) dxdz = \iint_{A'^*} (x^2 + z^2) dxdz \tag{G.8}$$

where A'^* is the circle defined by $A'^* = \{(x, z): x^2 + z^2 \leqslant cS/\pi\}$.

However, for any $A \in \Omega$ and its transform $A' \in \Omega'$, we have

$$\begin{aligned}\iint_{A} (x^2 + y^2) dxdy &= (1 + 1/c^2) \iint_{A} x^2 dxdy \\ &= \frac{(1 + 1/c^2)}{c} \iint_{A'} x^2 dxdz \\ &= \frac{(1 + 1/c^2)}{2c} \iint_{A'} (x^2 + z^2) dxdz\end{aligned} \tag{G.9}$$

Thus,

$$\begin{aligned}\min_{A \in \Omega} \iint_{A} (x^2 + y^2) dxdy &= \frac{(1 + 1/c^2)}{2c} \min_{A' \in \Omega'} \iint_{A'} (x^2 + z^2) dxdz \\ &= \frac{(1 + 1/c^2)}{2c} \iint_{A'^*} (x^2 + z^2) dxdz \\ &= \iint_{A^*} (x^2 + y^2) dxdy\end{aligned} \tag{G.10}$$

NOTES:

The parameter c defined in equation (G.2) is the measure of oblongity, so that the ellipse A^* is the closed simply connected plane set of oblongity c and area S with minimal standard distance, d_{A^*}. The squared minimal standard distance is given by

$$\begin{aligned}d_{A^*}^2 &= \frac{\iint_{A^*} (x^2 + y^2) dxdy}{\iint_{A^*} dxdy} \\ &= \frac{[(1 + 1/c^2)/(2c)] \iint_{A'^*} (x^2 + z^2) dxdz}{(1/c) \iint_{A'^*} dxdz} \\ &= \frac{1}{2}\left(1 + \frac{1}{c^2}\right) d_{A'^*}^2 = \frac{1}{2}\left(1 + \frac{1}{c^2}\right) \frac{cS}{2\pi} \\ &= \frac{1}{2}\left(c + \frac{1}{c}\right) \frac{S}{2\pi} = \frac{1}{2}\left(\frac{\sigma_y}{\sigma_x} + \frac{\sigma_x}{\sigma_y}\right) \frac{S}{2\pi}\end{aligned} \tag{G.11}$$

Appendix H: Measuring Central Asymmetry through the ASMD

Let ${}^{g}P$ denote the original geoset and ${}^{g}P'$ the same geoset rotated through 180°, in the same position, i.e., with the same center. Then the ASMD between ${}^{g}P$ and ${}^{g}P'$, ${}^{m}\ddot{D}^2(P, P')$, is a measure of the asymmetry. It has the value zero only in the case where ${}^{g}P$ and ${}^{g}P'$ coincide, i.e., when ${}^{g}P$ is symmetric. ${}^{m}\ddot{D}^2(P, P')$ can be written in the folloing alternative forms:

$$
\begin{aligned}
{}^{m}\ddot{D}^2(P, P') &= 2d^2 - 2[r^M(P, P')\sigma_M^2 + r^m(P, P')\sigma_m^2] \\
&= 2d^2 - 2\mathrm{cov}(P, P') \qquad \text{(H.1)}
\end{aligned}
$$

where r^M and r^m are the correlation coefficients for the major and the minor axes, respectively.

Therefore, the maximum value of ${}^{m}\ddot{D}^2(P, P')$ is $2d^2$, which is obtained when $\mathrm{cov}(P, P') = 0$. Thus, a relative measure of asymmetry is obtained as

$$
\begin{aligned}
\frac{{}^{m}\ddot{D}^2(P, P')}{2d^2} &= 1 - \frac{r^M(P, P')\sigma_M^2 + r^m(P, P')\sigma_m^2}{d^2} \\
&= 1 - \frac{\mathrm{cov}(P, P')}{d^2} = 1 - {}^{r,m}\mathrm{cov}(P, P') \qquad \text{(H.2)}
\end{aligned}
$$

where ${}^{r,m}\mathrm{cov}(P, P')$ is the relative covariance between ${}^{g}P$ and ${}^{g}P'$.

This measure of asymmetry lies between zero and one and the extreme values are obtained in cases of complete symmetry and complete asymmetry, respectively.

Appendix I: Numerical Example of Decomposition of the ASMD for Nonintersecting Geosets

To facilitate an understanding of the decomposition of ${}^{m}\ddot{D}^2(G, J)$, we give here a simple numerical example, shown in Tables I.1 and I.2 and Figure I.1: two nonintersecting geosets G and J, each of which includes three points. We can interpret the example of Figure I.1, as indicating the distribution of individuals belonging to two small ethnic groups in a city. As the number of permutations of $N = 3$ is $N! = 6$, there are six ways in which we can link each point in G with a point in J. Table I.1 and Figure I.1 show the linkages that minimize the total of the distances, where:

$$
{}^{m}\bar{D}(G, J) = 3.414 \qquad {}^{m}\ddot{D}^2(G, J) = 12 \qquad {}^{m}\ddot{D}(G, J) = 3.464 \qquad {}^{c}D^2(G, J) = 8
$$

Table I.1. Numerical Example for ${}^{m}\ddot{D}(G, J)$

Points G_i of geoset G	X_{G_i}	Y_{G_i}	Points J_i of geoset J	X_{J_i}	Y_{J_i}	Minimizing distance ${}^{m}d(G_i, J_i)$	${}^{m}d^2(G_i, J_i)$
G_1	1	1	J_1	1	4	3	9
G_2	7	7	J_2	10	7	3	9
G_3	7	1	J_3	10	4	4.243	18
Center	5	3	Center	7	5		

Let us refer again to equation (15.3) (Section 15.2):

$$ {}^{m}\ddot{D}^2(G, J) = {}^{c}D^2(G, J) + (d_G^2 + d_J^2) - 2[{}^{m}\text{cov}(X_G, X_J) + {}^{m}\text{cov}(Y_G, Y_J)] \qquad \text{(I.1)} $$

which shows that the average of the squares of the distances between minimizing distances can be obtained by adding: (a) the square of the distances ${}^{c}D^2(G, J)$ between the centers of the geosets G and J and (b) the sum of the distance variances $d_G^2 + d_J^2$, and by subtracting (c) twice the sum of the covariance between the abscissas of the elements of G_i and J_i linked in the minimizing way and (d) the covariance between the ordinates of G_i and J_i linked in the minimizing way. This sum will be

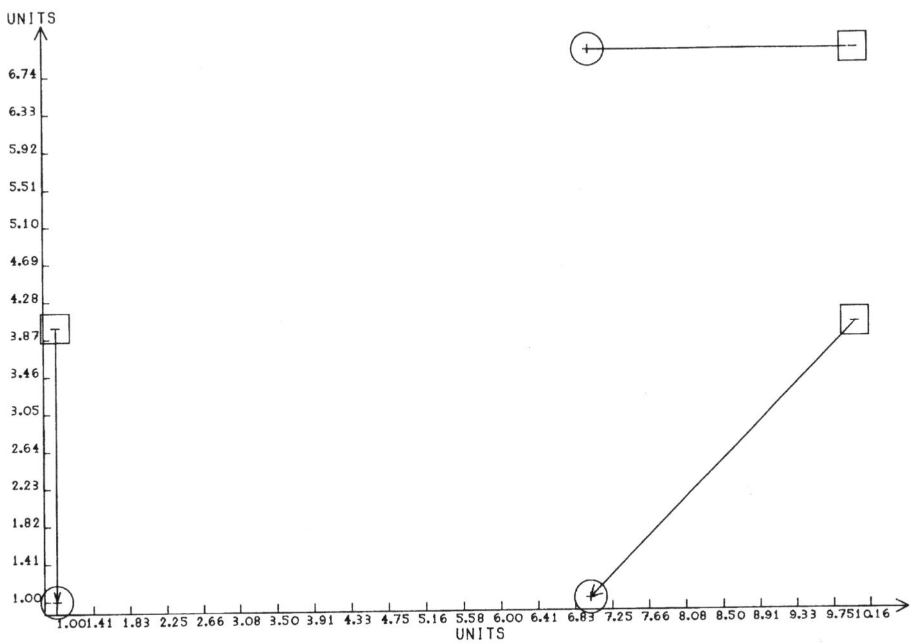

Figure I.1. The optimal transportation scheme (minimizing linkages) between G and J. Key: squares = origins; circles = destinations. The arrows indicate the direction of the shifted units; the areas of the circles and squares are proportional to the number of shifted units.

Table I.2. Details of Calculation of $^{m}\ddot{D}^2(G, J)$ for the Example of Table I.1

		X_{G_i}	Y_{G_i}	X_{J_i}	Y_{J_i}	$X^2_{G_i}$	$Y^2_{G_i}$	$X^2_{J_i}$	$Y^2_{J_i}$	$x_{G_i}x_{J_i}$	$y_{G_i}y_{J_i}$
G_1	J_1	−4	−2	−6	−1	16	4	36	1	24	2
G_2	J_2	+2	+4	+3	+2	4	16	9	4	6	8
G_3	J_3	+2	−2	+3	−1	4	4	9	1	6	2
		Ar. average				8	8	18	2	12	4

Squared distance between centers of G and $J[^{c}D^2(G, J)]$	8
Sums of distance-variances ($d_G^2 = 8 + 8 = 16$; $d_J^2 = 18 + 2 = 20$)	36
Minus twice the covariance between linked elements [2(12 + 4)]	−32
$^{m}\ddot{D}^2(G, J)$	12

indicated, for short, as covariance between linked elements:

$$^{m}\text{cov}(G, J) = (\Sigma\, x_{G_i}x_{J_i} + \Sigma\, y_{G_i}y_{J_i})/N \tag{I.2}$$

Table I.2 shows the calculation for the numerical example considered.

Appendix J: Geostatistical Methods and Related Computer Software*

J.1. Abstract

The research work described in this volume was accompanied and substantiated by intensive, on-going development of, and experimentation with, computer software. At various phases of the research different computer programs covered the entirety of the methods and statistical procedures described in the text. Both the graphical and the tabular illustrations to the text throughout the book were produced by that software.

Rapid, ongoing changes in technology render computer software obsolete at an incredible rate. In particular, scientists and scholars have, to a great extent, abandoned the use of mainframe computers and drifted toward desk-top solutions, where they are available. Concurrently with the compilation of this volume it was thought best to rewrite the entire suite of programs to accommodate the computing facilities typically at the disposal of today's reseacher.

A new, state-of-the-art comprehensive package known as *GeoStat* is now being constructed, to be released during 1999.

J.2. Contents

GeoStat offers several tools dealing with the analysis and display of spatial data. Normally the user would apply these tools to data representing spatial phenomena.

*Prepared by Michael Stier.

At this stage, GeoStat comprises the following features:

- Computation of statistics of geosets and their derivatives.
- Computation of the distance between two geosets (min D).
- Cartographic representation of the geosets as well as the statistics.
- Thematic mapping using graphical rational patterns.

No real world data are provided with the software package, except for illustration or clarification purposes. It is assumed that the user provides, on computer media, the following:

1. *Geosets*: See Section 3.1.1 for definition and contents.
2. *Outlines of territories, borders* (optional): See Section 3.2
3. *Landmarks, monuments, place-names* (optional).

Many of the cartographic products of GeoStat are very basic and can often be considered as overlays to actual, labeled, and enhanced maps of the territory. The package facilitates limited reproduction of cartographic features provided by the user, so that in certain cases the drawings can be considered maps.

Data formats, size limitations, and restrictions are discussed with the package's accompanying documentation.

J.3. System Requirements

To load and run the GeoStat package the following are required:

1. ***Hardware***
   - PC with 16 MB of RAM memory (32 MB preferred)
   - 20 MB free hard disk space
   - CD-ROM drive
   - Color ink-jet printer (optional)
   - Pen plotter, 1/1000 in. resolution (optional)
2. ***Software***
   Windows 95/98 (registered trademarks of Microsoft Corporation)—The package is self-contained and does not require any external libraries or supplements.
3. ***Required Skills***
   It is assumed that the user is familiar with the use of Windows' user-interface to the extent of installing a package (following basic instructions) and selecting data files.
4. ***GeoStat and the Internet***
   An improved version of GeoStat will be posted on the internet (on a try-till-you-buy basis).

Bibliography*

Alonso, W., 1960, A model of the urban land market: Location and densities of dwellings and businesses, unpublished Ph.D. thesis, University of Pennsylvania, Philadelphia.

Alonso, W., 1964, *Location and Land Use*, Harvard University Press, Cambridge, Mass.

Anderson, T.W., 1956, Potential models and spatial distribution of population, *Papers and Proceedings of the Regional Science Association* **2**.

Anderson, T.W., 1958, *An Introduction to Multivariate Statistical Analysis*, John Wiley and Sons, New York.

Anderson, T.W., and Egelard, J., 1961, Spatial aspects of social area analysis, *American Sociological Review* **26**: 392–399.

Andrews, D.F., 1972, Plots of high-dimensional data, *Biometrics* **28**: 125–136.

Arkin, H., and Colton, R., 1936, *Graphs: How to Make and Use Them*, Harper and Brothers, New York.

Austin, T.L., 1959, An approximation to the point of minimum aggregate distance, *Metron* **XIX** (3–4): 10–21.

Aviolo, P.J., 1976, Regional population distribution: *Application of the Lorenz Curve to the Description and Analysis of Spatial Concentration*, Center for Urban Development Research, Cornell University.

Bachi, R., 1932, La mobilita della popolazione all'interno delle grandi città Europle, Federazione della Proprietà Edilizia, Rome, LI:502.

Bachi, R., 1934, Metodi per lo studio della distribuzione topografica dei fenomeni: La distanza media, *Barometro Economico Italiano*, Aprile 10, 1934.

Bachi, R., 1955, The graphical representation of geographical series, *Bulletin of the International Statistical Institute* **34**(4) Rome.

Bachi, R., 1958, Statistical analysis of geographical series, *Bulletin de l'Institute International de Statistique* **36**: 229–240. Also in *Spatial Analysis*, 1968 (B.J.L. Berry and D.F. Marble, eds.) pp. 101–109, Prentice Hall, Englewood Cliffs, N.J.

Bachi, R., 1961, Some methods for the study of geographical distribution of internal migrations, Paper submitted to the International Population Union Conference, New York.

Bachi, R., 1962*a*, Some new methods for the graphical representation of statistical data, *Bulletin of the International Statistical Institute* **40**, Paris.

Bachi, R., 1962*b*, Standard distance measures and related methods for spatial analysis, *Papers and Proceedings of the Regional Science Association* **10**: 83–132.

Bachi, R., 1963, Standard distance measures for spatial analysis, *Regional Science Associated Papers* **X**: 83–132.

Bachi, R., 1965, Graphical representation and analysis of geographical statistical data, *Bulletin of the International Statistical Institute* **XLI**(1): 219–238, Belgrade.

Bachi, R., 1968, *Graphical Rational Patterns: A New Approach to Graphical Presentation of Statistics*, Israel Universities Press, Jerusalem.

Bachi, R., 1973, Geostatistical analysis of territories, *Bulletin of the International Statistical Institute* **45**(1): 121–33, Vienna.

*This bibliography was compiled with the help of Shalmit Bejarano.

Bachi, R., 1975, Graphical methods, achievements and challenges for the future, *Bulletin of the International Statistical Institute* **XLVI**(2): 441–455.

Bachi, R., 1976, Geostatistical analysis of internal migrations, *Journal of Regional Science* **16**: 1–19.

Bachi, R., 1978*a*, Geostatistical analysis of urban data, Proceedings of the 11th Meeting of the International Association for Regional and Urban Statistics, pp. 7–19.

Bachi, R., 1978*b*, Graphical statistical methodology in the automation era, in: *Graphic Presentation of Statistical Information* (Technical paper 43, pp. 3–6), US Department of Commerce, Bureau of the Census, Washington, DC.

Bachi, R., 1978*c*, Proposals for the development of selected graphical methods, in: *Graphic Presentation of Statistical Information* (Technical paper 43, pp. 23–67), US Department of Commerce, Bureau of the Census, Washington, DC.

Bachi, R., 1981, Mapping the main characteristics of distributions of populations over territories, Invited paper, *Bulletin of the International Statistical Institute* **XLIX**(2): 1003–1026, Buenos Aires.

Bachi, R., 1984, Parametri delle caratteristich generali di distribuzioni di popolazioni sul territorio, *Atti della XXXII Riunione Scientifica della Societa Italiana di Statistica* **II**: 149–169, Sorrento.

Bachi, R., 1989*a*, Rational maps and parameters of geographical–statistical data, *Espace, Populations, Sociétés* **3**: 337–348.

Bachi, R., 1989*b*, Graphical and geostatistical methods for presentation and analysis of territorial statistical data in urban research, Third World Congress of Regional Science Association, Jerusalem.

Bachi, R., 1989*c*, Rational maps and geostatistical parameters for presentation and analysis of official data detailed by small areas, The example of the *Statistical Atlas of Italy*, International congress on value of research data for government and business, 61 pp, mimeographed, Jerusalem.

Bachi, R., 1993, Geostatistical methods: The experience of the *Statistical Atlas of Italy*, *ISI Bulletin* **49**, Florence.

Bachi, R., and DeAre, D., 1978, Interim report on work carried out and first findings in the demographic field, Unpublished statistical paper, No. 14, Washington, DC.

Bachi, R., and Napolitano, P., 1988, Introduction to: *Atlante Statistico Italiano*, Istituto Centrale di Statistica, Rome.

Bachi, R., and Samuel-Cahn, E., 1976, Characterization of the shape of statistical distributions, *Regional Science and Urban Economics* **6**: 205–227.

Bachi, R., DellaPergola, S., and Klaff, V., 1988, Urban ecology of the Jews in the Diaspora, Unpublished report to the US–Israel Binational Science Foundation, Jerusalem.

Bachi, R., Klaff, V., and Ever-Hadani, S., 1988, Geostatistical methods in urban ecology: Application to Toronto religious subpopulations, Technical paper.

Balchin, W.G.V., and Coleman, A.M., 1966, Graphicacy should be the fourth ace in the pack, *The Cartographer* **3**: 23–28.

Barabba, V.P., 1978, Automating statistical graphics: A tool for communication, in: *Graphic Presentation of Statistical Information* (Technical paper 43, pp. 7–13), US Department of Commerce, Bureau of Census, Washington, DC.

Barabba, V.P., and Finkner, A.L.V., 1978, The utilization of primary printing colors in displaying more than one variable, in: *Graphic Presentation of Statistical Information* (Technical paper 43, pp. 14–22), US Department of Commerce, Bureau of Census, Washington, DC.

Bartlett, M.S., 1971, Physical nearest neighbour models and non-linear time series, *Journal of Applied Probability* **8**: 222–232.

Beniger, J.R., and Robyn, D.L., 1978, Quantitative graphics in statistics: A brief history, *American Statistician* **32**: 1–10, American Statistical Association, Alexandria, Va.

Berry, B.J.L., 1971, Internal structure of the city, in: *Internal Structure of the City* (L.S. Bourne, ed.), pp. 97–103, Oxford University Press, New York.

Berry, B.J.L., and Garrison, W.L., 1958, The functional bases of the central place hierarchy, *Economic Geography* **34**: 145–154.

Berry, B.J.L., and F. Horton, 1970, *Geographic Perspectives on Urban Systems*, Prentice-Hall, Englewood Cliffs, N.J.

Berry, B.J.L., and Marble, D.F. (eds.), 1968, *Spatial Analysis: A Reader in Statistical Geography*, Prentice-Hall, N.J.

Berry, B.J.L., Simmons, J.W., and Tennant, R.J., 1963, Urban population densities: Structure and change, *Geographical Review* **53** (July): 389–405.
Berry, B.J.L., Simmons, J.W., and Tennant, R.J., 1966, Urban population densities: Structure and change, *Geographical Review* **56** (April): 213–225.
Bertin, J., 1967, *Semiologie Graphique*, Mouton et Gauthier-Villars, Paris.
Bertin, J., 1973, *Le traitement graphique de l'information*, Laboratoire de Cartographie, Paris.
Bertin, J., 1981, *Graphics and Graphic Information Processing*, Walter De Gruyter, Berlin-New York.
Bertin, J., 1983, *Semiology of Graphics*, University of Wisconsin Press, Madison.
Blair, D.J., and Bliss, T.H., 1967, The measurement of shape in geography: An appraisal of methods and techniques, *Bulletin of Quantitative Data for Geographers*, Department of Geography, Nottingham University.
Boots, B.N., and Lamoureux, L.F., 1972, Working notes and bibliography on the study of shape in human geography and planning (No. 346), Council of Planning Librarians Exchange and Bibliography, Monticello, Il.
Bourne, L.S. (ed.), 1971, *Internal Structure of the City*, Oxford University Press, New York.
Boyce, R.R., and Clark, W.A.V., 1964, The concept of shape in geography, *Geographical Review* **54**: 561–572.
Brinton, W.C., 1939, *Graphic Presentation*, Brinton Associates, New York.
Bunge, W., 1962, *Theoretical Geography: Land Studies in Geography, 2nd ed.*, Serie CI, Lund: Geerup, London.
Bureau of the Census, 1978, *Graphic Presentation of Statistical Information*, Technical Paper No. 43, Washington, DC.
Burkhardt, F., 1929, Der statistische Schwerpunkt und seine Bedeutung für Theorie und Praxis, *Allgemeines Statistisches Archiv* **XIX**: 473–483.
Caprio, R.J., 1970, Centrography and geostatistics, *Professional Geographer* **XXII**: 15–19.
Carrothers, G.A.P., 1956, An historical review of the gravity and potential concepts of human interaction, *Journal of the American Institute of Planners* **XXII**: 94–102.
Chaddock, R.E., 1935, *Principles and Methods of Statistics*, Houghton-Mifflin, New York.
Chambers, J.M., Cleveland, W.S., Kleiner, B., and Tukey, P.A., 1983, *Graphical Methods for Data Analysis*, Wadsworth, Brooks, and Cole, Pacific Grove, Cal.
Chapin, F.S., Jr., and Weiss, S.F., 1962, *Urban Growth Dynamics in a Regional Cluster of Cities*, John Wiley and Sons, New York.
Chernoff, H., 1973, The use of lace to represent points in *k*-dimensional space graphically, *Journal of the American Statistical Association* **68**: 361–368.
Chernoff, H., 1978, Graphical representation as a discipline, in: *Graphical Representation of Multivariate Data* (P.C. Wang, ed.), pp. 1–11, Academic Press, New York.
Chernoff, H., 1979, Graphical representations as a discipline, in: *Information Linkage between Applied Mathematics and Industry* (P.C.C. Wang, A.L. Schoestadt, B.I. Russa, and C. Comstock, eds.), pp. 49–59, Academic Press, New York.
Chernoff, H., and Rizvi, M.H., 1975, Effect on classification error of random permutations of features in representing multivariate data by faces, *Journal of the American Statistical Association* **70**: 548–554.
Chorley, R., and Haggett, P., 1967, *Models in Geography*, Meuthen and Co., London.
Choynowski, M., 1959, Maps based on probabilities, *Journal of the American Statistical Association* **54**: 385–388.
Clark, C., 1951, Urban population densities, *Journal of the Royal Statistical Society* **114A**: 490–496.
Clark, W.A.V., and Gaile, G.L., 1973, The analysis and recognition of shapes, *Geografiska Appaler* **55**, Series B(2): 153–163.
Cleveland, W.S., 1984, Graphs in scientific publications, *The American Statistician* **38**: 261–269.
Cleveland, W.S., 1985, *The Elements of Graphing Data*, Wadsworth Advanced Book Program, Monterey, Calif.
Cleveland, W.S., 1993, *Visualizing Data*, AT&T Bell Laboratories, Murray Hill, New Jersey.
Cleveland, W.S., and McGill, R., 1984, Graphical perception: Theory, experimentation, and application to the development of graphical methods, *Journal of the American Statistical Association* **79**(387): 531–554.
Cliff, A.D., and Ord, J.K., 1981, *Spatial Processes: Models and Applications*, Pion, London.

Coffey, W.J. (ed.), 1988, *Geographical Systems and Systems of Geography*, Department of Geography, University of Western Ontario.

Coccossis, H.N., 1980, The shape and structure of metropolitan areas: An application of centrographic analysis to Atlanta, Georgia, 1940–1975, Ph.D. dissertation, Cornell University.

Cole, J.P., and King, C.A.M., 1966, *Quantitative Geography*, John Wiley and Sons, London.

Costigan-Eaves, P.A.J., 1984, Data graphics in the 20th century: A comparative and analytic survey, unpublished Ph.D. thesis, Rutgers, the State University of New Jersey, New Brunswick.

Court, A., 1964, The elusive point of minimum travel, *Annals of the Association of American Geographers* **54**: 400–403.

Crawford, P.V., 1976, Optimum spatial design for thematic maps, *Cartographic Journal* **13**: 134–144.

Cressie, N.A.C., 1991, *Statistics for Spatial Data*, John Wiley and Sons, New York.

Croxton, F.E., and Cowden, D.J., 1939, *Applied General Statistics*, Prentice-Hall, New York.

Crum, W.L., Patton, A.C., and Tebutt, A.R., 1938, *Introduction to Economic Statistics*, McGraw-Hill, New York.

Dacey, M.F., 1968*a*, A review on measures of contiguity for two and *k*-colour maps, in: *Spatial Analysis: A Reader in Statistical Geography* (B.J.L. Berry, and D.F. Marble, eds.), pp. 479–495, Prentice-Hall, Englewood Cliffs, N.J.

Dacey, M.F., 1968*b*, A model for the areal distribution of population in a city with multiple population centres, *Tijdschrift voor Economische en Sociale Geografiew* **59**: 232–236.

Danzig, G.B., 1951, Application of the simplex method to a tranportation problem, in: *Activity Analysis of Production and Allocation* (Tjalling Charles Koopmans, ed.), John Wiley and Sons, New York.

Degani, A., 1979, Methodological observations on the state of cartographic analysis in the context of automated spatial information systems, in: *Map Data Processing* (H. Freeman and G.G. Pieroni, eds.), pp. 207–221, Academic Press, London.

Dent, B.D., 1993, *Cartography—Thematic Map Design*, W.M.C. Brown Publishers, Dubuque, Iowa.

DePietri-Tonelli, A., 1949, *Un Sistema di Logo-Diagrammi Tipici*, CEDAM, Padova.

Dickinson, G.C., 1963, *Statistical Mapping and the Presentation of Statistics*, Edward Arnold Publishers, London.

Diggle, P.J., 1983, *Statistical Analysis of Spatial Point Patterns*, Academic Press, London.

Duncan, O.D., 1957, The measurement of population distributions, *Population Studies* **11**(1): 27–45.

Duncan, O.D., and Davis, B., 1953*a*, Contributions to the theory of segregation indexes, *Urban Analysis Report*, No. 14, University of Chicago.

Duncan, O.D., and Davis, B., 1953*b*, Measurement of population distribution, *Urban Analysis Report*, No. 16, University of Chicago.

Duncan, B., and Duncan, O.D., 1955, Analysis of segregation indexes, *American Sociological Review* **20**(2): 210–217.

Duncan, O.D., Cuzzort, R.P., and Duncan, B., 1961, *Statistical Geography: Problems in Analyzing Areal Data*, Glencoe Press, New York.

Dunn, E.S., 1956, The market potential concept, *Papers and Proceedings of the Regional Science Association* **2**.

Ebdon, D.S., 1976, On the underestimation inherent in the commonly used formulae, *Area* **8**: 165–169.

Ebdon, D.S., 1977, *Statistics in Geography*, Basil-Blackwell, Oxford.

Eells, W.C., 1925, The center of population—A prophecy and its fulfillment, *Scientific Monthly* **20** (January): 78–84.

Eells, W.C., 1930, A mistaken conception of the center of population, *Journal of the American Statistical Association* **XXV**: 33–40.

Eells, W.C., Development of interest in centrography in United States, unpublished mimeo.

Ever-Hadani, S., 1980, Applications of cluster analysis algorithm to geostatistical series, *Regional Science and Urban Economics* **19**: 123–151.

Feinberg, S.E., 1979, Graphical methods in statistics, *American Statistician* **33**: 165–178.

Feinberg, B.M., and Franklin, C.A. (eds.), 1975, *The Social Graphics Bibliography*, Bureau of Social Science Research, Washington, DC.

Fitzpatrick, P.J., 1962, The development of graphic presentation of statistical data in the United States, *Social Science* **37**: 203–214.

Fisher, W.D., 1958, On grouping for maximum homogeneity, *Journal of the American Statistical Association* **53**(284).

Flaskämper, P., 1962, *Bevölkerungsstatistik*, Verlag von Richard Meiner, Hamburg.
Flaskämper, P., 1963, Fläche und Bevölkerung eines Gebietes, in: *Zur Anwendung statistischer Methoden* (A. Heinze, ed.), pp. 351–359, Karl-Marx Universität, Leipzig.
Furfey, P.H., 1927, A note of Lefever's standard deviational ellipse, *American Journal of Sociology* **33**: 94–98.
Funkhouser, H.G., 1937, Historical development of the graphical representation of statistical data, *Osiris* **3**: 269–404.
Gabriel, K.R., 1971, The Biplot graphic display of matrices with applications to principal components analysis, *Biometrika* **58**: 453–467.
Gabriel, K.R., 1978, Least squares approximation of matrices by additive and multiplicative models, *Journal of the Royal Statistical Society* **B40**: 186–196.
Galvani, L., 1933, Sulla determinazione del centro di gravita e del centro mediano di una popolazione, *Metron* **XI**(1): 17–48.
Galvani, L., 1934, Sulla determinazione del centro di gravità e del centro mediano di una popolazione, con applicazione alle popolazione censita il 1 Dicembre, 1921, in: *Atti del Congresso Internazionale per gli Studi della Popolazione*, Roma, 1931, Vol. X, Instituto Poligrafica dello Stato, Rome.
Garrison, W.L., 1956, Problems of spatial interaction, *Papers and Proceedings of the Regional Science Association* **2**.
Getis, A., and B. Boots, 1978, *Models of Spatial Processes*, Cambridge University Press, Cambridge.
Ghindoli, G.B., 1950, Il centro della poplazione Italiana, *Statistica*, Ano **X**(4).
Gibbs, J.P., 1961, *Urban Research Methods*, D. Van Nostrand, New York.
Gini, C., 1914, Du una misura della dissomiglianza tra due gruppi di quantità e delle sue applicaziono allo studio delle relazioni statistiche, Atti del reale Instituto Veneto di Scienze, *Lettee ed Arti* **74**: 185–213.
Gini, C., 1930, Letter in editor's notes on the center of population and point of minimum travel, *Journal of the American Statistical Association* **XXV**: 447–448.
Gini, C. (ed.), 1934, *Atti del Congresso Internazionale per gli Studi Sulla Popolazione, Vol. X*, Istituto poligrafico dello stato, Rome.
Gini, C., and Galvani, L., 1929, Di Talune Estensioni dei concetti di media ai caratteri qualitative, *Metron* **VIII**: 136–138.
Gini, C., Boldrini, M., Galvani, and L., Venere, A., 1933, Sui Centri delle popolazioni e sulle loro applicazioni, *Metron* **XI**(2): 11–15.
Gini, C., 1939, Sur la méthode des profils et sur d'autres diagrammes a ordonées jointes, dans la cas de series non ordonées, *Revue de l'Institute International Statistique* **7**(1): 1–31, **7**(4): 225–246.
Glazer, N., and Moynihan, D.P., 1963, *Beyond the Melting Pot: The Negroes, Puerto Ricans, Jews and Irish of New York City*, MIT Press, Cambridge, Mass.
Gordon, D., Kellerman, A., and Waterman, S., 1982, On the computation of the cross of dispersion, *Cartographic Journal* **19**: 133–134.
Gould, P.R., 1969, Methodological developments since the fifties, *Progress in Geography* **1**: 1–50.
Gregory, S., 1963, *Statistical Methods and the Geographer*, Longmans, London.
Grenander, U., 1976, *Lectures in Pattern Theory I. Pattern Analysis*, Springer-Verlag, New York.
Grenander, U., 1978, *Lectures in Pattern Theory II. Pattern Synthesis*, Springer-Verlag, New York.
Griffin, F.L., 1933, The center of population for various continuous distributions of population over areas of various shapes, *Metron* **XI**(1).
Hagerstrand, T., 1967, *Innovation Diffusion as a Spatial Process*, University of Chicago Press, Chicago.
Haggett, P., 1966, *Locational Analysis in Human Geography*, St. Martin's Press, New York.
Haggett, P., Cliff, A.D., and Frey, A., 1977, *Locational Analysis in Human Geography*, 2nd ed., Arnold Publishers, London.
Haining, R., 1990, *Spatial Data Analysis in the Social and Environmental Sciences*, Cambridge University Press, Cambridge.
Haldane, J.B.S., 1948, Note on the median of a multivariate distribution, *Biometrica* **XXXV**: 414–415.
Hammer, C., 1952, A functional interpretation of the cost-utility index, Mimeographed seminar report, Columbia University, New York.
Hammond, R., and McCullagh, 1974, *Quantitative Techniques in Geography*, Clarendon Press, Oxford.
Harrison, R.J., and Weinberg, D.H., 1992, Changes in racial and ethnic residential segregation, 1980–1990, paper prepared for the American Statistical Association meetings, Boston, Mass.

Hart, J.F., 1954, Central tendency in areal distributions, *Economic Geography* **30**: 48–59.

Hartshorn, T.A., 1980, *Interpreting the City*, John Wiley and Sons, New York.

Harvey, D.W., 1968, Pattern, process and the scale problem in geographical research, *Transactions of the Institute of British Geographers* **45**: 71–78.

Hauser, P.M., Duncan, O.D., and Duncan, B.D., 1956, A summary report, A.F. Personnel and training research center, Texas.

Hawley, A.H., 1950, *Human Ecology*, Ronald Press, New York.

Hayford, J.F., 1902, What is the center of area, or the center of a population, *Journal of the Statistical Association* **VIII**: 47–58.

Herbert, J.D., and Stevens, B.H., 1960, A model for the distribution of residential activity in urban areas, *Journal of Regional Science* **II**(2) Fall: 21–26.

Hitchcock, F.L., 1941, Distribution of a product from several sources to numerous localities, *Journal of Mathematical Physics* **20**.

Hoover, E.M. Jr., 1941, Interstate distribution of population, 1850–1940, *Journal of Economic History* **1** (Nov.): 199–205.

Hoover, E.M., 1948, *The Location of Economic Activity*, McGraw-Hill, New York.

Hoover, E.M., and Vernon, R., 1959, *Anatomy of a Metropolis*, Harvard University Press, Cambridge, Mass.

Hoyt, H., 1936, City growth and mortgage risk, Insured Mortgage Portfolio, A publication of the Federal Housing Administration, Washington, I, Nos. 6–10 (December–April, 1937).

Huff, D., 1954, *How to Lie with Statistics*, Norton, New York.

Isard, W., 1956, *Location and Space Economy*, John Wiley and Sons, New York.

Isard, W., Bramhall, O.F., Carrothers, G.A.P., Cumberland, J.H., Moses, L.N., Price, D.O. and Schooler, E.W., 1960, *Methods of Regional Analysis*, MIT Press, Cambridge, Mass., and London.

ISTAT (Instituto Centrale di Statistica), 1988, *Atlante Statistico Italiano* (introduction by Bachi, R., and Napolitano, P), Rome.

ISTAT (Instituto Nazionale di Statistica), 1992, *Atlante Statistico Italiano, Analisi Geostatistica* (prepared by Bachi, R., Napolitano, P., Gargano, O., Pedachia, U., and Ever-Hadani, S.), Rome.

Izenman, R.J.K., 1980, Developments in statistical graphics 1960–1980, Bureau of the Census, Technical Paper 49: First General Conference on Social Graphics, pp. 51–113, US Government Printing Office, Washington, DC.

Jahn, J.A., 1950, The measurement of ecological segregation: Derivation of an index based on the criterion of reproducibility, *American Sociologic Review* **15** (February): 100–104.

Jakubs, J.F., 1981, A distance-based segregation index, *Socio-Economic Planning Sciences* **15**(3): 123–136.

Jefferson, M., 1909, The anthropography of some great cities: A study in the distribution of population, *Bulletin of the American Geographical Society* **41**: 551–566.

Jerome, H., 1924, *Statistical Methods*, Harper and Brothers, New York.

Johnston, R.J., 1970, On spatial patterns in the residential structure of cities, *Canadian Geographer* **14**: 361–367.

Johnston, R.J., 1978, *Multivariate Statistical Analysis in Geography*, Longmans, London.

Jones, B.G., 1960, The theory of the urban economy: Origins and development with emphasis on intraurban distribution of population and economic activity, Ph.D. dissertation University of North Carolina.

Jones, B.G., 1980, Applications of centrographic techniques to the study of urban phenomena: Atlanta, Georgia, 1940–1975, *Economic Geography* **56**(3): 201–222.

Jones, B.G., Manson, D.M., and Coccossis, H.N., 1978*a*, Task I report: Applications of centrographic techniques to the study of urban development: The growth of Atlanta, 1940–1975. Program in Urban and Regional Studies, Cornell University, Ithaca (September 15).

Jones, B.G., Manson, D.M., and Coccossis, H.N., 1978*b*, Task II report: Applications of centrographic techniques to the study of urban development: Theoretical implications of the expansion of Atlanta, Program in Urban and Regional Studies, Cornell University, Ithaca (October 1).

Jones, B.G., Manson, D.M., and Coccossis, H.N., 1978*c*, Task III report: Applications of centrographic techniques to the study of urban development: The expansion of metropolitan Atlanta, 1940–1975, Program in Urban and Regional Studies, Cornell University, Ithaca (December 20).

Jones, B.G., Manson, D.M., Jin, Y.H., and Corpuz, A.G., 1989, *The Evolution of Urban Form: Centrographic Measures Applied to Wichita, Kansas*, Cornell Institute for Social and Economic Research Program in Urban and Regional Studies, Ithaca.

Kadmon, N., 1968, The mapping of distribution parameters, *Cartographic Journal* **1968**: 64–69.

Kadmon, N., 1971, Comparing economic maps with the aid of distribution-parameter graphics, *International Yearbook of Cartography* **11**: 278–287.

Kadmon, N., 1975, Dynamic multi-dimensional mapping of diseases by computer, *Merhavim* **2**: 111–121.

Kadmon, N., 1983, Graphic communication and the importance of teaching cartographic principles in the computer age, in: *International Year Book of Cartography* **23**: 161–167.

Kain, J.F., 1962, The journey-to-work as a determinant of residential locations, *Papers and Proceedings of the Regional Science Association* **9**: 137–160.

Kain, J.F., 1975, *Essays on Urban Spatial Structure*, Ballinger Publishing, Cambridge, Mass.

Karsten, K.G., 1923, *Charts and Graphs*, Prentice-Hall, New York.

Kellerman, A., 1981*a*, *Centrographic Measures in Geography*, CATMOG 32, Geoabstracts, University of East Anglia, Norwich.

Kellerman, A., 1981*b*, The use of standard distance as an equity criterion for public facility location, *Area* **13**: 245–249.

Kelly, T.L., 1924, *Statistical Methods*, Macmillan & Co., New York.

Kemper, P., and Schmenner, R., 1974, The density gradient for manufacturing industry, *Journal of Urban Economics* **1**: 4110–4127.

King, L.J., 1969, *Statistical Analysis in Geography*, Prentice-Hall, Englewood Cliffs, N.J.

King, R., 1982, On geography, cartography and the fourth language, *Graphical Research Forum* **5**: 42–56.

Kosslyn, S.M., 1994, *Image and Brain: The Resolution of the Imagery Debate*, MIT Press, Cambridge, Mass.

Krishna Iyer, P.V., 1949, The first and second moments of some probability distributions arising from points on a lattice and their applications, *Biometrika* **36**: 135–141.

LaGory, M., and Pipkin, J., 1981, *Urban Social Space*, Wadsworth, Belmont, Cal.

Lee, D.B., Jr., 1967, Analysis and Description of Residential Segregation: An Application of Centrographic Techniques to the Study of the Spatial Distribution of Ethnic Groups in Cities. Center for Housing and Environmental Studies, Cornell University, Ithaca.

Lee, D.R., and Sallee, T.G., 1970, A method for measuring shape, *Geographical Review* **60**(4): 555–563.

Lefever, D.W., 1926, Measuring geographic concentration by means of the standard deviational ellipse, *American Journal of Sociology* **32**: 88–94.

Leti, G., 1961, Nouvi Tipi di distanze fra insiemi di punti e bro applicaziani alla statistica, *Metron* **21**: 1–33.

Leti, G., 1982, Distanze ed indici statistici, *La Goliardica* **1982**.

Lieberson, S., 1963, *Ethnic Patterns in American Cities*, Glencoe Press, New York.

Linders, F.J., 1933, Uber die Berechnung des Schwerpunkts und der Tragheitsellipse einer Bevolkerung, *Atti del Congresso Internazionale per lo Studio deo Problemi della Popolazione* **10** (Rome); and *Metron* **XI**(1): 3–10.

Lewis, P., 1977, *Maps and Statistics*, John Wiley and Sons, New York.

Loo, E.J.K., 1973, A spatial analysis of social structure by centrographics: Honolulu, Hawaii, unpublished Masters thesis, Cornell University, Ithaca.

Lorenz, M.O., 1904–1905, Methods of measuring the concentration of wealth, *Publications of the American Statistical Association* **9**.

Luu-Mau-Thanh, 1962, Distribution théoretiques des distances entre deux répartis uniformément sur une surface, in: *Les déplacements humains: Aspects méthologiques de leur mesure* (J. Sutter, ed.), pp. 173–184, Editions "Sciences Humanes," Monaco.

Maceachren, A.M., 1985, Compactness of geographic shape: Comparison and evaluation of measures, *Geografiska Annaler* **B67**(155): 53–67.

Mackay, J.R., 1951, Some problems and techniques in isopleth mapping, *Economic Geography* **XXVII**: 1–9.

Manson, D.M., 1979*a*, Spatial changes in a declining metropolitan area: Population and employment activities in Wilkes-Barre, Pennsylvania, 1900–1970, unpublished Ph.D. dissertation, Cornell University.

Manson, D.M., 1979*b*, The negative exponential density gradient and the standard radius: Mathematical and empirical relationships, Paper delivered at the Meeting of the Northeast Regional Science Association at Amherst, Massachusetts, May 1979.

Manson, D.M., 1981, The negative exponential density gradient and the standard radius: Mathematical and empirical relationships, *Environment and Planning A* **13**(9): 1053–1065.

March, L., 1911, *Bulletin de l'Institut International de Statistique* **19**: 118.

Marshall, A., 1885, On the graphic method of statistics, *Jubilee Volume of the Statistical Society*, pp. 251–260, Royal Statistical Society, London.

Massam, B.H., and Goodchild, M.F., 1971, Temporal trends in spatial organization of a service agency, *Canadian Geographer* **15**(3): 193–206.

Massey, D.S., and Denton, N.A., 1988, The dimensions of residential segregation, *Social Forces* **67**: 281–315.

Matérn, B., 1986, *Spatial Variations*, Springer-Verlag, Heidelberg.

Mather, E., 1994, A linear distance map of farm population in the United States, *Annals of the Association of American Geographers* **XXXIV**: 173–180.

Matheron, G., 1962, *Traité de geostatique appliqué*, Technip, Tome I, Paris.

Matheron, G., 1963, Principles of geostatistics, *Economic Geology* **58**: 1246–1266.

McCarty, H.H., 1956, *The Measurement of Association in Industrial Geography*, Department of Geography, State University of Iowa, Iowa City.

McElrath, D.C., 1962, The social area of Rome: A comparative analysis, *American Sociological Review* **27**: 376–391.

Mead, R., 1974, A test for spatial pattern at several scales using data from a grid of contiguous quadrants, *Biometrics* **30**: 295–307.

Merriam, D.F. (ed.), 1970, *Geostatistics*, Plenum Press, New York.

Meyer, M.A., Broome, F.R., and Schweitzer, R.H., Jr., 1975, Color statistical mapping by the US Bureau of Census, *American Geographer* **2** (1975): 100–117.

Meuriot, M.P., 1902, Du centre mathematique d'une population (de Villes), *Journale Societe Statistique de Paris* **XLIII**: 218–219.

Moellering, H., and Tobler, W., 1972, Geographical variances, *Geographical Analysis* **4**: 34–50.

Monkhouse, F.J., and Wilkinson, H.R., 1963, *Maps and Diagrams: Their Compilation and Construction*, Methuen & Co., London.

Monmonier, U., 1991, *How to Lie with Statistics*, University of Chicago Press, Chicago.

Monmonier, U., 1993, *Mapping It Out: Expository Cartography for the Humanities and Social Sciences*, University of Chicago Press, Chicago.

Morgan, B.S., 1982, The properties of a distance-based segregation index, *Socio-economic Planning Sciences* **16**(4): 167–171.

Morgan, B.S., 1983, Alternate approach to the development of a distance-based measure of racial segregation, *American Journal of Sociology* **88**: 1237–1249.

Morgan, B.S., and Norburg, J., 1981, Some further observations on the index of residential differentiation, *Demography* **18**: 251–256.

Morrill, R.L., and Symons, J., 1977, Efficiency and equity aspects of optimum location, *Geographical Analysis* **9**: 215–225.

Morris, A.E.J., 1972, *History of Urban Form*, John Wiley and Sons, New York.

Moses, L., and Williamson, H.F. Jr., 1967, The location of economic activity in cities, *American Economic Review, Papers and Proceedings* **1967** (May): 211–222.

Mudgett, B.D., 1930, *Statistical Tables and Graphs*, Houghton-Mifflin, Boston.

Murdie, R.A., 1969, Factoral ecology of metropolitan Toronto, 1951–1961, Research Papers **116**.

Murdie, R.A., 1970, A reply to 'On spatial pattern in the residential structure of cities,' *Canadian Geographer* **14**: 367–369.

Napolitano, P., 1988, Sul metodo di analisi delle distribuzioni di populazione su un territorio prefisso, *Genus* **XLIV**(3–4): 153–165.

Napolitano, P., 1989, Manuale di tecniche de indagine, *Le Rappresentazioni Grafiche di Dati Statistici*, ISTAT, Rome.

Napolitano, P., 1990, L'uso di un indice di dissomiglianza basata sulla distanza per l'analisi esplorativa di distribuzioni di popolazione sul territorio Italiano, *Atti delle Giornate di Studio Analisi dei Dati Spaziali*, Pescara.

Neft, D.S., 1966, *Statistical Analysis for Areal Distributions*, Monograph series 2 Regional Science Research Institute, Philadelphia.
Neprash, J.A., 1934, Some problems in the correlation of spatially distributed variables, *Journal of the American Statistical Association* **22** (Supplement): 167–168.
Newling, B., 1966, Urban growth and spatial structure: Mathematical models and empirical evidence, *Geographical Review* **56**: 213–225.
Norcliffe, G.B., 1977, *Inferential Statistics for Geographers*, Hutchinson, London.
Olbrechts, A., 1935, Le centre de population de la Belgique, *Melanges Ernest Mahain* (Liege) **1935**.
Olson, J.M., 1975, Auto correlation and visual map complexity, *Annals of the Association of American Geographers* **71**: 259–276.
Ord, J.K., 1972, Density estimation and tests for randomness using distance methods, in: *NATO Advanced Study Institute on Statistical Ecology*, Department of Statistics, Pennsylvania State University.
Ottensmann, J.R., 1975, *The Changing Spatial Structure of American Cities*, Lexington Books, Lexington, Mass.
Pedroni, F., 1953, *Corse di Rappresentazioni Statistiche*, Liberia Eredi Virgilio Veschi, Rome.
Pedroni, F., 1965, Rationalisation des representations statistiques a coordonnees orthogonales et des hachures des cartogrammes, *Bulletin de l'institut International de Statistique* **41**(2): 980–989.
Peguy, C.P., 1957, *Elements de statistique appliquee aux science geographiques*, Centre de Documentation Universitaire, Paris.
Peters, P.D., 1964, Residential Sectors and Urban Expansion, Center for Housing and Environmental Studies, Cornell University, Ithaca, New York.
Pielou, E.C., 1977, *Mathematical Ecology*, 2nd ed., John Wiley and Sons, New York.
Porter, P.W., 1963, What is the point of minimum aggregate travel? *Annals of the Association of American Geographers* **53**: 224–232.
Poulsen, T.M., 1959, Centrography in Russian Geography, *Annals of the Association of American Geographers* **49**: 326–327.
Quinn, J.A., 1943, The hypothesis of median location, *American Sociological Review* **VII** (April): 148–156.
Ravenstein, E.G., 1885, 1889, The laws of migration, *Journal of the Royal Statistical Society* **48**: 167–235; **52**: 241–305.
Riggelman, J.R., 1926, *Graphic Methods for Presenting Business Statistics*, McGraw-Hill, New York.
Ripley, B.D., 1981, *Spatial Statistics*, Wiley Interscience, New York.
Risser, R., 1937, Commenti su (2) e (3), *Journal de la Socièté de Statistique de Paris* **1937** (May).
Robinson, A.H., 1953, *Elements of Cartography*, John Wiley and Sons, New York.
Robinson, A.H., 1956, The necessity of weighting values in correlation analysis of areal data, *Annals of American Geographers* **XLVI**: 233–236.
Robinson, A.H., and Bryson, R.A., 1957, A method for describing quantitatively the correspondence of geographical distributions, *Annals of American Geographers* **XLVII**: 379–391.
Robinson, A.H., Sale, R.D., Morrison, J.L., and Muehrcke, P.C. (eds)., 1984, *Elements of Cartography*, John Wiley and Sons, New York.
Rogers, A., 1974, *Statistical Analysis of Spatial Dispersion*, Pion, London.
Rugg, H., 1925, *Primer of Graphics and Statistics for Teachers*, Houghton-Mifflin, Boston.
Salvemini, T., and Girone, G., 1981, *Lezioni de Statistica*, Cacucci, Bari.
Samuel, E., and Bachi, R., 1964, Measures of distances of distribution functions and some applications. *Metron* **23**(1–4): 83–122.
Scates, D.E., 1931, Notes on the center of minimum aggregate distance for a population distributed over a plane, unpublished paper for the Bureau of School Research, Cincinnati Public Schools (April 2).
Scates, D.E., 1933, Locating the median of the population in the United States, *Metron* **XI**, 1: 49–65.
Schmelz, O., 1959, Development of the Jewish population of Jerusalem during the last hundred years, Ph.D. thesis, Hebrew University of Jerusalem.
Schmid, C.F., 1978, The role of standards in graphic presentation, in *Graphic Presentation of Statistical Information* (Technical Paper 43), pp. 69–78, Bureau of the Census, Washington, DC.
Schmid, C.F., 1983, *Statistical Graphics: Design Principles and Practices*, John Wiley and Sons, New York.
Schmid, C.F., and Schmid, S.E., 1979, *Handbook of Graphic Presentation*, John Wiley and Sons, New York.
Schneider, J.B., 1967, Measuring the locational efficiency of the urban hospital, Health Services Research (Summer 1967).

Schneider, J.B., 1968, A new approach to area-wide planning of metropolitan hospital facilities, *Hospitals, Journals of the American Hospital Association* **42**: 79–83.

Secrist, H.D., 1920, *An Introduction to Statistical Methods*, Macmillan & Co., New York.

Seymour, D.R., 1965, IBM 7090 program for locating bivariate means and bivariate medians, in: Technical Report No. 16, ONR Task, No. 389-135, Contract Nonr 1228, C26, Office of Naval Research, Geography Branch, Northwestern University, Evanston, Ill.

Seymour, D.R., 1970, Note on Austin's 'An approximation to the point of minimum aggregate distance', *Metron* **28**(1–4): 412–421.

Shachar, A., 1967, Some applications of geo-statistical methods in urban research, *Papers and Proceedings of the Regional Science Association* **18**: 197–206.

Shachar, A., 1975, Patterns of population densities in the Tel-Aviv metropolitan area, *Environment and Planning* **A7**: 279–291.

Shimony, Z., 1963, Theoretical research in statistical-geographical series. Ph.D. thesis, mimeographed, Hebrew University of Jerusalem.

Shryok, H.S., 1959, The efficiency of internal migrations in the United States, International Population Conference, Vienna.

Shzyock, H.S., and Siegel, J.S., 1973, *The Methods and Materials of Demography*, US Department of Commerce, Washington.

Sloane, C.S., 1925, *Statistical Atlas of the United States—1924*. US Bureau of the Census, Department of Commerce, Washington, DC.

Smith, W.H., 1924, *Graphic Statistics in Management*, McGraw-Hill, New York.

Sonnino, E., 1968, Sulla distribuzione territoriale dei fenomeni demografici in Italia, Miscellanea di Studi Demografici, Instituto di Demografia, Facolta de Scienze Statistiche Demografiche ed Attuariali, Università di Roma **18**.

Soot, S., 1975, Methods and measures of centrography: A critical survey of geographic applications, Occasional publications of the Department of Geography, Paper No. 8, University of Illinois, Urbana-Champaign (April 1975).

Spear, M.E., 1952, *Charting Statistics*, McGraw Hill, New York.

Stephenson, L., 1972, Centrographical applications to intra-urban criminal distributions: Problems and promises, paper presented at the annual meeting of the Association of American Geographers, Kansas City.

Stewart, J.Q., 1942, A measure of the influence of population at a distance, *Sociometry* **5**: 63–71.

Stewart, J.Q., 1947*a*, Suggested principles of social physics, *Science* **106**.

Stewart, J.Q., 1947*b*, Empirical mathematical rules concerning the distribution and equilibrium of population, *Graphical Review* **XXVII**: 461–485.

Stewart, J.Q., 1948*a*, Concerning social physics, *Scientific American* **CLXXVIII** (May): 20–23.

Stewart, J.Q., 1948*b*, Demographic gravitation: Evidence and applications, *Sociometry* **XI**: 31–58.

Stewart, J.Q., 1950*a*, Potential of population and its relationship to marketing, in: *Theory in Marketing* (R. Cox and W. Alderson, eds), pp. 19–39, Richard D. Irwin, Chicago.

Stewart, J.Q., 1950*b*, The development of social physics, *American Journal opf Physics* **XVIII**: 239–253.

Stewart, J.Q., 1952, A basis for social physics, *Impact of Science on Society* **III**: 110–133.

Stewart, J.Q., 1953, Urban population densities, *Geographical Review* **XLIII**: 575–576.

Stewart, J.Q., 1958, Physics of population distribution, *Journal of Regional Science* **I**: 99–123.

Stewart, J.Q., 1959, Some parameters of the geographical distribution of population, *Geographic Review* **49**.

Stewart, J.Q., and Warntz, W., 1958, Macrogeography and social science, *Geographical Review* **48**: 167–184.

Sviatlovsky, E.E., and Eells, W.C., 1937, The centrographical method and regional analysis, *Geographical Review* **27**: 240–254.

Syslow, M.M., Deo, N., and Kowalik, J.S., 1983, Discrete Optimization Algorithms with Pascal Programs: Ch. 1.3, Transportation problem, pp. 54–71, Prentice-Hall, Englewood Cliffs, N.J.

Taeuber, K., and Taeuber, A.F., 1965, *Negroes in Cities*, Aldine Publishing Co., Chicago.

Taitel, M., 1960, On problems of measuring the distribution of population in an urban area, *Proceedings of the Social Statistics Section, American Statistical Association* **1960**: 160–165.

Taylor, P.J., 1971, Distances within shapes: An introduction to a family of finite frequency distributions, *Geografiska Annaler B* **53**: 40–53.

Taylor, P.J., 1977, *Quantitative Methods in Geography: An Introduction to Spatial Analysis*, Houghton-Mifflin, Boston.

Thomas, R.W., and Huggett, R.J., 1980, *Modelling in Geography*, Barnes and Noble, Totowa, N.J.

Thompson, Sir d'Arcy Wentworth, 1917, *On Growth and Form* (1942 ed.), Cambridge University Press, Cambridge, Mass.

Tufte, E.R., 1983, *The Visual Display of Quantitative Information*, Graphic Press, Cheshire, Conn.

Tukey, J.W., 1979, Methodology and the statisticians responsibility for both accuracy and relevance, *Journal of the American Statistical Association* **74**: 786–793.

Unwin, D., 1981, *Introductory Spatial Analysis*, Methuen & Co., London.

Upton, G.J.G., and Fingleton, B., 1989, *Spatial Data Analysis by Example: Point Pattern and Quantitative Data, Volume 2*, John Wiley and Sons, London.

Wainer, H., 1980, Making newspaper graphs fit to print, in: *Processing of Visible Language, Vol. 2* (P.A. Kolers, M.E. Wrolstad, and H. Bouma, eds.), pp. 125–142, Plenum Press, New York.

Wainer, H., 1984, How to display data badly, *American Statistician* **38**: 137–147.

Wainer, H., and Thissen, D., 1981, Graphical data analysis, *Annual Review of Psychology* **32**: 191–241.

Warntz, W., 1956, Measuring spatial association with special consideration of the case of market orientation of production, *Journal of the American Statistical Association* **LI**: 597–604.

Warntz, W., and Neft, D.S., 1960, Contributions to a statistical methodology for areal distributions, *Journal of Regional Science* **2**(1): 47–66.

Waterman, W., 1969*a*, Some aspects of the urban geography of Acre, Israel, unpublished Ph.D. dissertation, University of Dublin.

Waterman, W., 1969*b*, Some comments on standard distance: A new application to Irish population studies, *Irish Geography* **6**: 51–62.

White, M.J., 1983, The measurement of spatial segregation, *American Journal of Sociology* **88**: 1008–1018.

White, M.J., 1984, The measurement of spatial segregation, *American Journal of Sociology* **90**: 187–191.

White, M.J., 1986, Segregation and diversity measures in population distribution, *Population Index* **52**(2): 198–221.

Whittington, G., Beaver, K.S.O., and Mabin, A.S., 1972, Compactness of shape: Review, theory and application, Occasional Paper **7**.

Wright, J.K., 1937, Some measures of distributions, *Annals of the Association of American Geographers* **26**.

Yam, J., 1968, Concentration and segregation of geostatistical distributions: New methods and exemplary illustrations, publication 478, Szold Institute, Jerusalem.

Yam, J., 1969, The analysis of divergence in geostatistical distributions, Ph.D. thesis, Hebrew University of Jerusalem.

Yeates, M., 1974, *An Introduction to Quantitative Analysis in Human Geography*, McGraw-Hill, New York.

Yuill, R.S., 1971, The standard deviational ellipse: An updated tool for spatial description. *Geografiska Annaler* **53**(B): 28-39.

Zahl, S., 1974, Applications of the S-method to the analysis of spatial pattern, *Biometrics* **30**: 513–524.

Zerdy, E.T., 1976, Entrophy modeling: With application to urban population distributions, Ph.D. dissertation, Cornell University.

Zipf, G.K., 1946, The P_1P_2/D hypothesis: On the intercity movement of persons, *American Sociological Review* **XI**: 677–686.

Zipf, G.K., 1949, *Human Behavior and the Principle of Least Effort*, Addison-Wesley, Cambridge, Mass.

Index

Absolute squared oblongity, 27, 266
Affinal equivalent territory, 282
Affine equivalence, 405–406
Aggregation, 191, 346
  data handling, 70–71
Algorithm, 331–333
Alps, 7, 10
Anaglyph method, 132
Apennines, 7, 10
Areal measurement of dispersion, 263–264
Argentina, 238–240
Asymmetry, 30, 278, 312, 424
Atlanta, 100, 432–437
Austin, 432–437
Automation, 118–119
Average of squared minimizing distances (ASMD), 379, 405, 408, 426
  applications of, 383, 394
  decomposition of, 381–382
    equations for, 381–386
    examples, 394, 454–456
      city with concentric growth, 397–398
      Italy, 396–397
      United States, 394–396
    graphical presentation, 389–391
    interpretation of linkages, 380, 382
    nonnegative components, 383–394
  measurement of central asymmetry, 454
  overview of, 379–380
  role in joint dispersion, 384–386
Average of squared distances, 250
  with repetition, 250, 256
Averaging ratio, 194–195

Berlin, 287, 304
Bidimensionality, 247–248
Bipolarity, 234

Canada, 156
Cartesian coordinate system, 125–128, 367–368, 417
  application to geosets, 24–25
Cartesian coordinate system (*cont.*)
  general properties of, 126–127
  transformation between geosets, 399, 404–405
  use in statistical mapping, 127–128
    limitations of, 127–128
Cateris paribus, 288, 293, 295, 297
Central asymmetry, 454
Centrography, 40–42
Choropleth maps, 162–165, 178, 417
Clustering, 30, 245, 278, 305–311, 424
  examples for Jerusalem, 306–307
  examples of, 309–310
  impact of, 310–311
  parameters, 310
  study of large populations, 308
Coincidence, 314–316, 403, 420
Collinearity, 26, 347–349, 357–361
Colored maps, 174–178
Compactness, 30–31; *see also* Looseness
Computer software, 428–430, 456–457
Comuni, 3–4, 152
Concavity, 30, 108, 286–288, 292–294,
Concentration, 23, 29–30
Concentric growth, 397–398
Congruence, 314–316, 319, 403, 420
Continental Italy, 7, 30–31, 272, 293–294
Contour lines, 132
Convex hull, 93–96, 109

Density, 23, 29–30, 196, 419
  averaging ratios, 194–195
  net vs. gross, 191–193
  ratios, 194–195
  relative, 193
  surplus, 193, 198
Discrepancy, 32–34
Dispersion, 23, 25–29, 31, 267–270, 421
  areal measurement of, 263–264
  ethnic, 433
  joint, 384–386

Dispersion (*cont.*)
linear measure of, 256–257
measurement of, 247–249, 256
territorial, 25–29, 249
Disrupting factor, 257
Dissimilarity, 247, 328, 405
Distance averaging, 249–250, 256–257
Distance-deviance, 255–256
decomposition of, 270
actual fragmented territory, 272–273
habitat vs. nonhabitat issues, 273–274
primary to derived geosets, 274–275
territorial models, 270–272
Distance range, 247
Distance-variance, 22, 27–29, 31, 249–250, 279–282, 288, 297, 402, 408, 421, 423–424
for a circle, 431
values for selected countries, 280–281
Distance-moments, 211–212, 245
Dot maps, 128–129, 132
Duncan's concentration index, 11, 190–191, 419
Dynamics radius, 256

Eccentricity, 23, 25, 209–210, 309
Ellipse of equivalent scatter, 98–99
Equivalence, 405–406
Equivalent scatter, 98–99
Equivalent territory, 282
Euclidean distance, 248

Form, 422–424
Fragmentation, 30, 272–273, 296, 423
France, 152–153, 238–240, 339–340

Gamme naturalle, 159
General mean distance deviation, 250
Geosets, 10–22, 187–190
applications of, 90
Atlanta, 100
gross territory, 89–90
concentration of population, 97–98
convex hull, 93–96, 109
criteria for solutions, 92–93
ellipse as substitute for habitat, 99–104
ellipses of equivalent scatter, 98–99
handling unbounded population distributions, 90–98
problem description, 90–92
Italy, 103–104, 109
Jerusalem, 105, 110
predetermined territory, 43–88
Tel-Aviv, 91–97
territory without population reference, 105
United States, 102
Washington, DC, 101

Geosets (*cont.*)
averaging, 84–85
calculations for Italy, 261, 265
Cartesian coordinates in, 24–25
concept of, 10
density evaluation, 191–193
Italy, 195–199
Jerusalem, 195
United States, 199–207
derived, 10–11, 15–22, 25
discrepancy measurement between, 32–34
dispersion, 23, 25–29, 31
centers as averages of locations, 209–211
Argentina, 238–240
France, 238–240
map interpretations for Italy, 219–231, 238–240
average income by province, 230–231
citrus groves, 221, 225–226
illiterates, 225–228
sheep distribution, 221–224
subpopulations, 228–229
map interpretations for United States, 231–238, 241
subpopulations, 233–237
total population, 231–233
mean, 209
calculation of population geosets, 213–215
definition of, 212–213
median, 209, 242, 244
calculation of, 243
definition of, 242
properties of, 242
uses of, 243
modal, 245
properties of, 216–219
triangle of centers, 221, 231
vs. median for Italy, 244
calculation of territorial geosets, 215
clustering aspects, 305–311
examples for Jerusalem, 306–307
examples of, 309–310
impact of, 310–311
parameters, 310
study of large populations, 308
comparison of actual vs. uniform population distribution, 48–63
cell networks, 52
habitat vs. nonhabitat approach, 55–56, 60–63
mapping techniques, 46–63
comparison over same territory, 63–69
comparison using rates or ratios, 88
comparison of mean vs. median centers, 243–244
comparison via summary parameter, 325–327

Geosets (*cont.*)
compound territories, 107–108
external concavities, 108
formation by separate fragments, 107
internal gaps, 108
subterritories, 107
compound, 73–77
extension to approach on, 80
formulation of, 73
handling subpopulations, 82–84
concept of, 43–44
covariance between coordinates, 257–258
dealing with aggregated data, 70–71
decomposition of, 87
definition of territory, 48–51, 61–63
derived, 449–450
determination using measurement scales, 111
distance-deviance, 255–256
decomposition of, 270–275
distance-moments, 211–212
distance-variance, 251–257
calculation, 251
components vs. compound geosets, 254
properties of, 253–254
territories with regular shapes, 252–253
values for selected countries, 280–281
eccentricity, 23, 25
effect of asymmetry, 312
finite, 106–107
geostatistical parameters, 57
intermediate, 86
linear measures of dispersion, 256–257
role of mean center, 256
role of median center, 256
selected central point, 256
standard distance, 256
looseness of territories, 279–282, 302–303
analysis of factors, 289–291
applications to cities, 300–305
Jewish population, 304–305
derived, 296–299
analysis of factors, 299
examples for Italy, 286–292, 297–300
examples for United States, 299–301
extension to subterritories, 291–292
fragmented territories, 294–296
sensitivity to location, 296
habitat vs. nonhabitat, 289–291
parameters, 279–282, 290, 302–303
primary, 296–299
analysis of factors, 297
territories with external concavities, 292–294
territories with internal concavities, 286–288
mean center, 25–27, 449

Geosets (*cont.*)
measurement of dispersion, 247–249, 256
distance averaging, 249–250, 256–257
historical development of, 249
methods of comparison, 11
minimal distance, 32–34, 326, 334–335
aligned geosets, 327–329, 334
between territories, 343–346
comparison between populations and subpopulations, 340–342, 344
concept of, 327–329, 335
example for France, 339–340
example for Italy, 3, 337–339, 342–343, 345–346
example for United States, 339–340, 343–344
expression as pure numbers, 337
graphical presentation of, 335–336
influence of aggregation of input data, 346
notation convention, 334–335
relationships, 75–76
transport problem algorithm, 331–333
minimization, 331–337, 424–425
two-dimensional geosets, 329–337
notation convention, 77–80
oblongity, 22–23, 25–30
operations on territories, 108–111
operations using set theory, 72
overview of, 43–44
parameter evaluation, 28–29
parameters of form, 277
parameters of position, 313
coincidence, 314–316
comparison of, 316–318
congruence, 314–316, 319
description of, 320–321
overview, 313
principal position, 318–319
quadrant settings, 318
reduced principal coordinates, 321
role of coordinates, 314–315, 317, 320
similarity, 321–322
transformation, 316, 320
partitioning, 311–312
population, 45–48, 65, 213–215
large, 69–72
patterns, 213–215, 277–278
potentials, 212
primary, 10–11, 22, 25, 169–183, 449–450
relative scatter of territory, 282–286
example for Italy, 289
general considerations, 282–283
regular geometrical figures, 283
isosceles trapezoids, 284–286
polygons, 283–284
simply connected figures, 286

Geosets (*cont.*)
relative weighting of, 47, 87–88, 214
scatter, 22, 28, 30–31
selected parameters, 22
shape of territories, 278–279
territorial, 65–69, 215
territorial coverage, 23, 29–31, 195–199
territorial dispersion, 25–29
territorial range of, 90–99
transformation operations, 56–61
handling elements, 50–62
matching, 56–58, 65–66
surpluses, 56–60, 65, 67–68
union vs. averaging of, 85–86
use of coordinates, 53–54
use of principal axes in, 258–267
definition of, 258–259
effect of scatter, 263–265
equating axes, 262
identification of, 259–262
oblongity considerations, 263–267
orientation, 262–263
use of set theory in, 44–45, 72–73
weighted averaging, 84–85
weighting of elements, 251–252, 260
GeoStat, 456–457
Geostatistical methods, 35–38
analysis of geosets by, 36–40
applications of, 35–38
comparative urban research, 36
national statistics of Italy, 3–34
population distribution studies, 36–37, 39
approaches to, 414–415
historical review of, 40–42
mainframe computer programs, 427
mapping, 417–418, 427
general considerations, 417
graphical rational patterns, 417
other graphical techniques, 418
variability, 427
overview of, 413–414
parameters, 418–424
density, 419
dispersion, 421
form, 422–424
oblongity, 421–422
overview of, 418–419, 427
position, 420
ratios, 422
territorial coverage, 419
rational maps by, 39
region definition in, 39
related computer software, 456–457
role in quantitative geography, 39–40
Geostatistical methods (*cont.*)
role of minimal distance, 424–427
applications of, 426–427, 429
basic concepts of, 424–425
properties of, 425–426
software packages, 428–430
features of, 429–430
maps, 428–430
parameter comparison, 428–429
transformation of data, 416
types of territorial containers, 415–416
gross, 415
habitat, 415–416
other types, 416
use of parameters in, 37–39, 42
Geostatistical parameters, 11, 22
dispersion-territorial coverage comparison, 30–31
examples, 22–24
losseness, 30
mean center, 25–27
measurement of, 27–31
net density, 29–30
oblongity, 27–29
order of presentation, 22–24
superimposition on maps, 24–25
territorial coverage, 29–30
territorial dispersion, 27–29
Geostatistical set, 10; *see also* Geosets
Geostatistics, 42
Graphicacy, 123
Graphical presentation, 389–391
Graphical rational patterns, 4, 6, 35, 37, 42, 115, 129
applications of, 148–158
comparative maps, 169–183
ecological correlations, 182–183
individual population elements, 148–149
localities, 150–151
territorial units, 151–158
choropleth maps, 162–165
comparison with Cartesian system, 144–148
comparison with other scales, 158–163, 182–183
circles, 159–160
framed rectangles, 160–161, 164
squares and hexagons, 160, 162–163
construction principles, 135–148
areal aspects, 142–143
black vs. white scale, 138–139
linear aspects, 140–142
multiple representation, 143–144
use of circular units, 140
use of integers, 135–140
ensuring correspondence with data, 166–168
in colored maps, 174–178
overview of, 135
representation of rates or ratios, 178–182

Graphical rational patterns (*cont.*)
  use in mapping of geosets, 166–168
    selected examples, 169–170
Graphical semiology, 120
Gravity models, 41
Gross population density, 191–193, 419
Gross territory, 36–37, 89–90
  concentration of population, 97–98
  convex hull, 93–96, 109
  criteria for solutions, 92–93
  ellipse as substitute for habitat, 99–104
  ellipses of equivalent scatter, 98–99
  handling unbounded population distributions, 90–98
  problem description, 90–92

Habitat, 9, 37, 55–56, 97, 183, 188, 196, 199, 223, 226, 234, 273, 289, 297, 419

Inconsistency ratio, 350–353
Indexes, 188–190, 194
  interpretation of, 190
    cost/benefit approach, 190
    effects of aggregation, 191
    insensitivity to geography, 190–191
Integers, use in graphical rational patterns, 135–140
Internal gap, 108
Israel, 165
Italy, 3–34, 103–104, 109, 150, 154, 166, 169–182, 195–199, 219–231, 238–240, 244, 261, 265, 286–292, 297–300, 337–339, 342–343, 345–346, 352–353, 371, 396–397

Jerusalem, 46–61, 105, 110, 148–149, 195, 306–307, 335–336
  population density of, 195
Joint dispersion, 384–386

Linkages
  interpretation of, 380, 382
  optimal, 333
Lodz, 304
London, 304
Looseness, 23, 29–30, 277, 279–282, 286–291, 423–424
  in analysis of territories, 279–282, 302–303
    analysis of factors, 289–291
    applications to cities, 300–305
      Jewish population, 304–305
    derived, 296–299
      analysis of factors, 299
    examples for Italy, 286–292, 297–300
    examples for United States, 299–301
    extension to subterritories, 291–292

Looseness (*cont.*)
  in analysis of territories (*cont.*)
    fragmented territories, 294–296
      sensitivity to location, 296
    habitat vs. nonhabitat, 289–291
    parameters, 279–282, 290, 302–303
    primary, 296–299
      analysis of factors, 297
    territories with external concavities, 292–294

Mainframe computer programs, 427
Mapping, 417–418, 427
  choropleth, 162–165, 178, 417
  colored, 174–178
  comparative, 169–183
  construction principles, 135–148
  dot, 128–129, 132
  general considerations, 417
  geostatistical parameters on, 24–25
  graphical rational patterns, 417
  interpretations for Italy, 219–231, 238–240
    average income by province, 230–231
    citrus groves, 221, 225–226
    illiterates, 225–228
    sheep distribution, 221–224
    subpopulations, 228–229
  interpretations for United States, 231–238, 241
    subpopulations, 233–237
    total population, 231–233
  multi-colored, 178
  other graphical techniques, 418
  rational, 39
  selected examples, 169–170
  software packages, 428–430
  statistical
    Cartesian coordinate system in, 125–128
      general properties of, 126–127
    general criteria for, 133–134
    historical highlights of, 117–121
    Non-Cartesian methods in, 133
    overview of, 115–117
    principles of, 123–125
    recent scientific developments, 120–121
      state of the art, 121–123
    role of automation in, 118–119
    role of graphical rational patterns, 135–168
    role of software packages in, 118, 122–123
    standards for, 123–125
    stereogram construction in, 130–132
    use of contour lines in, 132
    use of dot maps, 128–129
      aggregates of location, 129
      population distribution, 128
  techniques in, 46–63

Mapping (*cont.*)
use of graphical rational patterns in, 166–168
role of integers, 135–140
variability, 427
Mean center, 23, 25–27, 209, 420
Median center, 209
Milan, 304
Minimal distance
as a metric, 347
between net geosets, 376–377
between three geosets, 347–356
measure of inconsistency, 347, 350–351
ethnic subpopulations in United States, 350–351
triangularity, 347–350
calculation of, 370–375
collinearity of three geosets, 357–361
comparison of geosets over different territories, 403
compound geoset, 365–366, 368
conversion to pure number, 406–409
dimensionless measures, 406–409
ideal solution, 406–407
coordinate transformation between geosets, 399
checking affine equivalence, 405
checking coincidence, 403
checking congruence, 403
comparison of geosets translated to their centers, 401–402
comparison of reduced geosets, 404–405
comparison of standardized geosets, 404–405
dilatation, 404
geosets in similar position, 403–404
parallel translation, 399–401
reflection, 402–403
rotation, 402–403
similar transformation, 404–405
translation to its center, 401
derived geosets, 361–363
extensions to other geosets, 370–372
geoset from two paradigmatic distributions, 351–352
example for Italy, 352–353, 371
joint analysis, 364
multi-ordered geosets, 355–357
nonintersecting components, 370
practical rules, 368–369
primary geosets, 361–365
role in geostatistical methods, 424–427
use of triangle of distances, 348, 362–363, 365–368
role of coordinates, 367–368
Modal center, 245
Models of potentials, 41
Multi-colored maps, 178

National Statistical Institute of Italy, 3
Net density, 191–193
New York, 446–447
Non-Cartesian methods, 133
Nonhabitat, 9, 55–56, 97, 183, 188, 199, 223, 226, 234, 273, 289, 419

Oblongity, 22–23, 25–30, 248–249, 263–267, 282–283, 267–270, 421–422, 452–453
absolute squared, 27, 266
principal axes in, 258–267
One-dimensional variables, 247–249
Optimal linkage, 333
Optimal transportation scheme, 331–337, 345, 400
Orthogonal axes, 258–259

Parallel translation, 399–401
Partitioning, 311–312
Pearson correlation coefficient, 183
Po Valley, 7, 12, 14
Population distribution
comparison with agriculture, 13–14, 18–19
comparison with uniform distribution, 5–10, 48–63
concentration, 97–98
density determinations, 433
Atlanta, 433
Austin, 433
Washington, DC, 433
ethnic dispersion, 433
Atlanta, 434–437
Austin, 434–437
New York, 446–447
Washington, DC, 434
formalization of geosets, 10–22
geosets in study of, 23–34
geostatistical parameters on, 22–31
religious-ethnic groups in Toronto, 444
effects of minimal distance, 445–446
effects of religious segregation, 445
studies, 36–37, 39
territorial distribution of Jews, 437
in European cities, 437–444
territory of Italy, 3–34, 220–231
cell network, 5
economic aspects, 12–14
employment in heavy industry and construction, 14
employment in petrol industry, 16
employment in advertising and public relations, 15
geoset formalization, 10–23
habitat vs. nonhabitat, 9
illiterates, 20
labor force, 21
relative weighting, 6
surplus, 8, 21
unbounded, 90–98
use of dot maps, 128–129

Population distribution (*cont.*)
  using ellipse, 431
    Atlanta, 432
    Austin, 432
    United States, 432
    Washington, DC, 432
Population potential, 212

Quadrant setting, 318, 403, 420
Quadratic average, 248, 329, 391–392, 404–406
Quantitative geography, 39–40

Rational mapping, 3–10, 39, 42, 156
Relative density, 193
Relative oblongity, 28, 267, 301, 422
Relative quadratic minimal distance (RQMD), 408
Relative scatter, 282–286, 289, 301, 422–423
Relative weighting, 6, 47, 87–88, 214
Rome, 75–83, 104, 300, 304, 355–361, 384, 387, 397–398

Sardina, 7, 30–31, 272, 293–294
Scatter plot, 126–127
Scatter, 22, 28, 30–31, 263–265, 422
Sea of Galilee, 286
Segregation, 23, 29–30, 189–190
Sets, 44
  operations using, 44–45
Sicily, 7, 30–31, 272, 293–294
Similarity, 247, 321–322, 420
Software packages, 428–430
Standard distance, 29, 256–257, 408, 421, 452–453
Statistical Atlas of Italy, 220–231
  applications of, 3–21
  population distribution, 4
  principles, 4
Statistical mapping,
  Cartesian coordinate system in, 125–128
    general properties of, 126–127
  general criteria for, 133–134
  historical highlights of, 117–121
  Non-Cartesian methods in, 133
  overview of, 115–117
  principles of, 123–125
  recent scientific developments, 120–121
    state of the art, 121–123
  role of automation in, 118–119
  role of graphical rational patterns, 135–168
  role of software packages in, 118, 122–123
  standards for, 123–125
  stereogram construction in, 130–132
  use of contour lines in, 132
  use of dot maps, 128–129
    aggregates of location, 129
    population distribution, 128
Stereograms, 130–132
Sunflower, 128–129
Surplus of population, 8, 21, 23
Symmetry, 312

Tel-Aviv, 91–97
Territorial coverage, 23, 29–31, 187–188, 196, 289; *see also* Territorial units
  application to cities, 206–207
  comparative aspects of geosets, 187–188
  concentration of population, 188–191
  effect of habitat on, 196–198
  effect of subterritories, 198–199
  effect of time on, 196–198
  evaluation of densities, 191–193
    averaging ratios, 194–195
    net vs. gross, 191–193
    surplus, 193
  habitat of Jewish population, 205
  Italy, 195–199
  Jerusalem, 195
  links between geosets, 188–190
  measurement of, 187–191
  segregation in, 189–190
  United States, 199–207
    ethnic variations, 201–206, 235–236
    habitat vs. nonhabitat, 200
    total population, 199
  use of common indexes in, 188–190, 194
    interpretation of, 190
      cost/benefit approach, 190
      effects of aggregation, 191
      insensitivity to geography, 190–191
Territorial dispersion, 25–29, 249; *see also* Dispersion
Territorial distribution, 129
Territorial units, 133–134, 151–158, 178; *see also* Territorial coverage
  Alps, 7, 10
  Argentina, 238–240
  Atlanta, 100, 432–437
  Austin, 432–437
  Berlin, 287, 304
  Canada, 156
  Continental Italy, 7, 30–31, 272, 293–294
  France, 152–153, 238–240, 339–340
  Israel, 165
  Italy, 3–34, 103–104, 109, 150, 154, 166, 169–182, 195–199, 219–231, 238–240, 244, 261, 265, 286–292, 297–300, 337–339, 342–343, 345–346, 352–353, 371, 396–397
  Jerusalem, 46–61, 105, 110, 148–149, 195, 306–307, 335–336
  New York, 446–447
  parameters of looseness, 279–282, 286–291

Territorial units (*cont.*)
  Po Valley, 7, 12, 14
  relative scatter, 282–286
  shape of, 278–279
  Tel-Aviv, 91–97
  Toronto, 304, 444
  types of, 151–152
  United States, 102, 155, 181, 199–207, 231–238, 241, 299–301, 339–340, 343–344, 350–351, 394–396, 432
  variability within, 152–156
  Vienna, 304
  Warsaw, 304
  Washington, DC, 101, 432–434
  Zurich, 304
Territory, definition of, 48–51, 61–63
Toronto, 304, 444
Transport problem, 331–333
  minimization, 331–337, 424–425
Triangle of centers, 27, 221, 231
Triangle of distances, 348, 362–363, 365–368
Triangular inequality, 348
Triangularity, 347–350
Trinaria, 31
Two-colored maps, 174–177

Uniform distribution, 5–10, 24, 32–34
United States, 102, 155, 181, 199–207, 231–238, 241, 299–301, 339–340, 343–344, 350–351, 394–396, 432

Variability, 247–249
Vector graphics, 119
Venn diagram, 44–45, 110
Vienna, 304

Warsaw, 304
Washington, DC, 101, 432–434
Weighted averaging, 84–85
Weighting of elements, 251–252, 260

Zurich, 304